Gene Probes for Bacteria

Gene Probes for Bacteria

Edited by

Alberto J. L. Macario

Everly Conway de Macario

Wadsworth Center for Laboratories and Research
New York State Department of Health
and
School of Public Health
State University of New York
Albany, New York

ACADEMIC PRESS, INC.
Harcourt Brace Jovanovich, Publishers
San Diego New York Berkeley Boston
London Sydney Tokyo Toronto

This book is printed on acid-free paper. ∞

Academic Press, Inc.
San Diego, California 92101

United Kingdom Edition published by
Academic Press Limited
24–28 Oval Road, London NW1 7DX

Library of Congress Cataloging-in-Publication Data

Gene probes for bacteria / edited by Alberto J. L. Macario, Everly Conway de Macario.
p. cm.
Includes bibliographical references.
ISBN 0-12-463000-6 (alk. paper)
1. Diagnostic bacteriology--Technique. 2. DNA probes--Diagnostic use. I. Macario, Alberto J. L. II. Conway de Macario, Everly.
[DNLM: 1. Bacteria--isolation & purification. 2. Nucleic Acid Probes. 3. Nucleic Acids--diagnostic use. QU 58 G326]
QR67.2.G46--dc20
DNLM/DLC
for Library of Congress 89-17733
CIP

Printed in the United States of America
90 91 92 93 9 8 7 6 5 4 3 2 1

To our collaborators, assistants, and students
of sundry places and times.

Contents

9 Nucleic Acid Probes for *Bacteroides* Species

David J. Groves

10 Nucleic Acid Probes for *Campylobacter* Species

Bruce L. Wetherall and Alan M. Johnson

11 Detection of *Leptospira, Haemophilus,* and *Campylobacter* Using DNA Probes

W. J. Terpstra, J. ter Schegget, and G. J. Schoone

12 Nucleic Acid Probes for the Identification of *Salmonella*

Fran A. Rubin

13 Gene Probes for Detection of Food-Borne Pathogens

K. Wernars and S. Notermans

14 The Use of Gene and Antibody Probes in Identification and Enumeration of Rumen Bacterial Species

J. D. Brooker, R. A. Lockington, G. T. Attwood, and S. Miller

15 Gene Probe Detection of Human and Cell Culture Mycoplasmas

Ram Dular

16 Detection of TEM β-Lactamase Genes Using DNA Probes

Kevin J. Towner

17 SIA Technology for Probing Microbial Genes

Robert J. Jovell, Everly Conway de Macario, and Alberto J. L. Macario

Contributors

Numbers in parentheses indicate the pages on which the authors' contributions begin.

Peter W. Andrew (179), Department of Microbiology, University of Leicester, Medical Sciences Building, Leicester LE1 9HN, England

G. T. Attwood (389), Department of Animal Sciences, Waite Agriculture Research Institute, Glen Osmond, South Australia 5064, Australia

Christiane Bebear (45), Laboratoire de Bactériologie, Hôpital Pellegrin, 33076 Bordeaux, France

Lawrence H. Bopp (69), Laboratories for Bacteriology, Wadsworth Center for Laboratories and Research, New York State Department of Health, Albany, New York 12201-0509

Graham J. Boulnois (179), Department of Microbiology, University of Leicester, Medical Sciences Building, Leicester LE1 9HN, England

J. D. Brooker (389), Department of Animal Sciences, Waite Agriculture Research Institute, Glen Osmond, South Australia 5064, Australia

Arunsri Chatkaeomorakot (95), Armed Forces Research Institute of Medical Sciences, Bangkok 10400, Thailand

Everly Conway de Macario (485), Wadsworth Center for Laboratories and Research, New York State Department of Health, and School of Public Health, State University of New York, Albany, New York 12201-0509

Hirofumi Danbara (167), Department of Bacteriology, The Kitasato Institute, Minato-ku, Tokyo 108, Japan

James J. Donegan (1), Enzo Biochem, Inc., New York, New York 10013

Ram Dular (417), Ontario Ministry of Health, Regional Public Health Laboratory, Ottawa, Ontario K2A 1S8, Canada

Brigitte Dutilh[1] (45), Laboratoire de Bactériologie, Hôpital Pellegrin, 33076 Bordeaux, France
Peter Echeverria (95), Armed Forces Research Institute of Medical Sciences, Bangkok 10400, Thailand
Sherry P. Goltz (1), Enzo Biochem, Inc., New York, New York 10013
Patrick A. D. Grimont (45), Unité des Entérobactéries, Unité 199 INSERM, Institut Pasteur, 75724 Paris, France
Roy Gross (205), Sclavo Research Center, 53100 Siena, Italy
David J. Groves (233), Department of Pathology, McMaster University, and Department of Laboratory Medicine, St. Joseph's Hospital, Hamilton, Ontario L8N 4A6, Canada
Alan M. Johnson (255), Department of Clinical Microbiology, School of Medicine, Flinders University and Flinders Medical Center, Bedford Park, South Australia 5042, Australia
Robert J. Jovell (485), Wadsworth Center for Laboratories and Research, New York State Department of Health, and School of Public Health, State University of New York, Albany, New York 12201-0509
Norman Kelker (1), Enzo Biochem, Inc., New York, New York 10013
R. A. Lockington (389), Department of Animal Sciences, Waite Agriculture Research Institute, Glen Osmond, South Australia 5064, Australia
Alberto J. L. Macario (485), Wadsworth Center for Laboratories and Research, New York State Department of Health, and School of Public Health, State University of New York, Albany, New York 12201-0509
S. Miller (389), Department of Animal Sciences, Waite Agriculture Research Institute, Glen Osmond, South Australia 5064, Australia
Margarita M. Molina (1), Enzo Biochem, Inc., New York, New York 10013
S. Notermans (353), Laboratory of Water and Food Microbiology, National Institute of Public Health and Environmental Protection, 3720 BA Bilthoven, The Netherlands
Linda M. Parsons (69), Laboratories for Bacteriology, Wadsworth Center for Laboratories and Research, New York State Department of Health, Albany, New York 12201-0509
Marjorie Pollice (1), Enzo Biochem, Inc., New York, New York 10013
Rino Rappuoli (205), Sclavo Research Center, 53100 Siena, Italy
Fran A. Rubin (323), Department of Bacterial Immunology, Walter Reed Army Institute of Research, Washington, D.C. 20307-5100
J. ter Schegget (295), Department of Medical Microbiology, University of Amsterdam, 1105 AZ Amsterdam, The Netherlands

[1] Present address: Laboratoire d'Analyses Médicales, 218, rue Mandron, 33000 Bordeaux, France.

G. J. Schoone (295), Department of Tropical Hygiene, Royal Tropical Institute, 1105 AZ Amsterdam, The Netherlands

Jitvimol Seriwatana (95), Armed Forces Research Institute of Medical Sciences, Bangkok 10400, Thailand

Orntipa Sethabutr (95), Armed Forces Research Institute of Medical Sciences, Bangkok 10400, Thailand

Mehdi Shayegani (69), Laboratories for Bacteriology, Wadsworth Center for Laboratories and Research, New York State Department of Health, Albany, New York 12201-0509

W. J. Terpstra (295), Department of Tropical Hygiene, Royal Tropical Institute, 1105 AZ Amsterdam, The Netherlands

John A. Todd (1), Enzo Biochem, Inc., New York, New York 10013

Kevin J. Towner (459), Department of Microbiology and PHLS Laboratory, University Hospital, Nottingham NG7 2UH, England

Jacob Victor (1), Enzo Biochem, Inc., New York, New York 10013

Alfred L. Waring (69), Laboratories for Bacteriology, Wadsworth Center for Laboratories and Research, New York State Department of Health, Albany, New York 12201-0509

K. Wernars (353), Laboratory of Water and Food Microbiology, National Institute of Public Health and Environmental Protection, 3720 BA Bilthoven, The Netherlands

Bruce L. Wetherall (255), Department of Clinical Microbiology, Flinders Medical Center, Bedford Park, South Australia 5042, Australia

Peter H. Williams (143), Department of Genetics, University of Leicester, Leicester LE1 7RH, England

Huey-Lang Yang (1), Enzo Biochem, Inc., New York, New York 10013

Preface

Molecular genetics has advanced enormously in the past decade. It is changing our views of physiological and pathological phenomena and, consequently, our strategies to study and control them. The construction of gene probes to analyze microbes is one example of these contemporary trends.

Clinical and epidemiological applications of nucleic acid probes for bacteria have just begun. This book attempts to furnish a comprehensive account of the state of the art of this new area. It complements the treatise on antibacterial monoclonal antibodies published earlier (see Introduction). Both present extensive reviews of probes for bacteria. The earlier work dealt with immunologic probes and methods; this one covers molecular genetics and nucleic acids, emphasizing diagnosis. The preparation and use of nucleic acid probes for identifying bacteria in clinical specimens and in other samples of practical or scientific interest are described.

The contributors are pioneers in their fields of expertise. They are affiliated with academe, government, or industry, thus representing all major interest groups. Their chapters—particularly their methods, conclusions, and suggestions for the future—should be of great use to professionals, technicians, and R & D directors in universities, in federal and state-dependent service and research institutes, and in private (e.g., hospital) and industrial laboratories.

The probes and procedures described are applicable to a variety of areas pertinent to medicine, veterinary sciences, environmental (sanitary) engineering, agronomy, zoology, and to other branches of science and biotechnology dealing with microbes. These probes and procedures have been standardized for detecting the presence of bacteria and for identifying and classifying them in clinical specimens, environmental samples, laboratory cultures, etc. The probes are for pathogens affecting the oral cavity, nasopharynx, gastrointestinal and genitourinary tracts (including those causing venereal diseases), eyes, blood and meninges, the cardiovascular, respiratory, and immune systems, and for bacteria found in water and foods, tissue cultures, and other ecosystems of importance.

Salient features of the book are highlighted in the Introduction. It calls the reader's attention to chapter sections covering laboratory applications and identifies other sections of general interest. To enhance the book's utility, the Introduction also comprises a bibliographic update.

We would like to thank our teachers, assistants, collaborators, and students for their continual support and for their help in the many tasks that directly or indirectly culminated in this book. We would also like to thank the contributors for their excellent manuscripts and the staff of Academic Press for expert advice, unfailing support, and encouragement.

Alberto J. L. Macario
Everly Conway de Macario

Introduction: Molecular Genetics in Diagnostic Bacteriology

Molecular genetics, including genetic engineering, has revolutionized biology. Genes have been and are being isolated and characterized. Currently, studies on how genes function and are regulated are progressing rapidly. Techniques for manipulating and modifying genes are becoming widespread.

These advances have permeated into virtually all branches of biology. Examining how cells and organisms work can now be done from the vantage point of molecular genetics. One of the many outcomes of these innovative ideas and techniques has been the generation of nucleic acid probes for identifying microbes. Thus, diagnostic microbiology has been endowed with new means to accomplish its primary goals, namely, to detect and identify viruses, bacteria, and other microorganisms in samples of scientific or practical interest.

Diagnostic bacteriology, the central theme of this volume, is entering a new era marked by the application of gene probes in addition to other classical identification methods, such as culture-isolation and immunoassays, based on polyclonal or monoclonal antibodies. This book was conceived along the lines of its predecessor "Monoclonal Antibodies against Bacteria" (Volumes I–III, 1985–1986. A. J. L. Macario and E. Conway de Macario, editors. Academic Press, Inc.). The main objective was to provide a comprehensive report on the current status of gene probes for bacteria useful in diagnostics. Which nucleic acid probes are available, how and when to utilize them, what to expect in terms of results obtained with their use, and how to prepare new probes are all questions addressed in the various chapters.

The contributors were asked to prepare comprehensive manuscripts covering the microorganisms of their expertise. A few bacterial species are treated in more than one chapter in order to provide a multidimensional picture of important microbes and to compile the knowledge gained from different laboratories using the same or different (e.g., radioactive vs. nonradioactive) probes or methods. To maintain unity of format and ensure full coverage, the chapters include a uniform series of sections.

Sections I, "Introduction," and II, "Background," present the main theme of the chapter, its origins, and pertinent problems and questions. A brief historical overview is included in Section II, with descriptions of microorganisms and diseases dealt with in the chapter. The purpose of this introductory material is to familiarize the nonspecialist with topics that need to be known to understand the subsequent sections, particularly Section III, "Results and Discussion," in which the expertise of the authors is displayed. Data are reported and analyzed to illustrate preparation and/or use of nucleic acid probes. A critical examination of the results leads to inferences and speculations which are summarized in Section IV, "Conclusions." The advantages and disadvantages of nucleic acid probes as opposed to antisera and monoclonal antibodies are discussed in Section V, "Gene Probes versus Antisera and Monoclonal Antibodies." Here the bibliography serves as a starting point in the search for additional information regarding modern serology and immunoassays based on monoclonal antibodies. As such, Section V may help those unfamiliar with immunoassays to venture into the advanced field of immunology and immunochemistry that originated only a few years ago with the widespread application of the hybridoma technology.

Section VI, "Prospects for the Future," contains opinions, speculations, and suggestions by the authors based on their invaluable experience. Suggestions relate to a variety of problems (methodological, biological, pathological, environmental, biotechnological) the authors deem important for research in the near future.

Section VII is a synopsis of the chapter's content. It leads the reader into Section VIII, "Materials and Methods," which is very important since it gives detailed instructions for the preparation and/or utilization of the nucleic acid probes specifically treated in the chapter. Procedures already published in manuals available in virtually all laboratories working with nucleic acids are only briefly mentioned or not described at all, unless modifications have been introduced. These modifications are given in detail and represent the authors' contribution to a field burgeoning with innovations. Some are improvements on the original methods; others are adaptations required to deal with unique technical difficulties posed by special characteristics of some microorganisms. Section VIII should be useful not only at the bench, but also as a source of information on reagents and their suppliers, on new techniques, and on critical evaluations of published methodologies. A comprehensive list of references closes each chapter.

The references which follow are a brief list of selected publications that have appeared recently. The list's purpose is to update the reader on material discussed in this volume. These publications pertain to general concepts and methodology (13,32,36,38), gonococcus (8,27), *Chlamydia* (4,15,37), *Escherichia coli* (11,20,28,31), mycobacteria (3,5,9,12,21,23,29), *Bacteroides* (19,34), *Campylobacter* (1,26), *Leptospira interrogans* (41), *Haemophilus actinomycetemcomitans* (34), *Staphylococcus aureus* (24), *Streptococcus pneumoniae* (2), *Legionella* species (6,14,33), *Yersinia* species (10,17,18,25), lactobacilli (22,35), *Pseudomonas aeruginosa* (30,40), *Clostridium difficile* (39), *Erwinia amylovora* (7), and *Borrelia burgdorferi* (16).

ACKNOWLEDGMENT

This work was supported in part by Grant No. 706IERBEA85 from GRI-NYSERDA-NY Gas.

Alberto J. L. Macario
Everly Conway de Macario

REFERENCES

1. Chevrier, D., Larzul, D., Megraud, F., and Guesdon, J-L. (1989). Identification and classification of *Campylobacter* strains by using nonradioactive DNA probes. *J. Clin. Microbiol.* **27,** 321–326.
2. Cooksey, R. C., Swenson, J. M., Clark, N. C., and Thornsberry, C. (1989). DNA hybridization studies of a nucleotide sequence homologous to transposon Tn1545 in the "Minnesota" strain of multiresistant *Streptococcus pneumoniae* isolated in 1977. *Diagn. Microbiol. Infect. Dis.* **12,** 13–16.
3. Cooper, G. L., Grange, J. M., McGregor, J. A., and McFadden, J. J. (1989). The potential use of DNA probes to identify and type strains within the *Mycobacterium tuberculosis* complex. *Letters Appl. Microbiol.* **8,** 127–130.
4. Dean, D., Palmer, L., Raj Pant, C., Courtright, P., Falkow, S., and O'Hanley, P. (1989). Use of a *Chlamydia trachomatis* DNA probe for detection of ocular *Chlamydiae*. *J. Clin. Microbiol.* **27,** 1062–1067.
5. Drake, T. A., Herron, R. M., Jr., Hindler, J. A., Berlin, O. G. W., and Bruckner, D. A. (1988). DNA probe reactivity of *Mycobacterium avium* complex isolates from patients without AIDS. *Diagn. Microbiol. Infect. Dis.* **11,** 125–128.

6. Ezaki, T., Dejsirilert, S., Yamamoto, H., Takeuchi, N., Liu, S., and Yabuuchi, E. (1988). Simple and rapid genetic identification of *Legionella* species with photobiotin-labeled DNA. *J. Gen. Appl. Microbiol.* **34,** 191–199.
7. Falkenstein, H., Bellemann, P., Walter, S., Zeller, W., and Geider, K. (1988). Identification of *Erwinia amylovora,* the fireblight pathogen, by colony hybridization with DNA from plasmid pEA29. *Appl. Environ. Microbiol.* **54,** 2798–2802.
8. Granato, P. A., and Roefaro Franz, M. (1989). Evaluation of a prototype DNA test for the noncultural diagnosis of gonorrhea. *J. Clin. Microbiol.* **27,** 632–635.
9. Hampson, S. J., Thompson, J., Moss, M. T., Portaels, F., Green, E. P., Hermon-Taylor, J., and McFadden, J. J. (1989). DNA probes demonstrate a single highly conserved strain of *Mycobacterium avium* infecting AIDS patients. *Lancet.* **1,** 65–68.
10. Jagow, J. A., and Hill, W. E. (1988). Enumeration of virulent *Yersinia enterocolitica* colonies by DNA colony hybridization using alkaline treatment and paper filters. *Molec. Cell. Probes* **2,** 189–195.
11. Karch, H., and Meyer, T. (1989). Evaluation of oligonucleotide probes for identification of shiga-like-toxin-producing *Escherichia coli. J. Clin. Microbiol.* **27,** 1180–1186.
12. Kawa, D. E., Pennell, D. R., Kubista, L. N., and Schell, R. F. (1989). Development of a rapid method for determining the susceptibility of *Mycobacterium tuberculosis* to isoniazid using the Gen-probe DNA hybridization system. *Antim. Agents Chemother.* **33,** 1000–1005.
13. Kennedy, K. E., Daskalakis, S. A., Davies, L., and Zwadyk, P. (1989). Nonisotopic hybridization assays for bacterial DNA samples. *Molec. Cell. Probes* **3,** 167–177.
14. Laussucq, S., Schuster, D., Alexander, W. J., Thacker, W. L., Wilkinson, H. W., and Spika, J. S. (1988). False-positive DNA probe test for *Legionella* species associated with a cluster of respiratory illnesses. *J. Clin. Microbiol.* **26,** 1442–1444.
15. LeBar, W., Herschman, B., Jemal, C., and Pierzchala, J. (1989). Comparison of DNA probe, monoclonal antibody enzyme immunoassay, and cell culture for the detection of *Chlamydia trachomatis. J. Clin. Microbiol.* **27,** 826–828.
16. LeFebvre, R. E., Perng, G. C., and Johnson, R. C. (1989). Characterization of *Borrelia burgdorferi* isolates by restriction endonuclease analysis and DNA hybridization. *J. Clin. Microbiol.* **27,** 636–639.
17. McDonough, K. A., Schwan, T. G., Thomas, R. E., and Falkow, S. (1988). Identification of a *Yersinia pestis*-specific DNA probe with potential use in plague surveillance. *J. Clin. Microbiol.* **26,** 2515–2519.
18. Miliotis, M. D., Galen, J. E., Kaper, J. B., and Morris, J. G., Jr. (1989). Development and testing of a synthetic oligonucleotide probe for the detection of pathogenic *Yersinia* strains. *J. Clin. Microbiol.* **27,** 1667–1670.
19. Moncla, B. J., Strockbine, L., Braham, P., Karlinsey, J., and Roberts, M. C. (1988). The Use of whole-cell DNA probes for the identification of *Bacteroides intermedius* isolates in a dot-blot assay. *J. Dent. Res.* **67,** 1267–1270.
20. Olive, D. M., Atta, A. I., and Setti, S. K. (1988). Detection of toxigenic *Escherichia coli* using biotin-labelled DNA probes following enzymatic amplification of the heat labile toxin gene. *Molec. Cell. Probes* **2,** 47–57.
21. Peterson, E. M., Lu, R., Floyd, C., Nakasone, A., Friedly, G., and De La Maza, L. M. (1989). Direct identification of *Mycobacterium tuberculosis, Mycobacterium avium,* and *Mycobacterium intracellulare* from amplified primary cultures in BACTEC media using DNA probes. *J. Clin. Microbiol.* **27,** 1543–1547.
22. Petrick, H. A. R., Ambrosio, R. E., and Holzapfel, W. H. (1988). Isolation of a DNA probe for *Lactobacillus curvatus. Appl. Environ. Microbiol.* **54,** 405–408.
23. Picken, R. N., Plotch, S. J., Wang, Z., Lin, B. C., Donegan, J. J., and Yang, H. L. (1988). DNA probes for mycobacteria. *Molec. Cell. Probes* **2,** 111–124.

24. Rifai, S., Barbancon, V., Prevost, G., and Piemont, Y. (1989). Synthetic exfoliative toxin A and B DNA probes for detection of toxigenic *Staphylococcus aureus* strains. *J. Clin. Microbiol.* **27,** 504–506.
25. Robins-Browne, R. M., Miliotis, M. D., Cianciosi, S., Miller, V. L., Falkow, S., and Morris, J., Jr. (1989). Evaluation of DNA colony hybridization and other techniques for detection of virulence in *Yersinia* species. *J. Clin. Microbiol.* **27,** 644–650.
26. Romaniuk, P. J., and Trust, T. J. (1989). Rapid identification of *Campylobacter* using oligonucleotide probes to 16S ribosomal RNA. *Molec. Cell. Probes* **3,** 133–142.
27. Rossau, R., VanMechelen, E., De Ley, J., and Van Heuverswijn, H. (1989). Specific *Neisseria gonorrhoeae* DNA-probes derived from ribosomal RNA. *J. Gen Microbiol.* **135,** 1735–1745.
28. Saez-Llorens, X., Guzman-Verduzco, L. M., Shelton, S., Nelson, J. D., and Kupersztoch, Y. M. (1989). Simultaneous detection of *Escherichia coli* heat-stable and heat-labile enterotoxin genes with a single RNA probe. *J. Clin. Microbiol.***27,** 1684–1688.
29. Saito, H., Tomioka, H., Sato, K., Tasaka, H., Tsukamura, M., Kuze, F., and Asano, K. (1989). Identification and partial characterization of *Mycobacterium avium* and *Mycobacterium intracellulare* by using DNA probes. *J. Clin. Microbiol.* **27,** 994–997.
30. Samadpour, M., Moseley, S. L., and Lory, S. (1988). Biotinylated DNA probes for exotoxin A and pilin genes in the differentiation of *Pseudomonas aeruginosa* strains. *J. Clin. Microbiol.* **26,** 2319–2323.
31. Scotland, S. M., Willshaw, G. A., Said, B., Smith, H. R., and Rowe, B. (1989). Identification of *Escherichia coli* that produces heat-stable enterotoxin ST_A by a commercially available enzyme-linked immunoassay and comparison of the assay with infant mouse and DNA probe tests. *J. Clin. Microbiol.* **27,** 1697–1699.
32. Siegler, N. (1989). DNA-based testing: A progress report. *ASM News* **55,** 308–312.
33. Starnbach, M. N., Falkow, S., and Tompkins, L. S. (1989). Species-specific detection of *Legionella pneumophila* in water by DNA amplification and hybridization. *J. Clin. Microbiol.* **27,** 1257–1261.
34. Strzempko, M. N., Simon, S. L., French, C. K., Lippke, J. A., Raia, F. F., Savitt, E. D., and Vaccaro, K. K. (1987). A cross-reactivity study of whole genomic DNA probes for *Haemophilus actinomycetemcomitans, Bacteroides intermedius,* and *Bacteroides gingivalis. J. Dent. Res.* **66,** 1543–1546.
35. Tannock, G. W. (1989). Biotin-labeled plasmid DNA probes for detection of epithelium-associated strains of lactobacilli. *Appl. Environ. Microbiol.* **55,** 461–464.
36. Tenover, F. (1988). DNA probes—Where are they taking us? *Eur. J. Clin. Microbiol. Infect. Dis.* **7,** 457–459.
37. Timms, P., Eaves, F. W., Girjes, A. A., and Lavin, M. F. (1988). Comparison of *Chlamydia psittaci* isolates by restriction endonuclease and DNA probe analyses. *Infect. Immun.* **56,** 287–290.
38. White, T. J., Arnheim, N., and Erlich, H. A. (1989). The polymerase chain reaction. *Trends. Genet.* **5,** 185–189.
39. Wilson, K. H., Blitchington, R., Hindenach, B., and Greene, R. C. (1988). Species-specific oligonucleotide probes for rRNA of *Clostridium difficile* and related species. *J. Clin. Microbiol.* **26,** 2484–2488.
40. Wolz, C., Kiosz, G., Ogle, J. W., Vasil, M. L., Schaad, U., Botzenhart, K., and Doring, G. (1989). *Pseudomonas aeruginosa* cross-colonization and persistence in patients with cystic fibrosis. Use of a DNA probe. *Epidem. Inf.* **102,** 205–214.
41. Zuerner, R. L., and Bolin, C. A. (1988). Repetitive sequence element cloned from *Leptospira interrogans* serovar hardjo type hardjo-bovis provides a sensitive diagnostic probe for bovine leptospirosis. *J. Clin. Microbiol.* **26,** 2495–2500.

1

The Use of Nonradioactive DNA Probes for Rapid Diagnosis of Sexually Transmitted Bacterial Infections

SHERRY P. GOLTZ, JAMES J. DONEGAN, HUEY-LANG YANG, MARJORIE POLLICE, JOHN A. TODD, MARGARITA M. MOLINA, JACOB VICTOR, AND NORMAN KELKER

Enzo Biochem, Inc.
New York, New York

I. INTRODUCTION

The incidence of sexually transmitted diseases (STD) has been steadily increasing over the last 20–30 years and is reaching alarming proportions. Diseases like genital herpes, gonorrhea, chlamydia, and most recently, AIDS are now household words. The development of new methods for the early detection of STD has been pinpointed as a major factor in the prevention of these diseases, as well as in the evolution of effective methods for their treatment (58).

Many of the tests that have historically been used to diagnose STD (i.e., growth of bacteria or viruses by culture or direct microscopic examination of specimens) are somewhat insensitive, nonspecific, lengthy, and cumbersome. They have proved to be only partially effective in curbing the spread of many STD because they are often unable to detect these diseases in their early stages. In the last few years methodologies for immunological and genetic analyses, such as those employing monoclonal antibodies (MAb) and DNA probes, have been applied to the diagnosis of infectious diseases with mixed results.

In many cases, the use of polyclonal or monoclonal antibodies, either for direct specimen analysis or in conjunction with standard culture assays, has resulted in enhanced diagnostic sensitivity, specificity, and accuracy. At other times, however, these antibody-based assays fail to offer substantial improvements over the standard culture or cytological tests for STD.

Although attempts to apply DNA hybridization techniques to the detection of anogenital pathogens are still in their preliminary stages, some progress, most notably in the area of virally induced STD, has been made. Enzo Biochem, Inc. has pioneered the use of rapid, nonradioactive *in situ* hybridization assays for the development of direct and culture confirmation assays for herpes simplex viruses (HSV types I and II) and for the detection of human papilloma viruses (HPV) in biopsy specimens. *In situ* (nonradioactive) and Southern blot (radioactive) assays for HPV are also being marketed by Life Biotechnologies Co. In addition, Enzo is developing a novel *in situ* DNA hybridization assay for AIDS in which human immunodeficiency virus type 1 (HIV-1) DNA is detected in interphase nuclei or metaphase spreads of infected lymphocytes by fluorescence microscopy. The Cetus Co. is applying its polymerase chain reaction (PCR) technique to the development of an AIDS diagnostic test, as well.

The development of DNA-based diagnostic assays for the identification of bacterial pathogens has been limited thus far because of difficulties in (a) isolating highly specific nucleic acid probes, (b) developing assay formats that are sufficiently rapid and simple to compete with MAb assays, and (c) devising nonradioactive detection systems that provide the requisite sensitivity.

This chapter focuses on the approaches we have taken to overcome the problems associated with isolating DNA probes for bacterially mediated STD, as well as on the use of nonradioactive DNA hybridization–detection systems to develop rapid, simple, sensitive, and specific DNA hybridization assays for these pathogens. The examples given specifically relate to the development of rapid DNA probe diagnostics for gonorrhea and chlamydia, although efforts by others relating to the isolation of DNA probes and diagnostic tests for syphilis will also be discussed.

II. BACKGROUND

A. Selection of DNA Probes

Procedures used to identify and to isolate DNA probes for the detection of pathogenic organisms must meet two important criteria in order to be clinically useful diagnostic reagents. First, the DNA probes must be totally specific for the pathogen of interest. They must be able to distinguish these pathogens absolutely from both closely related organisms and more distantly related commensal organisms with some degree of sequence homology. Second, the probes must be able to identify all known isolates of the pathogen of interest. The use of probes that do not identify all members of a bacterial or viral species will yield assays with reduced diagnostic sensitivity and high levels of false negative results. If several different pathogen-specific DNA probes meeting these criteria are obtained, then other factors such as the length of the probes and the repetitive nature of the DNA segment can be used to select for probes that will provide high levels of assay sensitivity, as well as assay specificity.

In some cases, the identification of pathogen-specific DNA probes is a relatively simple matter. Viruses tend to have small genomes, which are often sufficiently unique to serve as DNA probes. The many different types of HPV can, for example, be distinguished from one another using entire viral genomes as DNA probes for hybridization assays (12,71). Even the full complement of chromosomal material in some bacterial species, such as the *Mycoplasma,* can be used as specific DNA probes (10).

However, in other instances, pathogens can share varying degrees of DNA sequence homology with both closely and more distantly related organisms. The ease with which totally specific DNA probes can be isolated for a given pathogen will generally depend on the proportion of genomic or endogenous plasmid material that is unique to that particular pathogen.

B. DNA Hybridization Assay Formats

The four most commonly used types of DNA hybridization assay formats are (1) Southern blot, (2) dot or spot blot, (3) *in situ* hybridization, and (4) sandwich hybridization assays. As with the selection of appropriate DNA probes, the choice of a hybridization assay format for detecting a pathogen in a clinical specimen rests upon the degree of specificity and sensitivity that it provides for accurately identifying that particular program in a given type of specimen. In addition, factors such as the speed, reliability, cost, and ease of performance and interpretation of the assay will also figure strongly into the decision to commercialize one assay format over another.

In Southern blot assays, specimen DNA is isolated and purified prior to restriction endonuclease digestion, separation of the digestion products by electrophoresis on an agarose gel, denaturation of the DNA in the gel, and transfer of the denatured DNA fragments to a solid matrix such as nitrocellulose membranes. The DNA bound to the solid matrix is then hybridized in the presence of radioactively labeled DNA probes to establish homology between probe and target DNA and/or to assess the restriction endonuclease patterns obtained. Hybridization of the probes to the target DNA is detected by autoradiography and often requires several days or weeks of exposure. The Southern blot procedure provides a considerable amount of information about the genetic contents of a given organism and is the method of choice for applications such as analyses of chromosomal abnormalities. It is, nevertheless, too lengthy and cumbersome for routine large-scale diagnosis of infectious diseases.

The dot-blot procedure also requires that specimen DNA be isolated and purified before being denatured and applied to a suitable solid matrix (e.g., nitrocellulose). Hybridization to the matrix-bound DNA is then performed using target-specific probes. The hybridization of probe DNA to the target DNA is detected either by autoradiography (when radioactively labeled probes are used) or by visual inspection (when nonradioactive hybridization–detection procedures are employed, as described in Section II,C). The spot-blot assay format is similar to that described for dot-blot assays, except that specimens or specimen lysates are directly applied to the solid matrix without prior extraction of their DNA. This modification greatly simplifies the assay and increases the attractiveness of its use in clinical laboratory settings. Spot-blot hybridization assays are relatively simple to perform and do not require the use of any expensive or highly technical machinery. They also allow many different samples to be processed at one time and are readily amenable to future automation. However, these assays are often subject to high backgrounds that complicate the interpretation of assay results.

In situ hybridization represents a third standard DNA hybridization assay format. In these assays, the DNA or RNA in the cells of a fixed tissue section or in fixed cultured bacterial or animal cells are hybridized to DNA probes directly on a microscope slide. The results are determined by microscopy if nonradioactive detection systems such as those described in Section II,C are used and by autoradiography if DNA probes are labeled with radioisotopes. There are several advantages to using this assay format for diagnostic purposes. One is that cellular morphology is retained in *in situ* hybridization assays. This provides an added dimension to the interpretation of these assays and reduces the likelihood of making incorrect assessments of assay results. A second advantage is that these assays can be used for both perspective and retrospective analysis of biopsy specimens for research or diagnostic purposes. In addition, *in situ* hybridization assays have the unique capability of distinguishing a single bacterial pathogen or pathogen-containing cell in the midst of as many as 10^5–10^6 nonpathogenic or uninfected cells, as long as that single cell has a detectable level of pathogen DNA or RNA. However, other assay formats are better suited for detecting the presence of many cells that contain few copies of pathogen DNA. *In situ* hybridization–detection assays that employ biotin–streptavidin–enzyme colorimetric detection systems can usually detect cells carrying 50–100 copies of pathogen DNA per cell (H.-L. Yang, Enzo Biochem, Inc., unpublished observation). More recently, *in situ* hybridization–detection assays performed on metaphase spreads or interphase nuclei using nonradioactively labeled DNA probes and fluorescently labeled antibody detection systems have provided the means for detecting a single copy of integrated viral DNA in an infected cell (69, 74), but this methodology has not yet been applied to the detection of bacterial pathogens. Procedures for enhancing the sensitivity of *in situ* hybridization assays by target amplification techniques are also being developed (78). Aside from the possible sensitivity concerns, a major problem associated with the use of *in situ* hybridization assays is that they are not currently suitable for screening large numbers of specimens. However, attempts to automate the performance and the analysis of *in situ* hybridization assays are currently in progress (25).

Sandwich hybridization assays are extremely specific in that they require that at least two different pathogen-specific probes hybridize to the target DNA, rather than just one. In these assays, one probe (the capture sequence) is bound to a solid support and is allowed to bind (capture) specimen DNA. A second probe (the signaling probe), with a sequence that is adjacent or close to the capture sequence on the target DNA is then allowed to hybridize to the support-bound target DNA. This signaling probe can be labeled with either radioactive (^{3}H, ^{35}S, ^{32}P, ^{125}I) or nonradioactive molecules (biotin, sulfonates, digoxigenin, or fluorescent dyes). The use of multiple probes and the removal of nonspecific cellular material in the first step of the procedure enhances the specificity of hybridization assays by reducing the effects of contaminating microorganisms or tissue debris. This factor, along with the physical setup of the assay, permits larger volumes of clinical specimens to be accurately analyzed. The use of sandwich assays for analyzing larger samples can potentially result in greater assay sensitivity for direct detention in clinical specimens. Additional sensitivity would be particularly desirable for testing cervical specimens. Sandwich assays are also amenable to automation, making their use advantageous for the rapid, large-scale testing of clinical specimens.

Enzo has, for example, developed a nonradioactive sandwich-type oligonucleotide-based DNA hybridization test for confirmation of *Mycobacterium tuberculosis* cultures that is amenable to either partial or complete automation (6). Following lysis of cultured specimens, hybridization is carried out in two steps and can be accomplished in <2 h (20–30 min "hands-on" time), even when as many as 30–60 specimens are to be analyzed. To date the test has been found to be 100% sensitive and specific.

In a departure from these standard hybridization procedures, which are performed on solid matrices, the Gen-Probe company has introduced a novel hybridization assay system in which appropriately labeled DNA probes are hybridized to ribosomal RNA (rRNA) target sequences in solution, then magnetically separated from unhybridized probes. Hybridization proceeds more efficiently in solution than when targets are bound to a solid support. In addition, rRNA is found in greater copy numbers in cells than most other sequences. Together, these two factors result in enhanced assay sensitivity. Moreover, this type of hybridization assay system is desirable for use in clinical laboratories because it is rapid and can be readily automated.

C. Nonradioactive DNA Hybridization–Detection Systems

DNA probes labeled with radioisotopes such as phosphorus-32 (^{32}P) have been used to identify and to characterize regions of chromosomal DNA that are unique to an organism or that specify a particular gene product or genetic trait in that organism (79). The radioactive label attached to the probes is used to monitor or to quantify the ability of a DNA probe to hybridize to its complementary (target) DNA sequences.

DNA hybridization reactions using radioactively labeled DNA probes are extremely specific and sensitive. These systems are, however, subject to a number of limitations that diminish their usefulness in clinical and diagnostic laboratory settings. First, there are problems in the economy of using these short-lived isotopes on a large scale (the half-life of ^{32}P, for example, is only 14 days). ^{32}P-Labeled nucleotides are expensive to purchase, have a fairly short half-life, and require costly disposal procedures. Employees must be properly trained to handle these materials safely. Detailed records for receipt, usage, and disposal must be maintained.

Enzo Biochem has developed a novel technology for identifying and detecting DNA by nonradioactive methods. This technology is based on the findings of David Ward and colleagues at Yale University (47). Labeled DNA probes are prepared by incorporating biotin, in the form of the biotinylated nucleotide Bio-11-dUTP, into the DNA structure using either standard enzymatic procedures (i.e., nick translation or tailing) or a new light-catalyzed method for biotin addition (i.e., with PhotoProbe Biotin, Vector Laboratories). The binding of these biotinylated DNA probes to their targets can be monitored colorimetrically by using complexes consisting of avidin or streptavidin (i.e., proteins with a high affinity for binding biotin), which are linked to enzymes such as horseradish peroxidase (HRP) and alkaline phosphatase. The complexed enzymes are selected for their ability to catalyze the production of colored products upon reaction with colorless substrate molecules in the presence of a chromogen. Quantitation of bound (hybridized) probe DNA can therefore be made on the basis of this color reaction. We, as well as D. Ward's group (20,47), have shown that this detection system can, in many cases, be as sensitive or almost as sensitive as ^{32}P. Biotinylated probes are also extremely stable. DNA probes labeled with biotin have been stored at Enzo for at least 3 years at 4°C with no loss of activity. Others have found these probes to be indefinitely stable at −20°C (20).

Variations of this biotin–avidin (streptavidin)–enzyme colorimetric hybridization–detection system have been devised whereby avidin (or streptavidin) is bound to fluorescently labeled antibody molecules and used as a bridge to link these labeled antibodies to biotinylated DNA probes. Multiple layers of avidin (or streptavidin), biotin, and antibodies can be used for signal amplification (67).

The CHEMIPROBE system (FMC BioProducts), in which DNA probes are labeled by sulfonation rather than by biotinylation, then allowed to bind labeled antisulfonate antibodies following hybridization and washing, is still another nonradioactive detection system that has been successfully used for monitoring the hybridization of DNA probes to their targets. Similar systems, in which DNA probes are labeled with other nonradioactive compounds and detected either colorimetrically or fluorimetrically using the appropriate antibodies to bind the label attached to the probes (i.e., digoxigenin; Boehringer-Mannheim Biochemicals), have also been commercialized.

More recently, Gen-Probe has developed a novel and sensitive chemiluminescent DNA hybridization–detection system in which DNA probes labeled with acridinium esters are hybridized to their targets and detected by a light-generating chemical reaction between these esters and alkaline hydrogen peroxide. This detection system, combined with its in-solution hybridization and magnetic bead separation technology (see Section II,B), forms the basis of Gen-Probe's PACE (Probe Assay–Chemiluminescence Enhance) Assay System.

Another approach that has been used for labeling probes is one in which an enzyme such as HRP or alkaline phosphatase is directly attached to probes via a linker arm. This technique, which has been used by DuPont in its Snap Probe line of products and by the DiGene Company, appears to be particularly well suited for labeling oligonucleotide probes.

Procedures have also been developed for directly labeling DNA or oligonucleotide probes with fluorescent dyes (fluorescein, rhodamine, Texas red, etc.) (1,73). These fluorimetric hybridization–detection systems permit hybridization to be detected directly rather than by the indirect formation of a colored product, and are particularly useful for monitoring *in situ* hybridization assays by fluorescence microscopy. Thus, fluorimetric detection is potentially faster and easier than colorimetric detection systems, but may lack some of the sensitivity of these enzymatic or antibody systems because this type of direct-labeling system does not result in signal amplification.

D. Diagnosis of Bacterially Mediated Sexually Transmitted Diseases

The three major bacterially mediated STD are chlamydia, gonorrhea, and syphilis. Chlamydia and gonorrhea account for the largest proportion of STD worldwide. There are ~1 million cases of gonorrhea and >5–10 million cases of chlamydia-related STD in the United States alone, making the development of rapid and accurate diagnostics for these infections a major health concern (83). Approximately 35,000 cases of infectious syphilis (i.e., primary and secondary syphilis) were reported in the United States in 1987 (54). This represents a 25% increase in the incidence of this disease over the number of cases reported in 1986. Worldwide, syphilis represents a significant health risk to millions of individuals, particularly in developing nations (50).

1. Gonorrhea

Gonorrhea is a common STD that results from infection with the gram-negative bacterium, *Neisseria gonorrhoeae*. It is manifested as mild inflammations of the urogenital tract, rectum, or pharynx that are sometimes accompanied by a purulent discharge. If untreated, gonorrhea can result in more serious conditions like pelvic inflammatory disease (PID) and infertility in females. Diagnostic tests for *N. gonorrhoeae* are usually performed on swabs from urethral exudates (males), endocervical swabs (females), and swabs from the anal canal (males and females). The presence of *N. gonorrhoeae* culture-positive genital tract specimens in pregnant women is associated with an increased incidence of chorioamnionitis and premature rupture of the membranes in the mother and ophthalmia neonatorum in the newborn (9). Gonorrheal infections are usually susceptible to treatment by a variety of antibiotics, including the sulfonamides and penicillins (4).

A variety of cytological and culture tests have been used to diagnose gonorrhea. Gram staining of smears from genital or rectal exudates is a rapid and inexpensive method for detecting *N. gonorrhoeae*. The identification of gram-negative diplococci within polymorphonuclear leukocytes is highly predictive of gonorrhea in males with symptomatic urethritis (with a sensitivity and specificity of 95%, compared to culture tests), but is a less satisfactory test for detecting these bacteria in women or asymptomatic males, where the sensitivity of the test drops to 40–70% (15). In addition, Gram-staining tests exhibit poor sensitivity for detecting *N. gonorrhoeae* in rectal smears (30–65% sensitivity compared to culture tests) and cannot be used to examine pharyngeal smears because of the presence of commensal nonpathogenic *Neisseria* in these specimens (15).

Culture tests for *N. gonorrhoeae*, in which bacteria from genital, rectal, and pharyngeal swabs are grown on selective media, are considered to provide the most definitive diagnosis of gonorrhea to date (80). Despite their high sensitivity in symptomatic males (95–98%) and their relatively high sensitivity for detecting gonorrhea in females (85%), there are still some limitations to the use of culture tests for *N. gonorrhoeae* that have prompted the development of alternate diagnostic tests for this STD (75). One of the major drawbacks of using culture tests for *N. gonorrhoeae* revolves around the difficulty of maintaining the viability of these bacteria in specimens when the testing site is far removed from the sampling site (80). Widespread use of carbon dioxide-generating transport devices containing specialized selective media has alleviated, but not eliminated, this problem (23). Loss of *N. gonorrhoeae* viability results from exposure of specimens to temperature extremes and to lengthy transport times, for example. In addition, the growth of as many as 10% of all gonococcal isolates is inhibited by the antibiotic (vancomycin), which is commonly used in selective transport and growth media (23). Moreover, the sensitivity of culture tests for gonorrhea is severely diminished after the initiation of antibiotic therapy, so that the course of the therapy cannot be adequately monitored (75). Another limitation of culture tests relates to the lengthiness of these assays. At least 2–3 days are required to obtain the results of culture tests, during which time an infected individual can transmit the bacteria to his or her sex partners, who in turn pass the disease on to others before they are diagnosed and treated (75). Complications such as salpingitis may also occur during the 2- to 3-day delay in treatment.

Clearly, there is a need for more rapid, reliable, and sensitive methods for diagnosing gonorrhea, particularly in asymptomatic males and in women. Various immunoassays have been developed over the last few years that can be used either in conjunction with existing culture assays (culture confirmation tests) or for direct detection of *N. gonorrhoeae* in clinical specimens with the aim of improving upon the speed, sensitivity, or specificity of culture tests.

Of these immunoassays, the Phadebact Monoclonal GC OMNI Test (Pharmacia) for culture confirmation of *N. gonorrhoeae* and the Gonozyme test (Abbott Laboratories) for the direct detection of *N. gonorrhoeae* in clinical specimens offer the best antibody-based alternatives to standard culture assays for gonorrhea. The Phadebact OMNI test uses MAb to detect various epitopes on a gonococcus-specific membrane-bound protein, protein I. It is highly specific (100%) and sensitive (98.4%) (7). These antibodies do not react with closely related organisms like *Neisseria meningitidis*, *Neisseria lactamica*, *Neisseria cinerea*, and *Branhamella catarrhalis*.

The Gonozyme test for *N. gonorrhoeae* is a solid-phase enzyme immunoassay in which polyclonal antibodies are used to detect gonococcal antigens. It is the most widely used nonculture method for detecting gonorrhea (15). The Gonozyme test is more rapid than culture analyses by virtue of the fact that it can be used for the direct detection of *N. gonorrhoeae* in clinical specimens, rather than following a culture step. In addition, the Gonozyme test does not require that the organisms in a specimen survive. It can detect antigen from *N. gonorrhoeae* in specimens as long as 5 days after collection (80). Although the sensitivity and specificity of the Gonozyme test is similar to that of culture and Gram-staining methods in high-prevalence populations (75), its high false negative rate in low-risk populations (72,75) and its restricted application to urogenital specimens (80) limits its general utility. An additional disadvantage of using the Gonozyme test is its inability to determine the antibiotic sensitivity of the infecting gonococcal strain (80).

Some progress has been made in applying DNA hybridization techniques to the development of rapid, simple, and reliable diagnostic tests for gonorrhea, but until recently this effort has been hampered by the lack of appropriate probes for the sensitive and specific detection of *N. gonorrhoeae* nucleic acids. *Neisseria gonorrhoeae* is closely related to *N. meningitidis,* as well as to several other species of nonpathogenic neisseriae. Up to 90% of the *N. meningitidis* genome is homologous to DNA in *N. gonorrhoeae* chromosomal DNA (32,43,44). Over the years, researchers have reported the identification of families of repeated chromosomal nucleotide sequences in *N. gonorrhoeae,* such as the 26-bp sequence identified by Correia *et al.* (14), but these have not been shown to be specific for *N. gonorrhoeae*. Other workers have successfully demonstrated the potential utility of nucleic acid hybridization assays for the direct diagnosis of gonorrhea in clinical specimens by using the cryptic plasmid of *N. gonorrhoeae* as the probes in their assays (60,81). In these studies, cryptic plasmid probes labeled with ^{32}P were used to examine urethral specimens from males. The sensitivity of this assay for detecting *N. gonorrhoeae* DNA was sufficient to generate an overall assay sensitivity and specificity of 89% and 100%, respectively, compared to culture assay results. The sensitivity of this assay for detecting *N. gonorrhoeae* DNA in urethral exudates from symptomatic males approximated that of culture assays (60). These studies are extremely encouraging for two reasons. First, their sensitivity is relatively high even though the probes used in these experiments were suboptimal. The cryptic plasmid was found in only 91.2% of the *N. gonorrhoeae* isolates identified in Seattle, Washington, where the specimens used in this study were obtained (81). Thus, the sensitivity of these spot assays is necessarily lower than what would be expected if the assays were performed using probe sequences found in all strains of *N. gonorrhoeae* at levels that are comparable to that of cryptic plasmid (~30 copies per cell). The sensitivity of this assay was calculated to be 98% if only isolates containing cryptic plasmids were used in the study. In general, the incidence of cryptic plasmid-bearing *N. gonorrhoeae* is variable and depends on the geographic location of the sample population

(13,17,64,81). In addition, some sequence homology between the cryptic plasmids of *N. gonorrhoeae* and plasmids carried by pharyngeal isolates of *N. meningitidis* and *N. lactamica* has been noted and would tend to diminish the specificity of assays using *N. gonorrhoeae* cryptic plasmid probes (39). Second, the observed sensitivity of the assays—that is, 0.1 pg of purified gonococcal plasmid DNA, which is equivalent to the amount of plasmid DNA present in 100–5000 colony-forming units (CFU) of *N. gonorrhoeae*—suggests that these assays can provide the means for efficiently detecting *N. gonorrhoeae* in cervical specimens when they are performed using optimal probes. Culture-positive cervical specimens contain on the average 4×10^2–1×10^7 CFU, a quantity that falls within the detection sensitivity limits of this DNA hybridization assay (60). The average number of *N. gonorrhoeae* CFU in culture-positive male urethral exudates is even higher.

The problem of isolating probes for detecting *N. gonorrhoeae* that provide the sensitivity and the specificity of the cryptic plasmid is not trivial. We have identified several such probes using the novel selection scheme described by Donegan *et al.* (18) and are currently in the process of using these probes to develop rapid, sensitive and specific spot-blot tests for gonorrhea (see Section II,A).

Gen-Probe is also marketing a nonradioactive DNA probe test for the direct detection of *N. gonorrhoeae* in urogenital specimens using its Pace Assay System (see Section II,B and II,C). Although the specificity, speed, and performance characteristics of this assay are excellent, the actual sensitivity of the assay does not exceed that of the MAb tests for *N. gonorrhoeae* and is lower than that of standard culture assays for specimens from a mixed population. The PACE test exhibits high sensitivity and specificity (93% and 99%, respectively) for detecting *N. gonorrhoeae* in high-prevalence male and female populations, however (21,24). In addition, the test cannot be used on specimens from patients who are undergoing antibiotic therapy because antibiotics interfere with rRNA metabolism.

2. *Chlamydia*

The chlamydiae are a group of obligate, intracellular bacteria. *Chlamydia trachomatis* is the primary human chlamydial pathogen. Infections with *C. trachomatis* can result in trachoma, inclusion conjunctivitis, lymphogranuloma venereum, urethritis, epididymitis, cervicitis, and if left untreated, PID, perihepatitis, ectopic pregnancy, endometrial infections, and infertility in females (3,5,76). Infants born to women with *C. trachomatis* infections are often premature, and at least 50% of them suffer from neonatal conjunctivitis and/or infant pneumonia (76). Many chlamydial infections are asymptomatic. However, even symptomatic chlamydial infections are difficult to diagnose because they are manifested as fairly nonspecific inflammatory diseases, such as urethritis (76). *Chlamydia trachomatis* infections are amenable to treatment with common antibiotics such as tetracycline and erythromycin (5,42).

Chlamydia trachomatis exists in two forms: the infectious, extracellular elementary body (EB) and the noninfectious, intracellular reticulate body (RB). The EB is a nondividing cell with approximately equivalent amounts of DNA and RNA. It has a thick, impermeable cell wall that is relatively resistant to sonication and trypsin treatment. The metabolically active RB, on the other hand, has three to four times more RNA than DNA. The RB forms have nonrigid cell walls and are sensitive to sonication and to trypsin proteolysis. *Chlamydia trachomatis* RB are about three times larger than their corresponding EB (5). It should be noted that infected cells contain ~10–1000 infectious EB (depending on the strain) at 18–24 hr postinfection.

The cytological criteria for diagnosing *C. trachomatis* involves the identification of intracellular inclusions either in specimens or in cultured cells stained by iodine, Giemsa, or immunofluorescence techniques. Cytological tests on direct specimens are <50% as sensitive as culture assays for *C. trachomatis* (3). Culture tests for *C. trachomatis* currently represent the definitive means for diagnosing these infections and are the standard by which all other chlamydial diagnostic tests are measured (3). Nevertheless, culture tests are lengthy, expensive, and technically difficult to perform and interpret (76). Specimens must be refrigerated for storage and transport in order to maintain cell viability. Even so, specimens must be transported to the testing site within 24 hr after collection (30). If same-day transport is not possible, samples can be shipped to the laboratory on dry ice, but this is costly and inconvenient for small operations.

Two types of commercial immunoassays for *C. trachomatis* have recently become available, which are designed to offer rapid diagnosis in several hours rather than the 2–7 days required for standard tissue culture analyses. These tests are based on the identification of chlamydial antigens, thus eliminating the need to maintain cell viability (30). The MicroTrak test for *C. trachomatis* (Syva) and a similar test developed by the Genetic Systems Corp. are direct monoclonal fluorescent antibody slide tests. Chlamydiazyme (Abbott Laboratories) and the Pathfinder Chlamydia EIA (Kallestead Diagnostics) are enzyme-linked immunoassays.

In the MicroTrak test for chlamydia, the EB in smears from urogenital, nasopharyngeal, or ocular specimens are detected using fluorescently labeled MAb to EB membrane antigens. The sensitivity of the MicroTrak test varies from 70 to 98%, depending on the population tested. For males in a high-prevalence population, the sensitivity and specificity of the MicroTrak test for urogenital secretions (93 and 98%, respectively) approximates that of culture assays (35,59). Similar values were obtained when MicroTrak was used to test nasopharyngeal specimens. In low-prevalence or asymptomatic populations, the specificity and sensitivity of this test for urogenital specimens are somewhat lower: 87.5 and 97.4%, respectively (35). The ability of the MicroTrak test to detect *C. trachomatis* in cervical secretions (sensitivity of 70–76%) and in conjunctival specimens from adults with acute follicular conjunctivitis and babies suspected of having chlamydia-related ophthalmia neonatorum (80%), however, is much lower than that of culture assays (28,49). The advantages of the MicroTrak assay over culture tests for *C. trachomatis* is that the MicroTrak assay is an extremely rapid direct test that is as sensitive as culture assays for male patients in high-risk or symptomatic populations. These direct assays cannot, however, be used to replace culture assays for testing cervical or ocular specimens. Moreover, the skill and experience of the technicians performing these tests is a major factor in their accuracy. Errors in interpreting the MicroTrak test can easily be made by inexperienced personnel (76).

The Genetic Systems test uses a MAb to the principal membrane protein present in EB that is common to all serovars of human *C. trachomatis* (22). The sensitivity and specificity of this direct slide test is comparable to both culture and MicroTrak assays.

The sensitivity of the Chlamydiazyme test for detecting *C. trachomatis* antigens in swab specimens is somewhat lower than that of the MicroTrak and culture assays with the possible exception of tests performed on cervical specimens, for which the sensitivity of Chlamydiazyme may be better than that of the MicroTrak test (30). However, there may be some problems due to the presence of commensal bacteria in urogenital specimens when using the Chlamydiazyme assay (30). The sensitivity and specificity of this test are reported to be ~83 and 98%, respectively, compared to culture assays (2,30). Nevertheless, Chlamydiazyme is easier to perform and is less subject to interpretive errors than is the MicroTrak test for chlamydia and may offer comparable sensitivity in the hands of inexperienced laboratory personnel (76).

The Pathfinder Chlamydia EIA (Kallestead Diagnostics) appears to be easier, faster, and more sensitive than the Chlamydiazyme assay for the direct testing of urogenital specimens from high-prevalence populations of males and females (51). In addition, it does not seem to exhibit the cross-reactivity or specimen interference problems that occur with the Chlamydiazyme assay.

More recently, a novel direct assay for detecting *C. trachomatis* that uses the dot immunobinding technique (DIBT), MAb, and a nonradioactive biotin–avidin–enzyme detection system has been introduced and appears to have a high sensitivity (96.7% agreement with culture assays) for detecting chlamydial RB membrane antigens (62). This assay is able to detect as little as 75 pg of chlamydial antigens in specimens that have been stored as long as 1 week at room temperature.

Serological tests for detecting antibodies to *C. trachomatis* are also available and can be used to detect these antibodies in culture-positive individuals (3). The disadvantage of this test, however, is that it is unable to distinguish between current and past infections.

There is some reason to believe that DNA hybridization assays may provide a valuable alternative to the lengthy and cumbersome culture assays for *C. trachomatis* and to the faster, but suboptimal, direct antibody assays for this organism. A DNA probe test for trachoma, for example, exhibited a sensitivity and specificity of 86 and 91%, respectively, compared to culture assays, whereas a fluorescent-antibody assay had a sensitivity of only 45% for the same specimen population (16).

Palmer and Falkow (60) have cloned a 7-kb cryptic plasmid from *C. trachomatis* that is common to all known serovars of this organism but that does not hybridize to *Chlamydia psittacci* (another member of the chlamydiae that shares 10% sequence homology with *C. trachomatis* and is the causative agent for psittacosis). Spot hybridization assays performed on specimens using this plasmid labeled with ^{32}P as probe were only 82% as sensitive as culture assays in a random mixture of high- and low-titer specimens, whereas the sensitivity of these assays was equivalent to that of culture assays for heavily infected specimens from symptomatic individuals (60). Similar spot-blot assays were performed by Hyypia *et al.* (38) using probes prepared from both chromosomal and plasmid DNA.

An alternative DNA hybridization assay format, nonradioactive *in situ* hybridization, which uses *C. trachomatis* cryptic plasmid probes, has been developed at Enzo Biochem (see Section II,C) for confirmation of *C. trachomatis* cultures. Horn *et al.* (34) have also used *in situ* DNA hybridization assays with cryptic plasmid probes to examine cervical scrapings such as Pap smear specimens. The results of these studies were extremely encouraging, with a sensitivity of 91% and a specificity of 80%, compared to culture assays.

Gen-Probe has applied its PACE Assay technology to the direct detection of *C. trachomatis* in urogenital swabs. The sensitivity and specificity of this test compared to standard chlamydial culture tests are 88 and 95%, respectively, according to literature provided by the company. These values are comparable to those obtained by Gen-Probe in a similar assay that employs DNA probes labeled with iodine-125 (^{125}I). Gen-Probe also claims that the PACE test for *C. trachomatis* is able to detect <5 infectious units of *C. trachomatis* per assay using its chemiluminescent labeling system. This being the case, it is not clear why the assay sensitivity is so much lower than that of culture assays. Nevertheless, this is a rapid (2 hr), direct test for *C. trachomatis* that does not suffer from the transport problems associated with culture assays; for example, *C. trachomatis* rRNA can be detected in urogenital specimens maintained in the PACE Assay System transport media for as long as 7 days.

Another potentially useful approach has been taken by Palva *et al.* (61). These workers have devised a sandwich hybridization assay that is capable of detecting 10^6 molecules of *C. trachomatis* DNA using various fragments of ^{125}I-labeled *C. trachomatis*-specific chromosomal DNA as probes. This overnight assay was used to detect *C. trachomatis* in urogenital specimens. Three culture-positive and three culture-negative specimens obtained from females with cervicitis, and five culture-positive and five culture-negative specimens from males with urethritis were tested in this pilot study. The sensitivity and specificity of the test compared to culture assays was 100% in all cases. A larger clinical study must be performed in order to assess accurately the efficacy of this assay. Further commercialization of sandwich hybridization assays for *C. trachomatis* would also require that the assay be modified so as to reduce its complexity and lengthiness. The use of nonradioactively labeled probes in this type of assay would also be desirable.

3. Diagnosis of Syphilis

The causative agent of syphilis is the spirochete, *Treponema pallidum* (41). Pathogenic treponemes have many antigens in common. It has been shown that active infection with one strain or subspecies provides some protection against infection with another in humans and experimental animals. However, these strains and subspecies must exhibit significant antigenic differences because infection with one type of *T. pallidum* does not confer complete resistance to others (50).

Treponemal infections are generally transmitted by direct contact with skin or mucous membrane lesions and are characterized by two infectious stages (primary and secondary syphilis), long periods of latency, and a third active but noninfectious stage (tertiary syphilis). Between 30 and 60% of known contacts of patients with infectious lesions of venereal syphilis become infected, with 3–6 weeks elapsing between contact and development of a primary infection, which is usually characterized by the presence of chancre sores (50). These infectious lesions generally heal spontaneously after 1–6 weeks and are followed by a secondary disease stage characterized by widespread cutaneous and systemic symptoms that last for 2–10 weeks (37). Tertiary, or late, syphilis occurs in approximately one-third of the infected individuals and results in a variety of cardiovascular, neurological, cutaneous, and visceral diseases (37,57). Syphilis can be transmitted to a developing fetus at any stage and results in fetal death or in the development of congenital syphilis (9). Congenital syphilis often results in neurological disorders, which can either be seen at birth or be delayed for as long as 50 years or more (37). Syphilis may be treated with common antibiotics (i.e., penicillin, tetracycline, or erythromycin) (37).

Clinical diagnosis of syphilis is often difficult (33). Even in STD clinics, where there is a higher incidence of syphilis, the positive predictive value of clinical diagnosis of syphilis based on the presence and the appearance of rashes or lesions is only 78% (33).

The diagnosis of early or primary syphilis currently requires serological or dark-field microscopy to demonstrate infection with *T. pallidum*. Serological screening tests, such as the VDRL (Venereal Disease Research Laboratory) test, are positive in 70–90% of patients with primary syphilis (22). The VDRL test detects the presence of reagin, an antibodylike substance that reacts with several phospholipids, such as cardiolipin, phosphoinositol, cholesterol, and lecithin (34,45). Reagin is thought to result from the presence of breakdown products due to the interaction between *T. pallidum* and host tissues (37), but it is not specific to *T. pallidum* infections. The VDRL test is neither sensitive nor specific, but positivity with a high titer is generally indicative of active infections with *T. pallidum* (57). False positive reactions with the VDRL test and with another, similar serological test (the rapid reagin plasma or RPR test) have been associated with hepatitis, infectious mononucleosis, viral pneumonia, malaria, chickenpox, measles, smallpox vaccination, pregnancy, and more commonly with autoimmune and collagen vascular diseases such as systemic lupus erythematosus (SLE) (34,37,45,46). Other conditions that can produce false positive reactions include narcotic addiction, aging, leprosy, and malignancy (37). In addition, these tests often fail to detect serological responses to syphilis in individuals with HIV as a result of the immunosuppressing characteristics of this disease (29). The predictive value of the VDRL test is dependent on the stage of the disease (27). Both the VDRL and RPR tests work best for the secondary and early latent stages of syphilis. The relative insensitivity of these tests for detecting early and late syphilis and their lack of specificity limits their utility (33,37). Nevertheless, these tests are usually recommended for screening general populations (low-prevalence groups), for following posttreatment serum titers to detect recurrences of syphilis, and for diagnosing neurosyphilis [cerebrospinal fluid (CSF) VDRL].

Tests for serum antibodies to *T. pallidum,* such as the FTA-ABS (fluorescent treponemal antibody absorption) test, microhemagglutination assay for *T. pallidum* (MHA-TP), and *T. pallidum* immobilization (TI), are more specific than the VDRL and RPR tests, but otherwise suffer from many of the same disadvantages (22,37,53). The FTA-ABS test is the most commonly used member of this group.

The sensitivity of the FTS-ABS test for primary syphilis is 86%; for secondary syphilis it is 100%, for early latent syphilis, 99%, for latent syphilis 96%, and for late (tertiary) syphilis 95% (37). Generally, the FTS-ABS test is unable to detect antibodies to *T. pallidum* antigens in the first week of syphilitic infections. False positive results are seen primarily with autoimmune and drug-induced collagen vascular diseases. Thus, the FTA-ABS test has a higher sensitivity and specificity than the VDRL test and becomes positive earlier in primary syphilis. The disadvantages of this test are its lack of titer quantifaction so it cannot be used for posttreatment follow-up, its inability to distinguish between past and present *T. pallidum* infections, and its low predictive value when used to screen low-prevalence populations (37). In addition, it is more costly and difficult to perform than the VDRL and RPR tests (22).

The results of analyses performed using a new enzyme-linked immunoabsorbent assay (ELISA) for detecting the presence of antitreponemal antibodies in serum, the Syphilis Bio-EnzaBead Test (Litton Bionetics), were in good agreement (93%) with the FTA-ABS test. This assay uses *T. pallidum* antigens, which are bound to ferrous metal beads to capture (bind) anti-*T. pallidum* antibodies in serum specimens. The EnzaBead test is easy to perform and results in clear, unequivocal results (77).

Two variations of the FTA-ABS test have been developed for detecting IgM to treponemal antigens, rather than total immunoglobulins: the IgM fluorescent treponemal antibody absorption test (IgM-FTA-ABS) and the solid-phase hemadsorption assay (SPHA). Both IgM-FTA-ABS and SPHA tests can be used to demonstrate the presence of IgM antibodies in 83–100% of patients with active syphilis, depending on the stage of the infection. The advantage of this test over the FTA-ABS test is that IgM antibodies are absent in serum from patients who have been treated and cured. Therefore, these serological tests permit resolution of current and past infections. The overall sensitivity of the IgM-FTA-ABS test (92%) was comparable to that of the SPHA test (96%), but the SPHA test was more specific. Also, the SPHA test, but not the IgM-FTS-ABS test, detected IgM in instances of neurosyphilis. Both tests detected the reappearance of syphilis in patients who were reinfected with *T. pallidum* (53).

In cases where serological tests fail to detect syphilis, such as in the early primary stage, positive dark-field microscopy often provides the only evidence of *T. pallidum* infections (37). Dark-field microscopy for the detection of characteristic treponemal organisms in clinical specimens provides a method for rapid and simple diagnosis of syphilis, but it is not available to most clinical laboratories and is unreliable for the analysis of specimens from oral and rectal lesions because of the presence of commensal nonpathogenic organisms that closely resemble *T. pallidum* (33).

Antibody tests performed directly on specimens using polyclonal sera are not sufficiently sensitive or specific to permit accurate identification of *T. pallidum* to be made. However, the use of MAb probes for microscopic analysis of specimens in the diagnosis of early syphilis generally results in tests which are quite specific (33). Genetic Systems has developed a MAb test for the direct detection of *T. pallidum* in specimens by microscopy. This test employs a fluorescein-conjugated MAb directed against a 48,000 Da *T. pallidum* protein. This MAb recognizes *T. pallidum* and *Treponema pertenue,* but not *Treponema phagedenis* (biotype Reiter) or other commensal treponemes (22). The sensitivity of this test is comparable to dark-field and serological tests, but the MAb test is much more specific (33).

The diagnosis of neurosyphilis in the tertiary age of syphilis is difficult because of the lack of a good laboratory test, diversity of clinical manifestations, and a rising incidence of atypical forms of the disease (48). Traditionally, the VDRL test, cell count, and protein content have been used to diagnose neurosyphilis from CSF specimens. The CSF VDRL test is sufficiently specific, but it lacks sensitivity. Increases in the CSF cell count or protein content are not specific for syphilitic inflammation of the CNS (31). The FTA-ABS test for CSF had a sensitivity of 96%, compared to 79% for VDRL CSF tests (48).

The development of DNA probe tests for the detection of *T. pallidum* infections and, ultimately, for the diagnosis of syphilis is still in its early stages. A variety of *T. pallidum*-specific genes have been isolated, cloned, and expressed in *Escherichia coli* with the idea of using them to provide a source of *T. pallidum*-specific antigens for antibody production or as vaccines (8,11,26,36,65,66,68,82). Nevertheless, these cloned DNA sequences provide a ready source of DNA probes for the detection of *T. pallidum* by DNA or RNA hybridization assays. Realistically, these assays could only be developed for detecting *T. pallidum* from lesion specimens during primary or secondary syphilis, because lesions containing this organism are most accessible during these disease stages. The efficacy of these tests would, therefore, be dependent on the number of *T. pallidum* organisms in these specimens. In fact, the majority of patients with dark-field-positive tests for syphilis have only one *T. pallidum* cell or less per oil immersion field examined, and the number of treponemes in these specimens decreases with increasing progression of the disease; that is, fewer treponemes are detectable during latency or in tertiary visceral, cardiovascular, or neurological lesions than in primary and secondary syphilitic lesions (63).

Perine (63) has described studies in which the cloned *T. pallidum* probe pAW305 (82) was used to examine lesion exudates from six adult patients living in Seattle, Washington, who had primary syphilis that was identified by dark-field microscopy. The probe DNA used in these experiments was labeled with ^{32}P and could detect 0.1–1 pg of *T. pallidum* clone DNA, or as few as 10^4 *T. pallidum* bacteria using a dot-blot hybridization format. A 1- to 3-day exposure of the blot to X-ray film was necessary to achieve this degree of sensitivity. This assay permitted detection of syphilis in 66% (four of six) of the specimens tested. Similar assays performed by other workers (63) using two *T. pallidum* DNA probes, pAW305 and pAW327, to examine lesion exudates from syphilitic patients from Zambia had sensitivity and specificity levels of 75 and 70.5%, respectively.

Thus, the results of preliminary studies aimed at detecting *T. pallidum* in specimens from primary syphilitic lesions are not terribly encouraging. However, the use of alternate assay formats, specimen collection techniques, or target amplification techniques (e.g., the PCR technique) may result in clinically relevant assays for the diagnosis of syphilis during its early stages. These tests would be more sensitive, specific, and accurate than dark-field microscopy and, in addition, would not require that *T. pallidum* viability be maintained during specimen transport or analysis.

III. RESULTS AND DISCUSSION

A. Identification and Isolation of *Neisseria gonorrhoeae*-Specific DNA Probes

Studies performed by Palmer and Falkow (60), Perine *et al.* (64) and by Totten *et al.* (81) were aimed at developing DNA hybridization assays for detecting *N. gonorrhoeae* and successfully demonstrated the feasibility of using DNA probes to detect this organism in clinical specimens from males with urethritis. But the actual assays developed by these investigators are not clinically useful because they (1) involved the use of cryptic plasmid probes that are not present in all strains of this organism, (2) were lengthy and complex, and (3) used radioisotopic labels to monitor hybridization reactions.

As a first step toward developing clinically relevant nonradioactive DNA hybridization assays for gonorrhea, we undertook the formidable task of identifying and isolating *N. gonorrhoeae*-specific DNA probes. The isolation of probes for this bacterium was extremely difficult because of the high degree of sequence homology between *N. gonorrhoeae* and *N. meningitidis* DNA (32,43,44).

In order to isolate DNA probes for gonorrhea, it became necessary for us to develop a novel method for identifying segments of *N. gonorrhoeae* DNA that specifically hybridize to all strains of *N. gonorrhoeae*, but not to other neisseriae or to other unrelated commensal organisms (18). This technique involves the use of a sandwich-type hybridization assay format to screen recombinant phages containing the *Mbo*I restriction endonuclease digestion products of *N. gonorrhoeae* chromosomal DNA and M13mp8 vector DNA.

Two features of this screening process were crucial to its success. First, conditions were used that permitted specific hybridization to occur without the prior purification of phage DNA. The ease with which large numbers of recombinants could be screened using this approach greatly enhanced the feasibility of this undertaking. Second, the use of labeled vector sequences for detecting the hybridization of the cloned DNA to target sequences was important because it (1) amplified the signal that would normally be obtained if nick-translated clone DNA was used for screening, (2) facilitated the large-scale screening of clone DNA by reducing the number of probes requiring nick translation to one (^{32}P-labeled M13mp8 DNA), and (3) enabled us to quantify the specificity of the recombinant DNA by determining the counts per minute (cpm) of ^{32}P bound to the washed hybridized filters. In an attempt to maximize our chances of identifying genomic probes, rather than cryptic plasmid sequences, we selected a strain of *N. gonorrhoeae* that did not contain the cryptic plasmid to prepare this recombinant DNA library. We later discovered that this precaution was irrelevant to the success of the probe isolation procedure because *Mbo*I sites are not present on this plasmid and thus, no plasmid-derived *Mbo*I restriction fragments would have been included in our recombinant phage pool.

The screening procedure used to identify phages containing *N. gonorrhoeae*-specific DNA was developed by Donegan *et al.* (18). Initially, stocks were prepared from each of 3000 individual phage plaques. Hybridization was accomplished by transferring a portion of each phage stock to individual tubes containing hybridization solution and a nitrocellulose filter with dots of chromosomal DNA from *N. gonorrhoeae* and *N. meningitidis* fixed at separate sites, and incubating this mixture under the appropriate conditions. In the second stage of this procedure, the hybridization–detection sandwich was completed by allowing any filter-bound recombinant phage DNA to hybridize to denatured ^{32}P-labeled M13 replicative-form (RF) DNA to permit the detection of hybridized phage DNA. The portion of hybridizing DNA in this step was therefore homologous to the vector DNA attached to the insert, not to the cloned *N. gonorrhoeae* DNA. The filters were again washed, then dried, cut to separate the *N. gonorrhoeae* and *N. meningitidis* spots, and counted in a scintillation counter. Recombinant phages whose DNA hybridized to *N. gonorrhoeae* at a level that was at least 20 times higher than that at which they hybridized to *N. meningitidis* DNA (i.e., exhibited a ratio of at least 20 : 1 for cpm bound to *N. gonorrhoeae* vs. cpm bound to *N. meningitidis* DNA) were selected for further testing to ensure that they would not hybridize to other bacterial species.

Of the 3000 recombinant phages tested using this procedure, 50 were able to hybridize to the parent strain of *N. gonorrhoeae* (i.e., the strain used to prepare the *N. gonorrhoeae* recombinant phage library used in these studies), but not to the strain of *N. meningitidis* used in the primary screening procedure. These clones were subjected to a secondary screening procedure in which they were allowed to hybridize to filters containing DNA from representatives of all strains of *N. gonorrhoeae* and *N. meningitidis* (listed in Table I). Four recombinant phages whose DNA hybrid-

TABLE I

Representative *Neisseria gonorrhoeae* and *Neisseria meningitidis* Isolates Tested

Species	Strain	Serotype	Source
N. gonorrhoeae	7876		NRL
	8038		NRL
	7925		NRL
	7122		NRL
	6611		NRL
	5016		NRL
	7325		NRL
	7726		NRL
	7929		NRL
	34271		NRL
	33152		NRL
N. meningitidis	Y	Y	NRL
	W135	W135	NRL
	MCA4	A	E. Gotschlich, Rockefeller University
	ST10		E. Gotschlich, Rockefeller University
	1372		E. Gotschlich, Rockefeller University
	2D72		E. Gotschlich, Rockefeller University
	C2241		E. Gotschlich, Rockefeller University
	7185		E. Gotschlich, Rockefeller University
	7086		E. Gotschlich, Rockefeller University
	2B1		E. Gotschlich, Rockefeller University

ized to all 27 isolates of *N. gonorrhoeae,* but not to any of the *N. meningitidis* serotypes, were selected for further analysis, since these contained *N. gonorrhoeae* DNA that could be used in the development of clinically relevant diagnostic assays for gonorrhea. Cloned DNA that shared significant homology with the DNA in any of the strains of *N. meningitidis* tested or that failed to identify any of the *N. gonorrhoeae* isolates were not considered to be useful diagnostic tools.

Finally, it was necessary to determine if these four candidate clones could hybridize to DNA from bacteria that are commonly found in the urogenital tract (either as pathogens or as commensal organisms), since these would be likely contaminants of urogenital specimens that could interfere with the accuracy of subsequent diagnostic analyses. To test this possibility, RF DNA prepared from each of the four recombinant phages was labeled with ^{32}P by nick translation and used to probe nitrocellulose filters containing fixed spots of DNA from each of the organisms listed in Table II. Figure 1 illustrates the results of one such experiment. None of the four clones contained insert DNA that hybridized to DNA from these bacteria. These recombinant phages could therefore be used as *N. gonorrhoeae*-specific DNA probes.

The DNA from these phages was further characterized by restriction endonuclease mapping studies. The restriction maps of these recombinant plasmids are presented in Fig. 2. Two of the plasmids, GC155 and GC704, were separate isolates that contained the same 850-bp *Mbo*I fragment of *N. gonorrhoeae* DNA in opposite orientations. Therefore, only the map of GC155 is shown. The DNA inserts in the other two plasmids, GC1657 and GC801, are also small (i.e., 850 and 1300 bp, respectively). The small sizes of the inserts identified by this procedure may reflect (1) the use of a restriction endonuclease (*Mbo*I) that cuts DNA frequently to prepare the original recombinant phage library, (2) the use of M13 vectors, which favor the cloning and maintenance of small DNA inserts, or (3) the possibility that *N. gonorrhoeae*-specific DNA is present in short sequences that are interspersed throughout the genome of this bacteria. Southern blot analysis of *N. gonorrhoeae* chromosomal DNA indicates that the GC155 insert is repeated at least three or four times in the genome of this bacteria and that the inserts in GC1657 and GC801 are probably present in only one copy per genome. This observation is consistent with the observation that two of the four independent isolates contained the GC155 insert.

The *N. gonorrhoeae* DNA inserts in GC155 and GC1657 have subsequently been joined together and inserted into several different vectors (M13mp9, pUC9, and pIBI76). Dot-blot hybridization assays performed using ^{32}P-labeled probes consisting of the joined inserts in M13mp8 vectors (GC2B) were 40–50% more sensitive than the corresponding assays performed using probes with only the GC155 insert in M13mp8.

TABLE II

Bacteria Tested for Cross-Reactivity with *Neisseria gonorrhoeae*-Specific Probes

Strain	Origin
1. *Kingella kingae*	ATCC 23330
2. *Neisseria canis*	ATCC 14687
3. *Neisseria animalis*	ATCC 19573
4. *Neisseria mucosa*	ATCC 19696
5. *Neisseria subflava*	ATCC 14799
6. *Mycoplasma pneumoniae*	ATCC 15531
7. *Neisseria meningitidis*/Y	NRL
8. *N. gonorrhoeae*/33152	NRL
9. *N. meningitidis*/1665	E. Gotschlich, Rockefeller University
10. *Neisseria caviae*	ATCC 14659
11. *Klebsiella pneumoniae*	ATCC 13883
12. *Enterobacter aerogenes*	ATCC 13048
13. *Ureaplasma urealyticum*	ATCC 27618
14. *Mycoplasma columbinasale*	ATCC 33549
15. *Mycoplasma gallopovonis*	ATCC 33551
16. *Streptococcus* GpA	Phillip Tierno, New York University Medical Center
17. *Neisseria lactamica*	ATCC 23970
18. *Staphylococcus aureus*	ATCC 12600
19. *Lactobacillus casei*	ATCC 00393
20. *Legionella pneumophila*	ATCC 33152
21. *Streptococcus* GpB	Phillip Tierno, New York University Medical Center
22. *Streptococcus* GpC	Phillip Tierno, New York University Medical Center
23. *Streptococcus* GpD	Phillip Tierno, New York University Medical Center

B. Development of Spot-Blot Hybridization Assays for *Neisseria gonorrhoeae*

These probes were used in the development of a rapid (40-min) and simple nonradioactive dot-blot hybridization assay format for detecting *N. gonorrhoeae* DNA. This assay is directly applicable to the confirmation *N. gonorrhoeae* cultures and possibly to a direct diagnostic test for gonorrhea. It has considerable potential for overcoming the disadvantages of existing tests.

In the spot-blot assay that we have developed, bacterial colonies are collected by touching them to a wooden applicator stick. The stick is dipped into a solution consisting of 0.5 *M* sodium hydroxide and 0.5%

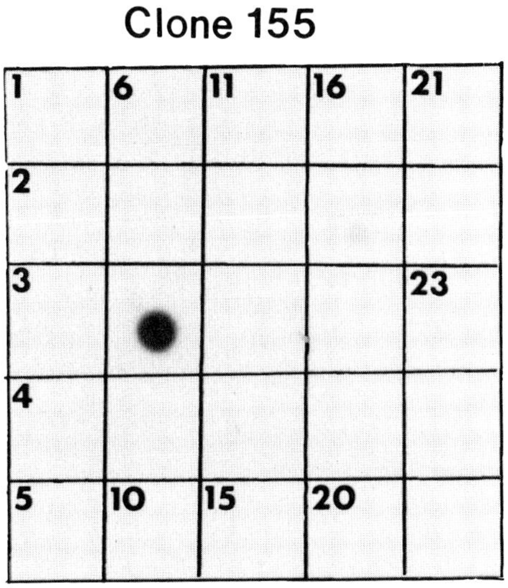

Fig. 1. Specificity of presumptive *Neisseria gonorrhoeae* DNA probes. DNA from each of the correspondingly numbered bacteria listed in Table II was spotted onto GeneScreen Plus filters and hybridized to ^{32}P-labeled pUC9-GC155 plasmid DNA. The autoradiogram of the hybridized, washed filter is shown.

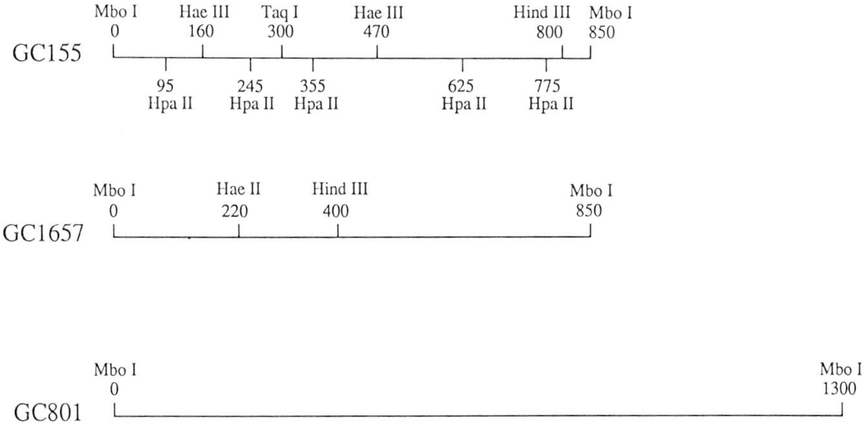

Fig. 2. Restriction endonuclease digestion maps of the isolated *Neisseria gonorrhoeae*-specific insert DNA. The insert DNA in GC155 and GC704 is identical, but occurs in opposite orientations. There are no restriction sites for *Ava*I, *Bam*HI, *Bgl*II, *Eco*RI, *Hae*II, or *Sal*I in the GC155 insert. Restriction enzymes *Ava*I, *Bam*HI, *Bgl*II, *Eco*RI, *Hae*III, *Hpa*II, *Sal*I, and *Taq*I do not cleave the GC1657 insert DNA. The GC801 insert cannot be cleaved by restriction endonucleases *Ava*I, *Bgl*II, *Eco*RI, *Hae*II, *Hae*III, *Hind*III, *Sal*I or *Taq*I.

sodium dodecyl sulfate (SDS) to lyse the attached cells and, at the same time, to denature their DNA. This treated material is then applied to a membrane (GeneScreen Plus) that has been fixed to a microscope slide. Cellular debris is removed by rigorous washing, while cellular DNA remains bound to the filter. Hybridization and detection steps are then performed using biotinylated *N. gonorrhoeae*-specific probe DNA (i.e., GC155 insert DNA) and a biotin–streptavidin–horseradish peroxidase (DETEKI-*hrp*) detection system. An example of the results obtained using this spot-blot assay is presented in Fig. 3. This assay is able to detect *N. gonorrhoeae* DNA specifically in 10^6–10^7 bacteria. No hybridization to any of the nongonococcal bacteria listed in Table II was detected.

C. Development of an *in Situ* Hybridization Assay for *Chlamydia trachomatis*

Recently, we developed a 45-min nonradioactive shell vial *in situ* hybridization assay for detecting *C. trachomatis* DNA in infected McCoy cells following a 24-hr culture period. The probe DNA used in these assays consists of a nick-translated, biotinylated *C. trachomatis*-specific cryptic plasmid DNA sequence. This probe is analogous to that described by Palmer and Falkow (60) and is capable of recognizing all 15 serotypes of *C. trachomatis,* but does not hybridize to *C. psittaci* DNA using the assay conditions described in Section VIII,F (M. Molina, Enzo Biochem, unpublished observations). An example of the kind of staining obtained when a biotin–avidin–HRP colorimetric detection system was used to detect hybridization of the *C. trachomatis*-specific probe DNA to its target sequences in the inclusions of the infected cells is shown in Fig. 4. The inclusions found in the cytoplasm next to the nuclei of infected cells are stained a strong, readily discernible brick red color (which appear as black areas in Fig. 4). Unlike the signal produced using fluorescence immunoassays, the reddish deposit in the inclusions forms a permanent record of the assay results that does not fade with time.

IV. CONCLUSIONS

The continued lack of simple, inexpensive tests for reliably detecting gonorrhea and chlamydia in women has precluded the effective control of these STD in most of the world (58). In response to this deficiency, we have initiated a series of experiments that will eventually result in clinically relevant DNA hybridization assays for these diseases.

As a first step toward developing a diagnostic assay for gonorrhea, we isolated DNA probes that are capable of specifically detecting all known

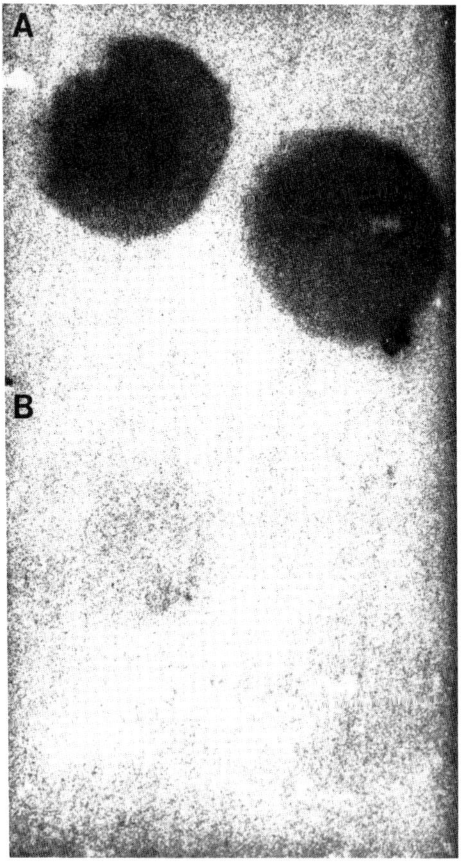

Fig. 3. Nonradioactive spot-blot hybridization assay for confirmation of *Neisseria gonorrhoeae* cultures. Ten colonies of either (A) *N. gonorrhoeae* strain GC33152 or (B) *Neisseria meningitidis* strain Y were treated with sodium hydroxide and SDS as described in Section VIII,E and applied in duplicate to a piece of GeneScreen Plus filter (attached to a microscope slide). Hybridization of the matrix-bound DNA to biotinylated GC155 insert DNA was visualized using DETEKI-*hrp*. Positive reactions with *N. gonorrhoeae* DNA are indicated by the formation of a precipitate (dark spot) at the site of sample application.

isolates of *N. gonorrhoeae*. We then used these probes in conjunction with a nonradioactive, colorimetric hybridization–detection system to develop a rapid spot-blot assay for *N. gonorrhoeae*. Despite the fact that these probes consisted of chromosomal DNA sequences that are repeated 1–4 times in the bacterial genome, compared to the 30 copies per cell at which the *N. gonorrhoeae* cryptic plasmids are present, and despite the fact that

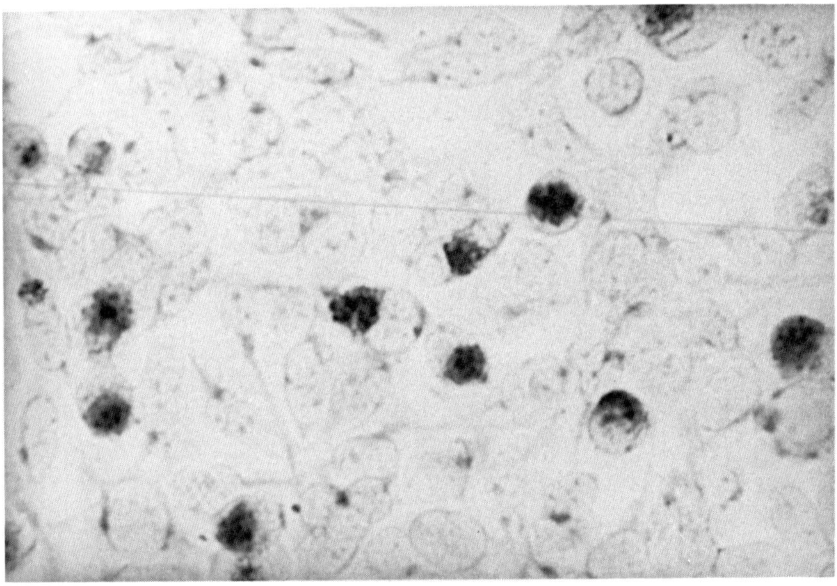

Fig. 4. Rapid, nonradioactive *in situ* DNA hybridization assay for *Chlamydia trachomatis*. McCoy cells infected with *C. trachomatis* serotype G were fixed onto a microscope slide and hybridized to biotinylated *C. trachomatis*-specific probe DNA in shell vials. The *C. trachomatis* DNA in the inclusions of the infected cells (darkly staining perinuclear regions) were visualized using a DETEKI-*hrp* detection system. The cells were counterstained with Evans blue prior to viewing. Magnification × 400.

nonradioactive detection systems were used in place of ^{32}P labels, the sensitivity of this assay was comparable to that of the radioactive dot-blot hybridization assay described by others (60,81) and is therefore likely to provide a clinically relevant alternative to these assays.

Although a variety of culture and direct assays for chlamydial infections are available, this disease remains the most prevalent STD worldwide. As with gonorrhea, infection with *C. trachomatis* is particularly difficult to diagnose in women. Unlike gonorrhea, however, DNA probes for the specific detection of *C. trachomatis* were easily obtainable (60), but sensitivity, performance, and cost considerations have hampered the development of clinically relevant diagnostic assays for chlamydia.

In situ hybridization assays can be satisfactorily used for *C. trachomatis* detection because *C. trachomatis* DNA can be found in relatively large, easily identifiable intracellular inclusions. Diffuse cytoplasmic or extracellular staining can be readily identified as background. In contrast, background hybridization in dot-blot, sandwich hybridization, or even *in situ* hybridization assays of "free-floating bacteria" like *N. gonorrhoeae,* can easily and erroneously be interpreted as a positive result. A second advantage of *in situ* hybridization assays for *C. trachomatis* is that the presence of a single cell with the appropriate staining characteristics permits positive identification of *C. trachomatis* infections to be made with good certainty. In sandwich and dot-blot assays, the presence of a single infected cell in the midst of many uninfected cells would be difficult to distinguish from background hybridization.

Clinical evaluations of the Enzo *in situ* hybridization test for confirmation of *C. trachomatis* cultures have not been performed. However, given the strength of the signal in the inclusions, it seems likely that this assay will provide a high degree of sensitivity and specificity in these analyses.

V. GENE PROBES VERSUS ANTISERA AND MONOCLONAL ANTIBODIES

DNA hybridization assays offer many advantages over other available assays for gonorrhea and chlamydia. Foremost among these is the extreme specificity offered by the use of DNA probes. DNA probes can be selected to be species-specific and to recognize an entire bacterial species, regardless of the number of serotypes. This was demonstrated by our ability to isolate DNA probes that are specific for all known isolates of *N. gonorrhoeae*. DNA hybridization assays can also be used to distinguish pathogenic bacteria from their nonpathogenic family members, as shown in this report and by S. Falkow and his colleagues in their work on the identification of enterotoxigenic *E. coli* in clinical specimens (55,56). Moreover, DNA probes offer the potential for assessing the antibiotic resistance properties of a given bacterial strain.

This last point is particularly significant because there has been a worldwide increase in *N. gonorrhoeae* antibiotic resistance (58). In the United States, increasing levels of *N. gonorrhoeae* plasmid-mediated resistance to tetracycline, β-lactamase-producing stains (plasmid-encoded penicillin resistance), and chromosomally mediated antibiotic resistance have been noted (58). Studies performed by Perine *et al.* (64), in which DNA spot hybridization assays were performed directly on urethral exudates of males with urethritis using cryptic plasmid probes to detect the presence of *N. gonorrhoeae* DNA and *N. gonorrhoeae* plasmids containing the TEM-type β-lactamase gene (which is present in the majority of penicillin-resistant gonococci) suggest that DNA hybridization assays will be extremely useful for determining bacterial antibiotic resistance characteristics. In these studies, which were performed using urethral exudates from high-prevalence male populations, the sensitivity and specificity of the DNA probe test for gonorrhea were 99 and 93%, respectively (compared to culture assays), and the sensitivity and specificity of β-lactamase detection in these bacteria were 91 and 96%, respectively, compared to electrophoretic detection of the β-lactamase plasmid. Thus, the results of these tests were comparable to those obtained by the gold standard tests, but were faster and easier to perform. In fact, a promising direct nonradioactive hybridization test for detecting the TEM-1 plasmid in *N. gonorrhoeae* has recently been developed by the Chiron Corp. (70). The major problem associated with the use of DNA probe tests to determine bacterial antibiotic resistance characteristics is the possibility of erroneous results due to the presence of commensal organisms in direct testing of clinical specimens. However, specimens from body fluids or organs that are normally bacteria-free, such as blood, CSF, and some genital specimens should provide reliable results with this type of antibiotic resistance test.

Polyclonal antisera are usually not able to achieve the degree of specificity that is characteristic of DNA probes. While MAb (for review, see 50a) are more specific than polyclonal antibodies and can be selected so as to be species-specific, often a panel of MAb must be employed to achieve the dual goal of recognizing all members of a given species, including new strain variants, and distinguishing these members from other closely related bacteria. Lack of reproducibility due to variations in the production of one or more of the MAb in a panel can be a serious problem for assays that employ multiple MAb probes.

In addition, MAb often recognize secondary, rather than primary, protein structures. Factors such as protein denaturation or degradation can severely diminish the ability of highly specific MAb to recognize antigenic determinants. In contrast, DNA probes specifically recognize homologous sequences on denatured nucleic acid targets. Absolute maintenance of the integrity of large stretches of these DNA or RNA targets is not a prerequisite for detection by DNA probes. Pathogen DNA, in contrast to protein antigens, is stable to harsh treatments such as formaldehyde, detergent, boiling temperatures, and strong base. Thus careful handling or transport procedures to maintain reactivity are minimized. In fact, the Viral Disease Laboratory of the California State Health Department has carried out studies that indicate that HSV DNA can be detected in human brain tissue that has been stored in the frozen state for at least 2 years (19). In these experiments, the DNA-based assay proved to be more reliable than a MAb-based immunoassay performed on the same tissues. We have performed retrospective studies in which DNA probes were used specifically to detect HPV DNA in sections from formalin-fixed, paraffin-embedded biopsy specimens that had been stored at room temperature for ≤10 years. We have also successfully performed DNA hybridization analyses on cultures of *M. tuberculosis* that had been maintained at 4°C for several years.

One of the most critical factors involved in the effective prevention and treatment of STD rests with the ability to detect these diseases in their early or latent stages. While the sensitivity of immunoassays is often enhanced by the high copy number of the protein antigens that they recognize, compared to the generally lower copy number of DNA or RNA targets, it is often necessary to wait for a period in the life cycle of the pathogen when this antigen is being produced. DNA probe assays can potentially recognize DNA sequences at any stage in the life cycle of a pathogen, including periods of latency, provided that sufficiently sensitive detection systems are used to monitor the hybridization reaction.

Currently, there are two areas in which MAb assays exhibit marked superiority to DNA probe assays. Immunoassays are, for the most part, faster and less complex to perform than DNA hybridization assays, primarily because the hybridization assays require a greater number of wash steps to ensure high specificity and low backgrounds. This makes the use of immunoassays more attractive for use in clinical laboratory settings. However, it is likely that DNA probe assays will be automated in the not too distant future so that they will become more acceptable.

Of greater significance is that DNA hybridization assays often provide lower sensitivity for a given disease than do the corresponding immunoassays. This is because the antibody probes used in these assays recognize proteins that are generally present in greater numbers than DNA targets.

The approach taken by Gen-Probe offers one solution to this problem. This company has enhanced the sensitivity of their hybridization assays by targeting their DNA probes to species-specific rRNA sequences, as well as by performing their hybridization reactions in solution (rather than on a solid matrix such as nitrocellulose membranes). For high-prevalence populations, the sensitivity and specificity levels of Gen-Probe's direct assays for gonorrhea and chlamydia are similar to those obtainable using MAb probes for the same diseases. However, by targeting their assays to rRNA, Gen-Probe has limited the utility of these assays to the detection of bacterial pathogens in individuals who are not receiving antibiotic treatments, since antibiotics often interfere with bacterial RNA metabolism. The Gen-Probe assays cannot therefore be used for virus detection or to monitor the course of therapeutic treatments.

The use of the target amplification techniques, such as the PCR procedure developed by Cetus, may also result in greater overall sensitivity for DNA hybridization assays by providing a means for amplifying the number of target molecules present in a given specimen. Hybridization–detection systems that provide higher sensitivity or the implementation of alternate hybridization assay formats may also be forthcoming, which will further enhance the applicability of DNA hybridization assays.

VI. PROSPECTS FOR THE FUTURE

There are clearly some areas of medicine, such as human genetic analyses and viral pathogen detection, for which nucleic acid hybridization assays represent the definitive diagnostic tool of the future. The case for using DNA probe assays for the detection of bacterial pathogens is, however, less obvious. DNA probes can provide greater specificity for distinguishing between closely related bacterial species and can identify a potentially unlimited number of isolates for bacterial species that frequently alter their surface antigens with greater facility than MAb probes. As importantly, direct or short-term culture DNA probe assays can yield this information in considerably less time than standard bacterial culture assays and can sometimes result in definitive diagnoses before the antigens recognized by MAb probes are detectable. In instances, however, where DNA probes and MAb provide similar information, detection sensitivity, and specificity within a comparable time period or disease stage, immunoassays often represent the preferred method because they can be performed with greater ease.

Easier and more sensitive DNA hybridization assay formats will almost certainly be developed in the next few years. Some attempts to automate existing hybridization assay formats are already under way (25). Alternate nonisotopic hybridization–detection systems are also being developed that will further enhance the sensitivity and, hence, the desirability of DNA probe assays. Assays that can provide increased sensitivity for detecting bacterially mediated STD in females will have a definite advantage over other types of diagnostic tests.

These new and improved hybridization assay methodologies will eventually result in the development of more rapid and reliable DNA-based diagnostic tests for STD. These advances will be particularly important for the detection of *N. gonorrhoeae*, because the strengths of the DNA probe assays most directly answer the requirements for diagnosis and treatment of gonorrhea. At present, the reasons for selecting DNA hybridization assays in preference to immunoassays for the diagnosis of chlamydial infections are less compelling than those cited for gonorrhea. Nevertheless, DNA hybridization assays will take precedence over immunoassays if they eventually prove to yield greater sensitivity for the detection of *C.*

trachomatis in females. At present, it seems unlikely that effective DNA hybridization assays for syphilis can be easily developed. The most likely stage for these assays would be during primary or early secondary syphilis, when bacteria are easily obtainable from lesion exudates, but even here the reported number of *T. pallidum* in these exudates is extremely low. Significant improvements in the sensitivity of DNA hybridization assays must be made before DNA probe diagnostics for syphilis become feasible.

VII. SUMMARY

The lack of rapid and reliable tests for diagnosing bacterially mediated STD is a major world health problem, particularly in that the incidence of these diseases is steadily increasing. The currently available cell culture methods for identifying the bacteria that cause gonorrhea, chlamydia, and syphilis (*N. gonorrhoeae, C. trachomatis,* and *T. pallidum*) are lengthy, cumbersome, and expensive. The use of MAb probes for direct specimen analysis has reduced the time necessary to obtain laboratory test results for these STD, but has not greatly enhanced our ability to detect these diseases. This is particularly true when these tests are applied to specimens from females, asymptomatic males, or low-risk populations. DNA hybridization assays that employ bacteria-specific DNA probes and nonradioactive hybridization–detection systems for the detection of *N. gonorrhoeae* and *C. trachomatis* in cultured clinical specimens are described here. DNA hybridization assays for the direct detection of these organisms in clinical specimens are under development. DNA hybridization assays are more rapid than culture assays for these bacteria and have the potential of being more sensitive, specific, and informative than the corresponding MAb-based immunoassays for these organisms.

VIII. MATERIALS AND METHODS

A. Bacteria and Bacteriophage Strains and Growth Conditions

Escherichia coli strain JM103 (Δ*lacpro, thi, str*A, *sup*E, *end*A, *sbc*B15, *hsd*R4, F′, *tra*D36, *pro*AB, *lac*I^{q}ZΔM15) was used as the host for growth of M13 bacteriophage and as the competent cells for transformation by recombinant phage and plasmid DNA.

The clinical isolates of *Neisseria gonorrhoeae* and *Neisseria meningitidis* used in this study are listed in Table I. Strains representing all known serotypes of these species were obtained from Joan Knapp, Neisseria Reference Laboratory (NRL), University of Washington, Seattle. Strains of other neisseriae and commensal bacteria are listed in Table II. These strains were obtained from the American Type Culture Collection (ATCC).

Neisseria gonorrhoeae and *N. meningitidis* were grown on chocolate II agar plates (BBL Microbiology Systems, Cockeysville, Maryland) at 37°C in an atmosphere of 5% carbon dioxide. Other bacterial species were grown on the respective media recommended by the ATCC.

Confluent monolayers of McCoy cells grown on coverslips in shell vials were infected with *C. trachomatis* at a multiplicity of infection of 0.1–0.25 and incubated at 37°C in minimal essential medium (MEM) with nonessential amino acids (Difco) supplemented with 2% heat-inactivated fetal calf serum, gentamicin, and glutamine. After addition of the bacteria, the shell vials were centrifuged at 3600 rpm for 45–60 min to pellet the *C. trachomatis* cells, then incubated for 45–90 min at 37°C. The supernatant was removed and replaced with an equivalent amount of the same medium already described, supplemented with 1 μg/ml cycloheximide.

Bacteriophage M13mp8 and M13 wild type were obtained from PL Biochemicals, Inc. (Milwaukee, Wisconsin) and were propagated on *E. coli* JM103 using the procedures described by Maniatis *et al.* (52).

B. Isolation and Labeling of DNA

Chromosomal DNA for screening purposes was prepared from cells grown on one to six plates containing the appropriate medium. These cells were resuspended in TE buffer [10 m*M* Tris-HCl, pH 7.6–1 m*M* ethylenediaminetetraacetate (EDTA)] and lysed by the addition of SDS to a final concentration of 1%. This mixture was incubated at 65°C for 10 min, then extracted four times with chloroform–isoamyl alcohol (24 : 1). Precipitation of the DNA was accomplished by the addition of two volumes of 95% ethanol, followed by incubation overnight at −20°C. After removing the ethanol, the DNA precipitate was collected by centrifugation, dried *in vacuo,* then resuspended in 1–3 ml of TE buffer. The concentration of the DNA solution was determined by UV spectroscopy (i.e., absorbance at a wavelength of 260 nm), then adjusted to a final concentration of ~1 mg/ml with TE buffer.

Plasmid and bacteriophage DNA were prepared using the procedures described by Maniatis *et al.* (52).

Wild-type double-stranded (RF) M13 and M13-GC2B DNA were labeled with [^{32}P]dCTP (New England Nuclear Co., Boston, Massachusetts) by nick translation using standard procedures (52).

Plasmid, phage DNA, and GC155 insert DNA were labeled with Bio-11-dUTP (Enzo Biochem, Inc., New York, New York) by nick translation according to the package insert instructions.

C. Preparation of the *Neisseria gonorrhoeae*–M13mp8 Recombinant Phage Library

The procedures used to prepare and to screen the *N. gonorrhoeae*–M13mp8 recombinant phage library were previously described by Donegan *et al.* (18).

Neisseria gonorrhoeae chromosomal DNA from strain 33152 was digested to completion with the restriction endonuclease *Mbo*I. The products of this digestion were ligated to M13mp8 vector DNA that had been linearized by digestion with *Bam*HI. This ligated recombinant DNA was used to transform competent JM103 cells. Recombinant phages were identified as colorless plaques on lawns of JM103 cells grown on agar plates containing 2× YT medium, 0.002% X-Gal (5-bromo-4-chloro-3-indolyl-β-galactoside, Boehringer-Mannheim) and 50 μ*M* IPTG (isopropyl-β-thiogalactopyranoside, Sigma). Single plaques were picked and used to infect cultures of JM103 to provide phage stocks. All of the procedures used to prepare this recombinant phage library were described by Maniatis *et al.* (52).

D. Identification of *Neisseria gonorrhoeae*-Specific Probes

Numbered pieces of nitrocellulose membranes (Schleicher and Schuell BA85) containing separate spots of *N. gonorrhoeae* (strain 33152) chromosomal DNA (1 μg/spot) and *N. meningitidis* (strain Y) chromosomal DNA (1–2 μg/spot) were used to identify phage-carrying recombinant DNA that could hybridize to *N. gonorrhoeae* DNA, but not to *N. meningitidis* DNA. The filters were prepared using chromosomal DNA, which was first denatured in 0.2 *N* sodium hydroxide, then neutralized by the addition of 3*M* ammonium acetate, pH 7, to a final concentration of 1 *M*, as described by Kafatos *et al.* (40). Approximately 1–2 μg of this denatured chromosomal DNA was applied to the nitrocellulose membranes using a dot-blot apparatus (Aquebogue Machine Shop, Aquebogue, New York). The filters were washed once with 0.6 *M* ammonium acetate–0.1 *M* Tris-Hcl, pH 8.0 following application of the DNA, then air-dried and baked *in vacuo* for 2 hr at 80°C. These filters were prehybridized in 1.5-ml Eppendorf tubes for 1–3 hr at 65°C in 0.4 ml of a solution containing hybridization solution: 5× SSC, 1.25× Denhardt's solution, 0.31% SDS, and 250 μg/ml sonicated, denatured calf thymus DNA (Sigma), where 1× SSC is 0.15 *M* sodium chloride–0.015 *M* sodium citrate and 1× Denhardt's solution is 0.02%

Ficoll (Sigma, MW 400,000), 0.02% polyvinylpyrrolidone (Sigma, MW 360,000), 0.02% bovine serum albumin (BSA): Sigma, Fraction V).

In the primary screening step, 0.1 ml of each of the 3000 phage stocks prepared from isolated recombinant phage plaques according to Maniatis *et al.* (52) was added to a separate Eppendorf tube containing a numbered filter and 0.4 ml of the hybridization solution described before. The hybridization solution and conditions used in these experiments resulted in the liberation of phage DNA in a hybridizable form without any additional treatment steps. These tubes were then incubated at 65°C for 12–16 hr to permit the recombinant phage DNA to hybridize to the filter-bound chromosomal DNA. The filters were then removed from each tube and washed twice in batches at 65°C for 30 min in 2× SSC–0.1% SDS.

The detection of hybridized recombinant phage DNA was accomplished using the following procedure. First, filters were incubated batchwise in a hybridization solution consisting of ^{32}P-labeled M13 wild-type DNA (specific activity, 1×10^8 cpm/ml), 4× SSC, 1× Denhardt's solution, 0.25% SDS, and 200 μg/ml sonicated, denatured calf thymus DNA for 12–16 hr at 65°C. The filters were washed three times in 1× SSC–0.1% SDS and once in 0.1× SSC–0.1% SDS at 65°C for 30 min per wash. The filters were then cut in half to separate the *N. gonorrhoeae* and *N. meningitidis* target DNA dots. The cpm ^{32}P bound to each filter was determined using a scintillation counter (Beckman LS 6800). The results obtained for each pair of *N. gonorrhoeae* and *N. meningitidis* filters were compared. Filters subjected to mock hybridization reactions (i.e., hybridization to *N. gonorrhoeae* or *N. meningitidis* DNA first in the presence of 10^{11} M13 wild-type phage, then in the presence of ^{32}P-labeled M13 DNA), were used to determine the background level of radioactivity on the filters.

In the secondary screening stage, DNA from phages that hybridized to *N. gonorrhoeae* (33152) chromosomal DNA, but not to *N. meningitidis* (strain Y) chromosomal DNA was hybridized to nitrocellulose filters with an array of target DNA dots containing denatured chromosomal DNA from representatives of all *N. gonorrhoeae* and *N. meningitidis* strains (see Table I). The filters were then cut apart to separate the dots. The ^{32}P cpm of each dot was determined by scintillography. The filter preparation, hybridization, detection, and counting procedures used in this stage of the study were identical to those used in the primary screening step.

Recombinants whose DNA hybridized to all strains of *N. gonorrhoeae*, but not to any of the *N. meningitidis* strains tested, were subjected to a final screening step that was designed to detect homology between the *N. gonorrhoeae* DNA in these phages and that of the other neisseriae or bacteria often present in urogenital specimens (commensal bacteria or pathogens), listed in Table II. In this tertiary screen, double-stranded RFI

DNA was prepared from each of the recombinants selected for further analysis following the secondary screening step. This phage DNA was labeled with ^{32}P, then used directly to probe nitrocellulose filters containing dots of chromosomal DNA from each of the organisms listed in Table II. The hybridization conditions used here were identical to those used in the first stages of the primary and secondary screening steps. The individual dots were separated and counted in a scintillation counter.

E. Rapid Spot-Blot Assay for Confirmation of *Neisseria gonorrhoeae* Cultures

1. Sample Preparation

Neisseria gonorrhoeae culture-positive urogenital specimens or specific isolates of *N. gonorrhoeae* were cultured on chocolate agar II plates for 18–26 hr at 37°C. Approximately 10 medium-sized colonies were collected onto the bottom of a wooden applicator stick. The stick was then dipped into 50 μl of solubilization solution (0.5 *M* sodium hydroxide–0.5% SDS) two times. The sample on the bottom of the wooden stick was applied to a strip of GeneScreen Plus membrane (DuPont), previously mounted onto a microscope slide with silicon glue (Sialastic Medical Adhesive Silicone, Dow Corning Co.), by firmly pressing the bottom of the stick onto the surface of the GeneScreen Plus matrix. Phosphate-buffered saline (PBS: 10 m*M* sodium phosphate buffer, pH 7.2–130 m*M* sodium chloride) containing 0.5% SDS was applied to each sample and allowed to sit on the matrix for 1 min at room temperature. The surface of the matrix was scrubbed with a cotton swab for ~5 sec. The matrix was then rinsed with a strong stream of deionized water from a squeeze bottle for 10–15 sec to ensure that all bacterial debris was removed. Failure to remove this debris results in high background staining. Slide results can be determined following completion of the assay using either wet or dry filters.

2. Hybridization and Detection

The procedure described here is intended for use with nitrocellulose membranes measuring 8 × 12 mm. The quantities of all reagents used for different-sized membranes should be adjusted accordingly. A 50-μl portion of hybridization solution containing biotinylated *N. gonorrhoeae* probe DNA (5 μg/ml of purified GC155 insert DNA), 50% formamide, 0.5× SSC, 5% dextran sulfate (Sigma), and sonicated; 0.4 mg/ml denatured salmon sperm DNA (Sigma) was placed on top of the matrix/slide. The sample was covered with a coverslip and incubated at 95°C for 3–4 min on a heating block to permit denaturation to occur. The matrix/

slide was removed from the heating block and allowed to hybridize at room temperature for 10 min. The coverslip was removed and the matrix was washed with a strong stream of distilled H_2O from a squeeze bottle for 5 sec. Excess water was removed by tapping the slide. Approximately 200–400 μl of a stringency wash solution (0.05× SSC–50% formamide) was applied to the hybridized matrix/slide and incubated for 10 min at room temperature. This solution was removed by tapping. The matrix/slide was washed with distilled H_2O for 5 sec.

Detection of the hybridized probe DNA was accomplished colorimetrically, using a complex consisting of streptavidin and biotinylated HRP marketed by Enzo Biochem, Inc. (DETEKI-*hrp*). A 200-μl aliquot of DETEKI-*hrp* diluted 1 : 200 in PBS containing 5% BSA was applied to the matrix/slide and incubated for 5 min at room temperature. The solution was removed from the matrix/slide by tapping. Following a 5-sec rinse with distilled H_2O, 500 μl of PBS–0.5% SDS was added to the slide, incubated for 10 min at room temperature, and washed off with water for 10 sec. Chromogen [aminoethylcarbazole (AEC)] and HRP substrate (hydrogen peroxide) were then added in buffer (100 m*M* sodium acetate, pH 4) and incubated for 5 min. The matrix/slide was washed with H_2O. Several drops of sodium azide (0.02%) were placed on the matrix/slide to stop the reaction. The matrix/slide was dried on a 95°C heating block, then visually inspected using a light microscope. Positive reactions are indicated by the formation of a red precipitate at the site of sample application.

F. *In Situ* Hybridization Assay for *Chlamydia trachomatis* Cultures

In situ hybridization assays for detecting *C. trachomatis* inclusions in infected cell cultures were performed in shell vials (purchased from Research Products, Inc.) containing confluent McCoy cells that were infected with *C. trachomatis* serotype G. Approximately 24 hr after infection, the media were removed from the shell vials by aspiration. The cells were rinsed once or twice with PBS, then fixed with cold anhydrous alcohol (Baker Co.) for 5 min. The alcohol was removed by aspiration or by decanting and the shell vials were allowed to air-dry at room temperature. It is essential that the cells and shell vials be completely dry before continuing with the assay.

The shell vial *in situ* hybridization assay for *C. trachomatis* was performed using the *C. trachomatis* BioProbe (Enzo Biochem, Inc.) and components from Enzo's ColorGene kits (i.e., wash buffer, posthybridization reagent, detection reagent, and chromogen/substrate reagent). Five drops of *C. trachomatis* BioProbe reagent were added to each shell vial containing the fixed cells. The vials were heated for 5 min on a heating block at 92°–95°C to denature probe and target DNA, and allowed to stand at room temperature for 10 min to permit hybridization to occur. The vials were washed three times by first filling the vials completely with wash buffer, then removing the wash solution by aspiration. The vials were inverted following the last rinse and allowed to drain. Posthybridization reagent (five to six drops) was added to each vial. The vials were gently agitated, then incubated at room temperature for 10 min. The posthybridization reagent was removed by aspiration. The vials were washed five to six times with wash buffer as described previously.

Detection of hybridized *C. trachomatis* probe DNA was accomplished by adding five to six drops of detection reagent and incubating in the presence of this reagent for 10 min at room temperature. The vials were again washed with wash buffer (five to six times). The vials were incubated at room temperature for 10 min following the addition of chromogen/substrate reagent. This reagent was removed by aspiration and the vials were washed three times with wash buffer. The coverslips were removed from the shell vials with the aid of a bent needle and a pair of forceps, placed on a drop of H_2O on a microscope slide, then viewed under a light microscope. The presence of perinuclear inclusions that are stained a brick red color in the cytoplasm of healthy cells is indicative of *C. trachomatis* infection (i.e., represents a positive reaction).

ACKNOWLEDGMENTS

This work was funded by NIH SBIR grants to H.-L. Yang and also by the New York State Science and Technology Foundation.

REFERENCES

1. Agrawal, S., Christodoulou, C., and Gait, M. J. (1986). Efficient methods for attaching non-radioactive labels to the 5′ ends of synthetic oligodeoxyribonucleotides. *Nucleic Acids Res.* **14,** 6227–6245.
2. Amortegui, A. J., and Meyer, M. P. (1985). Enzyme immunoassay for detection of *Chlamydia trachomatis* from the cervix. *Obstet. Gynecol. (N.Y.)* **65,** 523–526.
3. Batteiger, B. E., and Jones, R. B. (1987). Chlamydial infections. *In* "Infectious Disease Clinics of North America" (H. H. Handsfield, ed.), Vol. I, pp. 55–81. Saunders, Philadelphia, Pennsylvania.
4. Black, J. R., and Sparling, P. F. (1985). *Neisseria gonorrhoeae. In* "Principles and Practice of Infectious Diseases" (G. L. Mandell, R. G. Douglas, and J. E. Bennett, eds.), pp. 1195–1205. Wiley, New York.
5. Bowie, W. R., and Holmes, K. K. (1985). Chlamydial diseases. *In* "Principles and

Practice of Infectious Diseases'' (G. L. Mandell, R. G. Douglas, and J. E. Bennett, eds.), pp. 1047–1061. Wiley, New York.
6. Brakel, C. L., Donegan, J. J., Linn, C. -I. P., Molina, M., Pollice, M., Wang, Z., and Yang, H. -L. (1988). An automatable, colorimetric DNA hybridization test for *M. tuberculosis* confirmation. *Am. Soc. Microbiol., Abstr. Annu. Meet.,* p. 144.
7. Carlson, B. L., Calnan, M. B., Goodman, R. E., and George, H. (1987). Phadebact monoclonal GC OMNI test for confirmation of *Neisseria gonorrhoeae. J. Clin. Microbiol.* **25,** 1982–1984.
8. Chamberlain, N. R., Radolf, J. D., Hsu, P. -L., Sell, S., and Norgard, M. V. (1988). Genetic and physiochemical characterization of the recombinant DNA-derived 47-kilodalton surface immunogen of *Treponema pallidum* subsp. *pallidum. Infect. Immun.* **56,** 71–78.
9. Chapman, S. (1986). Bacterial infections in pregnancy. *Clin. Obstet. Gynecol.* **13,** 397–416.
10. Christiansen, G., Christiansen, C., and Freundt, E. A. (1985). Lack of genetic relatedness between *Mycoplasma alvi* and *Mycoplasma sualvi. Int. J. Syst. Bacteriol.* **35,** 210.
11. Coates, S., Sheridan, P. J., Hansen, D. S., Laird, W. J., and Erlich, H. A. (1986). Serospecificity of a cloned-protease resistant *Treponema pallidum*-specific antigen expressed in *Escherichia coli. J. Clin. Microbiol.* **23,** 460–464.
12. Coggin, J. R., and zur Hausen, H. (1979). Workshop on papillomavirus and cancer. *Cancer Res.* **39,** 545–546.
13. Copley, C. G., and Egglestone, I. (1982). Gonococci without plasmids. *Lancet* **1,** 1133.
14. Correia, F. F., Inouye, S., and Inouye, M. (1986). A 26-base pair repetitive sequence specific for *Neisseria gonorrhoeae* and *Neisseria meningitidis* genomic DNA. *J. Bacteriol.* **167,** 1009–1015.
15. Dallabetta, G., and Hook, E. W. (1987). Gonococcal infections. *In* ''Infectious Disease Clinics of North America'' (H. H. Handsfield, ed.), Vol. I, pp. 25–54. Saunders, Philadelphia, Pennsylvania.
16. Dean, D., and O'Hanley, P. O. (1988). High sensitivity of a *C. trachomatis* DNA probe compared with multiple culture passage and clinical exam as a valid test for trachoma. *Program Abstr., Intersci. Conf. Antimicrob. Agents Chemother., 28th* p. 268.
17. Dillon, J. R., and Pauze, M. (1981). Relationship between plasmid content and auxotype in *Neisseria gonorrhoeae* isolates. *Infect. Immun.* **33,** 625–628.
18. Donegan, J. J., Lo, A., Manwell, A., Picken, R. N., and Yang, H. -L. (1989). Isolation of a species-specific DNA probe for *Neisseria gonorrhoeae* using a novel technique particularly suitable for use with closely related species displaying high levels of DNA homology. *Mol. Cell. Probes* **3,** 13–26.
19. Forghani, B., Dupuis, K. W., and Schmidt, N. J. (1985). Rapid detection of herpes simplex virus DNA in human brain tissue by *in situ* hybridization. *J. Clin. Microbiol.* **22,** 656–658.
20. Garbutt, G. J., Wilson, J. T., Schuster, G. S., Leary, J. J., and Ward, D. C. (1985). Use of biotinylated probes for detecting sickle cell anemia. *Clin. Chem. (Winston-Salem, N.C.)* **31,** 1203–1206.
21. Gegg, C. C., Kranig-Brown, D., Gonzalez, C., You, M. S., Yang, Y. Y., and Harper, M. E. (1988). Direct Identification of *Neisseria gonorrhoeae* in urogenital specimens using the PACE DNA probe test: Confirmation of discrepant analysis. *Program Abstr., Intersci. Conf. Antimicrob. Agents Chemother, 28th,* p. 321.
22. Goldstein, L. C., and Tam, M. R. (1985). Monoclonal antibodies for the diagnosis of sexually transmitted diseases. *In* ''Clinics in Laboratory Medicine'' (W. -S. Lee, ed.), Vol. 5, pp. 575–588. Saunders, Philadelphia, Pennsylvania.

23. Granato, P. A., and Roefaro, M. (1985). Comparative evaluation of enzyme immunoassay and culture for the laboratory diagnosis of gonorrhea. *Am. J. Clin. Pathol.* **83,** 613–618.
24. Granato, P. A., and Franz, M. R. (1988). Evaluation of a prototype DNA probe test for the non-cultural diagnosis of gonorrhea. *Program Abstr., Intersci. Conf. Antimicrob. Agents Chemother., 28th,* p. 320.
25. Gualtieri, P., Benedetti, P. A., and Evangelista, V. (1988). A microprocessor-controlled microscope for diagnostic cytology. *Am. Biotechnol. Lab.* **6,** 36–40.
26. Hansen, E. B., Pedersen, P. E., Schouls, Severin, E., and van Embden, J. D. A. (1985). Genetic characterization and partial sequence determination of a *Treponema pallidum* operon expressing two immunogenic membrane proteins in *Escherichia coli. J. Bacteriol.* **162,** 1227–1237.
27. Hart, G. (1983). The role of treponemal tests in therapeutic decision making. *Am. J. Public Health* **73,** 739–743.
28. Hawkins, D. A., Wilson, R. S., Thomas, B. J., and Evans, R. T. (1985). Rapid, reliable diagnosis of chlamydial ophthalmia by means of monoclonal antibodies. *Br. J. Ophthamol.* **69,** 640–644.
29. Hicks, C. B., Benson, P. M., Lupton, G. P., and Tramont, E. C. (1987). Seronegative secondary syphilis in a patient infected with the Human Immunodeficiency Virus (HIV) with Kaposi Sarcoma. *Ann. Intern. Med.* **107,** 492–494.
30. Hipp, S. S., Han, Y., and Murphy, D. (1987). Assessment of enzyme immunoassay and immunofluorescence tests for detection of *Chlamydia trachomatis. J. Clin. Microbiol.* **25,** 1938–1943.
31. Hische, E. A. H., Tutuarima, J. A., Wolters, E. C., van Trotsenburg, L., van Eyk, R. V. W., Bos, J. D., Albert, A., and van der Helm, H. J. (1988). Cerebrospinal fluid IgG and IgM indexes as indicators of active neurosyphilis. *Clin. Chem. (Winston-Salem, N.C.)* **34,** 665–667
32. Hoke, C., and Vedros, N. A. (1982). Taxonomy of the *Neisseriae*. Deoxyribonucleic acid (DNA) base composition, interspecific transformation and hybridization. *Int. J. Syst. Bacteriol.* **32,** 57–66.
33. Hook, E. W., Roddy, R. E., Lukehart, S. A., Hom, J., Holmes, K. K., and Tam, M. R. (1985). Detection of *Treponema pallidum* in lesion exudate with a pathogen-specific monoclonal antibody. *J. Clin. Microbiol.* **22,** 241–244.
34. Horn, J. E., Hammer, M. L., Falkow, S., and Quinn, T. C. (1986). Detection of *Chlamydia trachomatis* in tissue culture and cervical scrapings by *in situ* DNA hybridization. *J. Infect. Dis.* **153,** 1155–1159.
35. Houssain, A. (1987). Rapid diagnosis of *Chlamydia trachomatis* infections by a monoclonal antibody direct immunofluorescence test. *J. Trop. Med. Hyg.* **90,** 307–310.
36. Hsu, P. -L., Qin, M., Norris, S. J., and Sell, S. (1988). Isolation and characterization of recombinant *Escherichia coli* clones secreting a 24-kilodalton antigen of *Treponema pallidum. Infect. Immun.* **56,** 1135–1143.
37. Hughes, G. B., and Rutherford, I. (1986). Predictive value of serologic tests for syphilis in otology. *Ann. Otol., Rhinol., Laryngol.* **95,** 250–259.
38. Hyypia, T., Jalava, A., Larsen, S. H., Terho, P., and Hukkanen, V. (1985). Detection of *Chlamydia trachomatis* in clinical specimens by nucleic acid spot hybridization. *J. Gen. Microbiol.* **131,** 975–978.
39. Ison, C., Bellinger, C. M., and Walker, J. (1986). Homology of cryptic plasmids of *N. gonorrhoeae* with plasmids from *N. meningitidis* and *N. lactamica. J. Clin. Pathol.* **39,** 1119–1123.
40. Kafatos, F. C., Jones, C. W., and Efstratiadis, A. (1979). Determination of nucleic acid

sequence homologies and relative concentration by a dot blot procedure. *Nucleic Acids Res.* **7,** 1541–1552.
41. Kampmeier, R. H. (1982). Syphilis. *In* "Bacterial Infections of Humans" (A. S. Evans and H. A. Feldman, eds.), pp. 553–577. Plenum, New York.
42. Kane, J. L., Woodland, R. M., Forsey, T. Darougar, S., and Elder, M. G. (1984). Evidence of chlamydial infection in infertile women with and without fallopian tube obstruction. *Fertil. Steril.* **42,** 843–848.
43. Kingsbury, D. T. (1967). Deoxyribonucleic acid homologies among species of the genus *Neisseria. J. Bacteriol.* **94,** 870–874.
44. Kingsbury, D. T., Fanning, G. R., Johnson, D. E., and Brenner, D. J. (1969). Thermostability of interspecies *Neisseria* DNA duplexes. *J. Gen. Microbiol.* **55,** 201–208.
45. Koike, T., Maruyama, N., Funaki, H., Tomioka, H., and Yoshida, S. (1984). Specificity of mouse hybridoma antibodies to DNA. II. Phospholipid reactivity and biological false positive serological test for syphilis. *Clin. Exp. Immunol.* **57,** 345–350.
46. Koike, T., Sueishi, H., Funaki, H., Tomioka, H., and Yoshida, S. (1984). Antiphospholipid antibodies and biological false positive serological test for syphilis in patients with systemic lupus erythematosus. *Clin. Exp. Immunol.* **56,** 193–199.
47. Langer, P. R., Waldrop, A. A., and Ward, D. C. (1981). Enzymatic synthesis of biotin-labeled polynucleotides: Novel nucleic acid affinity probes. *Proc. Natl. Acad. Sci. U.S.A.* **78,** 6633–6637.
48. Lee, J. B., Kim, S. C., Lee, S., Whang, K. H., and Choi, I. S. (1983). Symptomatic neurosyphilis. *Int. J. Dermatol.* **22,** 577–580.
49. Lipkin, E. S., Moncada, J. V., Shafer, M. -A, Wilson, T. E., and Schachter, J. (1986). Comparison of monoclonal antibody staining and culture in diagnosing cervical chlamydial infection. *J. Clin. Microbiol.* **23,** 114–117.
50. Lukehart, S. A. (1985). Prospects for development of a treponemal vaccine. *Rev. Infect. Dis.* **7,** Suppl. 2, S305–S313.
50a. Lukehart, S. A. (1986). Identification and characterization of *Treponema pallidum* antigens by monoclonal antibodies. *In* "Monoclonal Antibodies against Bacteria" (A. J. L. Macario and E. Conway de Macario, eds.), Vol. 3, pp. 1–27. Academic Press, Orlando, Florida.
51. Mach, P., Thompson, J., Ness, G., and Soule, H. (1988). Characterization of Pathfinder Chlamydia Enzyme Immunoassay. *Am. Soc. Microbiol.,Abstr. Annu. Meet.,* p. 370.
52. Maniatis, T., Fritsch, E. F., and Sambrook, J. (1982). "Molecular Cloning: A Laboratory Manual." Cold Spring Harbor Lab., Cold Spring Harbor, New York.
53. Merlin, S., Andre, J., Alacoque, B., and Paris-Hamelin, A. (1985). Importance of specific IgM antibodies in 116 patients with various stages of syphilis. *Genitourin. Med.* **61,** 82–87.
54. *Morbid. Mortal. Wkly. Rep.* (1988). **37,** 486–489.
55. Mosley, S. L., Huq, I., Ali, A. R. M. A., So, M., Samapour-Motalebi, M., and Falkow, S. (1980). Detection of enterotoxigenic *Escherichia coli* by DNA colony hybridization. *J. Infect. Dis.* **142,** 892–898.
56. Mosley, S. L., Echeverria, P., Seriwatana, J., Tirapat, C., Chaicumpa, W., Sakuldaipeara, T., and Falkow, S. (1982). Identification of enterotoxigenic *Escherichia coli* by colony hybridization using three enterotoxin gene probes. *J. Infect. Dis.* **145,** 863–869.
57. Musher, D. M. (1987). Syphilis. *In* "Infectious Disease Clinics of North America" (H. H. Handsfield, ed.), Vol. I, pp. 83–95. Saunders, Philadelphia, Pennsylvania.
58. 1986 NIAID Study Group Summary & Recommendations (1987). "Sexually Transmitted Diseases." National Institute of Allergy and Infectious Diseases, Bethesda, Maryland.

59. Paisley, J. W., Lauer, B. A., Melinkovich, P., Gitterman, B. A., Ferten, D., and Berman, S. (1986). Rapid diagnosis of *Chlamydia trachomatis* pneumonia in infants by direct immunofluorescence microscopy of nasopharyngeal secretions. *J. Pediatr.* **109,** 653–655.
60. Palmer, L., and Falkow, S. (1985). Selection of DNA probes for use in the diagnosis of infectious disease. *In* "Rapid Detection and Identification of Infectious Agents" (D. T. Kingsbury and S. Falkow, eds.), pp. 211–218. Academic Press, Orlando, Florida
61. Palva, A., Jousimies-Somer, H., Saikku, P., Vaananen, L. P., Soderland, H., and Ranki, M. (1984). Detection of *Chlamydia trachomatis* by nucleic sandwich hybridization. *FEMS Microbiol. Lett.* **23,** 83–89.
62. Patel, J. D., Joseph, J. M., and Falkler, W. A. (1988). Direct detection of *Chlamydia trachomatis* in clinical specimens by a dot-immunobinding technique using monoclonal antibody. *J. Immunol. Methods* **108,** 279–287.
63. Perine, P. L. (1987). Use of DNA probes for diagnosis of Treponemal diseases. *In* "Microbiology—1986" (L. Leive, ed.), pp. 129–130. Am. Soc. Microbiol., Washington, D. C.
64. Perine, P. L., Totten, P. A., Holmes, K. K., Sng, E. H., Ratnam, A. V., Widy-Wersky, R., Nsanze, H., Habte-Gabr, E., and Westbrook, W. G. (1985). Evaluation of a DNA-hybridization method for detection of African and Asian strains of *Neisseria gonorrhoeae* in men with urethritis. *J. Infect. Dis.* **152,** 59–63.
65. Peterson, K. M., Baseman, J. B., and Alderete, J. F. (1986). Isolation of a *Treponema pallidum* gene encoding immunodominant outer envelope protein P6, which reacts with sera from patients at different stages of syphilis. *J. Exp. Med.* **164,** 1160–70.
66. Peterson, K. M., Baseman, J. B., and Alderete, J. F. (1987). Cloning structural genes for *Treponema pallidum* immunogens and characterisation of recombinant treponemal surface protein P2*. *Genitourin. Med.* **63,** 289–96.
67. Pinkel, D., Straume, T., and Gray, J. W. (1986). Cytogenetic analysis using quantitative, high-sensitivity, fluorescence hybridization. *Proc. Natl. Acad. Sci. U.S.A.* **83,** 2934–2938.
68. Rodgers, G. C., Laird, W. J., Coates, S. R., Mack, D. H., Huston, M., and Sninsky, J. J. (1986). Serological characterization and gene localization of an *Escherichia coli*-expressed 37-kilodalton *Treponema pallidum* antigen. *Infect. Immun.* **53,** 16–25.
69. Rosenstraus, M., Spadoro, J., and Kelker, N. (1988). Ultra-sensitive, nonradioactive detection *in situ* detection of human papilloma virus DNA. *Am. Soc. Microbiol., Abstr. Annu. Meet.,* p. 318.
70. Sanchez-Pescador, R., Stempien, M., and Urdea, M. (1988). DNA hybridization assay for the detection of penicillin resistance (TEM-1) in *Neisseria gonorrhoeae. Am. Soc. Microbiol., Abstr. Annu. Meet.,* p. 338.
71. Schneider, A., Kraus, H., Schuhmann, R., and Gissmann, L. (1985). Papillomavirus infection of the lower genital tract: Detection of viral DNA in gynecological swabs. *Int. J. Cancer* **35,** 443–448.
72. Skeels, M. R., Matsuda, B., Horton, H., Sampson, J., Sawyer, A., and Mitchell, J. (1985). Evaluation of a modified enzyme immunoassay for *Neisseria gonorrhoeae* in high- and low-risk females. *Can. J. Microbiol.* **31,** 893–895.
73. Smith, L. M., Fung, S., Hunkapiller, M. W., Hunkapiller, T. J., and Hood, L. E. (1985). The synthesis of oligonucleotides containing an aliphatic amino group at the 5′ terminus: Synthesis of fluorescent DNA primers for use in DNA sequence analysis. *Nucleic Acids Res.* **13,** 2399–2412.
74. Spadoro, J. R., Lee, Y., Brakel, C. L., Laurence, J., and Rosenstraus, M. J. (1988).

Non-radioactive *in situ* detection of latent immunodeficiency virus DNA. *Program Abstr., Intersci. Conf. Antimicrob. Agents Chemother., 28th,* p. 258.
75. Stamm, W. E., Cole, B., Fennell, C., Bonin, P., Armstrong, A. S., Herrmann, J. E., and Holmes, K. K. (1984). Antigen detection for the diagnosis of gonorrhea. *J. Clin. Microbiol.* **19,** 399–403.
76. Stamm, W. E. (1988). Diagnosis of *Chlamydia trachomatis* genitourinary infections. *Ann. Intern. Med.* **108,** 710–717.
77. Stevens, R. W., and Schmitt, M. E. (1985). Evaluation of an enzyme-linked immunosorbent assay for treponemal antibody. *J. Clin. Microbiol.* **21,** 399–402.
78. Tecott, L. H., Barchas, J. D., and Eberwine, J. H. (1988). *In situ* transcription: Specific synthesis of complementary DNA in fixed tissue sections. *Science* **240,** 1661–1664.
79. Tenover, F. C. (1988). Diagnostic deoxyribonucleic acid probes for infectious diseases. *Clin. Microbiol. Rev.* **1,** 82–101.
80. Thomas, E., Scott, S. D., Grefkees, I., Hession, G., Pollock, R., Martin, T., and Albritton, W. (1986). Validity and cost-effectiveness of the Gonozyme test in the diagnosis of gonorrhea. *Can. Med. Assoc. J.* **134,** 121–146.
81. Totten, P. A., Holmes, K. K., Handsfield, H. H., Knapp, J. S., Perine, P. L., and Falkow, S. (1983). DNA hybridization technique for the detection of *Neisseria gonorrhoeae* in men with urethritis. *J. Infect. Dis.* **148,** 462–471.
82. Walfield, A. M., Hanff, P. A., and Lovett, M. A. (1982). Expression of *Treponema pallidum* antigens in *Escherichia coli. Science* **216,** 522–523.
83. Weisner, P. J., and Thompson, S. E. (1982). Gonococcal infections. *In* "Bacterial Infections of Humans" (A. S. Evans and H. A. Feldman, eds.), pp. 235–258. Plenum, New York.

2

Detection of *Chlamydia trachomatis* with DNA Probes

BRIGITTE DUTILH,*[,1] CHRISTIANE BEBEAR,* AND PATRICK A. D. GRIMONT†

**Laboratoire de Bactériologie*
Hôpital Pellegrin
Bordeaux, France

†Unité des Entérobactéries
Unité 199 INSERM
Institut Pasteur
Paris, France

[1] Present address: Laboratoire d'Analyses Médicales, 218, rue Mandron, 33000 Bordeaux, France.

I. INTRODUCTION

Chlamydia trachomatis, one of the most common of sexually transmitted pathogens, is responsible for a variety of clinical syndromes including urethritis, cervicitis, salpingitis, and lymphogranuloma venereum (LGV) (55). Besides genital infections, *C. trachomatis* is the causative agent of trachoma, a chronic conjunctivitis that is one of the major causes of blindness (52). Thus *C. trachomatis* infections are an important health problem in both industrialized and developing countries.

Chlamydia trachomatis is an obligatory intracellular parasite. The diagnosis of chlamydial infections usually relies on the multiplication of the organism in cell culture (36). This technique requires the presence of viable bacteria in the sample, the use of a special transport medium, and cell culture expertise in the clinical laboratory. Direct demonstration of chlamydial antigens without isolation became recently available by direct staining with fluorescent monoclonal antibodies (MAb) or by enzyme-linked immunosorbent assay (ELISA) (20,57). However, no technique is presently satisfactory for the detection of *C. trachomatis* associated with salpingitis, tubal infertility, Reiter's syndrome, or LGV.

Recent advances in DNA hybridization technology have made possible the specific detection of viruses, protozoans, and bacteria (58), including *C. trachomatis*. This technology is potentially specific, is sensitive, and can be automated. Though often confined to research laboratories, probe technology is at the beginning of its development, and technological improvements can be expected in a near future.

II. BACKGROUND

A. *Chlamydia trachomatis* Growth Cycle and Classification

Chlamydia trachomatis is a gram-negative bacterium, unable to grow on an artificial medium. It is characterized by its unique intracellular growth cycle. This cycle involves attachment and active penetration of the infectious particles—the elementary bodies (EB)—into the eukaryotic host cell. After entering the cytoplasm of the host cell, EB change to a metabolically active and multiplying form called the reticulate bodies (RB). Then, RB multiply by binary fission within a cytoplasmic vacuole. After 18–24 hr, the RB reorganize into infectious EB. The whole cycle lasts 48–72 hr. Cells are lysed and release numerous EB.

Chlamydia trachomatis is subdivided into 15 serovars defined by polyclonal antisera (62) and MAb (63). Serovars A, B, Ba, and C are the etiological agents of trachoma; serovars D, E, F, G, H, I, J, and K are associated with oculogenital infections, and serovars L1, L2, and L3 cause LGV. The specific determinants that characterize each serovar are associated with the major outer membrane proteins (MOMP).

B. Chlamydial Genetics

Progress in knowledge on chlamydial genetics has been hampered by the difficulties of cultivating the bacterium in large amounts and by the need for purifying the chlamydial components from host cell material.

The guanine + cytosine content of *C. trachomatis* DNA is 45 mol% (27). The chromosome is a closed, circular, double-stranded DNA molecule of MW 660 $\times$ 10^6 (50). This value is twice as large as that estimated previously (6 $\times$ 10^5 bp in size) (26) and is one of the smallest of bacterial genomes. The DNA relatedness among different *C. trachomatis* serovars is 96–97% (filter method), whereas relatedness between *C. trachomatis* and *Chlamydia psittaci* is <10% (27). DNA from *C. trachomatis* serovars B, C, D, E, F, L1, and L2 shows a single *Eco*RI restriction pattern when probed with ^{32}P-labeled 16 S rRNA or with an end-labeled 31-mer oligonucleotide complementary to a unique sequence on *C. trachomatis* 16 S rRNA (40).

The presence of a plasmid of ~7 kb was reported in *C. trachomatis* and *C. psittaci* strains (22,34,39). The plasmids present in both species were found to be distinct by restriction endonuclease analysis, DNA–DNA hybridization, and electron-microscopic heteroduplex analysis, although some homology was detected in Southern blots (24). The function of the plasmid is still unknown. Recently, the entire nucleotide sequences of the cryptic plasmid of serovar B (54) and L1 (15) were described. The exact size of the plasmid is 7.5 kb; its guanine + cytosine content is 36.3 mol%. The plasmid sequence contains four copies of a 22-bp repeated unit. Since the mean plasmid copy number per chlamydial cell was estimated to be between 7 and 10 (39,43), the plasmid represents a naturally amplified sequence, a suitable target for a probe.

A fragment of *C. trachomatis* chromosomal gene coding for the MOMP of serovars L1, L2, B, and C was cloned and the sequences compared (2,46,56). Gene sequences from four serovar MOMP showed several conserved domains and four variable regions. It was recently demonstrated, using amplification by polymerase chain reaction (PCR; Section VI), that a 129-bp fragment located in the conserved domain of the MOMP gene was conserved in the 15 *C. trachomatis* serovars (10).

The availability of *C. trachomatis* DNA sequences is essential for the development of specific probes.

C. Chlamydial Infections and Diagnosis

The 15 serovars of *C. trachomatis* are responsible for a wide variety of syndromes, including sexually transmitted diseases, eye infections, and newborn infections.

Chlamydia trachomatis is the most common sexually transmitted pathogen. Genital infections caused by *C. trachomatis* include urethritis, epididymitis, Reiter's syndrome in men, cervicitis, urethritis, endometritis, salpingitis followed in some cases by infertility and perihepatitis in women, proctitis, and LGV in both sexes. Some serovars cause oculogenital infections in adults. Chlamydial carriage occurs in asymptomatic patients (52,55). Trachoma, a chronic conjunctivitis leading to blindness, is a major public health problem in some developing countries (52). Newborns can be infected at birth by contact with their mother's genital exudate. Syndromes include conjunctivitis and pneumonia (14).

The microbiological diagnosis of *C. trachomatis* infection can be assessed by culture, antigen detection, and serology. Serological tests can be useful for invasive infections (e.g., salpingitis, perihepatitis, LGV), but are rarely useful for superficial genital diseases (55).

The standard method for *C. trachomatis* diagnosis is the isolation of the organism by cell culture followed by staining of the intracellular inclusions with iodine, Giemsa stain, or fluorescent antibodies (36). The specificity is excellent, but the sensitivity depends on proper storage and transport of specimens and on culture conditions and staining methodology. Identification by culture requires ~48 hr and is relatively expensive. A number of clinical laboratories do not have expertise in cell culture techniques and thus do not perform this isolation. Recently, an alternative to culture has been made available by the direct detection of chlamydial EB in smears by immunofluorescence (57,60) or ELISA (8,20) using monoclonal or polyclonal antibodies. *Chlamydia trachomatis* detection can now be done in many clinical laboratories using these techniques. However, both methods, culture and antigen detection, have limitations (51) especially in specimens such as fallopian tube exudate, cervical scraping, rectal tissue, or lymph nodes. Cytotoxic compounds in these samples may impair the isolation of *Chlamydia*. Antigenic detection using MAb can be prevented by the presence in the sample of antibodies that might bind antigenic sites *in vivo*. For these reasons, the development of alternate methods should be considered for the *Chlamydia*. Nucleic acid hybridization has a promising future in this area.

III. RESULTS AND DISCUSSION

Nucleic acid (NA) hybridization was applied for the first time to *C. trachomatis* detection by Hyypia (22). Since this first report, several different approaches have been used. These are summarized in this section.

A. Probe Preparation

The preparation of a probe requires several preliminary steps: large-scale *C. trachomatis* culture, purification of chlamydial EB free from host cell DNA, and extraction and purification of DNA. Total genomic DNA, cloned chromosomal fragments, or cryptic plasmid DNA can be used as a probe. Oligonucleotide probes could also be devised from published sequences of chlamydial genes.

1. Large-Scale Chlamydia trachomatis *Culture*

Mass production of *C. trachomatis* was obtained by chick embryo yolk sac culture or cell culture. The most widely used cell lines were HeLa 229 (28) and McCoy (48) treated by cycloheximide. For large-scale *C. trachomatis* DNA production, cultures were grown in 250-ml plastic bottles. Culture amplification required the following steps. Several passages in 2-ml vials were necessary before 80% of the cells in a vial were infected. Then the content of one vial was used to inoculate five vials. When 80% of the cells were infected, the content of the five vials was used to inoculate a 250-ml plastic bottle (29,59). After inoculation, plastic bottles were incubated at 37°C for 72 hr. The cells were removed and disrupted with sterile glass beads. Lymphogranuloma venereum strains L1 and L2 were the most widely used for the preparation of probes.

2. Purification of Chlamydial Elementary Bodies

Purification of EB was necessary to prevent contamination by host cell DNA. The most widely applied method was purification using differential centrifugation followed by Renografin (diatrizoate meglumine and diatrizoate sodium, 76% for injection, E.R. Squibb & Sons, Princeton, New Jersey) gradients (7,19,59).

3. Purification of DNA

Chromosomal DNA was extracted from the pellet of purified EB of *C. trachomatis* after lysis by addition of sarkosyl, sodium dodecyl sulfate (SDS), and ethylenediaminetetraacetic acid (EDTA). After treatment with proteinase K and then ribonuclease A, the DNA was purified by phenol and chloroform extractions, and ethanol precipitation (35).

Cryptic plasmid DNA was extracted from the EB by alkaline extraction procedure (3,39) and purified from chromosomal DNA by cesium chloride–ethidium bromide centrifugation (24,54).

4. *Preparation of DNA Probe*

The probe DNA may be either total genomic DNA or cloned chromosome or plasmid DNA. However, the use of genomic DNA caused nonspecific hybridizations in dot blots because of conserved sequences among bacterial strains. Cloned probes were prepared from a chromosomal fragment (21,41–43,45) or from the cryptic plasmid (18,21,25,43,45) present in all *C. trachomatis* serovars. The probes thus prepared were 2–10 kb long. The cloned cryptic plasmid DNA was used as a probe with or without the cloning vector.

5. *Labeling of DNA*

Radioactive labeling methods have been widely described (1). Three radioactive isotopes were used for labeling of *C. trachomatis* probes. Labeling of the DNA was by nick translation with [α-^{32}P]dCTP (21,22,25,41,45), [^{125}I]dCTP (32,42), or [^{35}S]dATP (17,18). ^{32}P and ^{125}I were used for hybridization on solid support or in solution and ^{35}S for *in situ* hybridization. Presently, radioactive probes are most sensitive but unfortunately unsuitable for most clinical laboratories.

Three nonradioactive commercially available systems of tagging probes were used. A biotin-labeled probe (Enzo Diagnostics, Inc., New York, New York) has recently been proposed for dot-blot and Southern blot analysis and for *in situ* hybridization (37). Another labeling technique using DNA sulfonation was applied for *in situ* hybridization on cell cultures and clinical specimens (9). A chemiluminescent *C. trachomatis* probe (Gen-Probe, Inc., San Diego, California) for liquid-phase hybridization was recently proposed (47).

B. Dot-Blot Hybridization

1. *Procedures*

Two different hybridization techniques were used. The first one was the classical dot-blot procedure, in which the target DNA is immobilized onto a filter and the probe is in solution in the hybridization mixture (1). The second one was the sandwich hybridization method (42–44). This method requires two separate specific NA reagents that are derived from two adjacent nonoverlapping regions on the bacterial plasmid or chomosome. One of the fragments is immobilized on a filter in a single-stranded form and serves to capture the homologous sample DNA; the other fragment, radiolabeled, is used as a probe. The test is positive when the sample DNA binds to both the membrane-bound fragment and the labeled probe. The sandwich approach is usually more specific (44) and less sensitive than the usual dot-blot hybridization.

2. *Treatment of Specimens*

Hybridization reactions can be performed on different samples: purified DNA, purified EB, infected cell culture, or clinical samples. The use of this method for detecting *C. trachomatis* in cell culture or clinical samples requires the extraction of DNA. The EB were lysed by SDS with or without sarkosyl and with 200 μg/ml of proteinase K at 37°C for at least 1 hr, and then DNA was precipitated with ethanol. In some cases the extracted DNA was purified by phenol and chloroform extraction and then by chromatography on diethylaminoethyl (DEAE) cellulose (41). The DNA was denatured before being spotted onto nitrocellulose or nylon filter by heating for 10 min at 100°C or by treating with NaOH.

3. *Hybridization Conditions*

A prehybridization was achieved by incubating the filter in the hybridization buffer containing polyvinylpyrrolidone, Ficoll 400, bovine serum albumin (BSA), and heterologous DNA (e.g., herring sperm) to prevent nonspecific binding of the probe to the filter. The probe was denatured by heating for 10 min at 100°C, then immersed in ice and added to the hybridization buffer. Final concentration of a radioactive probe should be ~10 ng/ml (1 to 4 $\times$ 10^5 dpm/ml). Nonradioactive probes were used in a higher concentration (~0.5 μg/ml).

The addition of polymers accelerated the hybridization reaction and simultaneously increased its sensitivity. The best results were obtained by adding 10% polyethylene glycol 6000 to the hybridization mixture (41). However, this cannot be used on a nitrocellulose filter at a temperature >42°C. Dextran sulfate at a final concentration of 10% also gave good results. Hybridization temperature was 39°–42°C in the presence of 50% formamide or 65°C in the absence of formamide. Generally 16 hr were allowed for the completion of hybridization.

After hybridization, the filters were washed at various temperatures according to levels of stringency. Stringency could be raised by raising the temperature from room temperature to 65°C and by lowering the sodium ion molarity from 0.4 to 0.02 *M* to eliminate the free probe and imperfect hybrids. The binding of probe DNA to the target was detected by autoradiography or by scintillation counting for radioactive probes and by enzyme-linked antibodies for nonradioactive probes.

4. Results

The results obtained by various techniques are summarized in Table I. Although different procedures were used, the results were fairly similar. For most of the experiments, the dot-blot hybridization allowed the detection of 100 pg of *C. trachomatis* DNA, which corresponds to $\sim 10^5$ EB. These results are comparable to those obtained with other bacteria (11). Cloned probes were more specific, especially when the vector plasmid was eliminated.

The dot-blot hybridization detection technique was used by several authors to detect *C. trachomatis* in genital specimens (21,41–43,45). The results of this procedure were compared with those obtained by culture. The samples found to be highly positive by culture (41) were all positive by hybridization. Only 42% of the samples found to be weakly positive by culture were found to be positive by hybridization. These results were improved by the sandwich hybridization technique using as a probe both chromosomal and plasmid DNA (21). Pao *et al.* (45) studied 317 endocervical specimens by slot-blot hybridization, culture, and direct detection of chlamydial antigen with enzyme-linked immunoassay. Compared to cell culture, sensitivity and specificity were, respectively, 91.7 and 95.3% for the slot-blot hybridization and 68.8 and 95.7% for the enzyme-linked immunoassay.

In summary, dot-blot hybridization was the first approach used for *C. trachomatis* detection by probe technology. The sensitivity of this technique was rather low, since the amount of organisms detected in the first reports ranged between 10^5 and 10^6. It is always difficult to evaluate the sensitivity of a dot-blot procedure using a radioactive probe, since increasing the amount of probe DNA or the autoradiography exposure time always increases sensitivity as well as cost and time.

C. Solution Hybridization

Solution hybridization was applied recently to *C. trachomatis* detection by Gen-Probe. The ^{125}I-labeled DNA probe is complementary to a specific part of a ribosomal RNA (rRNA) that is present in numerous copies per EB. The test is reported to be completed in 3 hr. The sensitivity and the specificity of the detection were 86.4 and 97%, respectively, when compared with culture. The DNA probe was able to detect as little as 1 inclusion-forming unit (IFU) (23). The probe was applied to 201 endocer-

TABLE I

Detection of *Chlamydia trachomatis* with DNA Probes by Dot-Blot Hybridization

Source of DNA probe	Hybridization method	Probe label	Limit of sensitivity	Number of clinical specimens tested	Reference
Genomic DNA or cloned plasmid	Filter	^{32}P	100 pg	0	(22)
Cloned chromosomal fragments	Sandwich	^{125}I	10^6 Genomes	16	(42)
Cloned chromosomal fragment	Filter	^{32}P	10^5 Genomes	231	(41)
Genomic DNA or cloned plasmid or cloned chromosomal fragment	Filter	^{32}P	10–100 pg DNA	76	(21)
Cloned chromosomal fragments and/or cloned plasmid	Sandwich	^{32}P	10^5 Genomes	268	(43)
Cloned plasmid or cloned chromosomal DNA	Filter	^{32}P	1 pg DNA	317	(45)

vical specimens tested at the same time by culture and MAb ELISA (32). Compared to cell culture, the sensitivities and specificities were, respectively, 88 and 99% for the Gen-Probe assay and 97 and 96% for the ELISA.

Recently a nonisotopic DNA probe (Gen-Probe) was developed using a chemiluminescent label. Detailed methodology had not been published prior to the completion of this chapter. The DNA probe was compared with antigen detection (ELISA) and culture on 220 endocervical specimens. Of 18 culture-positive samples, 16 were positive by hybridization and 17 by ELISA; and of 202 culture-negative samples, 191 were negative by hybridization and 195 by ELISA (47).

In preliminary reports (23,47), solution hybridization is fast (2 or 3 hr) and suitable for testing a large number of samples.

D. *In Situ* Hybridization

1. *Procedures*

The probes used for dot-blot hybridization can generally be used for *in situ* hybridization. These probes include genomic DNA (9), cryptic plasmid (17,18) and cloned DNA fragments (37). Probe labeling should be chosen for best *in situ* detection (Table II). If radiolabeling is used, ^{35}S

TABLE II

Detection of *Chlamydia trachomatis* with DNA Probes by *in Situ* Hybridization

DNA probe	Probe label	Method of visualization of hybrids	Specimens	References
Cloned plasmid	^{35}S	Autoradiography	Cell culture; cervical scraping	(18)
Genomic DNA	Sulfonation	Immunoenzymatic reaction	Cell culture; genital specimens	(9)
Cloned plasmid	^{35}S	Autoradiography	Cell culture; monkey tissue; rectal biopsy	(17)
Not disclosed (Enzo Diagnostics)	Biotinylation	Biotinylated enzyme reaction	Cell culture	(37)

would be preferred over ^{32}P, which is a too strong a β emitter. Any chemical labeling seems suitable for *in situ* hybridization.

The procedure used for the detection of *C. trachomatis* in cell cultures or genital specimens should permeabilize host and bacterial cells to allow hybridization with the probe while maintaining host cell morphology. The following steps are necessary: (1) pretreatment of glass slides to avoid nonspecific probe fixation; (2) sample fixation followed by permeabilization of host and bacterial cells; (3) DNA denaturation; (4) hybridization and washing; and (5) visualization of label on hybridized DNA.

2. *Sample Treatment*

Protocols for *in situ* hybridization of viral NA have been described with radioactive probes (4,5,13) and biotin-labeled probes (6,30,61). These protocols have been adapted to *C. trachomatis* detection (9,17,18,37).

Microscope glass slides were pretreated according to Brahic and Haase (4), then acetylated (16). The technique is detailed in Section VIII. Another protocol used treatment with poly-D-lysine (17,61).

The different samples studied by *in situ* hybridization included experimentally infected cell cultures (9,17,18,37), genital specimens (9,17,18), and tissue biopsy specimens (17). For cell cultures, three different procedures were used. Cells were grown and infected in tissue culture chamber slides (Lab-Tek chamber slide, Miles Scientific, Naperville, Illinois)(18), or infected cells (in a bottle) were trypsinized and deposited on a glass slide by use of a cytocentrifuge (9), or cells were cultured on glass coverslips (37). For endocervical specimens, samples were collected on a glass slide with an Ayre spatula (18) or cytocentrifuged from 2SP transport medium (9). Specimen fixation is a critical step in which cell loss should be avoided and cell morphology preserved. Fixatives included buffered formalin (17) or a mixture containing paraformaldehyde and glutaraldehyde (PFG) (5,9) or acetone (37).

Tissue biopsy samples were fixed in fresh buffered formalin (pH 7.4), embedded in paraffin, and sectioned to a thickness of 4 μm (17).

Fixed cells were permeabilized with HCl, and chlamydial EB were permeabilized with low proteinase K concentration (1 μg/ml) and SDS or saponin (9,17,18). When chemically labeled probes are used, permeabilization should allow the proteinaceous detection system to reach the target DNA (12). Samples embedded in paraffin were deparaffinized and treated with proteinase K (17).

3. *Hybridization Conditions*

For hybridization, permeabilized cells were covered with the hybridization mixture containing 50% formamide and the DNA probe, then with a treated coverslip. The probe DNA and the samples were denatured simultaneously by heating. The slides were then chilled on ice and sealed with rubber cement. Different incubation times were used: 24 hr in the dark at room temperature (17), 15 hr at 39°C (9), and 15 min at room temperature (37).

After hybridization, the coverslips were removed and the slides washed. Visualization of hybrids depended on the probe labeling. For ^{35}S labeling, slides were soaked in a gelatin solution and chromium potassium sulfate and autoradiographed by using nuclear track emulsion. The samples were counterstained with hematoxylin and examined at 400× magnification. The presence of EB in the samples was visualized by clusters or foci of silver grains over individual cells (17,18). When the Enzo biotinylated probe was used (37), hybrids were visualized with addition of a preformed streptavidin–biotinylated peroxidase complex, washed, and incubated with substrate. Chlamydial inclusions were dark, round or oval spots in the cells. For the sulfonated probe (9) an enzymatic immune reaction visualized the hybrids. The immunoenzymatic system included a mouse MAb specific for sulfonated cytosine and an anti-mouse polyclonal antibody conjugated to alkaline phosphatase (31). Chlamydial inclusions appeared as red grains in cells stained with hematoxylin (Fig. 1).

4. Results

Chlamydia trachomatis was detected by *in situ* hybridization in cell cultures, genital specimens, and tissue biopsy samples. This technique can visualize one EB under the microscope. Thus the actual sensitivity of the method depends on how infected cells and EB were concentrated on a slide. Comparison of *in situ* hybridization using a ^{35}S-labeled probe in cervical smears with cell culture of cervical specimens showed 91% sensitivity and 80% specificity (18). The Enzo probe was used to detect the organisms in infected cell culture in comparison with immunofluorescent-antibody staining. The sensitivity of *in situ* hybridization was 89.7% and specificity was 100% (37). In experimentally infected cell culture, a sulfonated total DNA probe detected intracellular inclusions as early as 8 hr after inoculation. Extracellular EB could also be detected by this probe (Fig. 1). When applied to five genital samples, the results were in agreement with those of culture (9). In *C. trachomatis*-infected monkey tissue, *in situ* hybridization with a ^{35}S-labeled probe showed 100% sensitivity and 100% specificity in comparison with culture (17). When the same probe was applied to human biopsy samples, five of six culture-positive specimens were found positive by DNA hybridization (17).

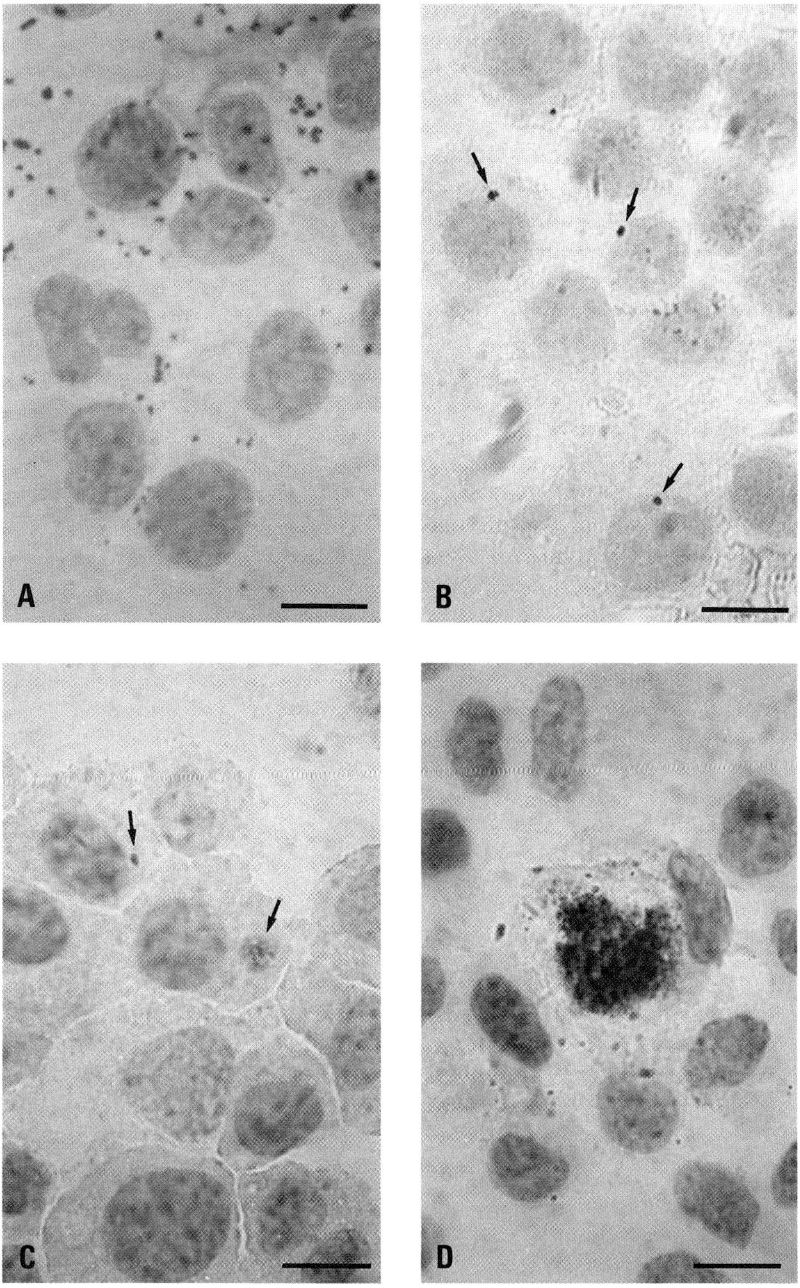

Fig. 1. *In situ* hybridization with sulfonated *C. trachomatis* total DNA (strain L2) grown in McCoy cells. Bacterial cells containing hybridized probe DNA are stained red. Cells were counterstained with hematoxylin. (A) 2 hr after infection; (B) 8 hr after infection; (C) 18 hr after infection; (D) 48 hr after infection (9). Arrows in B and C indicate chlamydial inclusions. Bar equals 10 μm.

IV. CONCLUSIONS

The application of NA hybridization to the detection of *C. trachomatis* in cell culture and in clinical samples has followed diverse strategies and has yielded diverse results. In dot-blot procedures, specificity is improved when cloned DNA fragments are selected for use as a probe (21,25,41–43,45) and when sandwich hybridization is used (41,43,44). Specificity has not been shown to be critical in *in situ* hybridization in which total DNA can be used as a probe (9). Sensitivity is more of a problem. Dot-blot hybridization can rarely detect $<10^5$ bacterial cells (21,22,41,43,45), and *in situ* hybridization, which theoretically can visualize one EB, can be more sensitive as long as microscopic fields are not obscured by cellular and other debris. Sensitivity can be improved by using probes whose targets are naturally amplified such as the multicopy cryptic plasmid or rRNA. Hybridization with a chemically labeled probe is usually less sensitive than that with a radioactive probe in the dot-blot procedure. Sensitivity is usually not affected by the choice of probe label in *in situ* hybridization. Improvement of sensitivity by gene amplification will be discussed in Section VI.

In situ hybridization is an interesting approach for the detection of pathogens, especially those that invade and multiply in host cells. In addition to a specific hybridization reaction, the method brings information on the shape of a bacterium and its relation to host cellular structures. A weak dot-blot hybridization signal is very difficult to interpret. A few EB or inclusions observed under the microscope after *in situ* hybridization are easier to interpret as proof of the presence of *Chlamydia*.

Speed and complexity of hybridization techniques are real problems. Time needed for hybridization is usually reduced when the concentration of target and probe DNA is raised, dextran sulfate or polyethylene glycol added, or target sequence reduced in size (oligonucleotide probe). The fastest procedure reported for detection of *C. trachomatis* by NA hybridization is the solution method of Gen-Probe (~3 hr), although methodological details have not been disclosed.

In situ hybridization presently requires too many critical steps to be of practical value in the clinical laboratory. More work is needed to simplify this method.

V. GENE PROBES VERSUS MONOCLONAL ANTIBODIES

Monoclonal antibody technology is now well established and rather well standardized for use in microscopic detection of bacteria (immunofluorescence or immunoenzymatic staining) or automated microtiter plate ELISA (34a). In contrast, NA probe technology is in its infancy and not ready for standardization. Thus there is little point in generalizing about specificity and sensitivity of various hybridization experiments (using different probes, different labeling, different protocols) compared to a rather standard immunological technique.

Both methodologies (MAb and NA) can be quite specific and both can detect nonviable organisms. Immunofluorescence detection of *C. trachomatis* requires trained personnel because the threshold between positive and negative reactions is not always obvious. Patients' antibodies might cover EB and thus prevent binding of MAb (17). In tissue sections, MAb are less efficient in visualizing *Chlamydia*. These different problems have stimulated research on *Chlamydia* probes. Although MAb are presently available for use by clinical microbiologists, probes are still confined to use in research laboratories. Probe technology is progressing fast and its future is very promising.

VI. PROSPECTS FOR THE FUTURE

As stated before, probe technology is in its infancy. New procedures are published each year that improve specificity, sensitivity, and speed. The latest breakthrough has been the availability of a gene amplification procedure called polymerase chain reaction (PCR). Until recently, 10^5 bacterial cells were generally the minimal number of target bacteria detected by DNA hybridization. The PCR method makes thousands of copies of a target sequence (49). This method can be used when the nucleotide sequence of a 100- to 2000-bp DNA fragment is known. The DNA fragment should be specific for the target bacterium. Two oligonucleotide primers are synthetized, which are complementary to short sequences on the ends of both DNA strands. These sequences flank the region to be amplified. The thermostable DNA polymerase of *Thermus aquaticus* (*Taq*) then copy both DNA strands by extension of the primers. Repeated cycles of denaturation, primer annealing, and primer extension result in amplification of the target sequence (49). The amplified DNA region can then be detected by hybridization with an oligonucleotide probe binding to the central region of the target sequence.

The PCR method has recently been applied to the detection of viral pathogens (33,38,53) and to the amplification of a 129-bp fragment (10) chosen within a MOMP gene of *C. trachomatis* (2,46,56). Two oligonu-

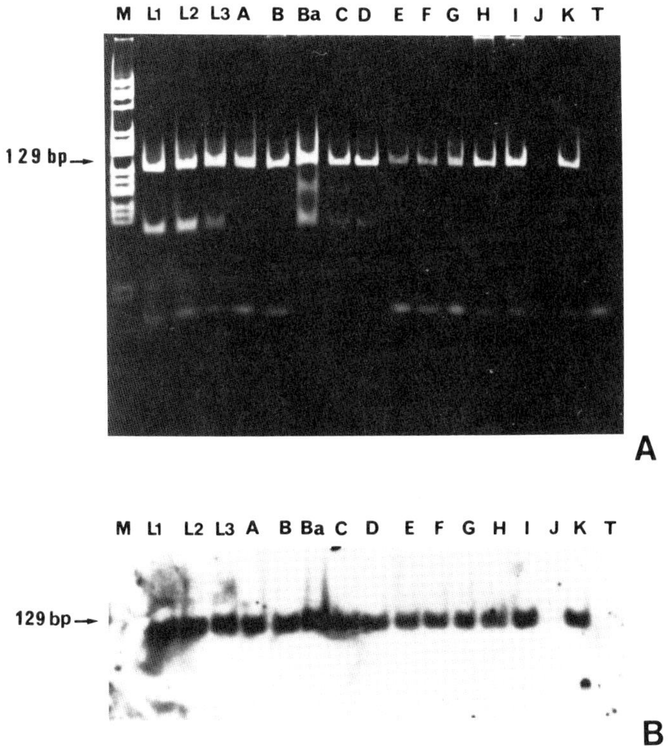

Fig. 2. PCR amplification (2 × 25 cycles) on each *C. trachomatis* serovar DNA and on 500 ng of McCoy cells DNA (line T). (A) Polyacrylamide gel electrophoresis of amplified DNA (molecular size marker was pBR-322 cleaved by *Hae*III). (B) Southern analysis of the gel with the ^{32}P-labeled internal oligonucleotide (10).

cleotide primers (21- and 23-mer) hybridized to sequences flanking the 129-bp region (Fig. 2). Pure chlamydial DNA or lysed EB were used as substrate for gene amplification. The amplified region was identified by agarose gel electrophoresis after *Eco*RI treatment followed by Southern transfer and hybridization with ^{32}P-end-labeled 21-mer oligonucleotide (recognizing a 21-bp portion of the 129-bp amplified sequence) (Fig. 3) (10).

Future work will include application of PCR to clinical specimens and assessment of sensitivity in the detection of *C. trachomatis*. Machines are now available to perform amplification cycles. When sensitivity is no longer a problem, simplification and automation of hybridization will need careful attention from developers.

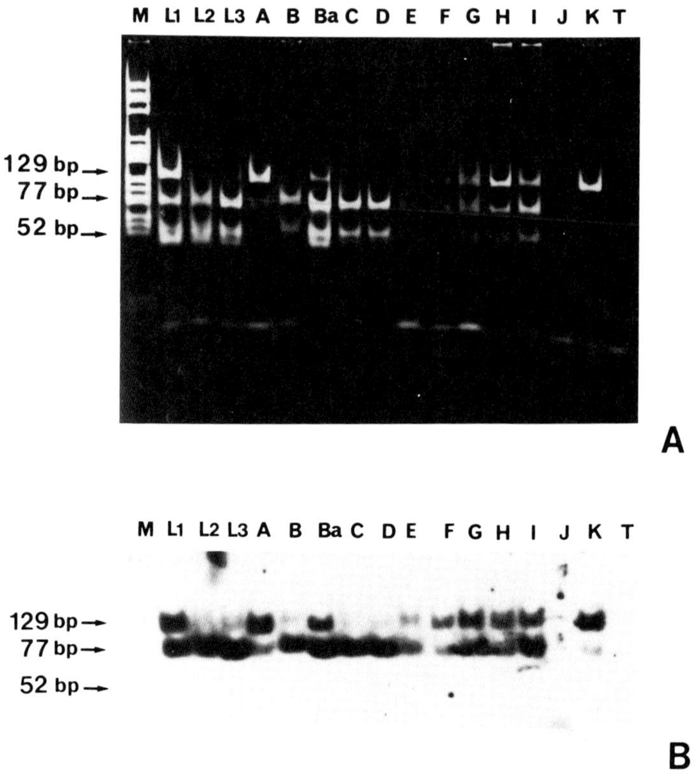

Fig. 3. *Eco*RI digestion of amplified DNA sequences from 15 *C. trachomatis* serovars and McCoy cells DNA. (A) Polyacrylamide gel electrophoresis of digested DNA (molecular size marker was pBR-322 cleaved by *Hae*III; line T is amplified McCoy cells DNA). (B) Southern analysis of the gel with the ^{32}P-labeled internal oligonuceotide. In this experiment, partial hydrolysis occurred for serovars L1, A, Ba, E, F, G, H, I, and K (10).

VII. SUMMARY

Chlamydia trachomatis is one of the most common of sexually transmitted pathogens and the causative agent of trachoma. The diagnosis of chlamydial infections relies usually on the multiplication of the organism in cell culture or on direct demonstration of chlamydial antigens. However, no technique is at present fully satisfactory for the detection of *C. trachomatis* in all specimens.

DNA hybridization technology was recently applied to *C. trachomatis* detection. Dot-blot hybridization was initially employed. The specificity of the detection was improved when cloned chromosomal or plasmid DNA fragments were used as a probe and the sandwich hybridization technique selected. However, dot-blot hybridization could rarely detect $<10^5$ bacterial cells. Solution hybridization was also applied, using a probe complementary to a specific part of rRNA. *Chlamydia trachomatis* could be located in tissue or cell culture by *in situ* hybridization with radioactive or nonradioactive probes. This technique reveals information on the pathogenesis of infections caused by this organism.

Probe technology is still in its infancy and should be simplified for use in the clinical laboratory. New procedures are proposed each year. The gene amplification procedure, called the polymerase chain reaction (PCR), is one of them and should be assessed for *C. trachomatis* detection.

VIII. MATERIALS AND METHODS

A. Culture and Purification of Chlamydial Elementary Bodies

1. Grow *C. trachomatis* L2 strain in McCoy cells. Culture is in 250-ml plastic flasks containing Eagle minimal essential medium supplemented with 10% (v/v) fetal calf serum, 2 m*M* glutamine, 10 μg/ml gentamicin, and 50 μg/ml vancomycin.
2. Inoculate 10 flasks each with 5 ml of bacterial suspension in cell culture medium consisting of $\sim 10^7$ IFU/ml, and incubate for 2 hr at 37°C with intermittent shaking.
3. Replace cell culture medium with 20 ml fresh medium supplemented with 1 μg/ml of cycloheximide (Sigma Chemical Co, St. Louis, Missouri) and 0.5% (w/v) glucose. Incubate at 35°C for 72 hr.
4. Remove and disrupt the cells with sterile glass beads and centrifuge the cell suspension at 500 *g* for 15 min.
5. Purification of EB from cell debris was achieved by differential centrifugation in Renografin (Section III,A,2). Collect the supernatants, centrifuge through an 8-ml cushion of 35% (v/v) Renografin in phosphate-buffered saline (8.0 g NaCl, 0.2 g KCl, 1.5 g NaH_2PO_4, 0.2 g K_2HPO_4 per liter of solution) for 1 hr at 30,000 *g* at 4°C in a JA-20 rotor (Beckman Instruments, Inc., Fullerton, California).
6. Collect the pellets in 1 ml cell culture medium and lay over a discontinuous Renografin gradient (13 ml of 40% Renografin, 8 ml of 44%, 5 ml of 52%, v/v). Centrifuge 1 hr at 43,000 *g* at 4°C in an SW27 rotor (Beckman Instruments).
7. Collect the EB band from the 44/52% interface. Dilute in 10 ml of cell

culture medium and pellet at 30,000 *g* for 30 min in the JA-20 rotor. Freeze the bacterial pellets at −20°C.

Note: Purification should be achieved in one step on fresh chlamydial culture without an intermediate freeze for maximum yield.

B. DNA Preparation

1. Lyse the pellet of purified EB with 5 ml of 1.5% (w/v) sarkosyl, 1% (w/v) SDS, and 50 m*M* EDTA. Incubate 1 hr at 37°C with intermittent shaking.
2. Add proteinase K (Boehringer, Mannheim, Federal Republic of Germany) to a final concentration of 100 μg/ml. Mix and incubate overnight at 37°C.
3. Extract the mixture twice with an equal volume of buffer-saturated (redistilled) phenol. Separate the phases by centrifugation at 3000 *g* after each extraction and transfer the aqueous layer to a clean tube.
4. Extract the aqueous phase twice with an equal volume of 24 : 1 chloroform–isoamyl alcohol. Transfer the aqueous layer to a clean tube.
5. Add ammonium acetate to final concentration of 2 *M*, add two volumes of ice-cold ethanol, mix well, and leave overnight at −20°C.
6. Recover the nucleic acids by centrifugation at 10,000 *g* for 30 min at 4°C. Discard the supernatant. Dry the pellet in a vacuum desiccator. Dissolve the pellet in TE buffer (10 m*M* Tris-HCl, pH 7.5, 1 m*M* EDTA).
7. Add RNase A (preincubated at 90°C for 10 min) to a final concentration of 50 μg/ml. Incubate 1 hr at 37°C.
8. Repeat steps 3–6. Store DNA at −20°C or at 4°C with a few drops of chloroform.

C. DNA Labeling by Sulfonation

1. Sonicate the DNA to a size of ~1 kb. The size can be checked by agarose gel electrophoresis.
2. Denature the DNA in a boiling-water bath for 10 min; chill in ice.
3. Label with DNA Chemiprobe Kit (Orgenics, Ltd, Yavne, Israel): to 100 μl of DNA solution add 50 μl of 2 *M* sodium bisulfite and 12.5 μl of 1 *M* *O*-methyl hydroxylamine. Incubate the mixture overnight at room temperature. DNA is modified and ready for use. A purification step is not necessary. Store the modified DNA at −20°C. According to the manufacturer, such a probe is supposedly stable for a year.

D. *In Situ* Hybridization of *Chlamydia trachomatis* in Cell Culture or Genital Specimens

1. Pretreatment of Slides before Hybridization

a. Incubate the slides in 1 *N* HCl for 30 min; wash three times in H_2O.
b. Transfer the slides to 95% ethanol for 30 min. Air-dry.
c. Soak the slides in Denhardt's solution (0.02% BSA, 0.02% polyvinylpyrrolidone 400, 0.02% Ficoll 400), 3× SSC (1× SSC 0.15 *M* NaCl, 0.015 *M* trisodium citrate) overnight at 65°C.
d. Wash the slides in distilled H_2O. Fix with 95% ethanol–acetic acid (3 : 1 v/v) 20 min. Air-dry.
e. Acetylate the slides: suspend in 0.1 *M* triethanolamine pH 8 and add, once only, acetic anhydride 5 ml per liter; mix vigorously for 10 min.
f. Wash the slides twice in H_2O, once in 95% ethanol. Air-dry.

2. Sample Treatment

a. Adjust infected McCoy cells suspension to ~10^5 cells/ml. Deposit 0.3 ml on the treated slides with the use of a cytocentrifuge at 500 rpm for 5 min. For genital specimens, this protocol is used with 2SP transport medium. Air-dry.
b. Incubate the slides in PFG: 0.5% paraformaldehyde, 0.5% glutaraldehyde, 0.1 *M* phosphate buffer pH 6.0, 1.6% glucose, 0.002% $CaCl_2$, 1% dimethyl sulfoxide for 20 min at 4°C.
c. Soak the slides in 0.15 *M* ethanolamine (pH 7.5) for 20 min at 4°C. Store the slides at 4°C.

3. Hybridization Conditions

Permeabilization step

a. Immerse the slides in 0.2 *N* HCl, 20 min. Wash twice in H_2O.
b. Incubate the slides in 50 m*M* Tris-HCl pH 7.5 and 5 m*M* EDTA containing 1 μg/ml proteinase K (Boehringer), 0.5% (w/v) saponin (Merck, Darmstadt, Federal Republic of Germany) at 37°C for 15 min (bacterial lysis step).
c. Wash the slides in 100 m*M* Tris-HCl pH 7.5, 100 m*M* NaCl, glycine 2 mg/ml. Dehydrate through a graded alcohol series (30, 60, 95%, absolute alcohol), 5 min each time. Air-dry.

Hybridization

a. Cover the cells with 20 μl of hybridization mixture containing 50% deionized formamide, 5× SSC, 25 m*M* sodium phosphate buffer pH 6.5, 1× Denhardt's solution, 10% (w/v) dextran sulfate, 50 μg/ml of sheared, denatured herring sperm DNA, and sulfonated *Chlamydia* DNA probe (10 μg/ml).

b. Cover the hybridization mixture with a treated coverslip.

c. Keep the slides on a heating rack 5 min at 95°C (infected cells and probe DNA are denatured). Chill in ice. Seal with rubber cement. Incubate 15 hr at 39°C.

d. Remove the coverslips. Wash the slides in 2× SSC for 20 min, and 1× SSC for 20 min at room temperature.

4. Immunological Visualization of Sulfonated DNA

a. Cover the slides with a blocking solution containing 350 mg/ml skimmed-milk powder, 37 mg/ml BSA, 0.05% (w/v) saponin, 3.5 mg/ml sodium heparin (Prolabo) in TEN buffer (50 m*M* Tris-HCl pH 8, 25 m*M* NaCl, 1 m*M* EDTA). Place the slides in a moist chamber for 45 min at room temperature.

b. Remove the blocking solution by a brief dip in a washing solution containing 1.5% (w/v) NaCl, 0.3% (v/v) Brij 35 solution (Sigma).

c. Incubate the slides for 1 hr with the blocking solution containing a 1 : 250 dilution of mouse MAb specific for sulfonated cytosine.

d. Wash the slides in the washing solution, with slow agitation, once for 3 min and three times for 5 min.

e. Incubate for 1 hr with alkaline phosphatase anti-mouse conjugate at a 1 : 125 dilution in the blocking solution. Wash again once for 3 min and three times for 10 min.

f. Prepare fresh substrate with 3 mg of Fast-Red TR salt (F 1500, Sigma) in 1 ml of naphthol solution (0.40 mg/ml naphthol MSX-phosphate (N 5000, Sigma) in 50 m*M* Tris-acetate pH 9.5, 10 m*M* Mg acetate).

g. Cover the slides with the substrate solution for 20 min in the dark at room temperature. Wash in distilled water. Air-dry.

h. Counterstain 5 min with Mayer's hematoxylin solution (Sigma). Rinse with distilled H_2O. Air-dry. Mount in buffered glycerin and examine by light microscopy (× 1000).

REFERENCES

1. Anderson, M. L. M., and Young, B. D. (1985). Quantitative filter hybridization. *In* "Nucleic Acid Hybridization: A Practical Approach" (B. D. Hames and S. J. Higgins, eds.), pp. 73–110. IRL Press, Washington, D. C.
2. Baehr, W., Zhang, Y.-X., Joseph, T., Su, T., Nano, F. E., Everett, K. D. E., and Caldwell, H. D. (1985). Mapping antigenic domains expressed by *Chlamydia trachomatis* major outer membrane protein genes. *Proc. Natl. Acad. Sci. U. S. A.* **85,** 4000–4004.

3. Birnboim, H. C., and Doly, J. (1979). A rapid alkaline extraction procedure for screening recombinant plasmid DNA. *Nucleic Acids Res.* **7,** 1513–1523.
4. Brahic, M., and Haase, A. T. (1978). Detection of viral sequences of low reiteration frequency by in situ hybridization. *Proc. Natl. Acad. Sci. U. S. A.* **75,** 6125–6129.
5. Brahic, M., Haase, A. T., and Cash, E. (1984). Simultaneous in situ detection of viral RNA and antigens. *Proc. Natl. Acad. Sci. U. S. A.* **81,** 5445–5448.
6. Brigati, D. J., Myerson, D., Leary, J. J., Spalholz, B., Travis, S. Z., Fong, C. K. Y., Hsiung, G. D., and Ward, D. C. (1983). Detection of viral genomes in cultured cells and paraffin-embedded tissue sections using biotin-labeled hybridization probes. *Virology* **126,** 32–50.
7. Caldwell, H. D., Kromhout, J., and Schachter, J. (1981). Purification and partial characterization of the major outer membrane protein of *Chlamydia trachomatis*. *Infect. Immun.* **31,** 1161–1176.
8. Caul, E. O., and Paul, I. D. (1985). Monoclonal antibody based ELISA for detecting *Chlamydia trachomatis*. *Lancet* **1,** 279.
9. Dutilh, B., Bébéar, C., Taylor-Robinson, D., and Grimont, P. A. D. (1988). Detection of *Chlamydia trachomatis* by *in situ* hybridization with sulphonated total DNA. *Ann. Inst. Pasteur/Microbiol.* **139A,** 115–128.
10. Dutilh, B., Bébéar, C., Rodriguez, P., Vekris, A., Bonnet, J., and Garret, M. (1989). Specific amplification of a DNA sequence common to all *Chlamydia trachomatis* serovars, using the polymerase chain reaction. *Res. Microbiol.* **140,** 7–16.
11. Edberg, S. C. (1985). Principles of nucleic acid hybridization and comparison with monoclonal antibody technology for the diagnosis of infectious diseases. *Yale J. Biol. Med.* **58,** 425–442.
12. Feldmann, G., Maurice, M., Bernau, D., Rogier, E., and Durand, A.-M. (1983). Penetration of enzyme-labeled antibodies into tissues and cells: A review of the difficulties. *In* "Immunoenzymatic Techniques" (S. Avrameas, P. Druet, R. Masseyeff, and G. Feldmann, eds.), pp. 3–15, Elsevier, Amsterdam.
13. Haase, A., Brahic, M., Stowring, L., and Blum, H. (1984). Detection of viral nucleic acids by in situ hybridization. *Methods Virol.* **7,** 189–225.
14. Harrison, H. R., and Alexander, E. R. (1984). *Chlamydia trachomatis* infections of the infant. *In* "Sexually Transmitted Diseases" (K. K. Holmes, P. A. Mårdh, P. F. Sparling, and P. J. Wiesner, eds.), pp. 270–280. McGraw-Hill, New York.
15. Hatt, C., Ward, M. E., and Clarke, I. N. (1988). Analysis of the entire nucleotide sequence of the cryptic plasmid of *Chlamydia trachomatis* serovar L1. Evidence for involvement in DNA replication. *Nucleic Acids Res.* **16,** 4053–4067.
16. Hayashi, S., Gillam, I. C., Delaney, A. D., and Tener, G. M. (1978). Acetylation of chromosome squashes of *Drosophila melanogaster* decreases the background in autoradiographs from hybridization with ^{125}I-labeled RNA. *J. Histochem. Cytochem.* **26,** 677–679.
17. Horn, J. E., Kappus, E. W., Falkow, S., and Quinn, T. C. (1988). Diagnosis of *Chlamydia trachomatis* in biopsied tissue specimens by using in situ DNA hybridization. *J. Infect. Dis.* **157,** 1249–1253.
18. Horn, J. E., Hammer, M. L., Falkow, S., and Quinn, T. C. (1986). Detection of *Chlamydia trachomatis* in tissue culture and cervical scrapings by in situ hybridization. *J. Infect. Dis.* **153,** 1155–1159.
19. Howard, L., Orenstein, N. S., and King, N. W. (1974). Purification on Renografin density gradients of *Chlamydia trachomatis* grown in the yolk sac of eggs. *Appl. Microbiol.* **27,** 102–106.
20. Howard, L. V., Coleman, P. F., England, B. J., and Herrmann, J. E. (1986). Evaluation

of Chlamydiazyme for the detection of genital infections caused by *Chlamydia trachomatis*. *J. Clin. Microbiol.* **23,** 329–332.
21. Hyypiä, T., Jalava, A., Larsen, S. H., Terho, P., and Hukkanen, V. (1985). Detection of *Chlamydia trachomatis* in clinical specimens by nucleic acid spot hybridization. *J. Gen. Microbiol.* **131,** 975–978.
22. Hyypiä, T., Larsen, S. H., Stahlberg, T., and Terho, P. (1984). Analysis and detection of chlamydial DNA. *J. Gen. Microbiol.* **130,** 3159–3164.
23. Jonas, V., Hogan, J. J., Young, K. M., and Bryan, R. N. (1988). Evaluation of an isotopic DNA probe assay for direct detection of *Chlamydia trachomatis* in urogenital specimens. *Am. Soc. Microbiol. Abstr. Annu. Meet.,* Abstr. C-232.
24. Joseph, T., Nano, F. E., Garon C. F., and Caldwell, H. D. (1986). Molecular characterization of *Chlamydia trachomatis* and *Chlamydia psittaci* plasmids. *Infect. Immun.* **51,** 699–703.
25. Kahane, S., and Sarov, I. (1986). Detection of Chlamydia by DNA hybridization with a native chlamydial plasmid probe. *In* ''Chlamydial Infections: Proceeding of the Sixth International Symposium on Human Chlamydial Infections'' (D. Oriel, G. Ridway, J. Schachter, D. Taylor-Robinson, and M. Ward, eds.), pp. 574–577. Cambridge Univ. Press, London.
26. Kingsbury, D. T. (1969). Estimate of the genome size of various microorganisms. *J. Bacteriol.* **98,** 1400–1401.
27. Kingsbury, D. T., and Weiss, E. (1968). Lack of deoxyribonucleic acid homology between species of the genus *Chlamydia*. *J. Bacteriol.* **96,** 1421–1423.
28. Kuo, C. C., Wang, S.-P., and Grayston, J. T. (1972). Differentiation of TRIC and LGV organisms based on enhancement of infectivity by DEAE–dextran in cell culture. *J. Infect. Dis.* **125,** 313–317.
29. Kuo, C. C., Wang, S.-P., and Grayston, J. T. (1977). Growth of trachoma organisms in HeLa 229 cell culture. *In* ''Nongonococcal Urethritis and Related Infections'' (D. Hobson and K. K. Holmes, eds.), pp. 328–336. Am. Soc. Microbiol., Washington, D. C.
30. Leary, J. J., Brigati, D. J., and Ward, D. C. (1983). Rapid and sensitive colorimetric method for visualizing biotin-labeled DNA probes hybridized to DNA or RNA immobilized on nitrocellulose. *Proc. Natl. Acad. Sci. U. S. A.* **80,** 4045–4049.
31. Lebacq, P., Squalli, D., Duchenne, M., Pouletty, P., and Joannes, M. (1988). A new sensitive nonisotopic method using sulfonated probes to detect picogram quantities of specific DNA sequences on blot hybridization. *J. Biochem. Biophys. Methods* **15,** 255–266.
32. Lebar, W., Herschman, B., Pierzchala, J., and Jemal, C. (1988). Comparison of DNA probe, monoclonal enzyme immunoassay and cell culture for the detection of *Chlamydia trachomatis*. *Program Abstr., Intersci. Conf. Antimicrob. Agents Chemother., 28th,* Abstr. No. 1185.
33. Loche, M. and Mach, B. (1988). Identification of HIV-infected seronegative individuals by a direct diagnostic test based on hybridization to amplified viral DNA. *Lancet* **2,** 418–421.
34. Lovett, M., Kuo, C. C., Holmes, K., and Falkow, S. (1980). Plasmids of the genus *Chlamydia*. *In* ''Current Chemotherapy and Infectious Disease'' (J. D. Nelson and C. Grassi, eds.), Vol. 2, pp. 1250–1252. Am. Soc. Microbiol., Washington, D. C.
34a. Macario, A. J. L., and Conway de Macario, E., eds. (1985–1986). ''Monoclonal Antibodies against Bacteria, Vols. 1–3. Academic Press, Orlando, Florida.
35. Maniatis, T., Fritsch, E. F., and Sambrook, J. (1982). ''Molecular Cloning: A Laboratory Manual.'' Cold Spring Harbor Lab., Cold Spring Harbor, New York.
36. Mardh, P.-A. (1984). Bacteria, chlamydiae and mycoplasmas. *In* ''Sexually Transmitted

Diseases" (K. K. Holmes, P. A. Mårdh, P. F. Sparling, and P. J. Wiesner, eds.), pp. 829–856. McGraw-Hill, New York.

37. Näher, H., Petzoldt, D., and Sethi, K. K. (1988). Evaluation of non-radioactive in situ hybridization method to detect *Chlamydia trachomatis* in cell culture. *Genitourin. Med.* **64,** 162–164.
38. Ou, C.-Y., Kwok, S., Mitchell, S. W., Mack, D. H., Sninsky, J. J., Krebs, J. W., Feorino, P., Warfield, D., and Schochetman, G. (1988). DNA amplification for direct detection of HIV-1 in DNA of peripheral blood mononuclear cells. *Science* **239,** 295–297.
39. Palmer, L. and Falkow, S. (1986). A common plasmid of *Chlamydia trachomatis*. *Plasmid* **16,** 52–62.
40. Palmer, L., Falkow, S., and Klevan, L. (1986). 16S ribosomal RNA genes of *Chlamydia trachomatis*. *In* "Chlamydial Infections: Proceeding of the Sixth International Symposium on Human Chlamydial Infections" (D. Oriel, G. Ridway, J. Schachter, D. Taylor-Robinson, and M. Ward, eds.), pp. 89–92. Cambridge Univ. Press, London.
41. Palva, A. (1985). Nucleic acid spot hybridization for detection of *Chlamydia trachomatis*. *FEMS Microbiol. Lett.* **28,** 85–91.
42. Palva, A., Jousimies-Somer, H., Saikku, P., Vaananen, P., Soderlund, H., and Ranki, M. (1984). Detection of *Chlamydia trachomatis* by nucleic acid sandwich hybridization. *FEMS Microbiol. Lett.* **23,** 83–89.
43. Palva, A., Korpela, K., Lassus, A., and Ranki, M. (1987). Detection of *Chlamydia trachomatis* from genito-urinary specimens by improved nucleic acid sandwich hybridization. *FEMS Microbiol. Lett.* **40,** 211–217.
44. Palva, A., and Ranki, M. (1985). Microbial diagnosis by nucleic acid sandwich hybridization. *Clin. Lab. Med.* **5,** 475–490.
45. Pao, C. C., Lin, S. S., Yang, T. E., Soong, Y. K., Lee, P. S., and Lin, J. Y. (1987). Deoxyribonucleic acid hybridization analysis for the detection of urogenital *Chlamydia trachomatis* infection in women. *Am. J. Obstet. Gynecol.* **156,** 195–199.
46. Pickett, M. A., Ward, M. E., and Clarke, I. N. (1987). Complete nucleotide sequence of the major outer membrane protein gene from *Chlamydia trachomatis* serovar L1. *FEMS Microbiol. Lett.* **42,** 185–190.
47. Putbrese, S. C., Meier, F. A., Johnson, B. A., Brookman, R. R., and Dalton, H. P. (1988). Comparison of a non-isotopic DNA probe and ELISA for detecting *Chlamydia trachomatis* in clinical samples. *Am. Soc. Microbiol., Abstr. Annu. Meet.,* Abstr. C-234.
48. Ripa, K. T., and Mårdh, P.-A. (1977). New simplified culture technique for *Chlamydia trachomatis*. *In* "Nongonococcal Urethritis and Related Infections" (D. Hobson and K. K. Holmes, eds.), pp. 323–327. Am. Soc. Microbiol., Washington, D. C.
49. Saiki, R. K., Gelfand, D. H., Stoffel, S., Scharf, S. J., Higuchi, R., Horn, G. T., Mullis, K. B., and Erlich, H. A. (1988). Primer-directed enzymatic amplification of DNA with a thermostable DNA polymerase. *Science* **239,** 487–491.
50. Sarov, I., and Becker, Y. (1969). Trachoma agent DNA. *J. Mol. Biol.* **42,** 581–589.
51. Schachter, J. (1985). Immunodiagnosis of sexually transmitted disease. *Yale J. Biol. Med.* **58,** 443–452.
52. Schachter, J. (1978). Chlamydial infections. *N. Engl. J. Med.* **298,** 428–435, 490–495, 540–549.
53. Shibata, D. K., Arnheim, N., and Martin, W. J. (1988). Detection of human papilloma virus in paraffin-embedded tissue using the polymerase chain reaction. *J. Exp. Med.* **167,** 225–230.
54. Sriprakash, K. S. and Macavoy, E. S. (1987). Characterisation and sequence of a plasmid from the trachoma biovar of *Chlamydia trachomatis*. *Plasmid* **18,** 205–214.
55. Stamm, W. E. and Holmes, K. K. (1984). *Chlamydia trachomatis* infections of the adult.

In "Sexually Transmitted Diseases" (K. K. Holmes, P. A. Mårdh, P. F. Sparling, and P. J. Wiesner, eds.), pp. 258–270. McGraw-Hill, New York.
56. Stephens, R. S., Pescador-Sanchez, R., Wagar, E., A., Inouye, C., and Urdea, M. S. (1987). Diversity of *Chlamydia trachomatis* major outer membrane protein genes. *J. Bacteriol.* **169,** 3879–3885.
57. Tam, M. R., Stamm, W. E., Handsfield, H. H., Stephens, R., Kuo, C. C., Holmes, K. K., Ditzenberger, K., Krieger, M., and Nowinski, R. C. (1984). Culture-independent diagnosis of *Chlamydia trachomatis* using monoclonal antibodies. *N. Engl. J. Med.* **310,** 1146–1150.
58. Tenover, F. C. (1988). Diagnostic deoxyribonucleic acid probes for infectious diseases. *Clin. Microbiol. Rev.* **1,** 82–101.
59. Terho, P. and Matikainen, M. T. (1982). Production of *Chlamydia trachomatis* antigen and antiserum: A review. *Scand. J. Infect. Dis.* **32,** 30–33.
60. Thomas, B. J., Evans, R. T., Hawkins, D. A., and Taylor-Robinson, D. (1984). Sensitivity of detecting *Chlamydia trachomatis* elementary bodies in smears by use of a fluorescein labelled monoclonal antibody: Comparison with conventional chlamydial isolation. *J. Clin. Pathol.* **37,** 812–816.
61. Unger, E. R., Budgeon, L. R., Myerson, D., and Brigati, D. J. (1986). Viral diagnosis by in situ hybridization; description of a rapid simplified colorimetric method. *Am. J. Surg. Pathol.* **10,** 1–8.
62. Wang, S.-P., and Grayston, J. T. (1971). Classification of TRIC and related strains with micro immunofluorescence. *In* "Trachoma and Related Disorders Caused by Chlamydial Agents" (R. L. Nichols, ed.), pp. 305–321. Excerpta Medica, Amsterdam.
63. Wang, S.-P., Kuo, C.-C., Barnes, R. C., Stephens, R. S., and Grayston, J. T. (1985). Immunotyping of *Chlamydia trachomatis* with monoclonal antibodies. *J. Infect. Dis.* **152,** 791–800.

3

Construction of DNA Probes for the Identification of *Haemophilus ducreyi*

LINDA M. PARSONS, MEHDI SHAYEGANI, ALFRED L. WARING, AND LAWRENCE H. BOPP

Laboratories for Bacteriology
Wadsworth Center for Laboratories and Research
New York State Department of Health
Albany, New York

I. INTRODUCTION

Haemophilus ducreyi, the fastidious gram-negative bacillus that causes chancroid, has traditionally been difficult to diagnose either by clinical symptoms or by laboratory testing. Symptoms may be mimicked by other organisms, and *H. ducreyi* is not easily cultured or identified biochemically. Nucleic acid (NA) probe technology offers microbiology laboratories an additional approach for the detection of organisms that are difficult or impossible to grow. This report describes the development and evaluation of DNA probes for the identification of *H. ducreyi.*

II. BACKGROUND

A. Chancroid

Chancroid or soft chancre is a sexually transmitted disease (STD) of humans characterized by painful ulcers of the genitalia, usually with inguinal lymph node involvement. The incubation period for the development of the ulcer tends to be 2–7 days (72,77,79). The ulcer crater, which can range from 1 mm to 2 cm in diameter, is usually purulent, and the edges are irregular and surrounded by an erythematous halo (33). Multiple ulcers often follow the appearance of the primary lesion, most likely caused by autoinoculation. This results in lesions of varying age, which naturally extend the duration of the illness to 1–3 months if appropriate treatment is not given.

In >50% of the cases, massive swelling of the lymph nodes (often with abscess formation) follows ulcer development by 1–2 weeks, and this chancroidal bubo can persist for many months (79). Systemic dissemination of *H. ducreyi* to other body sites is not seen.

The disease appears in men ~10 times more often than in women (79). Asymptomatic carriage of *H. ducreyi* has been described in both sexes, although it is more common in women (45).

B. History

Chancroid has been recognized as a clinical entity for centuries but was not differentiated from syphilis until 1852 (8). In 1889, Auguste Ducrey first saw the gram-negative bacillus in stained smears from genital ulcers (27). Around 1895 the organism was grown in culture, and in 1917 it was placed in the genus *Haemophilus*. Koch's postulates establishing the role of *H. ducreyi* as the causative agent of chancroid were fulfilled early in the twentieth century (32). In fact, patients' local skin reactions to heat-killed *H. ducreyi* (Ito–Reenstierna skin test) were often used to substantiate the clinical diagnosis (26,32).

C. Epidemiology

Gonorrhea, syphilis, and chancroid have traditionally been considered the three most commonly occurring STD. The incidence of chancroid is high in tropical countries and tends to decrease as the standard of living increases. In Europe, the incidence of chancroid had been decreasing since the mid-1800s until sporadic outbreaks occurred in times of war, beginning with World War I. An outbreak of the disease was also reported during the 1936 Olympic games, when athletes from around the world gathered in Berlin (79). In the early 1950s during the Korean War, chancroid was reported to occur in Korea 14–21 times more frequently than syphilis and almost twice as frequently as gonorrhea (7). Similarly, in the Vietnam conflict, Kerber (42) found chancroid to be second only to gonorrhea among venereal diseases. More recent outbreaks of the disease in England (33), France (52), the Netherlands (55), Germany (79), and Greenland (48) have been documented.

In 1984 it was reported that the worldwide incidence of chancroid exceeded that of syphilis (33). However, accurate numbers of cases of chancroid are difficult to determine, since most reports have listed cases based on clinical symptoms without cultural confirmation, even though other pathogens such as *Treponema pallidum,* herpes simplex virus (HSV), *Chlamydia trachomatis* (lymphogranuloma venereum), and *Calymmatobacterium granulomatis* (granuloma inguinale) can occasionally cause ulcerations similar to those of *H. ducreyi* (especially when polymicrobial flora exist in the lesion). The numbers have varied widely from country to country in reports from Africa and Asia (15,39,72,77). Several investigators from these areas of the world have attempted to confirm the presence of *H. ducreyi* in patients with clinically diagnosed chancroid using cultural methods. Results of these efforts have also demonstrated a wide variation in the numbers of positive specimens found: for example, the organism was isolated from 73% of the specimens cultured in New Delhi, India in 1981 (21), 8% in Singapore in 1982 (65), 38% in Bangkok, Thailand in 1982 (80), 67% in Nairobi, Kenya in 1983 (30), 40% in Durban, South Africa in 1984 (22), and 23% in Singapore in 1985 (84). These reports may indicate a true variation in the incidence of the disease, or they may suggest the lack of reliability of either clinical diagnosis or currently used cultural methods.

In North America, occasional isolation of *H. ducreyi* was reported in Canada in 1978 (34). In the United States, there were 87% fewer cases reported annually in the 1970s than in the 1940s (29). However, in the past decade outbreaks of chancroid have been reported from areas of both countries. Hammond *et al.* (35) reported an outbreak of 135 cases occurring in Winnipeg, Manitoba from July 1975 to September 1977. *Haemophilus ducreyi* was isolated from 19 patients. A smaller outbreak involving 14 cases of chancroid occurred in the same city from June 1987 to January 1988 (40). During May 1981 through February 1983, a large outbreak of chancroid involving 923 cases occurred in Orange County, California (10,20). *Haemophilus ducreyi* was recovered from 271 patients. No cases of chancroid had been reported in this county in the previous year. Forty-four cases were identified by laboratory or clinical data in West Palm Beach, Florida between August 1, 1982 and August 31, 1983 (9). In the Boston area, 53 patients with culture-confirmed or clinically suspected chancroid were treated between January 8 and September 30, 1985 (19). In the previous 2 years, only two cases had been diagnosed in all of Massachusetts. Outbreaks have been reported yearly in New York City since 1981 (87). In total, 6958 cases of chancroid have been diagnosed in New York City since 1984: 340 cases in 1984, 1323 cases in 1985, 2179 cases in 1986 (56), and 3116 cases in 1987 (88). Of the 5000 chancroid cases reported in the United States in 1987, 62.4% were from New York (62.3% from New York City), 16% from Florida, 8% from Texas, and 5% from Georgia (88).

D. Chancroid and Heterosexual Transmission of Human Immunodeficiency Virus

In Africa, a significant association has been established between past history of genital ulceration and heterosexual transmission of human immunodeficiency virus (HIV) (46,64,73). Presumably, STD causing genital ulcers that disrupt epithelial surfaces may enhance the transmission of HIV. Since chancroid has been reported to be the most common cause of genital ulcers in Africa (23), this disease has specifically been correlated with HIV seropositivity (73).

E. Therapy

Using *in vitro* testing of clinical isolates, it has been shown that the susceptibility of *H. ducreyi* to antimicrobial chemotherapy varies markedly from one geographic location to another (33,67,71). In addition, patterns of sensitivity have changed over the years in the same locations. Sturm (78) studied the antibiotic susceptibility patterns of 57 strains of *H. ducreyi* isolated in Amsterdam, the Netherlands, between 1978 and 1985. During the first 4 years, >30% of the isolates were β-lactamase-negative and sensitive to tetracycline. During the second 4 years, all strains except one produced β-lactamase and were tetracycline-resistant. Recent reports have described the widespread presence of plasmid-mediated resistance to antimicrobial agents (51). The spread of these plasmids in many areas of the world has made treatment with at least four important agents [tetracycline (5), sulfonamides (24), penicillins (13,24), and chloramphenicol (66)] ineffective against chancroid. In spite of the increased resistances seen, erythromycin continues to be active against *H. ducreyi* (75). Other newer antibiotics such as ciprofloxacin (54), cefriaxon (11), ceftriaxone (47), and cefotaxime have recently been used successfully (63,84).

F. Characterization of *Haemophilus ducreyi*

Haemophilus ducreyi is a slender gram-negative rod with bipolar staining, measuring 0.5–2 μm. In smears from ulcers or bubo aspirates, the organism often occurs in long, ropelike chains that resemble train tracks or shoals of fish. *Haemophilus ducreyi* is a fastidious, slow-growing organism, requiring 48–96 hr of incubation before colonies can be visualized. The organism grows best at 30°–34°C, with high humidity, under 5% CO_2 or anaerobic conditions. The colonies are small and smooth with a grayish-tan color, and grow in different sizes giving the culture a contaminated appearance. Colonies grown on solid media have a peculiar cohesive nature so that they slide over the agar surface intact when pushed with a loop. The colonies also remain tightly autoagglutinated when suspended in liquid.

Haemophilus ducreyi is biochemically less reactive than most other *Haemophilus* species. Still, its inclusion in this genus was based on biochemical similarities such as requirement for hemin, reduction of nitrate and alkaline phosphatase activity, and on the similarity of the guanine + cytosine (G+C) content of the DNA (38–39 mol%) (43). More recently, cross-reactivity has been shown between *H. ducreyi* and other *Haemophilus* species using human and animal polyclonal immune sera (69,70). However, using DNA hybridization *H. ducreyi* is genetically dissimilar from all other *Haemophilus* species (0 to 6% related) (6,17). In addition, a comparison of quinones extracted from bacterial membrane preparations demonstrated major physiological and chemotaxonomic differences between *H. ducreyi* and other *Haemophilus* species (16).

In 1945, Feiner and Mortara (28) described an assay for virulence of *H. ducreyi* that involved the production of lesions in rabbits following intradermal (id) injection. However, the lesions produced in the rabbit model differ from human lesions, since they are drier and usually do not ulcerate. More recently, Tuffrey *et al.* (86) reported that injections of virulent *H.*

ducreyi in mice resulted in the production of ulcerative lesions more similar to those seen in human disease.

Odumeru *et al.* (60) demonstrated that virulent *H. ducreyi* were resistant to killing by either human polymorphonuclear leukocytes or by the complement-mediated action of normal human and rabbit sera. In order to determine if virulent and avirulent strains of *H. ducreyi* can be differentiated on the basis of variations in cellular components, work on outer membrane proteins, surface-exposed proteins, and lipopolysaccharides has been reported from various laboratories (1–4,58,81). Results from these studies may lead to a better understanding of mechanisms of pathogenesis in *H. ducreyi*.

G. Laboratory Diagnostic Procedures

To maintain viability of the bacteria, *H. ducreyi* must be cultured within 4–6 hr after removal from the patient. Because of its fastidious nature, the organism does not grow well on most laboratory media and recovery rates from clinical specimens have traditionally been poor. Originating with Ducrey in the 1880s, the standard test for chancroid was autoinoculation of the patient's forearm or thigh with lesion material to see if a characteristic ulcer would result. Fortunately, this risky practice was eliminated as a diagnostic tool. In 1920, Teague and Deibert (82) used clotted rabbit blood heated to 55°C for 5 min as the primary culture medium. Although *H. ducreyi* grew well using this procedure, fresh rabbit blood was not readily available in most diagnostic laboratories. In 1978, Hammond *et al.* (34) reported the development of a selective enrichment agar consisting of gonococcal base medium with 1% hemoglobin, 1% IsoVitoleX, and 3 μg/ml vancomycin. Use of this medium with the addition of 5% fetal bovine serum, or adaptations of it using Mueller–Hinton agar base, chocolatized horse blood (in place of the hemoglobin), and the aforementioned supplements (76), has led to significantly higher recovery rates of *H. ducreyi*. However, isolation rates of only 60–70% are still the norm from patients with clinically diagnosed chancroid (53,70).

Since *H. ducreyi* is the only human *Haemophilus* species that requires X factor (hemin) but not V factor (NAD), the porphyrin test, which detects enzyme activities in the hemin-biosynthetic pathway, is most useful for identification (44). *Haemophilus ducreyi* is further differentiated from other *Haemophilus* species by its failure to produce H_2S, catalase, or indole, and its production of alkaline phosphatase (67). However, difficulties are often encountered in testing the biochemical reactivity of *H. ducreyi,* since the organism grows poorly in most biochemical test media and these tests require growing organisms for proper reactivity (18). Because of the poor growth of *H. ducreyi,* most strains are also biochemically inert when tested in the API-20E system (Analytab Products Inc., Plainview, New York) (67). The organisms have been reported to be appropriately reactive in the RapID NH system (Innovative Diagnostic Systems, Inc., Decatur, Georgia), which has been developed specifically for the identification of fastidious gram-negative bacilli (37).

Detection and identification methods using specific antisera have been described using polyclonal (25) or monoclonal antibodies (MAb) (38,70), but colonies of *H. ducreyi* grown on solid media are very cohesive and thus remain tightly autoagglutinated when suspended in liquid media. This cohesiveness interferes with serological tests based on agglutination, and the excessive clumping makes it difficult to interpret results obtained using fluorescent-antibody tests (25,70). In addition, cross-reactivity of polyclonal antisera with other *Haemophilus* species has been difficult to eliminate. Antisera specific for *H. ducreyi* are not commercially available at the present time.

Antibodies to *Haemophilus ducreyi* have been found in serum samples from patients with clinical chancroid. Approximately 40% of the serum samples obtained from 63 patients during outbreaks of chancroid in Florida and New York in 1984 and 1985 reacted strongly with outer-membrane preparations from *H. ducreyi,* while controls were negative (70). In a study reported in 1988 in which a sonicated whole-cell preparation of *H. ducreyi* was used as antigen, 89 and 55% (from Nairobi and Bangkok, respectively,) of men with proven chancroid had antibodies to *H. ducreyi,* compared with 2 and 17% of the controls (53).

No standard epidemiological typing system exists for *H. ducreyi.* Individual laboratories have attempted to develop procedures to differentiate within the species, which included: protein profiles of the outer membranes (57), indirect immunofluorescence assay using polyclonal rabbit antisera prepared against several strains of *H. ducreyi* (74), and enzyme profiles using the API-ZYM system (89). Plasmid profiles and restriction endonuclease fingerprinting of plasmids have also been used (14,36,85).

H. Background Summary

Problems exist in the diagnosis of chancroid because the symptoms of other STD can be similar to those seen with this disease. Also, *H. ducreyi,* the causative organism, is difficult to grow in culture unless plated immediately on special media, and once isolated, the organism is poorly reactive biochemically and thus difficult to identify. Detection methods for *H. ducreyi* using specific antisera have been hampered not only by the organism's tendency to autoagglutinate, but also by the cross-reactivity of the polyclonal antisera with other *Haemophilus* species. Although the use of polyclonal and monoclonal antisera has been described, neither is available for routine diagnostic use. Since the incidence of chancroid is increasing in the United States and Canada and this disease has been associated with heterosexual transmission of HIV in Africa, the problems involved in establishing an accurate diagnosis need to be resolved. To this end, we have developed DNA probes for use in the rapid and accurate identification of *H. ducreyi*.

III. RESULTS AND DISCUSSION

A. Construction of a *Haemophilus ducreyi* Genomic Library in λgt11

Figure 1 summarizes the procedures used in probe development (61). Initially, chromosomal DNA was isolated as previously described (12) from five different isolates of *H. ducreyi* obtained from the American Type Culture Collection, Rockville, Maryland [ATCC numbers 27721 (submitted to ATCC from the United States), 27722 (United States), 33921 (Canada), 33922 (Canada), and the type strain, 33940 (France)]. The thermal melt (T_m) values and G + C content of the DNA samples were determined by spectrophotometry during thermal denaturation. The results were as follows:

ATCC 27721, T_m = 85.6°C, G + C = 39.0 mol%
ATCC 27722, T_m = 85.5°C, G + C = 38.8 mol%
ATCC 33921, T_m = 85.1°C, G + C = 37.9 mol%
ATCC 33922, T_m = 85.5°C, G + C = 38.8 mol%
ATCC 33940, T_m = 85.2°C, G + C = 38.1 mol%

These results are comparable to those of Kilian (43), who determined the average G + C content for two strains of *H. ducreyi* to be 37.8 mol%, and with Piechulla *et al.* (62), who found T_m and G + C content values of 85.0°C and 37.6 mol%, and 85.2°C and 38 mol% for ATCC numbers 27722 and 33940, respectively. DNA from ATCC 33922 was selected for library construction because the organism grew fairly well and had been used previously in virulence and immunological assays (38,58,59). We confirmed the virulence of this strain by the production of necrotic lesions in rabbits following id injection (performed as previously described) (26,28,60).

An expression vector (λgt11, Promega Biotech, Madison, Wisconsin) was used for genomic library construction so that the recombinant phage

λ gt11
lacZ gene
cut with EcoRI
H. ducreyi (Hd)
chromosomal DNA
G AATTC
CTTAA G
(EcoRI sticky ends)
Anneal vector with Hd DNA fragments, seal gaps with ligase
Hd insert (lacZ gene interrupted)
In vitro packaging of phage in capsids
Hd genomic library
Infect E. coli host with packaged phage
Plate on agar containing β-gal inducer and substrate
Colorless plaques = recombinants
Blue plaques = nonrecombinants
Nitrocellulose filter gently placed on plate
Plaque lifts tested with Hd-specific antiserum (Hd antigens expressed by Hd insert in vector)
Obtain phage from original agar plate, isolate DNA from phage, cut with EcoRI, subclone Hd inserts into plasmid vector.

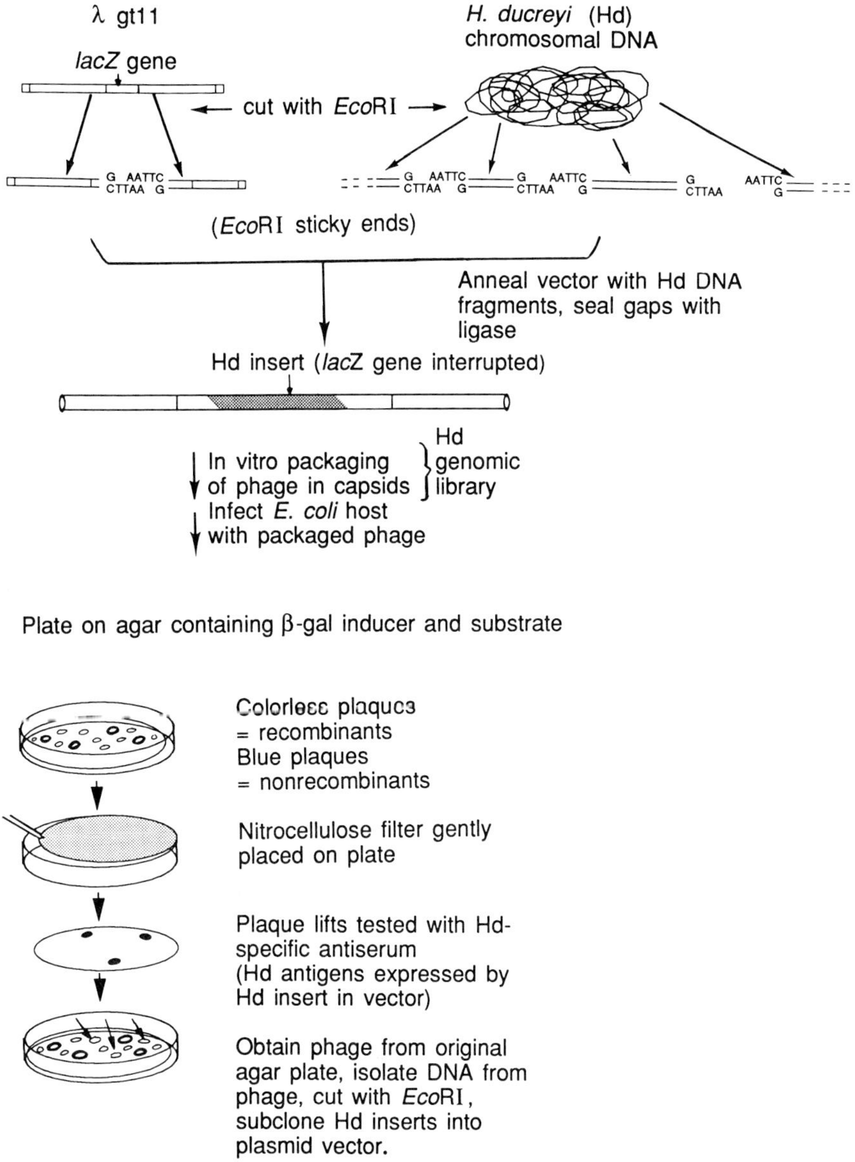

Fig. 1. DNA probe development: isolation of *H. ducreyi*-specific DNA sequences.

could be screened for the presence of *H. ducreyi* antigens using specific antiserum. Fragments of ATCC 33922 DNA previously digested with the restriction endonuclease, *Eco*RI, were inserted into the λ vector as previously described (41,61). The unique *Eco*RI site of λgt11 is located near the carboxyl terminus of the *lacZ* gene. Therefore, foreign DNA sequences inserted here are expressed (in a protease-deficient *Escherichia coli* host) as β-galactosidase (β-gal) fusion proteins under control of the *lacZ* gene promoter. Since insertion of the foreign DNA renders the bacteriophage gal^- while nonrecombinant λgt11 are gal^+, the presence of insert DNA is easily determined by growing the phage on a gal^- *E. coli* host in the presence of the β-gal inducer isopropyl-β-D-thiogalactopyranoside (IPTG) and substrate 5-bromo-4-chloro-3-indolyl-β-D-galactopyranoside (X-gal). The recombinant bacteriophages produce colorless plaques while the nonrecombinants produce blue plaques.

Polyclonal antiserum, prepared in rabbits by 6 months of biweekly injections of 10^8–10^9 formalin-killed *H. ducreyi* ATCC 33922 and absorbed with an *E. coli* cell lysate, was used to screen the recombinant phage for the expression of *H. ducreyi*-specific antigens using the Express Blot Assay Kit (Bio-Rad Laboratories, Richmond, California). Of the recombinant bacteriophages screened, 0.5% expressed antigens that reacted with the absorbed antiserum. Of the antiserum-positive plaques, 12 were purified for further study. DNA samples isolated from the bacteriophages were digested with *Eco*RI to release the *H. ducreyi* inserts, and the sizes of the fragments were determined by comparison with standards using agarose gel electrophoresis. Digested DNA was transferred to a nitrocellulose (NC) membrane by the method of Southern as described by Maniatis *et al.* (50), and hybridized with nick-translated ^{32}P-labeled *H. ducreyi* ATCC 33922 chromosomal DNA to verify the origin of the inserts.

B. Subcloning *Haemophilus ducreyi* Inserts into a Plasmid Vector

Because plasmid vectors are more convenient for laboratory manipulations than bacteriophage vectors, three inserts that hybridized with the chromosomal probe were subcloned into a pUC13 plasmid vector. The resulting recombinant plasmids (pLP1, 1.5-kb insert; pLP4, 5.4-kb insert; pLP8, 5.7-kb insert), which had been transferred to an *E. coli* host by transformation, were amplified in the bacterial host and purified by ultracentrifugation on two successive CsCl gradients to ensure removal of *E. coli* chromosomal DNA. Either purified plasmids containing the *H. ducreyi* inserts, or the inserts alone which had been recovered from agarose gels using the Gene Clean procedure (Bio 101, Inc., LaJolla, California), were labeled with ^{32}P by nick translation and used as probes.

C. Specificity of the Probes

A variety of bacterial species were obtained from the ATCC, the Centers for Disease Control (CDC) in Atlanta, Georgia, and from the lyophilized culture collection of the Laboratories for Bacteriology, Wadsworth Center for Laboratories and Research, New York State Department of Health, Albany, New York. Isolates of *H. ducreyi* recovered from patients in 1988 were obtained through the cooperation of the Bureau of Laboratories of the New York City Department of Health, New York, New York.

To prepare the bacterial cells for testing, the organisms were grown for 24–48 hr and then suspended in buffer either for direct use or for cell lysis and DNA isolation in solution. (For further details, see Section VIII.) The Minifold II Slot-Blotter apparatus (Schleicher and Schuell, Keene, New Hampshire) was used for filtration of the bacteria or DNA onto NC membranes. Also, bacterial colonies were transferred directly to NC from agar plates by gently placing the filter on the surface of the plate in contact with the colonies. The bacteria adhering to the NC were lysed and the DNA denatured as described by Maniatis *et al.* (50). Figure 2 illustrates the procedures used for testing bacterial and DNA suspensions with the probes. (See Section VIII.)

Table I lists the organisms tested and the results obtained using all three probes (61). All 16 strains of *H. ducreyi* reacted with the probes, as did the purified DNA from all 5 ATCC strains. When 37 other bacterial isolates, including organisms encountered in the urogenital tract, were tested with the probes, 28 were completely negative and 9 (6 belonging to other *Haemophilus* species and 3 belonging to *Pasteurella* species) reacted weakly when 10^7 colony-forming units (CFU) were used. No reactions were seen with these organisms when 10^5–10^6 CFU were tested. In contrast, this number of *H. ducreyi* organisms still reacted strongly with the probes. The presence of areas of homology in these organisms is not surprising, since antigenic cross-reactivity between *H. ducreyi* and other *Haemophilus* species has been reported (61,69,70). *Pasteurella* species may also share small areas of DNA relatedness with the cloned *H. ducreyi* sequences, since both *Pasteurella* and *Haemophilus* species are members of the family *Pasteurellaceae*. In contrast, the other significant genital pathogens, *Neisseria gonorrhoeae*, herpes simplex virus type 2 and *Treponema pallidum* were all completely nonreactive with the probes.

In addition to the bacteria and DNA that were tested in the slot-blot apparatus, nine recent isolates of *H. ducreyi* cultured from patients in 1988 in New York City were transferred directly to NC from growth on agar plates and were found to be positive when tested with the pLP8 probe whether present as pure cultures or mixed with other organisms (61).

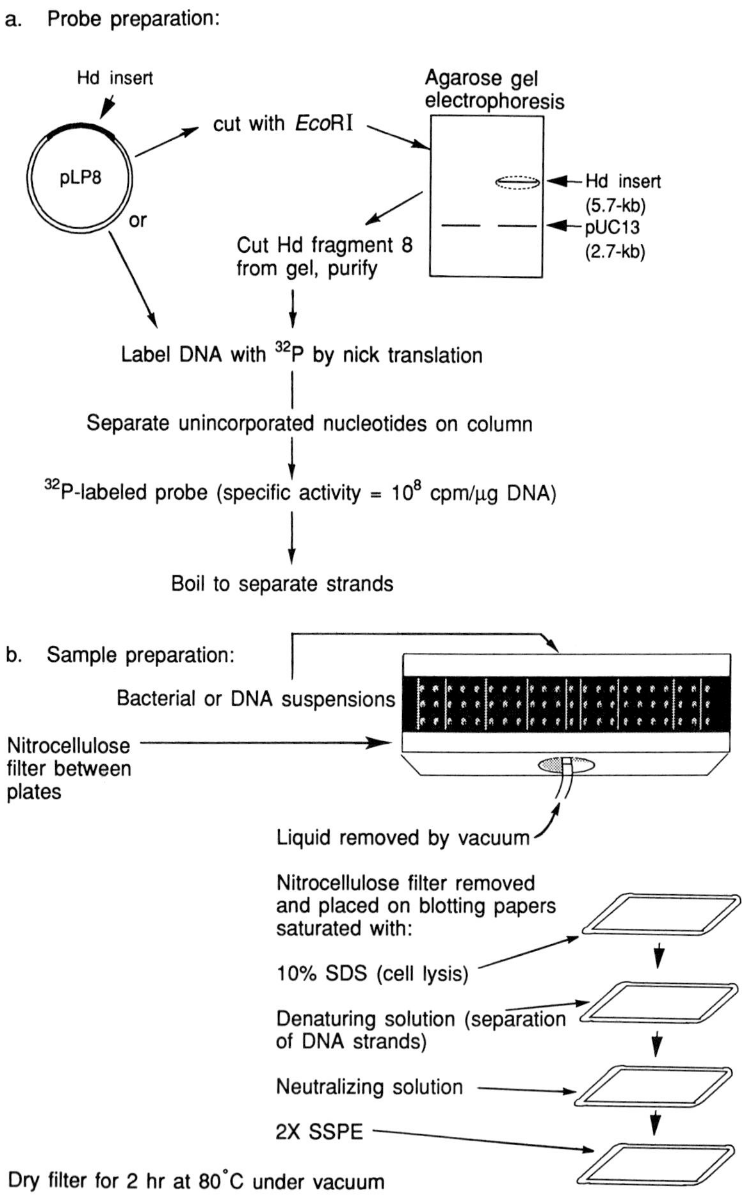

Fig. 2. Use of DNA probes for *H. ducreyi*.

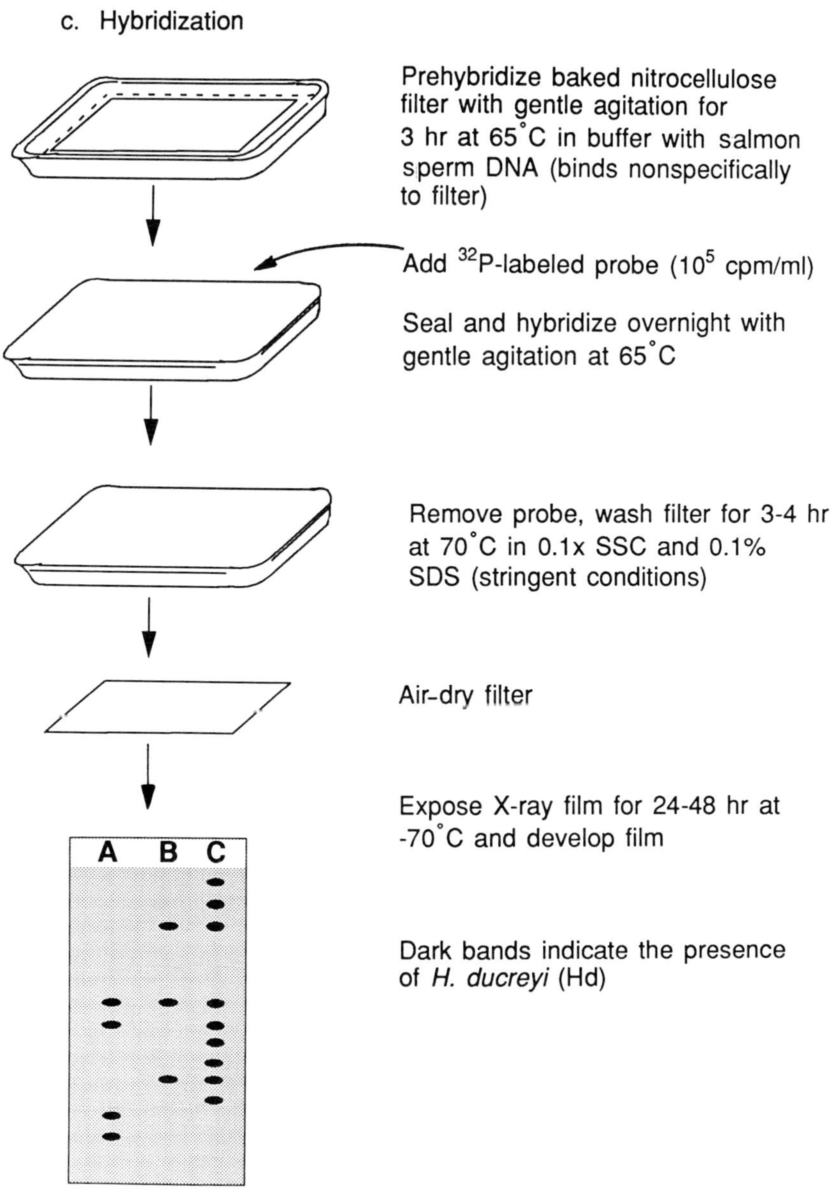

Fig. 2. (*Continued*)

TABLE I

Specificity of Probes for the Identification of *Haemophilus ducreyi*

Reaction	Organisms
Strong reaction with all probes (pLP1, pLP4, and pLP8, used separately)[a]	*Haemophilus ducreyi* (ATCC 27721, ATCC 27722, ATCC 33921, ATCC 33922, ATCC 33940, CDC 542, CDC 844, nine clinical isolates from New York City, 1988)
Negative with all probes[b]	*Acinetobacter calcoaceticus* subsp. *lwoffi* B277, *Actinobacillus actinomycetemcomitans* B1083, *Actinomyces* sp., strain B1228, *Alcaligenes faecalis* B38-78, *Bacteroides fragilis* ATCC 25285, *Clostridium perfringens* ATCC 13124, *Corynebacterium diphtheriae* C5703, *Escherichia coli* ATCC 25922, *Gardnerella vaginalis* ATCC 14018, (atypical) *G. vaginalis* B1905, *Lactobacillus acidophilus* ATCC 4962, *Moraxella osloensis* B1596, *Neisseria gonorrhoeae* 116, *Neisseria lactamica* B2159, *Neisseria meningitidis* W135, *Pasteurella haemolytica* M6169, *Peptostreptococcus anaerobius* A943, *Proteus mirabilis, Pseudomonas aeruginosa* ATCC 27853, *Staphylococcus aureus* ATCC 25923, *Staphylococcus,* coagulase negative strain B778, *Streptococcus agalactiae* B2545, *Streptococcus bovis* B1450, *Streptococcus faecalis* ATCC 29212, *Streptococcus mutans* B1254, *Streptococcus sanguis* II B1508, Yersinia *enterocolitica* ATCC 9610, *Yersinia rohdei* CDC 3022. Also herpes simplex virus type 2-infected cells and *Treponema pallidum* in rabbit testicular fluid.
Weak reactions with all probes[c]	*Haemophilus haemoglobinophilus* B1701, *Haemophilus influenzae,* types A and B, and nontypeable, biotypes III and V, *Haemophilus parainfluenzae* B1134, *Pasteurella gallinarum* B697, *Pasteurella multocida* B1221-76 and *Pasteurella pneumotropica* M5354.

[a] Approximately 1.6×10^6 CFU per well.
[b] Approximately 6.0×10^7 CFU per well.
[c] Weakly reactive at 6.0×10^7 per well, and negative at 10^5–10^6 CFU per well.

D. Sensitivity of the Probes

As previously reported (61), when dilutions of *H. ducreyi* were tested with the probes and plated to determine the number of CFU present, the three probes consistently detected 10^4 CFU of *H. ducreyi* in pure or mixed cultures. Because clinical specimens may not contain probe-detectable levels of *H. ducreyi,* we next attempted to amplify the organism by growth before testing. Known quantities of *H. ducreyi* were used to seed 2-ml aliquots of hemin broth, a highly enriched medium [prepared as previously described (4)], which supports the growth of this organism. Five dilutions of *H. ducreyi* ATCC 33922 (dilution 1, 4.5×10^4 CFU/ml; 2, 4.5×10^3 CFU/ml; 3, 4.5×10^2 CFU/ml; 4, 4.5×10^1 CFU/ml; and 5, 4.5×10^0 CFU/ml) were set up in triplicate and then subjected to the following four different temperature conditions for growth; condition A, 5 days at room temperature; condition B, overnight at 35°C–37°C followed by 4 days at room temperature; condition C, 5 days at 35°–37°C; and condition D, overnight at 35°–37°C followed by 4 days at 4°C. After the 5-day test period, 1.8 ml from each tube was centrifuged, the cells resuspended in 200 μl of harvesting buffer, and 100 μl filtered through the Bio-Rad Dot Blot apparatus. The hybridization results using fragment 8 (5.7 kb) as the probe are shown in Fig. 3. Following incubation under condition A, dilution 1 was weakly reactive; however, none of the lower dilutions was detectable. These results indicate that a detectable level of growth was not achieved during room temperature incubation. Conditions B, C, and D all began with an overnight incubation at 35°–37°C. The results for condition D (subsequently placed at 4°C) indicate that the cells in dilution tubes 1–3, but not 4 and 5, had grown to detectable levels during overnight incubation. The results for conditions B and C indicate that subsequent incubation at both room temperature and 35°–37°C (following an initial overnight incubation at 35°–37°C) will allow growth to detectable levels in all dilution tubes. A study using this broth to amplify *H. ducreyi* in clinical samples is currently under way in our laboratory.

E. Probing Lesion Material from Rabbits

Four strains of *H. ducreyi* (ATCC numbers 33922, 33940; CDC 542, and a recent clinical isolate from New York City, ME 38) were used to inject rabbits intradermally for the production of necrotic lesions, as previously reported (61). Exudates were obtained from the lesions 4–5 days after injection of 10^7 organisms (the lesions had healed in all rabbits by 3–4 weeks). From these exudates, only CDC 542 could be recovered by culture. However, using fragment 8 (5.7 kb) as a probe, *H. ducreyi* DNA could be detected as a weak to 1+ signal in material from lesions produced by all four strains. These results were interesting for two reasons: the *H. ducreyi* did not survive long in the lesions, and DNA hybridization was demonstrated to be a valuable tool for detection of *H. ducreyi* in the lesion exudate, even when the organisms were nonviable.

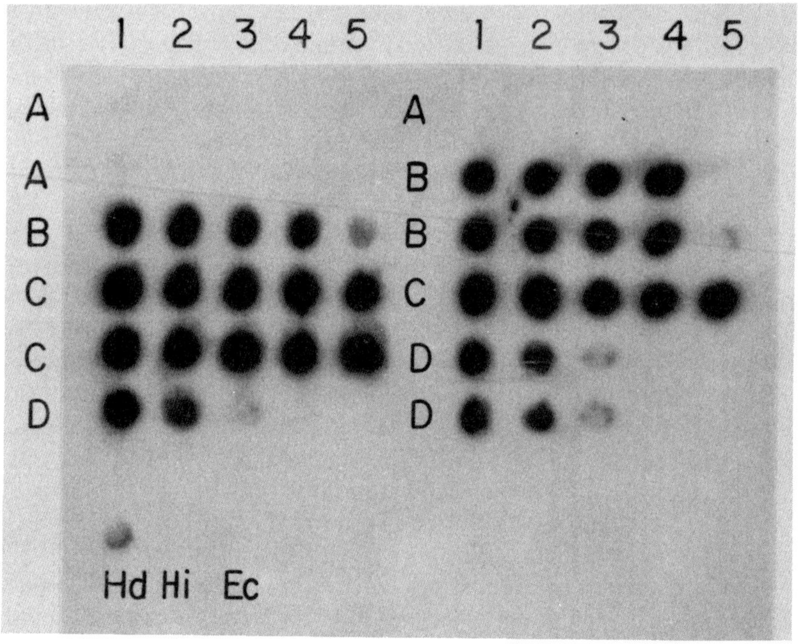

Fig. 3. Growth amplification of *H. ducreyi* to probe-detectable levels. The following dilutions of *H. ducreyi* ATCC 33922 were set up in triplicate and grown in hemin broth: (1) 4.5×10^4 CFU/ml; (2) 4.5×10^3 CFU/ml; (3) 4.5×10^2 CFU/ml; (4) 4.5×10^1 CFU/ml; (5) 4.5×10^0 CFU/ml. The growth conditions were as follows: (A) 5 days at room temperature; (B) overnight at 35°–37°C followed by 4 days at room temperature; (C) 5 days at 35°–37°C; (D) overnight at 35°–37°C followed by 4 days at 4°C. Control wells included $\sim 10^5$ CFU of *H. ducreyi* ATCC 33922 (Hd), *H. Influenzae* (Hi), and *E. coli* (Ec). Fragment 8 was used as the hybridization probe.

IV. CONCLUSIONS

In recent years, a need has been demonstrated for an accurate and more rapid diagnostic test for *H. ducreyi*. Outbreaks of chancroid are increasing in the United States and Canada. Because chancroid has traditionally been more of a problem in tropical and subtropical countries, clinicians in more temperate areas may be less familiar with the disease and hence may have more difficulty differentiating between the genital ulcerations caused by *H. ducreyi* and the lesions seen in other STD. Accurate diagnosis of the etiological agent is very important, since antimicrobial treatment against the different STD must be specific.

Even when diagnostic laboratory work is performed, problems exist. For successful recovery, *H. ducreyi* must be plated immediately on specialized media. The recent development of media specific for *H. ducreyi* has improved the recovery of this organism; however, isolation rates are still relatively low. Identification of the organism by biochemical reactivity is also difficult because the organism usually does not grow well enough to produce detectable reaction products. In addition, problems with autoagglutination of *H. ducreyi* and cross-reactivity of polyclonal antisera have limited the usefulness of identification by serological testing.

For these reasons, we developed DNA probes for the detection and identification of *H. ducreyi*. We have demonstrated that the probes are specific and sensitive. The use of these specific DNA probes in laboratory diagnostic tests should ensure a more rapid and accurate identification of *H. ducreyi*.

V. GENE PROBES VERSUS ANTISERA AND MONOCLONAL ANTIBODIES

An immunological approach to the detection of *H. ducreyi* was first reported by Denys *et al.* (25) in 1978. These investigators developed an indirect fluorescent-antibody (IFA) technique using rabbit polyclonal antisera. An IFA test was selected because excessive clumping of the organism ruled out serological agglutination tests. However, the authors reported that even the IFA test was difficult to interpret because of the clumped bacterial cells. The antiserum, produced against heat-killed bacteria, was reported to be specific for *H. ducreyi* only after extensive absorption with whole cells of *H. influenzae, H. parainfluenzae,* and *H. parahemolyticus.* Six strains of bacteriologically confirmed *H. ducreyi* were positive while 156 heterologous organisms were negative.

The IFA test using polyclonal antisera was the only experimental serological test for *H. ducreyi* until 1984, when Hansen and Loftus (38) isolated two MAb reactive with *H. ducreyi*. Using radioiodinated rabbit anti-mouse immunoglobulin as a probe for bound mouse MAb, the investigators demonstrated reactivity with 12 strains of *H. ducreyi* collected from six different countries. They also detected the organisms in necrotic lesion material from intradermally injected rabbits. The antibody probes were tested against five other organisms commonly found in the genital tract and were found to be nonreactive.

While investigating outbreaks of genital lesions in Florida and New York City in 1986, Schalla *et al.* (70) used MAb produced against *H. ducreyi* to test direct smears from genital lesions. Lesion material from six patients with clinically diagnosed chancroid were tested and three were positive using an IFA technique. Here again, interpretation was difficult because bacterial forms could not be distinguished clearly in the clumps that stained with the MAb.

Similar to the DNA probes described in this report, the immunological tests using MAb were shown to be specific for isolates of *H. ducreyi*. Both the probes and the MAb reacted only with the appropriate organism and with as great a diversity of the organism as was possible to test. As previously described (61), the sensitivity of the *H. ducreyi*-specific DNA probes was $\sim 10^4$ CFU. DNA probes developed for a variety of other organisms have previously been reported to detect homologous sequences in $\sim 10^2$–10^6 bacteria, and immunoassays have been reported to detect antigens in $\sim 10^3$–10^7 bacteria (31). The sensitivities of the two systems appear quite similar at this point. However, the development of new and innovative detection–visualization procedures (83) will determine the approach taken in future testing.

The time required for testing, either as time passed before the definitive report is available or as technician time required, is of real concern in the clinical laboratory setting. Antigen detection tests are currently less time-consuming than NA hybridization. However, a recent study using a DNA probe for the detection of *H. influenzae* in clinical specimens (49), indicates that by shortening the prehybridization and hybridization incubation times and by using a liquid scintillation counter for detecting hybridization instead of an autoradiograph, the time required for detection of *H. influenzae* was ~3 hr. Therefore, it appears possible, with the use of these or similar modifications, to reduce the time required to obtain a result using probe technology while still retaining the advantages of NA hybridization.

VI. PROSPECTS FOR THE FUTURE

Clinical studies are necessary to evaluate the efficacy of DNA probes as diagnostic tools for the detection and identification of *H. ducreyi*. To this end, a hemin broth (4) was evaluated for future use in transporting the clinical specimens from patients to testing laboratory, and for growth amplification of the *H. ducreyi* present to probe-detectable levels. Our results indicate that the hemin broth will enhance the growth of 450 CFU of *H. ducreyi* to probe-detectable levels in 1 day, and as few as 4–5 CFU to probe-detectable levels within 5 days. In collaborative studies with laboratories in New York City, we are presently using this broth to transport and growth-amplify the organisms from clinical specimens.

If growth in hemin broth and subsequent detection of the organism from clinical specimens is not satisfactory, a procedure for amplification of the target DNA (homologous to the *H. ducreyi*-specific probe) utilizing the polymerase chain reaction (PCR) will be developed. The PCR procedure will utilize specific primers and a thermostable DNA polymerase (Perkin Elmer Cetus, Norwalk, Connecticut) as previously described (68). The cloned *H. ducreyi* DNA fragments will be sequenced and from this sequence information, oligonucleotide primers will be synthesized for use in an automated thermal-cycler instrument to amplify homologous *H. ducreyi* sequences present in clinical specimens. The specimens will then be hybridized with the probe to determine whether homologous sequences (amplified to a detectable level) are present.

Genital ulcers may be caused by other infectious agents; in fact *H. ducreyi* may coexist in the same lesions with other pathogens, especially HSV and *T. pallidum*. It is therefore essential that specimens from genital ulcers be thoroughly examined for polymicrobial pathogenic flora. The development of these DNA probes specific for *H. ducreyi* is seen as the first step in the evolution of methods for testing a single clinical specimen from a genital lesion for the presence of *H. ducreyi*, HSV, and *T. pallidum*, the three agents most likely to cause genital ulceration.

VII. SUMMARY

Haemophilus ducreyi is the causative agent of chancroid, a sexually transmitted disease (STD) characterized by painful genital ulcers. In recent years recurring outbreaks of chancroid have been reported in Europe, Canada, and the United States. This disease can be troublesome to diagnose clinically, since chancroidal ulcers are often difficult to distinguish from lesions caused by other STD. In addition, *H. ducreyi* is a fastidious organism, and for successful recovery, must be cultured onto special media immediately after the specimen is obtained. The development of highly enriched, selective media has improved recovery of the organism, but isolation rates of only 60–70% are still the norm.

In this report, we describe the development and testing of DNA probes for the identification of *H. ducreyi*. The probes were selected from a genomic library on the basis of their ability to express antigens specific to *H. ducreyi*. Use of the probes with a wide variety of bacterial species has substantiated their specificity. We have also used the probes (1) to detect as few as 10^4 CFU of *H. ducreyi* in pure and mixed cultures, (2) to detect *H. ducreyi* in growth-amplified broth cultures seeded with as few as 4.5×10^0 CFU, and (3) to detect *H. ducreyi* that were no longer viable in specimens from lesions in intradermally infected rabbits. Our results indicate that these DNA probes are an important addition to the laboratory techniques available for the rapid and accurate identification of *H. ducreyi*.

VIII. MATERIALS AND METHODS

This section outlines the use of the DNA probes (see Fig. 2).

A. Probe Preparation

The *H. ducreyi* DNA fragment to be used for probing bacterial or DNA suspensions is maintained in a plasmid vector. For probe preparation, either the whole recombinant plasmid or the isolated fragment can be used. The *H. ducreyi* fragment can be excised from the plasmid by digestion with *Eco*RI followed by separation of the DNA vector and insert by agarose gel electrophoresis. The *H. ducreyi* fragment can then be cut from the gel and purified from the agarose using the Gene Clean procedure (Bio 101, Inc., LaJolla, California) according to the manufacturer's directions.

Approximately 200–600 ng of DNA is labeled with ^{32}P using [α-^{32}P]deoxycytidine triphosphate (Amersham Corporation, Arlington Heights, Illinois) and the BRL Nick Translation Kit [Bethesda Research Laboratories (BRL), Gaithersburg, Maryland]. Unincorporated nucleotides are removed on a NACS–PREPAC column (BRL). Both procedures are performed according to the manufacturer's directions. The specific activity of the probe is determined by counting a small portion of the probe in a liquid scintillation counter, and then calculating the counts per minute (cpm) per microgram of DNA. The results should be at least 10^7–10^8 cpm/μg of labeled DNA.

Just before use, the probe is placed in boiling water for 5 min to denature the double-stranded DNA molecules to single strands. Immediately placing the denatured probe on ice ensures that no renaturation will occur before use.

B. Sample Preparation

Bacterial suspensions of pure cultures can be suspended in harvesting buffer (50 m*M* Tris, pH 8.0–100 m*M* NaCl–50 m*M* EDTA–50 μg/ml freshly prepared pronase) and placed on NC filters (or other appropriate membrane filter) for probing. However, if growth is heavy (greater than a number 2 MacFarland turbidity standard) because of the presence of organisms other than *H. ducreyi,* as might be encountered in clinical specimens, the cells must first be lysed to separate the DNA from the cellular debris, which may clog the pores of the NC filter. This will ensure that an adequate amount of *H. ducreyi* DNA can bind. Cell lysis is performed by centrifuging 1.5 ml of the bacterial suspension and resuspending the cell pellet in 200 μl of 10 m*M* Tris-HCl, pH 8.0–1 m*M* EDTA. An equal volume of lysis buffer [0.4 *M* Tris-Cl, pH 8.0–0.1 *M* EDTA–1% (w/v) sodium dodecyl sulfate (SDS)–50 μg/ml freshly prepared pronase] is added. The cells are incubated for 1–2 hr in a 37°C water bath. The DNA in the samples is then purified by extraction with an equal volume of phenol–chloroform (1 : 1), followed by a second extraction with an equal volume of chloroform. One-half of the aqueous extract (which contains the DNA) can be used for blotting.

Either a slot-blot or dot-blot apparatus can be used to apply the bacterial or DNA suspensions to the NC filter, which has been prewet in 10× SSC (20× SSC is 3 *M* NaCl–0.3 *M* sodium citrate, pH 7.0) (see Fig. 2). Following assembly of the blotting apparatus according to the manufacturer's directions, vacuum is applied to the system and each well is rinsed with 20× SSC. The bacterial and DNA suspensions are added to the slots. When the slots are dry, the apparatus is disassembled and the filters are gently floated on the surface of a small amount of 10× SSC. The filters are placed DNA side up on a Whatman 3 MM filter paper saturated with 10% (w/v) SDS for 3 min, followed by transfer to another paper saturated with denaturing solution (0.5 *M* NaOH–1.5 *M* NaCl) for 5 min, then a third saturated with neutralizing solution (1.5 *M* NaCl–0.5 *M* Tris-Cl, pH 8.0) for 5 min, and finally to a fourth paper saturated with 2×-SSPE (20× SSPE is 3.6 *M* NaCl–200 m*M* $NaH_2PO_4 \cdot H_2O$–20 m*M* EDTA, pH 7.4) for 5 min. The filters are air dried and then baked in a vacuum oven at 80°C for 2 hr.

C. Hybridization

The baked NC filters are prehybridized for 3 hr at 65°C in 5× SSPE, 5× Denhardt's solution [50× Denhardt's is 1% (w/v) Ficoll 400, 1% (w/v) polyvinylpyrrolidone, 1% (w/v) bovine serum albumin], 0.1% SDS, and 100 μg/ml denatured salmon sperm DNA, and then hybridized overnight at 65°C with 10^5 cpm/ml ^{32}P-labeled probe. The filters are washed for 4 hr at 70°C in 0.1× SSC and 0.1% SDS. Bound probe is visualized by exposing X-ray film to the filters for 24–72 hr at −70°C, followed by film development.

ACKNOWLEDGEMENTS

We wish to thank Hillel Bercovier, David Odelson, Sho Ya Wang, Dianna Schoonmaker, Wendy Archinal, Arnold Steigerwalt, Yvonne Faur, Jill Robinson, and Konrad Wicher for their participation in parts of this study.

REFERENCES

1. Abeck, D., Johnson, A. P., and Taylor-Robinson, D. (1988). Antigenic analysis of *H. ducreyi* by Western blotting. *Epidemiol. Infect.* **101,** 151–157.
2. Abeck, D., Johnson, A. P., Wall, R. A., and Shah, L. (1987). *Haemophilus ducreyi* produces rough lipopolysaccharide. *FEMS Microbiol. Lett.* **42,** 159–161.
3. Abeck, D., and Johnson, A. P. (1987). Identification of surface-exposed proteins of *Haemophilus ducreyi. FEMS Microbiol. Lett.* **44,** 49–51.
4. Abeck, D., Johnson, A. P., Ballard, R. C., Dangor, Y., Fontaine, E. A., and Taylor-Robinson, D. (1987). Effect of cultural conditions on the protein and lipopolysaccharide profiles of *Haemophilus ducreyi* analysed by SDS–PAGE. *FEMS Microbiol. Lett.* **48,** 397–399.
5. Albritton, W. L., Maclean, I. W., Slaney, L. A., Ronald, A. R., and Deneer, H. G. (1984). Plasmid-mediated tetracycline resistance in *Haemophilus ducreyi. Antimicrob. Agents Chemother.* **25,** 187–190.
6. Albritton, W. L., Setlow, J. K., Thomas, M., Sottnek, F., and Steigerwalt, A. G. (1984). Heterospecific transformation in the genus *Haemophilus. Mol. Gen. Genet.* **193,** 358–363.
7. Asin, J. (1952). Chancroid. *Am. J. Syph., Gonorrhea, Vener. Dis.* **26,** 483.
8. Bassereau, P. I. A. L. (1852). "Traite des affections de la peau symptomatiques de la syphilis." Baillière, Paris.
9. Becker, T. M., DeWitt, W., and VanDusen, G. (1987). *Haemophilus ducreyi* infection in South Florida: A rare disease on the rise? *South. Med.J.* **80,** 182–184.
10. Blackmore, C. A., Limpakarnjanarat, K., Rigau-Perez, J. G., Albritton, W. L., and Greenwood, J. R. (1985). An outbreak of chancroid in Orange County, California: Descriptive epidemiology and disease-control measures. *J. Infect. Dis.* **151,** 840–844.
11. Bowmer, M. I., Nsanze, H., d'Costa, L. J., Dylewski, J., Fransen, L., Piot, P., and Ronald, A. R. (1987). Single-dose ceftriaxone for chancroid. *Antimicrob. Agents Chemother.* **31,** 67–69.
12. Brenner, D. J., McWhorter, A. C., Knutson, J. K. L., and Steigerwalt, A. G. (1982). *Escherichia vulneris:* A new series of *Enterobacteriaceae* associated with human wounds. *J. Clin. Microbiol.* **15,** 1133–1140.
13. Brunton, J., Meier, M., Erhman, N., Clare, D., and Almawy, R. (1986). Origin of small beta-lactamase-specifying plasmids in *Haemophilus* species and *Neisseria gonorrhoeae. J. Bacteriol.* **168,** 374–379.
14. Brunton, J., Meier, M., Ehrman, N., Maclean, I., Slaney, L., and Albritton, W. L. (1982). Molecular epidemiology of beta-lactamase-specifying plasmids of *Haemophilus ducreyi. Antimicrob. Agents Chemother.* **21,** 857–863.
15. Canziares, O. (1976). Dermatology in India. *Arch. Dermatol.* **112,** 93–97.
16. Carlone, G. M., Schalla, W. O., Moss, C. W., Ashley, D. L., Fast, D. M., Holler, J. S., and Plikaytis, B. D. (1988). *Haemophilus ducreyi* isoprenoid quinone content and structure determination. *Int. J. Syst. Bacteriol.* **38,** 249–253.
17. Casin, I. M., Grimont, F., Grimont, P. A. D., and Sanson-Le Pors, M. (1985). Lack of deoxyribonucleic acid relatedness between *Haemophilus ducreyi* and other *Haemophilus* species. *Int. J. Syst. Bacteriol.* **35,** 23–25.
18. Casin, I. M., Sanson-Le Pors, M. J., Gorce, M. F., Ortenberg, M., and Pérol, Y. (1982). The enzymatic profile of *Haemophilus ducreyi. Ann Inst. Pasteur/Microbiol.* **133B,** 379–388.
19. Centers for Disease Control (1985). Chancroid—Massachusetts. *Morbid. Mortal. Wkly Rep.* **34,** 711–718.

20. Centers for Disease Control (1982). Chancroid—California. *Morbid. Mortal. Wkly Rep.* **31,** 173–175.
21. Choudhary, B. P., Kumari, S., Bhatia, R., and Agarwal, D. S. (1982). Bacteriological study of chancroid. *Indian J. Med. Res.* **76,** 379–385.
22. Coovadia, Y. M., Kharsany, A., and Hoosen, A. (1985). The microbial aetiology of genital ulcers in black men in Durban, South Africa. *Genitourin. Med.* **61,** 266–269.
23. Dangor, Y., Miller, S. D., da la Exposto, F., and Koornhof, H. J. (1988). Antimicrobial susceptibilities of Southern African isolates of *Haemophilus ducreyi. Antimicrob. Agents Chemother.* **32,** 1458–1460.
24. Deneer, H. G., Slaney, L., Maclean, I. W., and Albritton, W. L. (1982). Mobilization of nonconjugative antibiotic resistance plasmids in *Haemophilus ducreyi. J. Bacteriol.* **149,** 726–732.
25. Denys, G. A., Chapel, T. A., and Jeffries, C. D. (1978). An indirect fluorescent antibody technique for *Haemophilus ducreyi. Health Lab. Sci.* **15,** 128–132.
26. Dienst, R. B. (1948). Virulence and antigenicity of *Haemophilus ducreyi. Am. J. Syph., Gonorrhea, Vener. Dis.* **32,** 289–291.
27. Ducrey, A. (1889). Experimentelle Untersuchungen über den Ansteckungsstoff des weichen Schankers and über die Bubonen. *Monatssch. Prakt. Dermatol.* **9,** 387–405.
28. Feiner, R. R., and Mortara, F. (1945). Infectivity of *Haemophilus ducreyi* for the rabbit and the development of skin hypersensitivity. *Am. J. Syph., Gonorrhea, Vener. Dis. 29,* 71–79.
29. Fiumara, N. J., Rothman, K., and Tang, S. (1986). The diagnosis and treatment of chancroid. *J. Am. Acad. Dermatol.* **15,** 939–943.
30. Fransen, L., Nsanze, H., Achola, J. N., D'Costa, L., Ronald, A. R., and Piot, P. (1987). A comparison of single-dose spectinomycin with five days of trimethoprim–sulfamethoxazole for the treatment of chancroid. *Sex. Transm. Dis.* **14,** 98–101.
31. Fung-Tomc, J. C., and Tilton, R. C. (1988). Application of nucleic acid hybridization in clinical bacteriology. *In* "Microbial Antigenodiagnosis" (K. Wicher, ed.), Vol. 1, pp. 237–244. CRC Press, Boca Raton, Florida.
32. Greenblatt, R. B., and Sanderson, E. S. (1938). The intradermal chancroid bacillary antigen test as an aid in the differential diagnosis of the venereal bubo. *Am. J. Surg.* **41,** 384.
33. Hafiz, S., Kinghorn, G. R., and McEntegart, M. G. (1984). *Haemophilus ducreyi* and chancroid. *In* "Medical Microbiology" Vol. 4 (C. S. F. Easmon and J. Jeljaszewicz, ed.), pp. 143–170. Academic Press, London.
34. Hammond, G. W., Lian, C. J., Wilt, J. C., and Ronald, A. R. (1978). Comparison of specimen collection and laboratory techniques for isolation of *Haemophilus ducreyi. J. Clin. Microbiol.* **73,** 39–43.
35. Hammond, G. W., Slutchuk, M., Scatliff, J., Sherman, E., Wilt, J. C., and Ronald, A. R. (1980). Epidemiologic, clinical, laboratory, and therapeutic features of an urban outbreak of chancroid in North America. *Rev. Infect. Dis.* **2,** 867–879.
36. Handsfield, H. H., Totten, P. A., Fennel, C. L., Falkow, S., and Holmes, K. K. (1981). Molecular epidemiology of *Haemophilus ducreyi* infections. *Ann. Intern. Med.* **95,** 315–318.
37. Hannah, P., and Greenwood, J. R. (1982). Isolation and rapid identification of *Haemophilus ducreyi. J. Clin. Microbiol.* **16,** 861–864.
38. Hansen, E. J., and Loftus, T. A. (1984). Monoclonal antibodies reactive with all strains of *Haemophilus ducreyi. Infect. Immun.* **44,** 196–198.
39. Hart, G. (1981). Chancroid. *In* "Communicable and Infectious Diseases" (P. F. Wehrle and F. H. Top, Sr., eds.), 9th ed., pp. 150–154. Mosby, St. Louis, Missouri.

40. Health and Welfare, Canada (1988). Chancroid outbreak—Winnipeg, Manitoba. *Can. Dis. Wkly. Rep.* **14,** 13–15.
41. Huynh, T. V., Young, R. A., and Davis, R. W. (1985). Constructing and screening cDNA libraries in lambda gt10 and lambda gt11. *In* "DNA Cloning: A Practical Approach" (D. M. Glover, ed.), Vol. 1, pp. 49–77. IRL Press, Washington, D.C.
42. Kerber, R. E., Rowe, C. E., and Gilbert, K. R. (1969). Treatment of chancroid. A comparison of tetracycline and sulfisoxazole. *Arch. Dermatol.* **100,** 604–607.
43. Kilian, M. (1976). A taxonomic study of the genus *Haemophilus* with the proposal of a new species. *J. Gen. Microbiol.* **93,** 9–62
44. Kilian, M. (1974). A rapid method for the differentiation of *Haemophilus* strains. The porphyrin test. *Acta Pathol. Microbiol. Scand., Sect. B* **82B,** 835–842.
45. Kinghorn, G. R., Hafiz, S., and McEntegart, M. G. (1983). Genital colonization with *Haemophilus ducreyi* in the absence of ulceration. *Eur. J. Sex. Transm. Dis.* **1,** 89–90.
46. Kreiss, J. K., Koech, D., Plummer, F. A., Holmes, K. K., Lightfoote, M., Piot, P., Ronald, A. R., Ndinya-Achola, J. O., D'Costa, L. J., Roberts, P., Ngugi, E. N., and Quinn, T. C. (1986). AIDS virus infection in Nairobi prostitutes: Spread of the epidemic to East Africa. *N. Engl. J. Med.* **314,** 414–418.
47. LeSaux, N. M., Slaney, L. A., Plummer, F. A., Ronald, A. R., and Brunham, R. C. (1987). In vitro activity of ceftriaxone, cefetamet (Ro 15-8074), ceftetrame (Ro 19-5247; T-2588), and fleroxacin (Ro 23-6240; AM-833) versus *Neisseria gonorrhoeae* and *Haemophilus ducreyi. Antimicrob. Agents Chemother.* **31,** 1153–1154.
48. Lykke-Olesen, L., Larsen, L., Pederson, T. G., and Gaarslev, K. (1979). Epidemic of chancroid in Greenland, 1977–78. *Lancet* **1,** 654–655.
49. Malouin, F., Bryan, L. E., Shewciw, P., Douglas, J., Li, D., Van Den Elzen, H., and LaPointe, J. R. (1988). DNA probe technology for rapid detection of *Haemophilus influenzae* in clinical specimens. *J. Clin. Microbiol.* **26,** 2132–2138.
50. Maniatis, T., Fritsch, E. F., and Sambrook, J. (1982). "Molecular Cloning: A Laboratory Manual." Cold Spring Harbor Lab., Cold Spring Harbor, New York.
51. McNicol, P. J., and Ronald, A. R. (1984). The plasmids of *Haemophilus ducreyi. J. Antimicrob. Chemother.* **14,** 561–573.
52. Morel, P., Casin, I., Gandiol, C., Vallet, C. and Civatte, J. (1982). Epidémie de chancre mou: Traitement de 587 malades. *Nouv. Presse Méd.* **11,** 655–656.
53. Museyi, K., Van Dyck, E., Vervoort, T., Taylor, D., Hoge, C., and Piot, P. (1988). Use of an enzyme immunoassay to detect serum IgG antibodies to *Haemophilus ducreyi. J. Infect. Dis.* **157,** 1039–1043.
54. Naamara, W., Plummer, F. A., Greenblatt, R. M., D'Costa, L. J., Ndinya-Achola, J. O., and Ronald, A. R. (1987). Treatment of chancroid with ciprofloxacin. A prospective, randomized clinical trial. *Am. J. Med.* **82,** Suppl. 4A, 317–320.
55. Nayyar, K. C., Stolz, E., and Michel, M. F. (1979). Rising incidence of chancroid in Rotterdam: Epidemiological, clinical, diagnostic, and therapeutic aspects. *Br. J. Vener. Dis.* **55,** 439–441.
56. New York City Department of Health (1987). Treatment of chancroid with ceftriaxone: New York City. *City Health Inf.* **6,** 1–2.
57. Odumeru, J. A., Ronald, A. R., and Albritton, W. L. (1983). Characterization of cell proteins of *Haemophilus ducreyi* by polyacrylamide gel electrophoresis. *J. Infect. Dis.* **148,** 710–714.
58. Odumeru, J. A., Wiseman, G. M., and Ronald, A. R. (1987). Relationship between lipopolysaccharide composition and virulence of *Haemophilus ducreyi. J. Med. Microbiol.* **23,** 155–162.
59. Odumeru, J. A., Wiseman, G. M., and Ronald, A. R. (1985). Role of lipopolysaccharide

and complement in susceptibility of *Haemophilus ducreyi* to human serum. *Infect. Immun.* **50,** 495–499.

60. Odumeru, J. A., Wiseman, G. M., and Ronald, A. R. (1984). Virulence factors of *Haemophilus ducreyi. Infect. Immun.* **43,** 607–611.
61. Parsons, L. M., Shayegani, M., Waring, A. L., and Bopp, L. H. (1989). DNA probes for the identification of *Haemophilus ducreyi. J. Clin. Microbiol.* **27,** 1441–1445.
62. Piechulla, K., Mutters, R., Burbach, S., Klussmeier, R., Pohl, S., and Mannheim, W. (1986). Deoxyribonucleic acid relationships of *"Histophilus ovis/Haemophilus somnus," Haemophilus haemoglobinophilus,* and *"Actinobacillus seminis." Int. J. Syst. Bacteriol.* **36,** 1–7.
63. Plummer, F. A., D'Costa, L. J., Nsanze, H., Karasira, P., Maclean, I. W., Piot, P., and Ronald, A. R. (1985). Clinical and microbiologic studies of genital ulcers in Kenyan women. *Sex. Transm. Dis.* **12,** 193–197.
64. Quinn, T. C., Mann, J. M., Curran, J. W., and Piot, P. (1986). AIDS in Africa: An epidemiologic paradigm. *Science* **234,** 955–963.
65. Rajan, V. S., Sng, E. H., and Lim, A. L. (1983). The isolation of *H. ducreyi* in Singapore. *Ann. N.Y. Acad. Med.* **12,** 57–60.
66. Roberts, M. C., Actis, L. A., and Crosa, J. H. (1985). Molecular characterization of chloramphenicol-resistant *Haemophilus parainfluenzae* and *Haemophilus ducreyi. Antimicrob. Agents Chemother.* **28,** 176–180.
67. Ronald, A. R., and Albritton, W. L. (1984). Chancroid and *Haemophilus ducreyi. In* "Sexually Transmitted Diseases" (K. K. Holmes, P. Mardh, P. F. Sparling, and P. J. Wiesner, eds.), pp. 385–393. McGraw-Hill, New York.
68. Saiki, R. K., Gelfand, D. H., Stoffel, S., Scharf, S. J., Higuchi, R., Horn, G. T., Mullis, K. B., and Erlich, H. A. (1988). Primer-directed enzymatic amplification of DNA with a thermostable DNA polymerase. *Science* **239,** 487–491.
69. Saunders, J. M., and Folds, J. D. (1986). Immunoblot analysis of antigens associated with *Haemophilus ducreyi* using serum from immunized rabbits. *Genitourin. Med.* **62,** 321–328.
70. Schalla, W. O., Sanders, L. L., Schmid, G. P., Tam, M. R., and Morse, S. A. (1986). Use of dot-immunobinding and immunofluorescence assays to investigate clinically suspected cases of chancroid. *J. Infect. Dis.* **153,** 879–887.
71. Schmid, G. P. (1986). The treatment of chancroid. *J. Am. Med. Assoc.* **255,** 1757–1762.
72. Sehgal, V. N., and Shyam Prasad, A. L. (1985). Chancroid or chancroidal ulcers. *Dermatologica* **170,** 136–141.
73. Simonsen, J. N., Ckameron, D. W., Gakinya, M. N., Ndinya-Achola, J. O., D'Costa, L. J., Karasira, P., Cheang, M., Ronald, A. R., Piot, P., and Plummer, F. A. (1988). Human immunodeficiency virus infection among men with sexually transmitted diseases: Experience from a center in Africa. *N. Engl. J. Med.* **319,** 274–278.
74. Slootmans, L., Vanden Berghe, D. A., and Piot, P. (1985). Typing *Haemophilus ducreyi* by indirect immunofluorescence assay. *Genitourin. Med.* **61,** 123–126.
75. Sng, E. H., Lim, A. L., Rajan, V. S., and Goh, A. J. (1982). Characteristics of *Haemophilus ducreyi.* A study. *Br. J. Vener. Dis.* **58,** 239–242.
76. Sottnek, F. O., Biddle, J. W., Kraus, S. J., Weaver, R. E., and Stewart, J. A. (1980). Isolation and identification of *Haemophilus ducreyi* in a clinical study. *J. Clin. Microbiol.* **12,** 170–174.
77. Stamps, T. J. (1974). Experience with doxycycline (vibramycin) in the treatment of chancroid. *J. Trop. Med. Hyg.* **77,** 55–60.
78. Sturm, A. W. (1987). Comparison of antimicrobial susceptibility patterns of fifty-seven

strains of *Haemophilus ducreyi* isolated in Amsterdam from 1978 to 1985. *J. Antimicrob. Chemother.* **19,** 187–191.

79. Stüttgen, G. (1981). "Ulcus molle: Chancroid." Grosse Verlag, Berlin.
80. Taylor, D. N., Duangmani, C., Suvongse, C., O'Connor, R., Pitarangsi, C., Panikabutra, K., and Echeverria, P. (1984). The role of *Haemophilus ducreyi* in penile ulcers in Bangkok, Thailand. *Sex. Transm. Dis.* **11,** 148–151.
81. Taylor, D. N., Echeverria, P., Hanchalay, S., Pitarangsi, C., Slootmans, L., and Piot, P. (1985). Antimicrobial susceptibility and characterization of outer membrane proteins of *Haemophilus ducreyi* isolated in Thailand. *J. Clin. Microbiol.* **21,** 442–444.
82. Teague, O., and Deibert, O. (1920). The value of cultural method in diagnosis of chancroid. *J. Urol.* **4,** 543–550.
83. Tenover, F. C. (1988). Diagnostic deoxyribonucleic acid probes for infectious diseases. *Clin. Microbiol. Rev.* **1,** 82–101.
84. Thirumoorthy, T., Sng, E. H., Doraisingham, S., Ling, A. E., Lim, K. B., and Lee, C. T. (1986). Purulent penile ulcers of patients in Singapore. *Genitourin. Med.* **62,** 253–255.
85. Totten, P. A., Handsfield, H. H., Peters, D., Holmes, K. K., and Falkow, S. (1982). Characterization of ampicillin resistance plasmids from *Haemophilus ducreyi. Antimicrob. Agents Chemother.* **21,** 622–627.
86. Tuffrey, M., Abeck, D., Alexander, F., Johnson, A. P., Ballard, R. C. and Taylor-Robinson, D. (1988). A mouse model of *Haemophilus ducreyi* infection (chancroid). *FEMS Microbiol. Lett.* **50,** 207–209.
87. U.S. Department of Health and Human Services, Public Health Service, Centers for Disease Control (1985). "Sexually Transmitted Disease Statistics," Issue No. 135. CDC, Atlanta, Georgia.
88. U.S. Department of Health and Human Services, Public Health Service, Centers for Disease Control (1987). "Sexually Transmitted Disease Statistics," Issue No. 136. CDC, Atlanta, Georgia.
89. Van Dyck, E., and Piot, P. (1987). Enzyme profile of *Haemophilus ducreyi* strains isolated on different continents. *Eur. J. Clin. Microbiol.* **6,** 40–43.

4

Detection of Diarrheogenic *Escherichia coli* Using Nucleotide Probes

PETER ECHEVERRIA, JITVIMOL SERIWATANA, ORNTIPA SETHABUTR, AND ARUNSRI CHATKAEOMORAKOT

Armed Forces Research Institute of Medical Sciences
Bangkok, Thailand

I. INTRODUCTION

Escherichia coli that cause diarrhea have been classified into four major categories: enterotoxigenic (ETEC), enteroinvasive (EIEC), enteropathogenic (EPEC), and enterohemorrhagic (EHEC) (80). These diarrheogenic *E. coli* cause diarrhea by at least one or more identified pathogenic mechanisms: enterotoxin production, enteroinvasion, enteroadherence, or cytotoxin production (79,80). Enteroadherence is the principal pathogenic mechanism identified in EPEC (79), although EPEC O26 and O111 produce cytotoxins and other EPEC serotypes cause disease by mechanisms that remain to be identified (135). These diarrheogenic *E. coli* are usually identified by *in vitro* or *in vivo* biological assays or by serotyping (23,128). An alternative approach to identify these *E. coli* using recent developments in molecular genetic technology is to identify nucleotide sequences of pathogenic determinants directly in specimens. Nucleotide probes have been especially useful in identifying diarrheogenic *E. coli* indistinguishable phenotypically from nonpathogenic *E. coli*. The use of nucleotide probes to identify these *E. coli* is based on a combination of gene cloning and nucleic acid (NA) hybridization techniques. Gene cloning provides probes for the specific polynucleotide sequence of an *E. coli* pathogenic determinant, and NA hybridization with either radiolabeled or nonradiolabeled probes involves the use of these probes to search for homologous sequences in the total NA present in a specimen.

Nucleotide probes have been constructed to identify plasmid-encoded genes coding for heat-labile and heat-stable enterotoxin (LT and ST), plasmid-encoded genes involved in enteroinvasion, plasmid-encoded genes coding for adherence to tissue cultures and human intestinal cells, and bacteriophage-encoded and chromosomally encoded genes for *Shiga*-like enterotoxin (SLT). Probe DNA refers to specific DNA fragments of plasmids containing cloned genes coding for enteropathogenic determinants or synthetic oligonucleotides constructed from the nucleotide sequences of these fragments.

These probes are used to search for homologous DNA sequences in specimens. Hybridization occurs when there is sufficient homology between single strands of probe DNA and single strands of target DNA to permit stable hydrogen binding between nucleotide bases and the formation of a double-stranded complex. The polynucleotide gene probes for *E. coli* enteropathogenic determinants range from 154 to 17,000 bp long and hybridize under stringent conditions with target DNA that is 80% homologous with these probes (40). Synthetic oligonucleotide probes to detect genes coding for LT, ST, and SLT are 20–50 bases long and may not hybridize to target sequences when only one or two nucleotides are mismatched within the sequence (8). Oligonucleotide probes are therefore very specific in detecting specific DNA sequences (11), but may not detect single-base mutations in target cell DNA, changes that might still preserve toxin production. Nucleotide probes to detect rRNA are sensitive and specific in identifying bacterial species (71), but have not been used to differentiate between diarrheogenic and nondiarrheogenic *E. coli*.

II. BACKGROUND

A. Enterotoxigenic *Escherichia coli*

Enterotoxigenic *E. coli* is one of the most common causes of diarrhea in infants and young children in tropical developing countries (4,25,127) and is one of the leading causes of diarrhea in travelers (26,91). The pathogenesis of ETEC has been frequently reviewed (78,80,129). In order to cause diarrhea, ETEC require two virulence factors: colonization adhesins and enterotoxin production. First, ETEC must adhere to the proximal small bowel, a process that is mediated by protein fimbrial adhesins termed colonization factor antigens (CFA) or *E. coli* surface (CS) fimbriae. A number of adhesins have been recognized including CFA I (41), CS1, CS2, CS3, CS4, CS5, CS6 (18,88,147), and PCF O159:H4 (156). CS1, CS2, and CS3 in various combinations were first reported as a single factor CFA II (18,147), and CS4, CS5, and CS6 were reported as colonization factor PCF 8775 (88), now called CFA IV (89). Genes coding for CFA I and CFA II are plasmid-encoded and are closely associated with genes coding for ST (33). Additional colonization factors have been described, but it is not clear from the published data if they are distinct from previously reported adhesins (59). These include CFA III [presumably CS6 (21,22)] and CFA V [presumably CS5 (85)]. Many ETEC strains have been isolated that do not have any of these CFA. Additional CFA of ETEC presumably exist and remain to be identified. DNA probes for adhesins of ETEC have not been evaluated in studies of human ETEC infections.

The second virulence factor is enterotoxin, which causes fluid and electrolyte secretion. Enterotoxigenic *E. coli* produce LT, ST, or both (129). The LT form is immunogenic and is composed of an A subunit of 25,500 Da and five B subunits of 11,500 Da each (16,19,73). The A subunit of LT acts on small-intestinal cells by stimulating adenyl cyclase with a resultant increase in intracellular cAMP (101) and net secretion of electrolytes and water (47). The B subunit of LT that attaches to ganglioside receptors on intestinal epithelial cells is involved in the binding of the toxin to intestinal cells and the delivery of the enzymatically active A subunit into the cell cytosol. *Escherichia coli* LT and its subunits (A and B) cross-react serologically with the closely related enterotoxin of *Vibrio cholerae* (CT) (47). The LT produced by strains of *E. coli* from humans (LT-H) and pigs (LT-P) have common and unique antigenic determinants (56,58,164). Antibodies of CT can neutralize LT-H and LT-P, which have been designated LT I (119). LT-H ETEC have only been isolated from humans, and LT-P ETEC have been isolated mostly from pigs. Genes coding for LT-H and LT-P are very similar in nucleotide sequence and cannot be differentiated by DNA hybridization. LT II, originally described in *E. coli* SA 53 isolated from a water buffalo in Thailand (49), resembles LT I with respect to several of its biological properties, but is not neutralized by antibodies to CT, LT-H, or LT-P (49). Genes coding for LT I and ST are plasmid-mediated, while genes coding for LT II are chromosomally encoded (49). DNA sequences coding for LT II do not hybridize with genes coding for LT I under hybridization conditions that allow for up to 20% base pair mismatch (119). Two different types of LT II have been identified serologically: LT IIa from *E. coli* SA 53 and LT IIb from *E. coli* 41 isolated from beef in Brazil (49,52,53,56).

Heat stable toxin is thermostable, poorly immunogenic, and has a MW <5000 (1). One form of ST, ST I, is soluble in methanol (9), acts by stimulating guanylate cyclase (43,60), and causes distension of suckling mouse intestine (23). The ST I forms are heterogeneous (63,97,98,151). Enterotoxigenic *E. coli* that produce ST Ia (ST-P) have been isolated from humans, pigs, and cows (97,98,151,168). A second ST I, designated ST Ib (ST-H), was originally isolated from a person with diarrhea in Bangladesh (96,98). Almost always, ETEC that produce ST Ib are isolated from humans. Genes coding for ST Ia are located on a transposon (150), and genes coding for ST Ib are often associated with plasmids coding for CFA I and II (33). Genes coding for ST Ia and ST Ib have a sequence divergence of 31% (98) and are thus distinct when examined by DNA hybridization under stringent hybridization conditions. ST II (STB) is a methanol-insoluble toxin that is active in the pig jejunal loop (9) but negative in the suckling mouse assay (23). Neither the DNA nor the amino acid sequences of ST II (77,118) have any resemblance to ST Ia or ST Ib. Enterotoxigenic *E. coli* have been described that contain genes coding for LT I, ST Ia, ST Ib, and ST II in different combinations (29,31,37,99).

B. Enteroinvasive *Escherichia coli*

Enteroinvasive *E. coli* and *Shigella* cause dysentery by invading epithelial cells of the colon (24,45,74,78). The former are difficult to identify because they may be confused with nonpathogenic *E. coli*. Enteroinvasive

E. coli may be lactose-fermenting or non-lactose-fermenting, and while they usually decarboxylate lysine and are nonmotile (143,162), no biochemical test is entirely specific in identifying these organisms. They belong to a number of *E. coli* serogroups: O28, O29, O124, O136, O143, O144, O147, O152, O164, and O167 (42,162,163). *Escherichia coli* O124 agglutinate in *Shigella dysenteriae* serotype 3 antisera (42).

The genetics of the virulence of *Shigella* and EIEC are similar (54,131,132). Both *Shigella* and EIEC contain plasmids of 120–140 MDa that are necessary for virulence. It is possible to identify EIEC by examining *E. coli* for 120- to 140-MDa plasmids. In a study of dysentery in Thailand, 364 (13%) of 2758 *E. coli* isolated from children with dysentery contained plasmids of ~140 MDa, but only 64 (18%) of these 364 *E. coli* were positive in the Sereny test (141). Thus screening *E. coli* for plasmids was too time-consuming and nonspecific to be practical.

The homology between the 120- to 140-MDa plasmids of EIEC and *Shigella* suggested that these plasmids shared DNA sequences that could be used as a specific probe to identify EIEC (132). A 17-kb *Eco*RI digestion fragment of pWR 100, the 140-MDa plasmid of *S. flexneri* 5 (M90T), was shown to be specific in differentiating EIEC from non-EIEC (5,141).

Both EIEC and *Shigella* have been detected by an enzyme-linked immunosorbent assay (ELISA) using antisera raised to a virulent EIEC O143 (116). This antiserum, subsequently absorbed with an avirulent derivative of the immunizing strain, contains ELISA-reactive antibodies specific for a unique antigen on the virulent strain, termed virulent marker antigen (VMA). This ELISA distinguishes virulent *Shigella* and EIEC from their avirulent derivatives. Avirulent derivatives of EIEC or *Shigella* that are probe-positive and Sereny test-negative (136) have been identified (117). Additional probes to identify *Shigella* and EIEC are other fragments of the large virulence plasmid of *Shigella* (144) including a probe coding primarily for invasion plasmid antigen *a* (Ipa) (169).

C. Enteroadherent *Escherichia coli*

Histopathological studies in infants and animals with EPEC demonstrated that these strains were adherent to the small-bowel mucosa, and close adherence was suggested as important for the induction of diarrhea (15,94,124,125,166) although cytotoxin production has also been identified in EPEC (65,135). Eighty percent of EPEC examined in the United Kingdom adhered to HEp-2 cells in tissue culture, and HEp-2 adherence was significantly more common among EPEC than among ETEC and normal flora *E. coli* (17). HEp-2 cell adherence was associated with 50- to 70-MDa plasmids in EPEC (2), and with a 60-MDa plasmid (pMAR 2) in EPEC

2348 (serotype O127:H5) (79). The presence of this plasmid correlated with the ability of *E. coli* 2348 to cause diarrhea in adult volunteers (79). The genes for HEp-2 adherence have been cloned and used to identify *E. coli* that adhere to tissue culture cells and human intestinal enterocytes in a localized pattern (68,69,104). In addition to localized adherence, EPEC strains also adhere intimately, attaching to projections of tissue culture and human enterocytes and causing localized destruction of microvilli (68,69). This lesion has been referred to as effacement.

The term EPEC adherent factor (EAF) has been applied to the adhesin factor that is specified by pMAR 2 and that confers mannose-resistant localized adherence of *E. coli* to HEp-2 cells and human intestinal cells (69,104), and enteropathogenicity in volunteers (79). A DNA probe for EAF constructed from pMAR 2 was both sensitive and specific in identifying EPEC that adhered to HEp-2 cells in a localized pattern (104). Strains of EPEC that hybridized with the EAF probe were more frequent in serogroups O55, O111, O119, O127, and O128 (referred to as class I EPEC serogroups) than in serogroups O44, O86, and O114 (referred to as class II EPEC) (104). Sixty-two percent of class I and none of class II EPEC hybridized with the EAF probe, but both classes of EPEC were associated with diarrhea in children in Peru—implying that EPEC causes diarrheal disease by an alternative mechanism (105).

In addition to EAF *E. coli* that adhere to HEp-2 and human intestinal cells in a localized pattern, *E. coli* also adhere in either an aggregative or diffuse adherence (DA) pattern (106,133). A DNA probe to identify diffusely adherent *E. coli* has been constructed (100). This probe is derived from a chromosomal gene encoding a protein associated with *E. coli* expression of diffuse adherence in the HEp-2 cell assay.

Escherichia coli of EPEC serotypes that adhere in a localized pattern cause effacement of human enterocytes (69). When plasmids encoding for EAF were transferred to *E. coli* K12, recipients adhered in a localized pattern but did not efface intestinal epithelial cells, implying that genes responsible for effacement are located on the chromosome in EPEC (68).

D. Enterohemorrhagic *Escherichia coli*

Shiga-like toxin (SLT)-producing *E. coli* [or verocytotoxin (VT)-producing *E. coli*] associated with the hemorrhagic colitis and hemolytic uremic syndrome are referred to as enterohemorrhagic *E. coli* (EHEC) (62,66,114,115,120,122,126). These organisms produce one or both of two antigenically distinct toxins, SLT I and SLT II, which are encoded by two different bacteriophages (933J and 933W) in *E. coli* 933 (107). Originally SLT *E. coli* strains were identified by testing sterile-culture supernatants of EPEC and other *E. coli* for cytotoxicity to Vero or HeLa cells (72,86,135). The genes for SLT I and SLT II have subsequently been cloned and sequenced, and polynucleotide DNA probes specific for each have been constructed (108,109,153). Genes coding for VT1 and VT2 have also been cloned and sequenced, and are similar in nucleotide sequence to genes coding for SLT I and SLT II (172,173).

Shiga-like toxin-producing *E. coli* O157:H7 also possess plasmid-encoded fimbriae that promote attachment to intestinal epithelial cells in tissue culture (64). A DNA probe consisting of a 3.4-kb *Hin*dIII cryptic fragment of pCVD 419, referred to as the "EHEC" probe hybridized with 99% of EHEC O157:H6, 77% of EHEC O26:H11, and 81% of EHEC of other serotypes (81) isolated from patients with hemolytic uremic syndrome or hemorrhagic colitis.

III. RESULTS AND DISCUSSION

A. Enterotoxigenic *Escherichia coli*

Biochemical tests cannot be used to identify ETEC. Testing pools of *E. coli* for enterotoxin production (10,92) or screening *E. coli* for serotypes commonly found among ETEC (93) are methods that have been used to identify this enteric pathogen. *Escherichia coli,* however, produce colicins that can lyse ETEC and lead to falsely negative results, especially if isolates are stored as pools (102); and not all ETEC are of ETEC serotypes (27). The most reliable way of identifying ETEC has been to test culture supernatants of individual colonies in bioassays (23,128) or immunoassays (50,55,155,175), both of which are time-consuming, expensive, or require specific antisera.

1. Polynucleotide Probes

Once genes coding for enterotoxins had been successfully cloned (20,76,148,149), searching for DNA sequences coding for enterotoxins directly became an alternative approach to identifying ETEC. The first use of DNA probes to identify diarrheogenic *E. coli* was performed in Bangladesh in 1981 with DNA sequences encoding LT I and ST Ia (ST-P) (96). Enterotoxin gene probes for LT I, ST Ia, and ST Ib were compared to testing *E. coli* for enterotoxin production with isolates from Asia (137) (see Table I) and Africa (46). Genes coding for other enterotoxins, LT II (119) and ST II (77,118), have subsequently been cloned and used as probes to identify other types of enterotoxin-producing *E. coli.* DNA fragments of plasmids containing cloned enterotoxin genes used as probes to identify ETEC are listed in Table II.

TABLE I

***Escherichia coli* Isolated[a] from Children in Asia with Diarrhea Tested for Enterotoxin Production and Colony Hybridization with Polynucleotide Enterotoxin Gene Probes**

Toxin Produced	Number	Number of isolates that hybridized with polynucleotide DNA probes coding for: LTST Ia ST Ib	LTST Ia	LTST Ib	LT I	ST Ia	ST Ib
LTST	246	44	46	156	0	0	0
LT	401	0	0	0	401	0	0
ST	328	0	0	0	0	84	244
None	733	0	0	0	0	0	0

[a] One isolate per child.

Plasmids containing the cloned genes for the enterotoxin of interest are isolated by ethidium bromide–cesium chloride ultracentrifugation (148). The enterotoxin gene probes are endonuclease digestion fragments of plasmids containing the whole or part of the cloned enterotoxin genes. For example, the LT I probe is an 850 bp *Hin*cII digestion fragment of pEWD 299 that codes for the A and part of the B subunit of the gene coding for LT I, while the DNA probes for ST Ia and ST Ib contain the entire sequences encoding ST I. Other polyclonal plasmid DNA probes for genes coding for LT I, ST Ia, and ST Ib have been constructed that contain the same sequences but are easier to prepare (152). The digested plasmid DNA is separated by electrophoresis on a polyacrylamide gel (Fig. 1). The appropriate DNA fragments are removed from the gel by electroelution. After

TABLE II

Polynucleotide DNA Probes Used to Identify Enterotoxigenic *Escherichia coli*

Toxin	Plasmid[a]	Digestion fragment
LT I	pEWD 299 (19)	850-bp *Hin*cII
LT II	pCP 2725 (119)	800-bp *Hin*dIII-*Pst*I
ST Ia (ST-P)	pRIT 10036 (76)	154-bp *Hin*fI
	or	or
	pDAS 101 (152)	157-bp *Eco*RI-*Bam*HI
ST Ib (ST-H)	pSLM 004 (98)	215-bp *Hpa*II-*Eco*RI
	or	or
	pDAS 100 (152)	215-bp *Bam*HI-*Pst*I
ST II	pCHL 6 (77)	460-bp *HinfI*

[a] Numbers in parentheses are references.

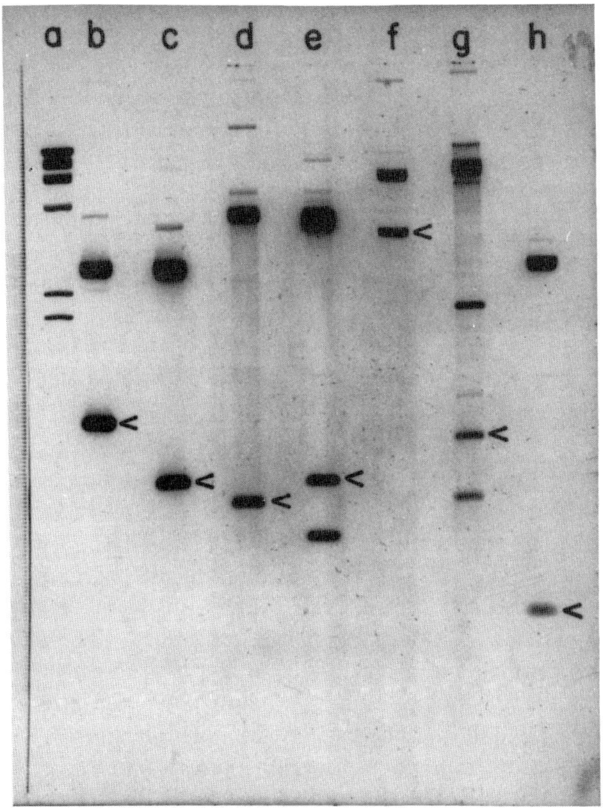

Fig. 1. Endonuclease digestion fragments of plasmids containing cloned genes coding for enterovirulent determinants used as polynucleotide probes separated on a polyacrylamide gel: (a) *Hin*dIII digests of λ DNA used as size markers; (b) 1154-bp *Bam*HI digestion fragment of pJN 37-19 used as the probe for genes coding for *Shiga*-like toxin (SLT I); (c) 842-bp *Pst*I–*Sma*I digestion fragment of pNN 110-18 used as the probe for genes coding for SLT II; (d) 750-bp *Hin*dII digestion fragment of p 746 used as the probe for genes coding for verocytotoxin I; (3) 850-bp *Ava*I–*Pst*I digestion fragment of p 363 used as the probe for genes coding for verocytotoxin II; (f) 3400-bp *Hin*dIII digestion fragment of pCVD 419 used as the probe for genes coding for "enterohemorrhagic *E. coli*"; (g) 1000-bp *Bam*HI–*Sal*I digestion fragment of pMAR 22 used as the probe for genes coding for enteropathogenic *E. coli* adhesin factor; (h) 450-bp *Pst*I digestion fragment of pSLM 852 (pW 22) used as the probe for genes coding for diffuse adherence.

ethanol precipitation the DNA fragments are labeled *in vitro* with [α-^{32}P]deoxynucleotides by nick translation (84).

Hybridization of specimens was first performed using a modification of a colony hybridization method (51) on nitrocellulose (NC) paper. After processing, filters are baked overnight at 65°C or for 2 hr at 85°C in a vacuum oven.

Only 18% of stool specimens that contain ETEC spotted at a field laboratory in rural Thailand hybridized with the enterotoxin gene probes, suggesting DNA was not retained on NC filters as effectively as in the laboratory in Bangkok (31). This was presumably due to the difficulties in fixing DNA onto NC paper in the field without a constant electrical power supply for the oven.

An alternative method has been described that is more suitable under field conditions (83). Specimens are spotted directly onto MacConkey agar and fixed on Whatman 541 filters. This procedure does not require baking and retains DNA on a more durable support. Alternative methods of retaining DNA on solid supports have been described (90).

Examining stool blots (stools streaked over a 1 × 1 cm area of a MacConkey plate, incubated at 37°C overnight, and transferred to Whatman 541 filters) identified 77–80% of ETEC infections identified by testing 10 *E. coli* colonies per patient for hybridization with the enterotoxin gene probes (34,36). In a recent study, 10 and 300 *E. coli* colonies per patient, as well as stool blots, were examined with enterotoxin gene probes (39). Infections with ETEC were identified in 38 (9%) of 416 children with diarrhea by examining 10 colonies, 52 (12%) by examining 300 colonies, and 42 (10%) by examining stool blots. Fewer ETEC infections were identified by examining stool blots than 300 well-separated colonies, presumably because bacterial cell remnants interfered with access of probe to target cell DNA. Previous studies in which 5–10 *E. coli* colonies per patient were examined for ETEC presumably underestimated the incidence of ETEC infections. Since examining 300 *E. coli* colonies per patient is not practical, alternate methods of detecting genes encoding enterotoxin directly in stools need to be developed (161). Whether this will change our understanding of the epidemiology of ETEC remains to be determined.

Radiolabeled enterotoxin gene probes have been used to identify ETEC infections worldwide (46,137,151). The major advantage of enterotoxin gene probes has been in testing large numbers of isolates. *Escherichia coli* can be fixed on a solid support and mailed to a reference laboratory to be examined with radiolabeled probes. Polynucleotide enterotoxin gene probes have been used to test large numbers of isolates to identify sources of ETEC in homes of children in Bangkok (29,37) and to define the epidemiology of ETEC in a longitudinal study of villages in Thailand (31) (Table III and Fig. 2). Examining *E. coli* with polynucleotide enterotoxin gene probes has been used by our department and the Division of Medical Sciences of the Thai Ministry of Health to identify ETEC infections throughout Thailand. Enterotoxin gene probes have also been used to identify ETEC in situations without access to a microbiology laboratory (a refugee camp on the Thai–Laotian border) (158) and in studies of diarrheal

TABLE III

Enterotoxigenic *Escherichia coli* Detected by DNA Hybridization in Environmental Specimens from the Homes of Children with ETEC Diarrhea, from Their Neighbors, and Homes Not Associated with ETEC Infections

	Number of specimens positive/number examined		
Specimen	42 Homes of children with ETEC diarrhea	42 Homes of their neighbors	866 Homes not associated with ETEC diarrhea
Source of drinking water	1/38	0/33	0/529
Drinking water in home	0/41	0/42	1/866
Food	0/84	0/84	2/1732
Mothers' hands	6/42	0/42	2/866
Childrens' hands	5/37	0/43	01/966

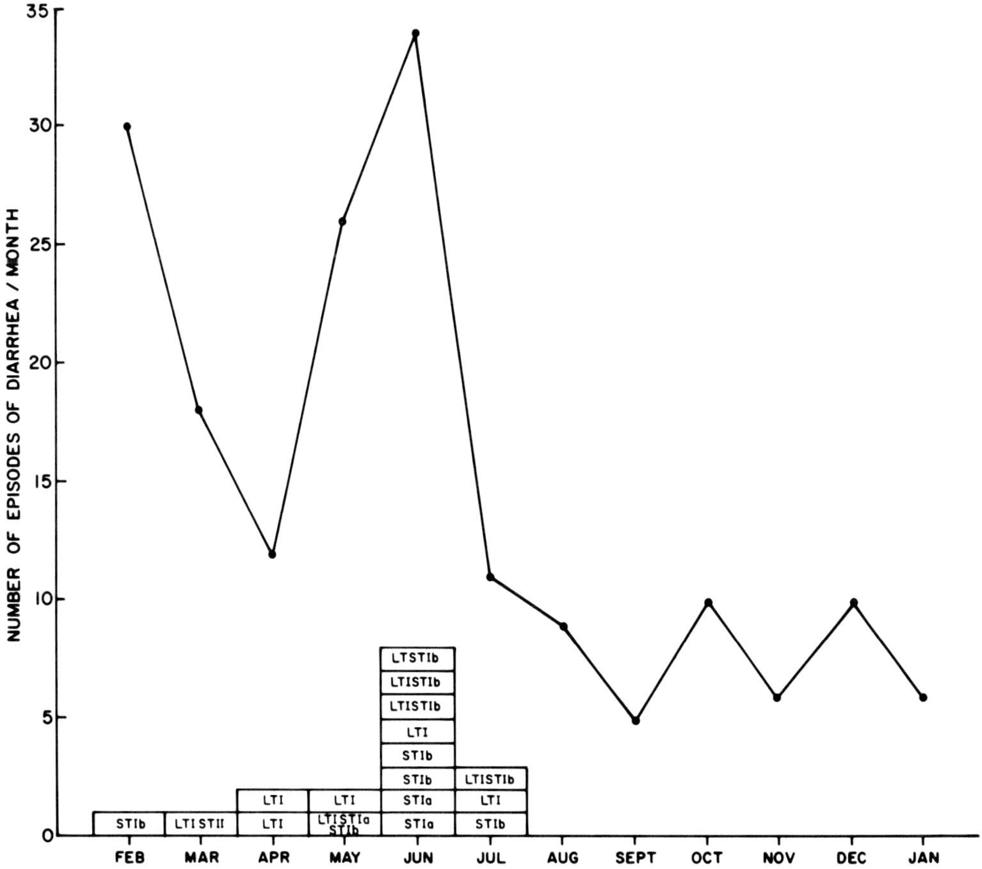

Fig. 2. Episodes of diarrhea associated with enterotoxigenic *Escherichia coli* identified with enterotoxin gene probes in a year-long study of a village in Thailand.

disease in children in Asia (137), South America (82), and Africa (46). The major advantage of this approach as compared with testing isolates for enterotoxin production lies in the number of isolates that can be examined expeditiously before isolates lose plasmids coding for enterotoxin. Enterotoxin gene probes are also sensitive tools to identify ETEC in food and water (28,29,37) (Table IV).

A DNA probe for genes coding for ST II was used to search for ST II *E. coli* in 3127 pigs, 465 people, and 312 water specimens at pig farms (30). This would have been extremely difficult to perform by isolating *E. coli* from each specimen and testing culture supernatants of isolates in pig ileal loops. DNA probes have also been used to characterize ETEC. In surveys of animals and humans in Thailand, STI ETEC isolated from animals hybridized with the ST Ia probe (168) (Table V), while ST I ETEC from humans hybridize with either the ST Ia or ST Ib probes (137) (Table I). With a specific probe for genes coding for LT II, it was determined that although LT II ETEC are isolated from buffalo and cows, they are isolated infrequently from people (140). Enterotoxin gene probes have also been used to demonstrate that plasmids coding for CFA I and CFA II carry genes coding for ST Ib (33). Enterotoxin gene probes have also been used to identify ETEC containing genes coding for different types of enterotoxin (32) (Table VI).

TABLE IV

Comparison of Identification of Enterotoxigenic *Escherichia coli* by Testing 10 Lactose-Fermenting Colonies for Enterotoxin Production versus Probing Filtrates of 100 ml of Water with the Radiolabeled Enterotoxin Gene Probes

Number of LT I ST Ib ETEC per ml of canal water[a]	Testing 10 colonies for enterotoxin productive[b]	Hybridization of filtrates with enterotoxin gene probe[c]
10^8	+	+
10^7	+	+
10^6	+	+
10^5	+	+
10^4	−	+
10^3	−	+
10^2	−	+
10	−	−
0	−	−

[a] Canal water contained 5.4×10^5 *A. hydrophila*, 5×10^4 non-ETEC, 2×10^4 *Escherichia agglomerans*, and 1×10^2 *Klebsiella pneumoniae*.

[b] As identified with the Y-1 adrenal and suckling mouse assays.

[c] Enterotoxin gene probes are radiolabeled polynucleotide probes coding for LT I and ST Ib.

TABLE V

Enterotoxigenic *Escherichia coli* Identified in Pigs by DNA Hybridization in 87 Litters of Pigs with Diarrhea at 10 Farms in Thailand in 1981

Farm (Month)	Number of pigs	Number of litters with ETEC/number of litters examined	Number of litters with type of ETEC infection
A (Jan.)	12	3/3	3 ST Ia
B (Feb.)	5	2/2	2 ST Ia
C (Feb.)	25	4/4	4 ST Ia
D (Feb.)	24	2/3	2 ST Ia
E (March)	133	2/17	2 LT I ST Ia
F (April)	144	6/15	1 LT ST Ia, 2 LT I, 3 ST Ia
G (April)	103	2/7	2 ST Ia
B (April)	95	2/7	2 ST Ia
H (April)	64	1/8	1 ST Ia
I (June)	72	2/5	2 ST Ia
H (Aug.)	37	0/7	—
I (Sept.)	29	0/3	—
J (March)	50	2/5	1 LT I ST Ia, 1 ST Ia
I (Oct.)	10	0/1	—

2. *Oligonucleotide Probes*

The DNA sequences encoding LT I, ST Ia, and ST Ib have been determined and synthetic oligonucleotide probes have been constructed and used to identify ETEC (34,36,103). The DNA sequence of the 26-mer LT

TABLE VI

Examination of Enterotoxigenic *Escherichia coli*-Containing Genes Encoding Enterotoxins as Identified by Southern Blot Analysis

Source	Serotype	Plasmids hybridizing with enterotoxin gene probes
Human	O88:H^-	LT I, ST II
Human	O9ab:H19	ST Ia, ST II
Buffalo	O7:H4	LT I, ST II
Human	O?	LT I, ST II
Human	O78:H12	LT I, ST Ia, ST Ib
Human	O112ab:H9	LT I, ST Ia
Human	O128:H12	LT I, ST Ib
Human	O159:H^-	ST Ia, ST Ib

oligo probe spans the A1 and A2 region of the toxin A gene, is homologous with the LT-H gene, and differs from the LT-P gene by a single base. The 26-mer ST Ia oligo probe is homologous with the ST Ib DNA sequence and differs from the ST Ia sequence by two base changes. The DNA sequences of the oligonucleotide probes that have been used to identify ETEC are listed in Table VII. Using the 26-mer oligo probes to examine *E. coli* isolated from 2626 children in Thailand and the Philippines, 92% of ETEC infections, as identified by testing isolates in the Y-1 adrenal and suckling mouse assays (23,128), were identified with polynucleotide plasmid probes and 89% with the oligonucleotide probes (36).

In another comparative study of cloned polynucleotide and synthetic 30- and 34-mer oligonucleotide probes all 74 reference strains of ETEC were identified with both types of probes (152). Either cloned polynucleotide or synthetic oligonucleotide probes can be used to identify ETEC. Since most laboratories do not have the equipment to construct specific polynucleotide probes, oligonucleotide probes that are commercially available (Molecular Biosystems) are more practical. This will enable laboratories with only minimal equipment to use DNA hybridization assays to identify ETEC.

There are *E. coli* that contain DNA sequences that hybridize with polynucleotide and oligonucleotide probes that do not produce enterotoxin (36,113). For example, *E. coli* H 10407P, the CFA I-negative deriv-

TABLE VII

Synthetic Oligonucleotide Probes Used to Identify Enterotoxigenic *Escherichia coli*

LT I
  34-mer 5′ A CGT TCC GGA GGT CTT ATG CCC AGA GGG CAT ATT 3′
  26-mer 5′ C ACC TCT AAG TAG TTG TTG TTA ATG T 3′
  20-mer 5′ GCG AGA GCA ACA CAA ACC G G 3′
  ST Ia
    5′ TGA CGA CAC TTG AAA CAA CAT TAG G A 3′
ST Ib
  5′ TGA CGA CAC TTA ACA CAA CAT TAG G A 3′
ST Ia
  30-mer 5′ GAA CTT TGT AAT CCT GCC TGT GCT GGA TGT 3′
  26-mer 5′ CAA CAG TGA AAA AAA ATC AGA AAA T 3′
  22-mer 5′ GCT GTG AAC TTT GTT GTA ATC C 3′
ST Ib
  30-mer 5′ GAA TTG TGT AAT CCT GCT TGT ACC GGG TGC 3′
  26-mer 5′ A AGT AAT AAA AGT GGT CCT GAA AGC 3′
  22-mer 5′ GCT GTG AAT TGT GTT GTA ATC C 3′

ative of LT ST CFA I *E. coli* H 10407, contains the genes coding for ST Ia, but is negative in the suckling mouse assay (99). Ten percent (19/198) of *E. coli* from Thailand and the Philippines that hybridized with both the radiolabeled oligonucleotide and cloned polynucleotide probes did not produce enterotoxins (36). Single amino acid substitutions in the ST Ia molecule have been shown to decrease or eliminate ST production (111) and nitrosoguanidine-induced mutants of ETEC that produced decreased amounts of LT have been constructed (165).

3. *RNA Transcript Probes*

To develop a probe to identify genes encoding ST Ib that was not dependent on large-scale plasmid purification procedures, electrophoresis equipment, or the expense of obtaining oligonucleotides, the genes encoding ST Ib were cloned into pSP 64 and RNA transcripts were constructed to detect ST Ib ETEC by colony hybridization (14). Briefly, the 215-bp DNA fragment encoding ST Ib from pSLM 004 was ligated with pSP 64 and transformed into *E. coli* K12. *Eco*RI digests of plasmid DNA from these transformants were used as templates for RNA transcription. RNA transcripts hybridized with ST Ib ETEC, but not with non-ETEC and ETEC producing ST Ia or LT. Enterotoxin gene RNA transcript probes offer minimally equipped laboratories an alternative method of identifying ETEC by nucleotide hybridization.

4. *Nonradioactive Probes*

The greatest technical obstacle to the use of DNA probes is the detection of the target gene sequences. Radiolabeled DNA probes are inconvenient for most laboratories, because radioisotopes require special handling and because of the short half-life of ^{32}P (14 days). Alkaline phosphatase conjugated (121) or biotinylated (3) DNA probes have been developed as practical alternatives. Although nonradioactive probes are less sensitive than radiolabeled probes and require extensive digestion of colony blots with proteinase K and Triton X-100, biotinylated polyclonal plasmid DNA probes were shown to be comparable to α-^{32}P probes in the identification of ETEC (3,67).

A nonradioactive DNA probe was constructed by attaching alkaline phosphatase directly onto a DNA probe (121). Nonradioactive oligonucleotide probes were constructed by covalently linking alkaline phosphatase directly to the C-5 position of a thymidine base through a 12-atom spacer arm (61). The alkaline phosphatase-conjugated oligonucleotides were as sensitive as the α-^{32}P-labeled probes in identifying homologous genes encoding LT I and ST Ib, but were less sensitive in detecting genes coding ST Ia, which differed by 2 bp (138); see Fig. 3. These nonradioac-

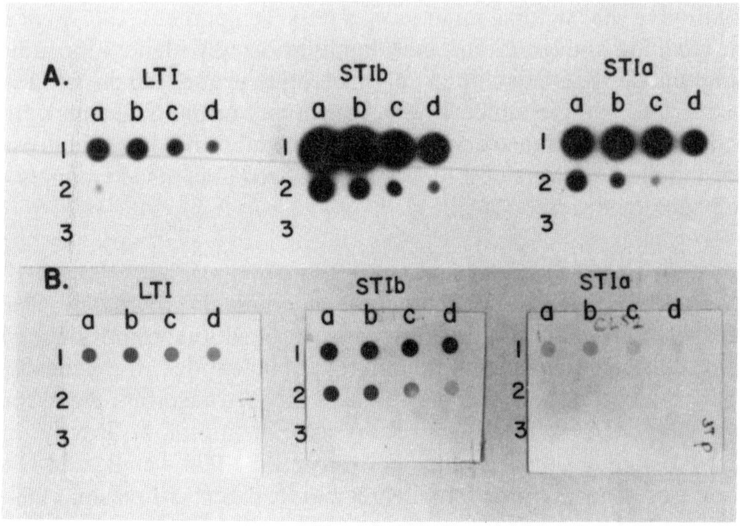

Fig. 3. Detection of decreasing concentrations of pEWD 299 (LT I), pSLM 004 (ST Ib), and pRIT 10036 (ST Ia) spotted on nitrocellulose paper: (A) hybridized with 10^5 cpm of radiolabeled oligonucleotide probes for 16 hr and exposed to X-Omat X-ray film for 16 hr at −70°C; (B) hybridized with alkaline phosphatase-conjugated oligonucleotide probes for 30 min and exposed to substrate for 4 hr. Concentrations of DNA: 1a, 50 ng; 1b, 25 ng; 1c, 12.5 ng; 1d, 6.25 ng; 2a, 312 ng; 2b, 1.56 ng; 2c, 0.78 ng; 2d, 0.39 ng; 3a, 0.19 ng; 3b, 0.095 ng; 3c, 0.047 ng; 3d, 0.023 ng.

tive probes identified a similar number of ETEC as radiolabeled probes and should facilitate the use of DNA probes in the identification of ETEC (110,113,138).

B. Enteroinvasive *Escherichia coli*

1. Polynucleotide Probes

Three different DNA probes have been used to identify EIEC (141, 144,157,169) (Table VIII), which are similar in sensitivity and specificity (174). Although EIEC identified with the 17-kb fragment of pRM 17 and confirmed in the Sereny test were infrequently isolated from Thai (38,159) and Chilean (82) children <2 years of age, the isolation rate of *Shigella* and EIEC increased in older Thai children (38). Of 410 children in Thailand, *Shigella* was isolated from 23% and EIEC from 4% of children with diarrhea (159). The isolation rate of both pathogens was highest in children

TABLE VIII

Polynucleotide DNA Probes Used to Identify *Shigella* and Enteroinvasive *Escherichia coli*[a]

1. 17-kb *Eco*RI fragment of pRM 17 (5,141)
2. 2.5-kb *Hin*dIII fragment of pSF 55 (144)
3. 1750-bp *Eco*RI fragment of pW 22 (169)

[a] Numbers in parentheses are references.

3–5 years of age from whom *Shigella* was isolated from 38% and EIEC from 9% (Fig. 4). *Shigella* was isolated from 52% and EIEC from 7% of 91 children with bloody diarrhea.

Enteroinvasive *E. coli* are found when *Shigella* isolation rates are high. In Thailand and Brazil EIEC strains were more often lactose-fermenting,

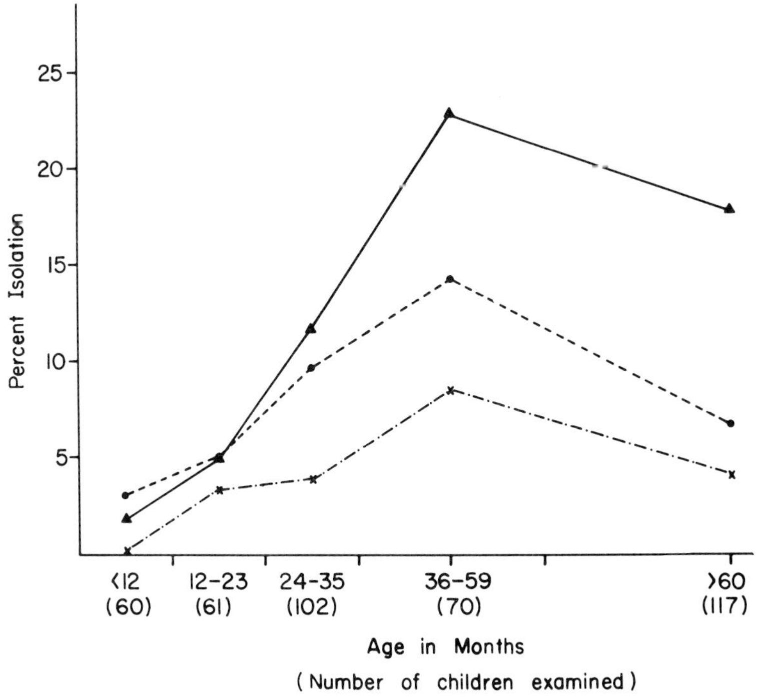

Fig. 4. Age-specific isolation rates of *Shigella* species (▲, *S. flexneri;* ●, *S. sonnei*) and enteroinvasive *Escherichia coli* (x) from 410 children with diarrhea in Children's Hospital, Bangkok, Thailand, January through June, 1985.

and most do not decarboxylate lysine (157,159,163). Of 58 EIEC isolated in Thailand, 10% of strains were of serotypes not previously associated with EIEC. It is possible to detect EIEC in large numbers of patients and to detect unrecognized bioserotypes with a specific probe. The percentage isolation rate of EIEC from different populations identified by DNA hybridization varied from 3% of children with diarrhea in Chile (82) to 2–9% of children with diarrhea in Thailand (38,157,159) depending on their age and clinical symptoms. DNA hybridization has permitted identification of EIEC in 2% of travelers with diarrhea in Nepal (160) and 5% of travelers with diarrhea in Mexico (171).

2. *Nonradioactive Probes*

A biotinylated DNA probe to identify *Shigella* and EIEC has been constructed by incorporating biotin-11-dUTP into the 17-kb *Eco*RI fragment of pRM 17 (142). In a case–control study of endemic diarrhea in Thai children the biotinylated DNA probe hybridized with the same 110 *E. coli* as the radiolabeled probe. Of the strains that hybridized with the DNA probes, 99% (109/110) were positive in the Sereny test. The one probe-positive Sereny test-negative isolate was of the same serotype (O164:H$^-$) as the two other Sereny test-positive *E. coli* isolated from the same patient.

The results of hybridization of stool blots from patients with diarrhea and controls are shown in Table IX. Examining stool blots with the radio-

TABLE IX

Detection of *Shigella* and Enteroinvasive *Escherichia coli* by Examining Stool Blots with Radiolabeled and Biotinylated 17-kb *Eco*RI Fragment of pMR 17

Condition of children	Number infected	Probe: Radiolabeled	Probe: Biotinylated	p-value[a]
With diarrhea				
Shigella	155	94	59	<.001
EIEC	19	13	10	NS[c]
None[b]	1056	14	7	<.025
Without diarrhea				
Shigella	3	0	0	NS
EIEC	6	4	2	NS
None[b]	1221	4	0	NS

[a] As determined by the McNemar test.
[b] No *Shigella* or EIEC identified.
[c] Not significant.

labeled probe identified more *Shigella* infections than the less sensitive biotinylated probe, although the difference between radiolabeled and biotinylated probes in identifying EIEC by examining colony blots was similar. Since only 68% of EIEC and 61% of *Shigella* infections were identified in children with diarrhea by probing stool blots with the radiolabeled probe, this procedure, as currently performed, should not be used to identify *Shigella* and EIEC infections in clinical specimens.

C. Enteroadherent *Escherichia coli*

1. Polynucleotide Probes

Between 91 and 100% of *E. coli* that adhere to HEp-2 (or HeLa) cells in a localized pattern hybridized with the EAF probe, and isolates that do not exhibit localized adherence rarely hybridized with the EAF probe (13,35,82,104,105). In Peru 83% of EAF *E. coli* strains were EPEC serogroups O111 and O119 (105). In Brazil localized adherence (LA) was observed much more frequently in *E. coli* considered to be EPEC serotypes (93%) than in other serotypes (14%) (134). The serotypes of LA *E. coli* isolated in Brazil included O55:H6, O86:H34, O111ab:H$^-$, O111ab:H2, O119:H$^-$, O119:H6, and O127:H$^-$. In a study of diarrheal disease in children in Chile, *E. coli* exhibiting LA to HEp-2 cells was found in 28% of 154 children with diarrhea and 11% of 66 children without diarrhea ($p = .005$) (82). In Mexico LA *E. coli* was isolated from 13% of 154 children with diarrhea and 0.7% of 137 controls ($p < .001$) (87).

In the same population in Chile, *E. coli* that adhered in an aggregative pattern was found in 36% of 154 patients and 23% of 66 controls of ($p <$ .03); with one exception (O148:H28), the serotypes of aggregative *E. coli* strains were not characteristic of other diarrheogenic *E. coli* (170). Preliminary results with a candidate DNA probe for aggregative *E. coli* showed high specificity and sufficient sensitivity to detect 49% of aggregative strains and suggested some heterogeneity between strains.

A DNA probe to identify *E. coli* that adhere in a diffuse pattern was moderately sensitive (75%) and highly specific (96%) (82). The probe correctly detected 33 of 44 strains that adhered to HEp-2 cells in a DA pattern. Although the probe used to identify *E. coli* that adhered to HEp-2 cells in a DA pattern did not hybridize with LA *E. coli,* 8 of 110 aggregative and 11 of 262 nonadherent *E. coli* hybridized with the diffuse-adherence probe. *Escherichia coli* that adhered to HEp-2 cells in a diffuse pattern or hybridized with the diffuse-adherent probe were not associated with diarrhea in Chile (82); however, in Mexico diffuse-adherent *E. coli* as identified in a HEp-2 adherence assay, was isolated from 21% of patients and 7% of controls ($p < .001$) (87). This discrepancy may be due to the assay employed in identifying diffusely adherent *E. coli;* isolates considered diffusely adherent in Mexico might have been considered aggregatively adherent in the assay used to examine strains from Chile. Development of probes for genes coding for different types of adherence would be a more specific way to identify isolates that adhere in certain patterns and would help define the enteropathogenicity of these adherent isolates.

Strains having the combination of EPEC serogroups, LA, and hybridization with the EAF probe were found in 19% of 154 children with diarrhea but in only 1.5% of 66 controls ($p < .001$) (82). *Escherichia coli* strains of non-EPEC serotypes that hybridized with the EAF probe were recovered with similar frequency from patients and controls. In studies of diarrheal disease in Thailand, *E. coli* strains that hybridized with the EAF probe and adhered to HeLa cells in a localized pattern were isolated from 7% of 272 children with diarrhea and 3% of controls <6 months of age and was the most common pathogen isolated from hospitalized children (38). Only half of the EAF *E. coli* belonged to EPEC serotypes (Table X). The illness was similar in patients with EAF *E. coli* regardless of serotype. *Escherichia coli* that hybridized with the EAF probe of EPEC and non-EPEC serotypes isolated in Thailand were positive in the immunofluorescent assay to measure effacement (70) and were subsequently shown to cause effacing lesions in human enterocytes.

Although LA *E. coli* have been categorized as EPEC, EAF probe-positive (and LA *E. coli*) are not always of EPEC serotypes. The use of

TABLE X

Serotypes of *Escherichia coli* That Hybridized with the EPEC Adhesin Factor Probe Isolated from Children <5 Years of Age in Thailand[a]

O119:H6 (16)[b]	O86:H$^-$ (1)
O127:H12 (12)[b]	O119:H2 (1)
O2:H21 (5)	O142:H$^-$ (1)
O142:H6 (3)[b]	O15:H18 (1)
O157:H19 (3)	O157:H45 (1)
O123:H21 (3)	O162:H? (1)
O53:H3 (2)	O78:H24 (1)
O51:H15 (2)	O51:H3 (1)
O86:H19 (1)	O?:H1, 12 (1)

[a] Numbers in parentheses are references.
[b] EPEC serotypes.

specific DNA probes and the tissue culture adherence assays have identified enteroadherence as an diarrheogenic determinant of *E. coli.*

2. *Nonradioactive Probes*

A biotinylated DNA probe to identify EAF *E. coli* was as sensitive and as specific in identifying *E. coli* that adhered to tissue culture cells in a localized pattern as a radiolabeled probe (13). Once nonradioactive probes are developed, different patterns of enteroadherent *E. coli* can be identified in minimally equipped microbiology laboratories. Effacement also appears to be a pathogenic mechanism of diarrheogenic *E. coli* of EPEC and non-EPEC serotypes. A DNA probe to identify these effacing *E. coli* could be useful to define the enteropathogenicity of these strains.

D. Enterohemorrhagic *Escherichia coli*

1. *Polynucleotide Probes*

Enterohemorrhagic *E. coli* strains were identified with DNA probes in 39% of patients with hemorrhagic colitis in Great Britain (145). In 12 of 32 samples, <10% of the coliforms examined hybridized with DNA probes for genes coding for VT (SLT). Specific probes have been used to identify genes coding for SLT (139) and VT (146) in *E. coli* and *Shigella*. Specific DNA probes were used to screen for SLT I- and II-producing *E. coli* in children <5 years old with bloody diarrhea, in infants <6 months of age, and in age-matched controls without diarrhea in Thailand (7). *Escherichia coli* producing SLT was identified in 4 of 54 children with bloody diarrhea in whom other enteropathogens were not identified, and in 3 of 50 controls. When SLT *E. coli* strains were identified, these bacteria composed 0.3–4.0% of the 300 coliforms screened. In 115 infants with diarrhea and 119 infants without diarrhea, SLT *E. coli* was identified. However, in one child the "EHEC" probe hybridized with *E. coli* O76:H7 lacking the genes for SLT I or SLT II. SLT I *E. coli* was found in one clone with diarrhea and one control without diarrhea in a longitudinal study to identify enteropathogenic determinants in serotyped *E. coli* isolated from infants with diarrhea and age-matched controls in Thailand. In this study 300 colonies per patient from 500 children <6 months of age and 500 age-matched controls were examined with the SLT probes; less than 10% of the 300 colonies examined per child hybridized with the SLT I probe. Enterohemorrhagic *E. coli* of serogroups other than O157 have been isolated from 5% of adults with diarrhea and 4% of controls without diarrhea. In contrast to the low frequency of EHEC in patients with diarrhea, these organisms have been identified with SLT probes in livestock and foods of animal origin in local markets in Bangkok.

In Thailand 35% (14/40) of SLT-producing *E. coli* from humans and 45% (129/285) of SLT-producing *E. coli* from animals hybridized with the "EHEC" probe. All of these SLT-producing *E. coli* were of non-O157 serogroups. Only 48% (14/29) of the *E. coli* from humans and 60% (129/196) of *E. coli* from animals that hybridized with the "EHEC" probe hybridized with the SLT probes. Before genes coding for SLT were cloned, the "EHEC" probe was used to search for SLT-producing *E. coli* in children in Chile (81). DNA probes that are more specific have now been constructed to identify SLT-producing *E. coli* (109). The VT1 and VT2 DNA probes used to identify EHEC (172,173) identified the same Thai isolates as the probes for SLT I and SLT II.

2. *Oligonucleotide Probes*

Synthetic oligonucleotide probes constructed from the A and B subunits of SLT I and SLT II have been constructed (7a). These oligonucleotide probes have been used at different degrees of stringency to identify SLT I, SLT II, and pig edema variants of SLT II-producing *E. coli*. Using these oligonucleotide probes, *E. coli* producing SLT II variants were identified in humans and animals with diarrhea. Using DNA hybridization with SLT oligonucleotides is considerably easier than performing neutralization assays of culture filtrates with SLT I and SLT II antisera.

IV. CONCLUSIONS

DNA probes for ETEC, EIEC, enteroadherent *E. coli,* and EHEC have been used to identify diarrheogenic *E. coli* worldwide. DNA probes have been used to search for ETEC in large numbers of patients and to identify potential environmental sources of infection. Probes for genes coding for different enterotoxins have been used to characterize *E. coli* that produce phenotypically similar enterotoxins and have subsequently been useful in defining the epidemiology of these infections. DNA probes to identify EIEC have been used to identify these infrequently identified pathogens in children and travelers with diarrhea and to appreciate further the serological and biochemical differences exhibited by these organisms. The DNA probe for EAF *E. coli,* originally identified in EPEC, has lead to a greater appreciation of the importance of enteroadherent *E. coli* of both EPEC and non-EPEC serotypes as a cause of diarrhea in young children and travelers. The SLT DNA probes enable investigators to determine the importance of *E. coli* containing genes coding for a defined toxin, and oligonucleotides constructed from the nucleotide sequences of these cloned genes have been used as probes to differentiate toxin types and variants. Identifying genotype, rather than phenotype, has been an important tool in identifying and characterizing diarrheogenic *E. coli* and has been the model on which other bacterial pathogens have been identified by DNA hybridization. Nonradioactive markers for these probes should increase their use for the clinical diagnostic laboratory.

V. GENE PROBES VERSUS IMMUNOLOGICAL ASSAYS

Immunological assays using either polyclonal or monoclonal antibodies have been developed and proposed as another method of identifying infectious agents. These immunological assays are sensitive and specific, and with diarrheogenic *E. coli* are directed to the same protein antigens encoded by the DNA sequences identified with specific nucleotide probes. Both immunological assays and DNA probes are similar in sensitivity and specificity in identifying diarrheogenic *E. coli* isolates, and both are less sensitive in identifying *E. coli* infections directly in stools (95). Streptavidin has been an important addition to immunoassays because it increases the signal generated by the reporting antibody. Immunoassays, however, detect only those antigens that are produced or survive enzymatic degradation (which in stool can be considerable). On the other hand, bacterial wall remnants interfere with access of probe to target cell DNA.

Nucleotide probes, however, detect nucleotide sequences within bacteria that encode for these enterovirulent determinants. This method is not dependent on the media and growth conditions of the bacteria and is therefore much more sensitive and specific. A number of bacterial species have been reported to produce diarrheogenic determinants, as identified with polyclonal or monoclonal antibodies, which do not hybridize with DNA probes used to identify these determinants in *E. coli* (e.g., SLT in *Shigella* and vibrios; LT in *Aeromonas* and *Salmonella*). Immunological probes are also used in identifying exoproteins encoded by genes that are too homologous in nucleotide sequence to be differentiated with specific nucleotide probes.

A. Enterotoxigenic *Escherichia coli*

Immunoassays that do not require tissue culture facilities or *in vivo* animal assays have been used to identify LT-producing *E. coli*. GM_1

gangliosides (155) or burro anticholera toxin (LT) (175) have been used to capture the LT molecule, which is subsequently detected with polyclonal or monoclonal antibody to CT. An ELISA or radioimmunoassay (RIA) was as sensitive (10 pg) as tissue culture assays in detecting LT (50,175). A modified Elek plate system (the Biken test) (57), latex agglutination (44), staphylococcal coagglutination (6), and passive immune hemolysis assays (12) are other immunological assays that have been used to identify LT-producing ETEC. In a collaborative evaluation of different methods of detecting LT ETEC conducted by the World Health Organization, the Biken test, the GM_1 ELISA, the Y-1 adrenal, the Chinese hamster ovary tissue culture cell assays, and the LT DNA probe were equally sensitive and specific in identifying LT ETEC (154). The choice between these assays is dependent on the laboratory facilities, the number of specimens to be examined, the frequency of testing organisms, and the availability of reagents.

In addition to diagnostic kits that use polyclonal antibodies to identify LT ETEC, monoclonal and polyclonal antibodies have been produced to characterize different epitopes of LT (56). Antibodies have been used to differentiate between LT-H and LT-P in immunodiffusion assays (58). Since these LT differ by only one of two nucleotide substitutions, their nucleotide sequences cannot be differentiated with polynucleotide and existing oligonucleotide probes. Antibodies to two different epitopes of LT II have also been used to characterize LT II (53). Gene sequences of different epitopes of LT II cannot be differentiated by DNA hybridization

Both ELISA and RIA have been constructed to identify ST I (123), and an ELISA has been developed to identify ST II ETEC (55). Since ST is nonantigenic, antibodies to ST were produced by immunizing rabbits with ST conjugated with human thyroglobulin or bovine serum albumin (BSA). An indirect ELISA technique was used to identify ST I and ST II.

The tissue culture, suckling mouse, and immunological assays detect enterotoxin in culture supernatants. Enterotoxin production is dependent on the composition of the culture media and growth conditions of the isolate (1). Identifying nucleotide sequences coding for enterotoxin are not.

B. Enteroinvasive *Escherichia coli*

Enteroinvasive *E. coli* can be identified with a number of DNA probes that are equally sensitive and specific in identifying EIEC and *Shigella* (48). Sereny test-positive *Shigella* and EIEC can develop deletions in the 120- to 140-MDa plasmid and become Sereny test-negative, but remain probe positive. This is rare with fresh specimens, however.

The VMA ELISA, which detects an outer-membrane protein antigen involved in enteroinvasion, identifies bacteria that produce keratoconjunctivitis in the Sereny test. Expression of these outer-membrane antigens is, however, also dependent on media composition and growth conditions (116). In a comparative study of the VMA ELISA and the 17-kb DNA probe to identify EIEC, there was complete agreement (157). Others have mentioned lack of correlation between the results of the VMA ELISA and the Sereny test, presumably because they did not grow isolates in the indicated media under the appropriate conditions.

C. Enteroadherent *Escherichia coli*

Although EPEC serogroups can be identified with polyclonal antisera by slide agglutination, no immunological assay has been developed to identify localized, diffuse, or aggregative adherent *E. coli*. Both ELISA and slide agglutination methods with polyclonal antibodies have been compared to DNA probes in the identification of K88 and K99 ETEC (75). Both methods were comparable in the identification of K99 ETEC. *Escherichia coli* strains were found that contained genes coding for K88 that were expressed poorly. Agreement was improved when strains were grown on blood agar.

VI. PROSPECTS FOR THE FUTURE

All of the DNA hybridization experiments discussed require incubating specimens on media to increase the number of copies of DNA to be detected with nucleotide probes. Examining the total bacterial mass of stools cultured on MacConkey agar (stool blots) identified fewer infections than testing 300 individual colonies, presumably because of the interference of bacterial cell wall remnants with the access of probe to target cell DNA. A procedure for amplifying specific DNA sequences within a sample by using a pair of primers and DNA polymerase has been developed (130). In this procedure specific primers are added that bind on either end of the nucleotide sequence to be amplified. On addition of DNA polymerase the target sequence is replicated first in one direction from one primer and then in the other direction from the opposite primer making multiple copies of the DNA sequence. This procedure has been used to increase the amount of nucleotide sequences coding for hepatitis B in sera (167) and has been used to identify LT ETEC in stools (112). Since the genes coding for LT, ST, SLT I, and SLT II have been sequenced, it should be possible to construct primers for DNA sequences encoding for these EPEC determinants. By extracting DNA from stool, augmenting the DNA sequence of interest with specific primers and DNA polymerase, it should be possible to increase the amount of target cell DNA of intestinal pathogens to quantities readily detectable with nonradioactive nucleotide probes.

This method will afford a method of examining stools directly for diarrheogenic *E. coli*. In developing countries this procedure may result in the identification of multiple infections in a single patient. As we have increased the sensitivities of our diagnostic microbiological methods in the study of enteric pathogens in Thailand, we have found that 20–30% of patients with diarrhea are infected with multiple enteric pathogens (38,157). It will be of interest to determine the clinical outcome of these multiple enteric infections.

VII. SUMMARY

Nucleotide probes have been used to identify diarrheogenic *E. coli*—ETEC, EIEC, enteroadherent *E. coli,* and EHEC—isolated during studies of diarrheal disease. By identifying genes coding for specific enteropathogenic determinants, it has been possible to define the epidemiology of the *E. coli* infections. This method has been used to determine the relative importance of *E. coli* with different enteropathogenic determinants.

As currently employed, DNA probes for *E. coli* diarrheogenic determinants have been used as epidemiological tools and to identify the enteropathogenic determinants of different isolates. Since treatment of most *E. coli* diarrhea is supportive, the decision to treat a specific patient with antibiotics is dependent more on the clinical status of the patient than on identifying the etiology of his or her diarrhea. We have not therefore tried to use nucleotide probes as a rapid way to identify diarrheogenic *E. coli*. Nucleotide probes have been proposed as a rapid means of identifying infectious agents and have been useful in investigating outbreaks. Once the *in vitro* biological assays have been established, examining a small number of isolates (<200) for hybridization with specific probes is no faster than performing biological assays.

These biological assays are expensive and labor-intensive; moreover, they require tissue culture capabilities and maintenance of large numbers of laboratory animals. Examining specimens with DNA probes would be more suitable for clinical laboratories. Because radioisotopes are difficult to obtain and use safely, nonradioactive probes have been developed that are sensitive and specific in examining *E. coli* isolates. Methods of augmenting target cell DNA in stool are being developed that will make it possible to identify *E. coli* infections directly by probing stools with nonradioactive probes.

VIII. MATERIALS AND METHODS

A. Polynucleotide Probes

Polynucleotide probes are endonuclease digestion fragments of plasmids containing cloned enterovirulent determinants.

1. Storage and Selection of Clones

For long-term storage *E. coli* K12-containing plasmids with cloned enterovirulent determinants are stored lyophilized or frozen in skimmed milk at −70°C. For short-term storage, isolates are kept on Dorset egg yolk slants. Careful storage of isolates is important for consistent results.

Escherichia coli K12 containing the cloned enterovirulent determinants are inoculated onto Mueller–Hinton agar; an antibiotic disk is placed on the agars and agar and disk are incubated at 37°C overnight. This is to select for isolates that contain the plasmid with the cloned determinant. (See tabulation below).

Determinant	Plasmid	Disk[a]	Broth (μg/ml)
LT I	pEWD 299	Ap	50
LT II	pCP 2725	Tc	10
ST Ia	pRIT 10036	Tc	10
	or		
	pDAS 101	Ap	50
ST Ib	pSLM 004	Tc	10
	or		
	pDAS 100	Ap	50
ST II	pCHL 6	Tc	10
DA	pW 22	Ap	50
EAF	pMAR 22	Ap	50
EHEC	pCVD 419	Ap	50
SLT I	pJN 37-19	Ap	50
SLT II	pNN 110-18	Ap	50
VT 1	p 746	Km	30
VT 2	p 363	Tc	10
EIEC	pRM 17	Tc	10

[a] Ap, Ampicillin (10 μg); Tc, tetracycline (30 μg); Km, kanamycin (30 μg).

2. *Plasmid Purification*

Method I

a. Colonies that grow near the antibiotic disk are inoculated into tubes of Mueller–Hinton broth containing the appropriate antibiotic concentration (see above tabulation) and incubated at 37°C overnight. One milliliter of the overnight growth is inoculated in 1 liter of M9 broth in 2-liter Erlenmeyer flasks (see Appendix, Section VII,C). *Escherichia coli* K12 pW 22 (DA) is grown in LB broth. Flasks are incubated at 200 rpm at 37°C until the optical density reaches 0.3. Chloramphenicol dissolved in ethanol is then added (final concentration 300 μg/ml), the shaking is reduced to 50 rpm, and incubation continued for another 16 hr. For each plasmid purification, 10 liters of bacteria are routinely prepared.
b. Centrifuge at 8000 *g* for 15 min at 4°C (6000 rpm in a GSA head of a Sorvall centrifuge).
c. Wash bacteria in TES.
d. Resuspend in 8 ml of 2.5% sucrose (Sigma S 9378) in TE.
e. Add 1 ml of 1% lysozyme in TE.
f. Add 2 ml 0.5 *M* EDTA pH 8 (Sigma, E 5134), mix with a glass rod, and incubate for 5 min at 25°C.
g. Add 13 ml of Triton lytic mix (see Section VIII,C, Appendix) slowly with gentle stirring and inversion until lysis is complete.
h. Centrifuge at 20,000 *g* (17,000 rpm in a GSA head) for 40 min at 4°C.
i. Add 25 g CsCl and mix gently until in solution.
j. Adjust density to 1.58–1.60 with additional CsCl or TE. This is done by weighing 10 ml of the suspension.
k. Transfer to VTi 50 or Ti 60 tubes.
l. Add 1 ml of ethidium bromide (Sigma, E 8751) (10 mg/ml in TE saturated with CsCl) and mix by inversion.
m. Centrifuge at 40 K for 40 hr (60 Ti) or 50 K for 12 hr (VTi 50).
n. Vent the tube (with sealed tubes with an 18-gauge needle) and collect the lower band with a syringe.
o. Extract ethidium bromide with isopropanol saturated with 5 *M* NaCl in TE four times (one time after the red color is gone).

Method II

Used routinely with pRM 17 (EIEC).

a. Inoculate 5 ml of starting culture into 500 ml of LB broth in 1-liter Erlenmeyer flasks, and incubate at 200 rpm at 37°C until the cell density reaches 5×10^8 cells/ml. Chloramphenicol dissolved in ethanol is then added (final concentration 150 μg/ml). The shaking is reduced to 50 rpm and incubation continued for another 16 hr. The volume of bacteria routinely prepared for each plasmid purification is 1 liter.

b. Centrifuge at 8000 *g* for 15 min at 4°C (6000 rpm in a GSA head of a Sorvall centrifuge).
c. Wash bacteria in TES.
d. Resuspend in 40 ml of solution I (see Section VIII,C, Appendix) and incubate for 40 min at 0°C.
e. Add 80 ml of solution II (see Section VIII,C, Appendix), invert gently, and incubate at 0°C for 5 min.
f. Add 60 ml of solution III (see Section VIII,C, Appendix), invert gently, and incubate for >1 hr at 0°C.
g. Centrifuge at 14,000 *g* for 15 min at 4°C and retain supernatant.
h. Add two volumes of 95% ethanol and incubate 1 hr at −70°C to precipitate the plasmid DNA.
i. Centrifuge at 14,000 *g* for 20 min at 4°C, dissolve the pellet in 27 ml of TES, and follow steps i–o in Method I. From a 500-ml culture, 300 μg of plasmid DNA are obtained.

3. *Isolation of DNA Fragment*

a. Digest 1 μg of plasmid DNA with 1 unit of the indicated endonuclease in the appropriate buffer at the indicated temperature (usually 37°C) for 1 hr, and then heat at 65°C for 10 min (see Section VIII,C, Appendix).
b. Separate the digested DNA by electrophoresis on the percentage polyacrylamide or low-melting agarose gel indicated in Tris-borate buffer at 5–10 V/cm of gel for 2 hr.

DNA fragments separated by electrophoresis are cut out of the gel with a razor blade and placed in dialysis tubing. Both ends of the tubing are closed with plastic clips and placed in an electroelution chamber (see Section VIII,C, Appendix) filled with Tris-borate buffer and electroeluted at 100 V for 1 hr. Alternately the DNA fragments are cut out of a 1% low-melting agarose gel (FMC) and heated at 65°C for 20 min to melt the agarose. Two volumes of 20 m*M* Tris pH 8–1 m*M* EDTA are added, extracted twice with phenol–chloroform, and the DNA is precipitated from the aqueous phase with two volumes of absolute ethanol in the presence of 0.3 *M* sodium acetate.

The electroeluted DNA within the dialysis bag is then ethanol-precipitated in dry ice–ethanol for 5 min. Further purification is not necessary.

Determinant	Plasmid	Endonuclease	Gel[a]	Fragment (bp)
LT I	pEWD 299	*Hinc*II	5% (P)	850
LI II	pCP 2725	*Hind*III/*Pst*I	5% (P)	800
ST Ia	pRIT 10036	*Hinf*I	10% (P)	154
	or	or		
	pDAS 101	*Eco*RI/*Bam*HI	10% (P)	157
ST Ib	pSLM 004[b]	*Hind*III/*EcoRI* and *Hpa*II	10% (P)	215
	or	or		
	pDAS 100	*Bam*HI/*Pst*I	10% (P)	215
ST II	pCHL 6[c]	*Hind*III and *Hinf*I	10% (P)	460
EHEC	pCVD 419	*Hind*III	1% (A)	3.4 (kb)
SLT I	pJN 37-19	*Bam*HI	5% (P)	1154
SLT II	pNN 110-8	*Pst*I/*Sma*I	5% (P)	842
VT 1	p 746	*Hinc*II	1% (A)	750
VT 2	p 363	*Ava*I/*Pst*I	1% (A)	850
EAF	pMAR 22	*Bam*HI/*Sal*I	5% (P)	1000
DA	pW 22	*Pst*I	5% (P)	450
EIEC	pRM 17	*Eco*RI	1% (A)	17 (kb)

[a] (P), Polyacrylamide gel; (A), 1% low-melting agarose gel.

[b] pSLM 004 is initially digested with *Hind*III/EcoRI, generating two fragments. The lower fragment is digested with *Hpa*II and the 215-base fragment is used as the ST Ib probe.

[c] pCHL 6 is initially digested with *Hind*III generating two fragments. The lower fragment is digested with *Hinf*I and the 460-bp fragment is used as the ST II probe.

4. Radiolabeling

Long digestion fragments (>500 bp) are nick-translated to incorporate [α-^{32}P]nucleotides. The reaction is performed in a 1.5-ml Eppendorf tube and the reaction is given for 25 μl.

a. 20 μl of double-distilled H_2O)–volume of DNA (usually 5–7 μl).
b. 2.5 μl of NT buffer (see Section VIII,C, Appendix).
c. 2.5 μl of 0.02 m*M* of each dNTP–dNTP containing α-^{32}P (usually [α-^{32}P]dCTP).
d. 0.5 μg of DNA in 5–7 μl (see Step a).
e. 0.5 μl DNase I (see Section VIII,C, Appendix).
f. Transfer above mixture to 50 μCi [α-^{32}P]dNTP (Sp act. $\geq$ 800 Ci/mmol) that has been dried in a 1.5-ml ethanol-washed Eppendorf. [α-^{32}P]Deoxynucleotides supplied in 50% ethanol are dried by covering the tube with parafilm, punching three or four holes in the parafilm, and drying under vacuum at room temperature (10 min).
g. 0.1 μl of 2 mg/ml of *E. coli* DNA polymerase I.

h. Incubate at 14°C for 1.5 hr.
i. Add 5 μl of 0.02 *M* Na_2EDTA.
j. To separate α-^{32}P-labeled probe from unincorporated nucleotides, pass the reaction through a G-50 column (5 Prime 3 Prime Inc.).

End-labeling of short (<500 bp) polynucleotide probes and oligo probes:

a. Mix: 5′-OH DNA* 50 pmol of 5′ ends (0.4 μg)
10× Kinase buffer, 2.5 μl
[γ-^{32}P]ATP (sp. act. ≥3000 Ci/mmol), 15 μl
T4 Polynucleotide kinase, 10 units
Distilled H_2O to 25 μl
b. Incubate at 37°C for 30 min.
c. Add 2 μl 0.5 *M* EDTA (pH 8.0).
d. Separate labeled DNA from unincorporated dAT γ-^{32}P by absorption and washing with 2× SSC on a NACS–Prepac column (see Section VIII,C, Appendix).

*One mole of dephosphorylated (OH) ends = the molecular weight of a single strand in grams. For example, a 22-bp-long single-stranded probe has a molecular weight of ~22 bp × 330 g/mole/bp = 7260 g/mol; 7260 pg/pmol = 7.26 ng/pmol; 50 pmol = 50 pmol × 2.26 ng/pmol = 363 ng = 0.36 μg.

5. *Fixation of Specimens for Hybridization*

Nitrocellulose filters

a. Place 0.45 μg nitrocellulose (NC) paper (Schleicher and Schuell, BA-85) in 4 × 5 cm pieces cut from a roll of NC paper on MacConkey agar.
b. Isolates or stools are inoculated onto these filters and incubated at 37°C overnight (12 hr).
c. Filters are placed on Whatman no. 3 paper saturated with 0.5 *N* NaOH for 10 min.
d. Filters are transferred to Whatman paper saturated with 1.0 *M* ammonium acetate and 0.02 *N* NaOH for 1 min.
e. Repeat four times; after the last transfer the filters are kept on the saturated Whatman paper for 10 min.
f. Air-dry and bake overnight at 65°C or 2 hr at 85°C in a vacuum oven.

Whatman 541 filters

a. Stools or isolates are inoculated directly onto MacConkey agar and incubated at 37°C overnight (12 hr).
b. Whatman 541 filters (7 cm diameter) are pressed evenly over this bacterial growth.

c. Place 541 filter in a glass petri dish on Whatman no. 3 paper saturated with 0.5 *N* NaOH and 1.5 *M* NaCl
d. Steam for 3 min in an autoclave *or* over a pan positioned over a beaker of boiling water *or* in a microwave oven for 30 sec.
e. Immerse 541 filter in 1 *M* Tris–2 *M* NaCl pH 7 for 4 min and air-dry.

6. *Hybridization with Radiolabeled Polynucleotide Probes*

a. Filters are incubated at 37°C for 3 hr in 10× Denhardt's solution, 4× SET, 0.5% SDS, and 10 μg/μl heat-denatured calf thymus DNA (see Section VIII,C,Appendix).
b. Filters are transferred to 50% formamide, 2× Denhardt's, 4× SET, 0.5% SDS, 10 μg/ml heat-denatured calf thymus DNA, 6% polyethylene glycol 8000, 500 μg/ml heparin, and 10^5–10^6 cpm/ml DNA probe. Hybridization is routinely done for 24 hr at 37°C.
c. Wash in 1× SSC and 0.1% SDS for 10 min at room temperature, and then in 1× SSC, 0.1% SDS for 45 min at 65°C; 0.1× SSC, 0.1% SDS for 45 min at 65°C; rinse twice in 2× SSC at room temperature, and air-dry.
d. Filters are exposed to X-Omat X-ray film (Eastman Kodak) with a Cronex Lightning Plus Screen (DuPont) for 24 hr at −80°C. The X-ray films are developed according to the manufacture's instructions (Eastman Kodak).

7. *Biotinylated Polynucleotide Probe*

Labeling with biotinylated nucleotides is performed by substituting biotin-11 dUTP [Bethesda Research Laboratories (BRL) 9507 SA] for radioactive nucleotides in a nick-translation reaction. To prepare 541 filters for hybridization with biotinylated probes, the filters—prepared as for examination with radiolabeled probes—are immersed in 0.01 *M* Tris (pH 7.8)–0.005 *M* EDTA–0.5% SDS for 10 min at 25°C, transferred to the same prewarmed solution containing 1 mg/ml proteinase K (BRL 5530 UA), and incubated at 50°C for 1 hr. Filters are then washed in the same buffer without proteinase K, washed once in 2× SSC, and air-dried. For hybridization with biotinylated probes, the digested filters are presoaked in 4× SET for 10 min at 25°C and prehybridized in 10× Denhardt's solution–4× SET–0.5% SDS–10 μg/ml heat-denatured calf thymus DNA at 42°C for 2 hr. Hybridization is performed at 42°C in 45% formamide–2× Denhardt's solution–4× SET–0.5% SDS–10 μg/ml heat-denatured calf thymus DNA–50 ng/ml heat-denatured biotinylated DNA probe for 24 hr. Filters are washed twice in 2× SSC–0.1% SDS for 5 min at 25°C, washed twice in 0.2× SSC–0.1% SDS for 5 min each at 25°C, then washed twice in 0.16× SSC–0.1% SDS for 15 min at 50°C, and rinsed in 2× SSC–0.1% SDS at 25°C. After washing, detection of biotinylated DNA probe–target hybrids with biotinylated alkaline phosphatase is carried out according to the manufacturers' instructions (DNA Detection Systems, BRL).

B. Oligonucleotide Probes

1. Construction of Oligonucleotides

The sequences used in the construction of synthetic oligonucleotides are shown in Table VII. The oligonucleotides are constructed by phospharamidite chemistry on a DNA synthesizer. These oligonucleotides are end-labeled with γ-^{32}P by T4 polynucleotide kinase (see above), and passed through a G-25 column (5 Prime 3 Prime).

2. Hybridization with Radiolabeled Oligo Probes

Colonies are spotted and processed on NC paper as previously described. Filters are prehybridized for 2 hr at 50°C in 6× SSC–5× Denhardt's–1 m*M* EDTA–0.2 μg/ml heat-denatured calf thymus DNA. Hybridization is carried out at 50°C overnight in the same buffer plus 10^6 cpm/ml of radiolabeled oligo probe. After hybridization the filters are washed twice in 6× SSC at 25°C for 10 min and then three times in 6× SSC for 30 min at 50°C. The filters are then air-dried and exposed to X-ray film with a Cronex Lightning Plus Screen (DuPont) at −70°C for 24 hr.

3. Hybridization with Alkaline Phosphatase-Conjugated Oligo Probes

The 26-base LT and ST oligonucleotide probes are synthesized using phosphoramidite chemistry on a DNA synthesizer (Applied Biosystems). Alkaline phosphatase is convalently linked directly to the C-5 position of a thymidine base through a 12-atom spacer arm (Molecular Biosystems). The alkaline phosphatase-conjugated oligo probes are commercially available.

For hybridization with alkaline phosphatase-conjugated oligonucleotide probes, bacteria fixed on NC filters are placed in 1× SSC with 2 mg/ml proteinase K at 50°C for 15 min and then washed in 1× SSC at 25°C. Filters are prehybridized in 5× SSC–0.5% BSA–0.5% polyvinylpyrrolidone–1% SDS for 30 min at 50°C. Filters are hybridized with 50 ng of alkaline phosphatase-conjugated probes per milliliter for 15 min at 60°C with the ST and for 30 min at 50°C with the LT oligo probe. This amount of probe was sufficient to hyridize 100 cm^2 of NC filter. Filters were washed twice for 5 min each in 1× SSC–1% SDS at 25°C, twice for 5 min in 1× SSC–1% SDS at 50°C for the LT and at 60°C for the ST oligo probes. Washed filters are then immersed in 7.5 ml of a solution containing 0.1 *M* Tris-HCl–0.1 *M* NaCl–0.05 *M* $MgCl_2$–0.1 m*M* $ZnCl_2$–0.02% sodium azide, pH 8.5, with

33 μl of 7% diethylformamide and 25 μl of 5-bromo-4-chloro-3-indolyl phosphate (50 mg/ml in dimethylformamide), and incubated for 4 hr.

C. Appendix

1. Media

LB (Luria–Bertani) Medium

Per Liter: 10 g Bacto-tryptone (Difco, 0123-01)
5 g Yeast extract (BBL, 11929)
5 g NaCl (Sigma, S 3014)

M9 Medium

Per Liter: 6 g Na_2HPO_4 (Sigma, S 3264)
3 g KH_2PO_4 (Sigma, P 0662)
0.5 g NaCl (Sigma, S 3014)
1 g NH_4Cl (Sigma, A 5666)

Autoclave and then add sterile:
1 *M* $MgSO_4$, 1 ml (Sigma, M 9397)
20% Glucose, 10 ml (Sigma, G 5767)
1% Thiamine, 0.05 ml (Sigma, T 4625)

2. Buffers and Solutions

TES
30 m*M* Tris pH 8 (Sigma, T 8524)
5 m*M* EDTA (Sigma, E 5134)
50 m*M* NaCl (Sigma, S 3014)

TE
1 m*M* EDTA (Sigma, E 5134)
10 m*M* Tris pH 8 (Sigma, T 8524)

Triton lytic mix
0.1% Triton X-100 (Sigma, T 6878)
50 m*M* EDTA (Sigma, E 5134)
50 m*M* Tris pH 8 (Sigma, T 8524)

Solution I
20% Lysozyme (Sigma, L 6876)
50 m*M* Glucose (Sigma, G 5000)
10 m*M* EDTA (disodium salt) (Sigma, E 5134)
25 m*M* Tris pH 8 (Sigma, T 8524)
Prepare immediately before use and store at 0°C.

Solution II
0.2 *N* NaOH (Sigma, S 5881)
1% SDS (Sigma, L 5750)

Solution III

3 *M* Sodium acetate pH 4.8 (Sigma, S 8750)

Tris-borate

89 m*M* Tris pH 8.3 (Sigma, T 8524)
89 m*M* Boric acid (Sigma, B 6768)
2.5 m*M* EDTA (Sigma, E 5134)

NT Reaction buffer

0.5 *M* Tris pH 7.5 (Sigma, T 8524)
0.1 *M* $MgSO_4$ (Sigma, M 9397)
10 m*M* Dithiothreitol (DTT) (Sigma, D 9779)
500 μg/ml BSA (Sigma, B 2518)

SSC

150 m*M* NaCl (Sigma, S 3014)
15 m*M* Sodium citrate pH 7.0 (Sigma, C 8532)

SET

150 m*M* NaCl (Sigma, S 3014)
30 m*M* Tris pH 8 (Sigma, T 8524)
1 m*M* EDTA (Sigma, E 5134)

Denhardt's solution

0.02% BSA fraction V (Sigma, A 4503)
0.02% Ficoll 400 (Sigma, F 4375)
0.02% Polyvinylpyrrolidone (Sigma, PVP-360)

10× Kinase buffer

0.5 *M* Tris-HCl pH 7.6 (Sigma, T 7149)
0.1 *M* $MgCl_2$ (Sigma, M 8266)
−50 m*M* DTT (Sigma, D 9779)
1 m*M* Spermidine (Sigma, S 2626)
1 m*M* EDTA (Sigma, E 5134)

Restriction endonucleases

Enzyme	Salt	Temperature (°C)
*Ava*I	Medium	37 (Bio Lab, 152)
*Bam*HI	Medium	37 (BRL, 5201 LA)
*Eco*RI	Medium	37 (BRL, 5202 SH)
*Hin*cII	Medium	37 (BRL, 5206 SB)
*Hin*dIII	Medium	37 (BRL, 5207 LA)
*Hin*fI	Medium	37 (Bio Lab, 155)
*Hpa*II	Low	37 (BRL, 5209 SA)
*Kpn*I	Low	37 (BRL, 5232 SA)
*Pst*I	Medium	30 (BRL, 5215 SC)
*Sal*I	Low	37 (BRL, 5217 SA)
*Sma*I	Special	37 (BRL, 5228 SA)

Buffers:	NaCl	Tris	$MgSO_4$	DTT
Low	0	10 m*M* (pH 7.4)	10 m*M*	1 m*M*
Medium	50 m*M*	10 m*M* (pH 7.4)	10 m*M*	1 m*M*
High	100 m*M*	50 m*M* (pH 7.4)	10 m*M*	0
*Sma*I	20 m*M* KCl	10 m*M* (pH 8)	10 m*M*	1 m*M*

DNase I

1. Stock solution of DNase I (Sigma, D 4527) is 1 mg/ml in 50 m*M* Tris (pH 7.5), 10 m*M* $MgSO_4$, 1 m*M* DTT, and 50% glycerol (Sigma, G 5516) stored at −20°C.
2. 0.5 μl of DNase I stock is diluted in 100 μl of 50 m*M* Tris (pH 7.5), 10 m*M* $MgSO_4$, 1 m*M* DTT, and 50 mg/ml of BSA fraction V on ice (0°C).
3. 0.5 μl of this dilution is used in a nick-translation reaction.

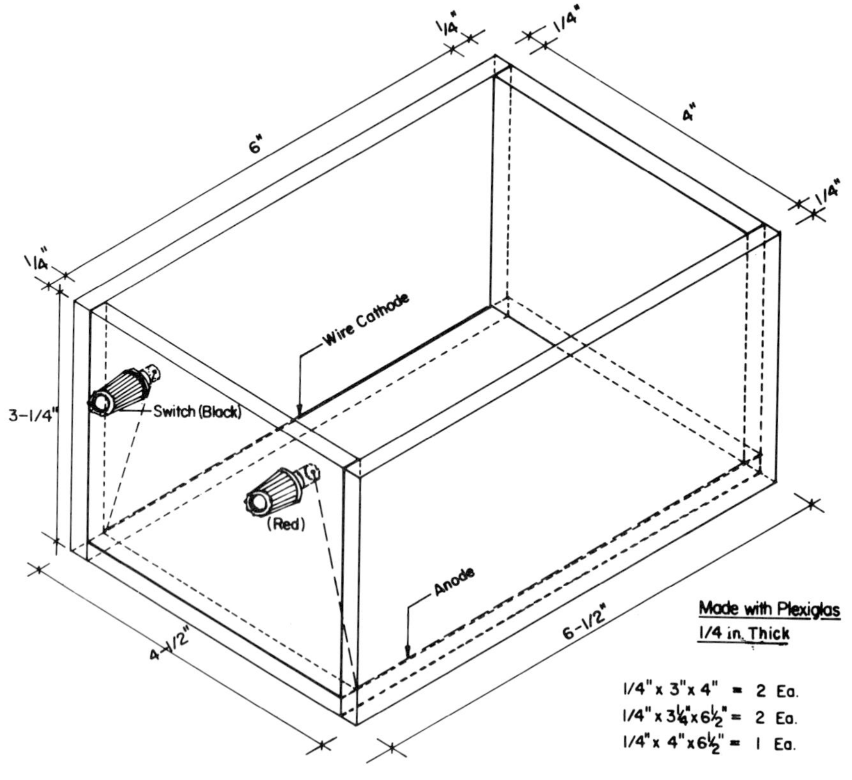

Fig. 5. Electroelution chamber made of Plexiglas in the maintenance shop at AFRIMS.

Electroelution chamber

Electrolution chambers are constructed locally with Plexiglas sheets and choroform (Fig. 5).

Investigators are encouraged to refer to Davis *et al.* (22a) for more information.

ACKNOWLEDGMENTS

We thank Chittima Pitarangsi, Prani Ratarasarn, Apichai Srijan, and Orapan Chivaratanond for performing the clinical bacteriology; Pradith Nabumrung and Sawat Boonnak for preparing media; Suchitra Changchawalit and Thamma Sakulkaipeara for performing enterotoxin assays and purifying plasmid DNA; Ladaporn Bodhidatta, Sajee Pinnoi, Ovath Thonglee, Nisara Wongkamhaeng, and Chuanchom Pravichpram for collecting specimens; Songmuang Piyaphong and Vitaya Khungvalert for fixing samples on Whatman 541 paper; and Nattakarn Siripraivan for preparing this chapter for publication.

REFERENCES

1. Alderete, J. F., and Robinson, D. C. (1978). Purification and chemical characterization of the heat-stable enterotoxin produced by porcine strains of enterotoxigenic *Escherichia coli*. *Infect. Immun.* **19,** 1021–1030.
2. Baldini, M. M., Kaper, J. B., Levine, M. M., Candy, D. C. A., and Moon, H. W. (1983). Plasmid-mediated adhesion in enteropathogenic *Escherichia coli*. *J. Pediatr. Gastroenterol. Nutr.* **2,** 534–538.
3. Bialkowska-Hobrzanska, H. (1987). Detection of enterotoxigenic *Escherichia coli* by dot-block hybridization with biotinylated DNA probes. *J. Clin. Microbiol.* **25,** 338–343.
4. Black, R. E., Merson, M. H., Rahman, A. S. M. M., Yuunus, M., Alim, A. R. M. A., Huq, I., Yolken, E. H., and Curlin, G. T. (1980). A two-year study of bacterial, viral, and parasitic agents associated with diarrhea in rural Bangladesh. *J. Infect. Dis.* **142,** 660–664.
5. Boileau, C. R., d'Hauteville, H. M., and Sansonetti, P. J. (1984). DNA hybridization techniques to detect *Shigella* species and enteroinvasive *Escherichia coli*. *J. Clin. Microbiol.* **20,** 959–961.
6. Brill, B. M., Wasilavskas, B. L., and Richardson, S. H. (1979). Adaptation of the staphylococcal coagglutination technique for detection of heat-labile enterotoxin of *Escherichia coli*. *J. Clin. Microbiol.* **9,** 45–48.
7. Brown, J. E., Echeverria, P., Taylor, D. N., Seriwatana, J., Vanapruks, V., Lexomboon, U., Neill, R. N., and Newland, J. W. (1989). Determination by DNA hybridization of Shiga-like toxin-producing *Escherichia coli* in children with diarrhea in Thailand. *J. Clin. Microbiol.* **27,** 291–294.

7a. Brown, J. E., Settabuth, O., Echeverria, P. (1989). *Infect. Immun.* (in press).

8. Bryan, R. N., Ruth, J. L., Smith, R. D., and Le Bon, J. M. (1986). Diagnosis of clinical samples with synthetic oligonucleotide hybridization probes. *In* "Microbiology—1985" (L. Leive, ed.), pp. 113–116. Am. Soc. Microbiol., Washington, D. C.
9. Burgess, N. M., Bywater, R. J., Cowley, C. M., Mullan, N. A., and Newsome, P. M. (1978). Biological evaluation of a methanol-soluble, heat-stable *Escherichia coli* enterotoxin in infant mice, pigs, rabbits, and calves. *Infect. Immun.* **21,** 526–531.
10. Byers, P. A. and DuPont, H. L. (1979). Pooling method for screening large numbers of *Escherichia coli* for production of heat-stable enterotoxin, and its application in field studies. *J. Clin. Microbiol.* **9,** 541–543.
11. Caruthers, M. H., Beaucage, S. L., Becker, C., Efcavitch, W., Matteveci, M., and Stabinsky, Y. (1982). New methods of synthesizing deoxyoligonucleotides. *In* "Genetic Engineering Principles and Methods" (A. Setlow and S. K. Hollaender, eds.), Vol. 4, pp. 1–17. Plenum, New York.
12. Castro, A. F. P., Sarafim, M. B., Gomes, J. A., and Gatti, M. S. (1980). Improvements in the passive immune hemolysis test for assaying enterotoxigenic *Escherichia coli. J. Clin. Microbiol.* **12,** 714–717.
13. Chatkaeomorakot, A., Echeverria, P., Taylor, D. N., Bettelheim, K. A., Blacklow, N. R., Sethabutr, O., Seriwatana, J., and Kaper, J. (1987). HeLa cell adherent *Escherichia coli* in children with diarrhea in Thailand. *J. Infect. Dis.* **156,** 669–672.
14. Chityothin, O., Sethabutr, O., Echeverria, P., Taylor, D. N., Vongsthongsri, V., and Tharavanij, S. (1987). Detection of heat-stable enterotoxigenic *Escherichia coli* by hybridization with an RNA transcript probe. *J. Clin. Microbiol.* **25,** 1572–1573.
15. Clausen, C. R., and Christie, P. L. (1982). Chronic diarrhea in infants caused by adherent enteropathogenic *Escherichia coli. J. Pediatr.* **100,** 358–361.
16. Clements, J. D., Yancey, R. J., and Finkelstein, R. A. (1980). Properties of homogenous heat-labile enterotoxins from *Escherichia coli. Infect. Immun.* **29,** 91–97.
17. Cravioto, A., Gross, R. J., Scotland, S. M., and Rowe, B. (1979). An adhesive factor found in strains of *Escherichia coli* belonging to the traditional infantile enteropathogenic serotypes. *Curr. Microbiol.* **3,** 95–99.
18. Cravioto, A., Scotland, S. M., and Rowe, B. (1982). Hemagglutination activity and colonization factor antigens I and II in enterotoxigenic and non-enterotoxigenic strains of *Escherichia coli* isolated from humans. *Infect. Immun.* **36,** 189–197.
19. Dallas, W. S., and Falkow, S. (1979). The molecular nature of heat-labile enterotoxin (LT) of *Escherichia coli. Nature (London)* **277,** 406–407.
20. Dallas, W. S., Gill, D. M., and Falkow, S. (1979). Cistrons encoding *Escherichia coli* heat-labile toxin. *J. Bacteriol.* **139,** 850–858.
21. Darfueille, A., Lafeuille, B., Joly, B., and Cluzel, R. (1983). A new colonization factor antigen (CFA/III) produced by enteropathogenic *Escherichia coli* O128:B12. *Ann. Microbiol. (Paris)* **134A,** 53–64.
22. Darfeuille-Michaud, A., Forester, C., Joly, B., and Cluzel, R. (1986). Identification of a non-fimbrial adhesive factor of an enterotoxigenic *Escherichia coli* strain. *Infect. Immun.* **52,** 468–475.
22a. Davis, R. W., Botstein, D., and Roth, J. R. (1980). "Advanced Bacterial Genetics: A Manual for Genetic Engineering." Cold Spring Harbor Lab., Cold Spring Harbor, New York.
23. Dean, A. G., Ching, Y. C., Williams, R. G., and Harden, L. B. (1972). Test for *Escherichia coli* enterotoxin using infant mice: Application in a study of diarrhea in children in Honolulu. *J. Infect. Dis.* **125,** 407–411.
24. DuPont, H. L., Formal, S. B., Hornick, R. B., Snyder, M. J., Libonati, J. P., Sheahan, D. G., LaBrec, E. H., and Kalas, J. P. (1971). Pathogenesis of *Escherichia coli* diarrhea. *N. Engl. J. Med.* **285,** 1–9.

25. Echeverria, P., Blacklow, N. R., Vollet, J. L., Uylangco, C. V., Cukor, G., Soriano, V. B., DuPont, H. L., Cross, J. H., Orskov, F., and Orskov, I. (1978). Reovirus-like agent and enterotoxigenic *Escherichia coli* infections in pediatric diarrhea in the Philippines. *J. Infect. Dis.* **138,** 326–332.
26. Echeverria, P., Blacklow, N. R., Sanford, L. B., and Cukor, G. G. (1981). Travelers' diarrhea among American Peace Corps volunteers in rural Thailand. *J. Infect. Dis.* **143,** 767–771.
27. Echeverria, P., Orskov, F., Orskov, I., and Plianbangchang, D. (1982). Serotypes of enterotoxigenic *Escherichia coli* in Thailand and the Philippines. *Infect. Immun.* **851–856.**
28. Echeverria, P., Seriwatana, J., Chityothin, C., Chaicumpa, W., and Tirapat, C. (1982). Detection of enterotoxigenic *Escherichia coli* in water by filter hyrbidization with three enterotoxin gene probes. *J. Clin. Microbiol.* **16,** 1086–1090.
29. Echeverria, P., Seriwatana, J., Leksomboon, U., Tirapat, C., Chaicumpa, W., and Rowe, B. (1984). Identification by DNA hybridization of enterotoxigenic *Escherichia coli* in homes of children with diarrhea. *Lancet* **1,** 63–65.
30. Echeverria, P., Seriwtana, J., Patamaroj, U., Moseley, S. L., McFarland, A., Chityothin, O., and Chaicumpa, W. (1984). Prevalence of heat-stable II enterotoxigenic *Escherichia coli* in pigs, water, and people at farms in Thailand as determined by DNA hybridization. *J. Clin. Microbiol.* **19,** 489–491.
31. Echeverria, P., Seriwatana, J., Taylor, D. N., Tirapat, C., Chaicumpa, W., and Rowe, B. (1985). Identification by DNA hybridization of enterotoxigenic *Escherichia coli* in a longitudinal study of villages in Thailand. *J. Infect. Dis.* **151,** 124–130.
32. Echeverria, P., Taylor, D. N., Seriwatana, J., Tirapat, C., and Rowe, B. (1985). *Escherichia coli* contains plasmids coding for heat-stable b and other enterotoxins and antibiotic resistance. *Infect. Immun.* **43,** 843–848.
33. Echeverria, P., Seriwatana, J., Taylor, D. N., Changchawalit, S., Smyth, C. J., Twohig, J., and Rowe, B. (1986). Plasmids coding for colonization factor antigens I and II, LT and STA-2 in *Escherichia coli*. *Infect. Immun.* **51,** 626–630.
34. Echeverria, P., Taylor, D. N., Seriwatana, J., Chatkaeomorakot, A., Khungvalert, V., Sakuldaipeara, T., and Smith, R. A. (1986). A comparative study of enterotoxin gene probes and tests for toxin production to detect ETEC. *J. Infect. Dis.* **153,** 255–260.
35. Echeverria, P., Taylor, D. N., Bettelheim, K. A., Chatkaeomorakot, A., Changchawalit, S., Thongcharoen, A., and Leksomboon, U. (1987). HeLa cell adherent enteropathogenic *Escherichia coli* in children under 1 year of age in Thailand. *J. Clin. Microbiol.* **25,** 1472–1475.
36. Echeverria, P., Taylor, D. N., Seriwatana, J., and Moe, C. (1987). Comparative study of synthetic oligonucleotide and cloned polynucleotide enterotoxin gene probes to identify enterotoxigenic *Escherichia coli*. *J. Clin. Microbiol.* **25,** 106–109.
37. Echeverria, P., Taylor, D. N., Seriwatana, J., Leksomboon, U., Chaicumpa, W., Tirapat, C., and Rowe, B. (1987). Potential sources of enterotoxigenic *Escherichia coli* in homes of children with diarrhea in Thailand. *Bull. W. H. O.* **65,** 207–215.
38. Echeverria, P., Taylor, D. N., Lexomboon, U., Bhaibulaya, M., Blacklow, N. R., Tamura, K., and Sakazaki, R. (1989). Case-control study of endemic diarrheal disease in Thai children under 5 years of age. *J. Infect. Dis.* **159,** 543–548.
39. Echeverria, P., Taylor, D. N., Seriwatana, J., Brown, J. E., and Lexomboon, U. (1989). Examination of colonies and stool blots for the detection of enteropathogens by DNA hybridization with eight DNA probes. *J. Clin. Microbiol.* **27,** 331–334.
40. Eisenstein, B. I., and Engleberg, N. C. (1986). Applied molecular genetics: New tools for microbiologists and clinicians. *J. Clin. Microbiol.* **153,** 416–430.
41. Evans, D. G., Silver, R. D., Evans, D. J., Chase, D. G., and Gorbach, S. L. (1975).

Plasmid-controlled colonization factor associated with virulence in *Escherichia coli* enterotoxigenic in humans. *Infect. Immun.* **12,** 656–667.
42. Ewing, W. H. (1986). "Edward's and Ewing's Identification of Enterobacteriaceae," 4th ed., pp. 49, 120–124. Am. Elsevier, New York.
43. Field, M., Graf, L. H., Laird, W. J., and Smith, P. L. (1978). Heat-stable enterotoxin of *Escherichia coli: In-vitro* effects on guanylate cyclase activity, cyclic AMP concentration, and ion transport in the small intestine. *Proc. Natl. Acad. Sci. USA* **75,** 2800–2804.
44. Finkelstein, R. A., and Yang, Z. (1983). Rapid test for identification of heat-labile enterotoxin-producing *Escherichia coli* colonies. *J. Clin. Microbiol.* **18,** 23–28.
45. Formal, S. B., and Hornick, R. B. (1987). Invasive *Escherichia coli. J. Infect. Dis.* **137,** 641–644.
46. Georges, M. C., Wachmuth, I. K., and Birkness, K. A. (1983). Genetic probes for enterotoxigenic *Escherichia coli* isolated from childhood diarrhea in the Central African Republic. *J. Clin. Microbiol.* **18,** 199–202.
47. Gill, D. M., and Richardson, S. H. (1980). Adenosine diphosphate-ribosylation of adenylate cyclase catalyzed by heat-labile enterotoxin of *Escherichia coli:* Comparison with cholera toxin. *J. Infect. Dis.* **141,** 64–70.
48. Gomes, T. A. T., Toldeo, M. R. F., Trabulsi, L. R., Wood, P. K., and Morris, J. G. (1987). DNA probes for the identification of enteroinvasive *Escherichia coli. J. Clin. Microbiol.* **25,** 2025–2027.
49. Green, B. A., Neill, R. J., Ruyechan, W. T., and Holmes, R. K. (1983). Evidence that a new enterotoxin which activates adenyl cyclase in eucaryotic target cells is not plasmid mediated. *Infect. Immun.* **41,** 383–390.
50. Greenberg, H. B., Sack, D. A., Rodriguez, W., Sack, R. B., Wyatt, R. G., Kalicia, A. R., Horswood, R. L., Chanock, R. M., and Kapikian, A. Z. (1977). Microtiter solid-phase radioimmunoassay for detection of *Escherichia coli* heat-labile enterotoxin. *Infect. Immun.* **17,** 541–545.
51. Grunstein, M., and Hogness, D. (1975). Colony hybridization: A method for the isolation of cloned DNAs that contain a specific gene. *Proc. Natl. Acad. Sci. U.S.A.* **72,** 3961–3965.
52. Guth, B. E. C., Pickett, C. L., Twiddy, E. M., Holmes, R. K., Gomes, T. A. T., Lima, A. A. M., Guerrant, R. L., Franco, B. D. G. M., and Trabulsi, L. R. (1986). Production of type II heat-labile enterotoxin by *Escherichia coli* isolated from food and human feces. *Infect. Immun.* **54,** 587–589.
53. Guth, B. E. C., Twiddy, E. M., Trabulsi, L. R., and Holmes, R. K. (1986). Variations in chemical properties and antigenic determinants among type II heat-labile enterotoxins of *Escherichia coli. Infect. Immun.* **54,** 529–536.
54. Hale, T. L., Sansonetti, P. J., Schad, D. A., Austin, S., and Formal, S. B. (1983). Characterization of virulence plasmids and plasmid-associated outer membrane proteins in *Shigella flexneri, Shigella sonnei,* and *Escherichia coli. Infect. Immun.* **40,** 340–350.
55. Handl, C., Ronnberg, B., Nilsson, B., Olsson, E., Jonsson, H., and Flock, J. I. (1988). Enzyme-linked immunosorbent assay for *Escherichia coli* heat-stable enterotoxin type II. *J. Clin. Microbiol.* **26,** 1555–1560.
56. Holmes, R. K., Twiddy, E. M., and Bramucci, M. G. (1983). Antigenic heterogeneity among heat-labile toxins from *Escherichia coli. In* "Advances in Research on Cholera and Related Diarrheas" (S. Kuwahara and N. F. Pierce, eds.), Vol. 1, pp. 293–300. Martinus Nihoff Publishers, Boston, Massachusetts.
57. Honda, T., Taga, S., Takeda, Y., and Miwatani, T. (1981). A modified Elek test for detection of heat-labile enterotoxins of enterotoxigenic *Escherichia coli. J. Clin. Microbiol.* **13,** 1–5.

58. Honda, T., Tsuji, T., Takeda, Y., and Miwatani, T. (1981). Immunological non-identity of heat-labile enterotoxins from human and porcine enterotoxigenic *Escherichia coli*. *Infect. Immun.* **34,** 337–340.
59. Honda, T., Arita, M., and Miwatani, T. (1984). Characterization of new hydrophobic pili of human enterotoxigenic *Escherichia coli:* A possible new colonization factor. *Infect. Immun.* **43,** 959–965.
60. Hughes, J. M., Murad, F., Chang, B., and Guerrant, R. L. (1978). Role of cyclic GMP in the action of heat-stable enterotoxin of *Escherichia coli. Nature (London)* **271,** 755–756.
61. Jablonski, E., Moomaw, E. W., Tullis, R., and Ruth, J. (1986). Preparation of oligodeoxynucleotide–alkaline phosphatase conjugates and their use as hybridization probes. *Nucleic Acids Res.* **14,** 6115–6128.
62. Johnson, W. M., Lior, H., and Bezanson, G. S. (1983). Cytotoxic *Escherichia coli* O157:H7 associated with hemorrhagic colitis in Canada. *Lancet* **1,** 76.
63. Kapitany, R. A., Scott, A., Forsyth, G. W., McKenzie, S. L., and Worthington, R. W. (1979). Evidence for two heat-stable enterotoxins produced by enterotoxigenic *Escherichia coli. Infect. Immun.* **24,** 965–966.
64. Karch, H., Hessemann, J., Lauf, R., O'Brien, A. D., Tucker, C. O., and Levine, M. M. (1987). A plasmid of enterohemorrhagic *Escherichia coli* O157:H7 is required for expression of a new fimbrial antigen adhesion to epithelial cells. *Infect. Immun.* **55,** 455–461.
65. Karch, H., Hessemann, J., and Laufs, R. (1987). Phage-associated cytotoxin production by and enteroadhesiveness of enteropathogenic *Escherichia coli* isolated from infants with diarrhea in West Germany. *J. Infect. Dis.* **155,** 707–715.
66. Karmali, M. A., Petric, M., Lim, C., Fleming, P. C., Arbus, G. S., and Lior, H. (1985). The association between idiopathic hemolytic uremic syndrome and infection by verocytotoxin-producing *Escherichia coli. J. Infect. Dis.* **151,** 775–782.
67. Kirii, Y., Danbara, H., Komase, K., Arita, H., and Yoshikawa, M. (1987). Detection of enterotoxigenic *Escherichia coli* by colony hybridization with biotinylated enterotoxin probes. *J. Clin. Microbiol.* **25,** 1962–1965.
68. Knutton, S., Baldini, M. M., Kaper, J. B., and McNeish, A. S. (1987). Role of plasmid-encoded adherence factors in adhesion of enteropathogenic *Escherichia coli* to HEp-2 cells. *Infect. Immun.* **55,** 78–85.
69. Knutton, S., Lloyd, B. R., and McNeish, A. S. (1987). Adhesion of enteropathogenic *Escherichia coli* to human intestinal enterocytes and cultured human intestinal mucosa. *Infect. Immun.* **55,** 69–77.
70. Knutton, S., Baldwin, T., Williams, P. H., and McNeish, A. S. (1988). New diagnostic test for enteropathogenic *Escherichia coli. Lancet* **1,** 1337.
71. Kohne, D., Hogan, J., Jones, V., Dean, E., and Adams, T. H. (1986). Novel approach for rapid and sensitive detection of microorganisms: DNA probes to rRNA. *In* "Microbiology—1985" (L. Leive, ed.), pp. 110–112. Am. Soc. Microbiol., Washington, D. C.
72. Konowalchuk, J., Speirs, J. I., and Stavric, S. (1977). Vero response to a cytotoxin of *Escherichia coli. Infect. Immun.* **18,** 775–779.
73. Kunkel, S. L., and Robertson, D. C. (1979). Purification and chemical characterization of the heat-labile enterotoxin produced by enterotoxigenic *Escherichia coli. Infect. Immun.* **25,** 586–596.
74. LaBrec, E. H., Schneider, H., Magnani, T. J., and Formal, S. B. (1964). Epithelial cell penetration is an essential step in the pathogenesis of bacillary dysentery. *J. Bacteriol.* **88,** 1503–1518.
75. Lanser, J. A., and Anargyros, P. A. (1985). Detection of *Escherichia coli* adhesins with DNA probes. *J. Clin. Microbiol.* **22,** 425–427.

76. Lathe, R., Hirth, P., DeWilde, M., Harford, N., and LeCocq, J.-P. (1980). Cell-free synthesis of enterotoxin of *E. coli* from a cloned gene. *Nature* (*London*) **284,** 473–474.
77. Lee, C. H., Moseley, S. L., Moon, H. W., Whipp, S. C., Gyles, C. L., and So, M. (1983). Characterization of the gene encoding heat-stable toxin II and preliminary molecular and epidemiological studies of enterotoxigenic *Escherichia coli* heat-stable toxin II. *Infect. Immun.* **42,** 264–268.
78. Levine, M. M., Kaper, J. B., Black, R. E., and Clements, M. L. (1983). New knowledge on pathogenesis of bacterial enteric infections as applied to vaccine development. *Microbiol. Rev.* **47,** 510–550.
79. Levine, M. M., Nataro, J. P., Karch, H., Baldini, M. M., Kaper, J. B., Black, R. E., Clements, M. L., and O'Brien, A. D. (1985). The diarrheal response of humans to some classic serotypes of enteropathogenic *Escherichia coli* is dependent on a plasmid encoding an enteroadhesive factor. *J. Infect. Dis.* **152,** 550–559.
80. Levine, M. M. (1987). *Escherichia coli* that cause diarrhea: Enterotoxigenic, enteropathogenic, enteroinvasive, enterohemorrhagic, and enteroadherent. *J. Infect. Dis.* **155,** 377–389.
81. Levine, M. M., Xu, J.-G., Kaper, J. B., Lior, H., Prado, V., Tall, B., Nataro, J., Karch, H., and Wachsmuth, K. (1987). A DNA probe to identify enterohemorrhagic *Escherichia coli* of O157:H7 and other serotypes that cause hemorrhagic colitis and hemolytic uremic syndrome. *J. Infect. Dis.* **156,** 175–181.
82. Levine, M. M., Prado, V., Robins-Browne, R. M., Lior, H., Kaper, J. B., Moseley, S., Gicquelais, K., Nataro, J. P., Vial, P., and Tall, B. (1988). DNA probes and HEp-2 cell adherence to detect diarrheagenic *Escherichia coli. J. Infect. Dis.* **158,** 224–228.
83. Maas, R. (1983). An improved colony hybridization method with significantly increased sensitivity for detection of single genes. *Plasmid* **10,** 296–298.
84. Maniatis, T., Jeffrey, A., and Kleid, D. G. (1975). Nucleotide sequence of the rightward operator of the phage lambda. *Proc. Natl. Acad. Sci. U.S.A.A* **72,** 1184–1188.
85. Mannig, P. A., Higgins, G. D., Lump, R., and Lanser, J. A. (1987). Colonization factor antigens and a new fimbrial type (CFA/V) on O115:H4 and H-strains of enterotoxigenic *Escherichia coli* in Central Australia. *J. Infect. Dis.* **156,** 841–844.
86. Marques, L. R. M., Moore, M. A., Wells, J. G., Wachsmuth, I. K., and O'Brien, A. D. (1986). Production of Shiga-like toxin by *Escherichia coli. J. Infect. Dis.* **154,** 338–341.
87. Matthewson, J. J., Oberhelman, R. A., DuPont, H. L., Javier de la Cabada, F., and Garibay, E. V. (1987). Enteroadherent *Escherichia coli* as a cause of diarrhea among children in Mexico. *J. Clin. Microbiol.* **25,** 1917–1919.
88. McConnell, M. M., Thomas, L. V., Day, N. P., and Rowe, B. (1985). Enzyme-linked immunosorbent assays for the detection of adhesion factor antigens of enterotoxigenic *Escherichia coli. J. Infect. Dis.* **152,** 1220–1227.
89. McConnell, M. M., Thomas, L. V., Willshaw, G. A., Smith, H. B., and Rowe, B. (1988). Genetic control and properties of coli surface antigens of colonization factor antigen IV (PCF 8775) of enterotoxigenic *Escherichia coli. Infect. Immun.* **56,** 1974–1980.
90. Meinkoth, J., and Wahl, G. (1984). Hybridization of nucleic acids immobilized on solid supports. *Annal. Biochem.* **138,** 267–284.
91. Merson, M. H., Morris, G. K., Sack, D. A., Wells, J. C., Feeley, R. B., Sack, R. B., Creech, W. B., Kapikian, A. Z., and Gangarosa, E. J. (1976). Traveler's diarrhea in Mexico: A prospective study of physicians and family members attending a congress. *N. Engl. J. Med.* **294,** 1299–1305.
92. Merson, M. H., Sack, R. B., Golam Kibriya, A. K. M., Mahamound, A. A., Adamed, Q.-S., and Huq, I. (1979). Use of colony pools for diagnosis of enterotoxigenic *Escherichia coli* diarrhea. *J. Clin. Microbiol.* **9,** 493–497.

93. Merson, M. H., Rowe, B., Black, R. E., Huq, I., Gross, R. J., and Eusof, A. (1980). Use of antisera for identification of enterotoxigenic *Escherichia coli*. *Lancet* **2,** 222–224.
94. Moon, H. W., Whipp, S. C., Argenzio, R. A., Levine, M. M., and Gianella, R. A. (1983). Attaching and effacing activities of rabbit and human enteropathogenic *Escherichia coli* in pigs and rabbits intestines. *Infect. Immun.* **41,** 1340–1351.
95. Morgan, D. R., DuPont, H. L., Wood, L. V., and Ericson, C. D. (1983). Comparison of methods to detect *Escherichia coli* heat-labile enterotoxin in stool and cell free culture supernatant. *J. Clin. Microbiol.* **18,** 798–802.
96. Moseley, S. L., Huq, I., Alim, A. R. M. A., So, M., Samadpour-Motalebi, M., and Falkow, S. (1980). Detection of enterotoxigenic *Escherichia coli* by DNA colony hybridization. *J. Infect. Dis.* **142,** 892–898.
97. Moseley, S. L., Echeverria, P., Seriwatana, J., Tirapat, C., Chaicumpa, W., Sakuldaipeara, T., and Falkow, S. (1982). Identification of enterotoxigenic *Escherichia coli* using three enterotoxin gene probes. *J. Infect. Dis.* **145,** 863–869.
98. Moseley, S. L., Hardy, J. N., Huq, M. I., Echeverria, P., and Falkow, S. (1983). Isolation and nucleotide sequence determinations of a gene encoding a heat-stable enterotoxin *Escherichia coli*. *Infect. Immun.* **39,** 1167–1174.
99. Moseley, S. L., Smadpour-Motalebi, M., and Falkow, S. (1983). Plasmid association of nucleotide sequence relationship of two genes encoding heat-stable enterotoxin production in *Escherichia coli* H10407. *J. Bacteriol.* **156,** 441–443.
100. Moseley, S. L., Clausen, C. R., and Smith, A. L. (1985). A new bacterial adhesin associated with enteritis in infants. *Program Abstr., Intersci. Conf. Antimicrob. Agents Chemother. 25th* Asbtr. No. 1128.
101. Moss, J., and Richardson, S. H. (1978). Activation of adenylate cyclase by heat-labile *Escherichia coli* enterotoxin: Evidence for ADP-ribosyl transferase activity similar to that of choleragen. *J. Clin. Invest.* **62,** 281–285.
102. Murray, B. E., Seriwatana, J., and Echeverria, P. (1981). Toxin detection after storage or cultivation of enterotoxigenic with colicinogenic *Escherichia coli:* A possible mechanism for toxin-negative pools. *J. Clin. Microbiol.* **13,** 179–183.
103. Murray, B. E., Mathewson, J. J., DuPont, H. L., and Hill, W. E. (1987). Utility of oligodeoxyribonucleotide probes for detecting enterotoxigenic *Escherichia coli*. *J. Infect. Dis.* **155,** 809–811.
104. Nataro, J. P., Scaletsky, I. C. A., Kaper, J. B., Levine, M. M., and Trabulsi, L. R. (1985). Plasmid-mediated factors conferring diffuse and localized adherence of enteropathogenic *Escherichia coli*. *Infect. Immun.* **48.** 378–383.
105. Nataro, J. P., Baldini, M. M., Kaper, J. B., Black, R. E., Bravo, N., and Levine, M. M. (1985). Detection of an adherence factor of enteropathogenic *Escherichia coli* with a DNA probe. *J. Infect. Dis.* **152,** 560–565.
106. Nataro, J. P., Kaper, J. B., Robins-Browne, R., Prado, V., Vial, P. A., and Levine, M. M. (1987). Patterns of adherence of diarrheagenic *Escherichia coli* to HEp-2 cells. *Pediatr. Infect. Dis.* **6,** 829–831.
107. Newland, J. W., Strockbine, N. A., Miller, C. F., O'Brien, A. D., and Holmes, R. K. (1985). Cloning of Shiga-like toxin structural genes from a toxin-converting phage of *Escherichia coli*. *Science* **230,** 179–181.
108. Newland, J. W., Strockbine, N. A., and Neill, R. J. (1987). Cloning of genes for production of *Escherichia coli* Shiga-like toxin type II. *Infect. Immun.* **55,** 2675–2680.
109. Newland, J. W., and Neill, R. J. (1988). DNA probes for Shiga-like toxins I and II and for toxin-converting bacteriophages. *J. Clin. Microbiol.* **26,** 1292–1297.
110. Nishibuchi, M., Arita, M., Honda, T., and Miwatani, T. (1988). Evaluation of a nonisotopically labeled oligonucleotide probe to detect the heat-stable enterotoxin gene of *Escherichia coli* by the DNA colony hybridization test. *J. Clin. Microbiol.* **26,** 784–786.

111. Okamoto, K., Okamoto, K., Yukitake, J., and Miyama, A. (1988). Reduction of enterotoxin activity of *Escherichia coli* heat-stable enterotoxin by substitution of an asparagine residue. *Infect. Immun.* **56,** 2144–2148.
112. Olive, D. M. (1989). Detection of enterotoxigenic *Escherichia coli* after polymerase chain reaction amplification with a thermostable DNA polymerase. *J. Clin. Microbiol.* **27,** 261–265.
113. Oprandy, J. J., Thornton, S. A., Gardiner, C. H., Burr, D., Batchelor, R., and Bourgeois, A. L. (1988). Alkaline phosphatase-conjugated oligonucleotide probes for enterotoxigenic *Escherichia coli* in travelers to South America and West Africa. *J. Clin. Microbiol.* **26,** 92–95.
114. Pai, C. H., Gordon, R., Sims, H. V., and Bryan, L. E. (1984). Sporadic cases of hemorrhagic colitis associated with *Escherichia coli* O157:H7: Clinical epidemiological and bacteriologic features. *Ann. Intern. Med.* **101,** 738–742.
115. Pai, C. H., Ahmed, N., Lior, H., Johnson, W. H., Sims, H. W., and Woods, D. E. (1988). Epidemiology of sporadic diarrhea due to verocytotoxin-producing *Escherichia coli:* A two-year prospective study. *J. Infect. Dis.* **157,** 1054–1057.
116. Pal, T., Pacsa, A. S., Emody, L., Voros, S., and Selley, E. (1985). Modified enzyme-linked immunosorbent assay for detecting enteroinvasive *Escherichia coli* and virulent *Shigella* strains. *J. Clin. Microbiol.* **21,** 415–418.
117. Pal, T., Echeverria, P., Taylor, D. N., Sethabutr, O., and Hanchalay, S. (1985). Identification of enteroinvasive *Escherichia coli* by modified enzyme-linked immunosorbent and DNA hybridization assays. *Lancet* **2,** 785.
118. Picken, R. N., Mazaistis, A. J., Maas, W. K., Rey, M., and Heyneker, H. (1983). Nucleotide sequence of the gene for heat-stable enterotoxin II of *Escherichia coli.* *Infect. Immun.* **42,** 269–275.
119. Pickett, C. L., Twiddy, E. M., Belisle, B. W., and Holmes, R. K. (1986). Cloning of genes that encode a new heat-labile enterotoxin of *Escherichia coli.* *J. Bacteriol.* **165,** 348–352.
120. Remis, R. S., MacDonald, K. L., Riley, L. W., Duhr, N. D., Wells, J. G., Davis, B. R., Blake, P. A., and Cohen, M. L. (1984). Sporadic cases of hemorrhagic colitis associated with *Escherichia coli* O157:H7. *Ann. Intern. Med* **101,** 624–626.
121. Renz, M., and Kurz, C. (1984). A colorimetric method for DNA hybridization. *Nucleic Acids Res.* **12,** 3435–3444.
122. Riley, L. W., Remis, R. S., Helgerson, S. D., McGee, H. B., Wells, J. G., Davis, B. R., Herbert, R. J., Olcott, E. S., Johnson, L. M., Hargrett, N. T., Blake, P. A., and Cohen, H. L. (1983). Hemorrhagic colitis associated with a rare *Escherichia coli* serotype. *N. Engl. J. Med.* **708,** 681–685.
123. Ronnberg, B., Carlson, J., and Wadstrom, T. (1984). Development of a new enzyme-linked immunosorbent assay for detection of *Escherichia coli* heat-stable enterotoxin. *FEMS Microbiol. Lett.* **23,** 275–279.
124. Rothbaum, R. A., McAdams, A. J., Gianella, R., and Partin, J. C. (1982). A clinicopathologic study of enterocyte-adherent *Escherichia coli:* A cause of protracted diarrhea in infants. *Gastroenterology* **83,** 441–454.
125. Rothbaum, R. J., Partin, J. C., Saulfeld, K., and McAdams, A. J. (1983). An ultrastructural study of enteropathogenic *Escherichia coli* infections in human infants. *Ultrastruct. Pathol.* **4,** 291–304.
126. Ryan, C. A. (1986). *Escherichia coli* O157:H7 diarrhea in a nursing home. *J. Infect. Dis.* **154,** 631–638.
127. Ryder, R. W., Sack, D. A., Kapikian, A. Z., McLaughlin, J., Chakraborty, J., Rahman, R. J. M. M., Merson, M. H., and Wells, J. G. (1976). Enterotoxigenic *Escherichia coli* and reovirus-like agent in rural Bangladesh. *Lancet* **1,** 659–662.

128. Sack, D. A., and Sack, R. B. (1975). Test for enterotoxigenic *Escherichia coli* using Y-1 adrenal cells in miniculture. *Infect. Immun.* **11,** 334–336.
129. Sack, R. B. (1975). Human diarrheal disease caused by enterotoxigenic *Escherichia coli*. *Annu. Rev. Microbiol.* **29,** 333–353.
130. Saiki, R. K., Gelfand, D. H., Stoffel, S., Scharf, S. I., Higuchi, R., Horn, G. T., Mullis, K. B., and Erlich, H. A. (1988). Primer-directed enzymatic amplification of DNA with a thermostable DNA polymerase. *Science* **239,** 487–491.
131. Sansonetti, P. J., Kopecko, D. J., and Formal, S. B. (1982). Involvement of a plasmid in the invasive ability of *Shigella flexneri*. *Infect. Immun.* **35,** 852–860.
132. Sansonetti, P. J., d'Hauteville, H., Formal, S. B., and Toucas, M. (1982). Plasmid-mediated invasiveness of *Shigella*-like *Escherichia coli*. *Ann. Microbiol.* (*Paris*) **132A,** 351–355.
133. Scaletsky, I. C. A., Silva, M. L. M., and Trabulsi, L. R. (1984). Distinctive patterns of adherence of enteropathogenic *Escherichia coli* to HeLa cells. *Infect. Immun.* **45,** 534–536.
134. Scaletsky, I. C. A., Silva, M. L. M., Toledo, M. R. F., Davis, B. R., Blake, P. A., and Trabulski, L. R. (1985). Correlation between adherence to HeLa cells and serogroups, serotypes, and bioserotypes of *Escherichia coli*. *Infect. Immun.* **49,** 528–532.
135. Scotland, S. M., Day, N. P., and Rowe, B. (1980). Production of a cytotoxin affecting Vero cells by strains of *E. coli* belonging to traditional enteropathogenic serotypes. *FEMS Microbiol. Lett.* **7,** 15–17.
136. Sereny, B. (1955). Experimental *Shigella* conjunctivitis. *Acta Microbiol. Acad. Sci. Hung.* **2,** 293–296.
137. Seriwatana, J., Echeverria, P., Escamilla, J., Glass, R., Huq, I., Rockhill, R., and Stoll, B. L. (1983). Identification of enterotoxigenic *Escherichia coli* in patients with diarrhea in Asia with three enterotoxin gene probes. *Infect. Immun.* **42,** 152–154.
138. Seriwatana, J., Echeverria, P., Taylor, D. N., Sakuldalpeara, T., Changchawalit, S., and Chivoratanond, O. (1987). Identification of enterotoxigenic *Escherichia coli* with synthetic alkaline-phosphatase-conjugated oligonucleotide DNA probes. *J. Clin. Microbiol.* **25,** 1438–1441.
139. Seriwatana, J., Brown, J. E., Echeverria, P., Taylor, D. N., Suthienkul, O., and Newland, J. (1988). DNA probes to identify Shiga-like toxin I and II producing enteric bacterial pathogens isolated from patients with diarrhea in Thailand. *J. Clin. Microbiol.* **26,** 1614–1615.
140. Seriwatana, J., Echeverria, P., Taylor, D. N., Rasrinaul, L., Brown, J. E., Peiris, J. S. M., and Clayton, C. L. (1988). Type II heat-labile enterotoxin-producing *Escherichia coli* isolated from animals and humans. *Infect. Immun.* **56,** 1158–1161.
141. Sethabutr, O., Echeverria, P., Taylor, D. N., Pal, T., and Rowe, B. (1985). DNA hybridization in the identification of enteroinvasive *Escherichia coli* and *Shigella* in children with dysentery. *In* "Infectious Diarrhea in the Young" (S. Tsipori, ed.), pp. 350–356. Elsevier (Biomedical Division), Amsterdam.
142. Sethbutr, O., Hanchalay, S., Echeverria, P., Taylor, D. N., and Leksomboon, U. (1985). A non-radioactive DNA probe to identify *Shigella* and enteroinvasive *Escherichia coli* in stools of children with diarrhea. *Lancet* **2,** 1095–1097.
143. Silva, R. M., Toledo, M. R. F., and Trabulsi, L. R. (1980). Biochemical and cultural characteristics of invasive *Escherichia coli*. *J. Clin. Microbiol.* **11,** 441–444.
144. Small, P. L. C., and Falkow, S. (1986). Development of a DNA probe for the virulence plasmid of *Shigella* spp. and enteroinvasive *Escherichia coli*. *In* "Microbiology—1986" (L. Leive, ed.), pp. 121–124. Am. Soc. Microbiol., Washington, D. C.
145. Smith, H. R., Rowe, B., Gross, R. J., Fry, N. K., and Scotland, S. M. (1987). Haemor-

rhagic colitis and verocytotoxin producing *Escherichia coli* in England and Wales. *Lancet* **1,** 1062–1065.
146. Smith, H. R., Scotland, S. M., Chart, H., and Rowe, B. (1987). Vero cytotoxin production and the presence of VT genes in strains of *Escherichia coli* and *Shigella*. *FEMS Microbiol. Lett.* **42,** 173–177.
147. Smyth, C. J. (1982). Two mannose-resistant hemagglutinins on enterotoxigenic *Escherichia coli* of serotype O6:K15:H16 or H- isolated from travelers' and infantile diarrhea. *J. Gen. Microbiol.* **128,** 2081–2096.
148. So, M., Boyer, H. W., Betlach, M., and Falkow, S. (1976). Molecular cloning of an *Escherichia coli* plasmid determinant that encodes for the production of heat-stable enterotoxin. *J. Bacteriol.* **128,** 403–412.
149. So, M., Dallas, W. S., and Falkow, S. (1978). Characterization of an *Escherichia coli* plasmid encoding for synthesis of heat-labile toxin: Molecular cloning of the toxin determinant. *Infect. Immun.* **21,** 405–411.
150. So, M., Heffron, F., and McCarthy, B. S. (1979). The *E. coli* gene encoding heat-stable toxin is a bacterial transposon flanked by inverted repeats of ISI. *Nature (London)* **277,** 453–456.
151. So, M., and McCarthy, B. J. (1980). Nucleotide sequence of the bacterial transposon Tn 1681 encoding a heat-stable (ST) toxin and its identification in enterotoxigenic *Escherichia coli* strains. *Proc. Natl. Acad. Sci. U.S.A.* **77,** 4011–4015.
152. Sommerfelt, H., Kalland, K. H., Raj, P., Moseley, S. L., Bhan, M. K., and Bjorvatn, B. (1988). Cloned polynucleotide and synthetic oligonucleotide probes used in colony hybridization are equally efficient in the identification of enterotoxigenic *Escherichia coli*. *J. Clin. Microbiol.* **26,** 2275–2278.
153. Strockbine, N. A., Jackson, M. P., Sung, L. M., Holmes, R. K., and O'Brien, A. D. (1988). Cloning and sequencing of the genes for Shiga toxin from *Shigella dysenteria* type I. *J. Bacteriol.* **170,** 1116–1122.
154. Sutton R. G. A., Merson, M., Craig, J. P., Escheverria, P., Moseley, S., Rowe, B., Trabulsi, L. R., Honda, T., and Takeda, Y. (1985). Evaluation of the Biken test for the detection of LT-producing *Escherichia coli*. *In* "Bacterial Diarrheal Disease" (Y. Takeda and T. Miwatani, eds.), pp. 209–218. KTK Scientific Publishers, Tokyo.
155. Svenneholm, A., and Holmgren, J. (1978). Identification of *Escherichia coli* heat-labile enterotoxin by means of a ganglioside immunosorbent assay (GM-1 ELISA) procedure. *Curr. Microbiol.* **1,** 19–23.
156. Tacket, C. O., Maneval, D. R., and Levine, M. M. (1987). A new fimbrial putative colonization factor of enterotoxigenic *Escherichia coli* serotype O159:H4, purification, morphology and genetics. *Infect. Immun.* **55,** 1063–1069.
157. Taylor, D. N., Echeverria, P., Pal, T., Sethabutr, O., Wankitcharoen, S., Sricharmorn, S., Rowe, B., and Cross, J. H. (1986). The role of *Shigella* spp. enteroinvasive *Escherichia coli* and other enteropathogens as causes of childhood dysentery in Thailand. *J. Infect. Dis.* **153,** 1132–1138.
158. Taylor, D. N., Echeverria, P., Pitarangsi, C., Seriwatana, J., Sethabutr, O., Bodhidatta, L., Brown, C., Herrman, J. E., and Blacklow, N. R. (1987). Application of DNA hybridization techniques in the assessment of diarrheal disease among refugees in Thailand. *Am. J. Epidemiol.* **127,** 179–187.
159. Taylor, D. N., Echeverria, P., Sethabutr, O., Pitarangsi, C., Leksomboon, U., Blacklow, N. R., Rowe, B., Gross, R., and Cross, J. (1988). Clinical and microbiological features of *Shigella* and enteroinvasive *Escherichia coli* infections detected by DNA hybridization. *J. Clin. Microbiol.* **26,** 1362–1366.
160. Taylor, D. N., Houston, R., Shlim, D. R., Bhaibulaya, M., Ungar, B. L. P., and

Echeverria, P. (1988). Etiology of diarrhea among travelers and foreign residents in Nepal. *JAMA, J. Am. Med. Assoc.* **260,** 1245–1248.
161. Tenover, F. C. (1988). Diagnostic deoxyribonucleic acid probes for infectious diseases. *Clin. Microbiol. Rev.* **1,** 82–101.
162. Toledo, M. R. F., Reis, M. H. L., Almeida, R. G., and Trabulsi, L. R. (1979). Invasive strains of *Escherichia coli* belonging to O group 29. *J. Clin. Microbiol.* **9,** 288–289.
163. Toledo, M. R. F., and Trabulsi, L. R. (1983). Correlation between biochemical and serological characteristics of *Escherichia coli* and results of the Sereny test. *J. Clin. Microbiol.* **17,** 419–421.
164. Tsuji, T., Taga, S., Honda, T., Takeda, Y., and Miwatani, T. (1982). Molecular heterogeneity of heat-labile enterotoxins from humans and porcine enterotoxigenic *Escherichia coli. Infect. Immun.* **38,** 444–448.
165. Tsuiji, T., Yoshida, S., Honda, T., and Miwatani, T. (1988). Isolation and characterization of enterotoxigenic *Escherichia coli* mutants that produce abnormal heat-labile enterotoxins. *FEMS Microbiol. Lett.* **51,** 67–72.
166. Ulsen, M. H., and Rollo, J. L. (1980). Pathogenesis of *Escherichia coli* gastroenteritis in man: Another mechanism. *N. Engl. J. Med.* **302,** 99–101.
167. Urdea, M. S., Running, J. A., Horn, T., Clyne, J., Ku, L., and Wagner, B. (1987). A novel method for the rapid detection of specific detection of specific nucleotide sequences in crude biological samples without blotting or radioactivity: Application to the analysis of hepatitis B virus in human serum. *Gene* **38,** 253–264.
168. Utomporn, P., Seriwatana, J., and Echeverria, P. (1983). Identification of enterotoxigenic *Escherichia coli* isolated from swine with diarrhea in Thailand by colony hybridization using three enterotoxin gene probes. *J. Clin. Microbiol.* **18,** 1424–1431.
169. Venkatesan, M., Buyssee, I. M., Vandendrics, E., and Kopecko, D. J. (1988). Development and testing of invasion-associated DNA probes for detection of *Shigella* spp. and enteroinvasive *Escherichia coli. J. Clin. Microbiol.* **26,** 261–266.
170. Vial, P. A., Robins-Browne, R., Lior, H., Prado, V., Kaper, J. B., Nataro, J. P., Maneval, D., and Elsayed, A. (1988). Aggregative *Escherichia coli* a putative agent of diarrheal disease. *J. Infect. Dis.* **158,** 70–79.
171. Wanger, A. R., Murray, B. E., Echeverria, P., Matthewson, J. J., and DuPont, H. L. (1988). Enteroinvasive *Escherichia coli* in travelers with diarrhea. *J. Infect. Dis.* **158,** 640–642.
172. Willshaw, G. A., Smith, H. R., Scotland, S. M., and Rowe, B. (1985). Cloning of genes determining the production of verocytotoxin by *Escherichia coli. J. Gen. Microbiol.* **131,** 3047–3053.
173. Willshaw, G. A., Smith, A. R., Scotland, S. M., Field, A. M., and Rowe, B. (1987). Heterogeneity of *Escherichia coli* phages encoding vero cytotoxins: Comparison of cloned sequences determining VT1 and VT2 and development of specific gene probes. *J. Gen. Microbiol.* **133,** 1309–1317.
174. Wood, P. K., Morris, J. G., Jr., Small, P. L. C., Sethabutr, O., Toledo, M. R. F., Trabulsi, L., and Kaper, J. B. (1986). Comparison of DNA probes with the Sereny test for the identification of invasive *Shigella* and *Escherichia coli* strains. *J. Clin. Microbiol.* **24,** 498–500.
175. Yolken, R. H., Greenberg, H. B., Merson, M. H., Sack, R. D., and Kapikian, A. Z. (1977). Enzyme-linked immunosorbent assay for detection of *Escherichia coli* heat-labile enterotoxin. *J. Clin. Microbiol.* **6,** 439–444.

5

DNA Probes for *Escherichia coli* Isolates from Human Extraintestinal Infections

PETER H. WILLIAMS
Department of Genetics
University of Leicester
Leicester, England

I. INTRODUCTION

Although *Escherichia coli* is a major aerobic component of the fecal flora of a healthy individual, some strains are pathogenic for humans and animals (28,38). They may exert their effects either within the intestine, causing diarrheal or dysenterylike diseases, or at a variety of extraintestinal sites. Systemic colibacillosis, for example, is a condition that affects a number of domestic animals, particularly calves and lambs, and less frequently piglets (28). The pathogenesis of the disease is unclear, and although infecting bacteria are thought to enter the bloodstream directly from the alimentary tract, there is in fact no experimental evidence that strains associated with the condition are able to penetrate target cells. Systemic *E. coli* infections are also a problem in poultry, where bacteria are inhaled in fecally contaminated dust, colonize the lungs, and then spread systemically to other organs. In some cases bacterial infection may be opportunistically related to virus or mycoplasma infection. Septicemia in humans is relatively uncommon but *E. coli* is the most frequently isolated pathogen when it occurs. It is also a major cause of human neonatal meningitis (38). Septicemia is usually preceded by severe infection elsewhere in the body (e.g., in the urinary tract), or results from trauma leading to contamination by gut contents; isolates from human disease are probably unable to translocate directly from the gut to cause systemic infection, as is thought to be the case in animals. The route of infection in neonatal meningitis is not known, but it has been speculated that the capsular antigen K1, which is most frequently associated with these strains, exhibits some kind of organotropism for the meninges.

Escherichia coli is also the most common cause of uncomplicated human urinary tract infections (UTI) (39). Because the serotypes of strains most frequently isolated from UTI are also the most common serotypes among fecal isolates, infection is generally considered to ascend the urinary tract following colonization of the periurethral area by fecal bacteria. The greater length of the male urethra presumably reduces the chances of such infection, so that there is a lower incidence of UTI in males than in females. However, in the first few months of life UTI is actually more common among boys than girls (15); thus, there is a high probability of fecal contamination of the prepuce of infant boys [which is significantly reduced in areas where infant circumcision is a widespread practice (52)], but preputial colonization by bacteria is likely to be rare in older boys, and the frequency of male UTI is therefore extremely low in the absence of predisposing factors. In females UTI occurs across the age range, and is often recurrent (38,39).

Several clinical forms of UTI are defined (39), the most common of which is asymptomatic (or covert) bacteriuria (ABU). While ABU is not clinically apparent to the patient, it may represent a silent (benign) phase of a primary or recurrent symptomatic infection. Cystitis, the most common symptomatic form of UTI, is characterized by inflammation and consequent oversensitivity of the bladder and by frequent and painful passage of urine. In some cases bacteria may ascend from the bladder into the upper urinary tract to infect the kidneys, causing pyelonephritis, although this is relatively uncommon in the absence of predisposing anatomical abnormalities. Chronic pyelonephritis, which is usually a consequence of UTI in very early life, is a very serious condition in which sufficient kidney tissue may be destroyed to cause renal failure.

The simplest explanation for UTI, encapsulated in the "prevalence theory" of *E. coli* pathogenicity, is that certain strains are most frequently isolated from UTI merely because they are the most prevalent strains in the gut flora, and therefore the most likely to contaminate the urinary tract. The obvious complexity of the pathogenesis of UTI and other extraintestinal infections, however, and the operation of a range of host defense mechanisms protecting the urinary tract and other extraintestinal sites, make the alternative "special pathogenicity theory" more attractive. This proposes that certain strains of fecal origin most often cause disease because they possess special characteristics—virulence determinants—that allow successful invasion of the well-defended tissues and body fluids of the host animal. It is the identification of these virulence determinants that forms the basis of diagnostic tests and provides opportunities for specific therapies. Since assessment of the role of potential virulence determinants is now most often approached by the techniques of gene cloning, it is natural that cloned DNA should be used to probe for the presence of the genetic capacity to express essential virulence determinants. The potential of such probes for epidemiology and diagnosis of extraintestinal infections, particularly UTI, by *E. coli* is the subject of this chapter.

II. BACKGROUND

Several potential virulence determinants have been suggested for uropathogenic *E. coli* (UPEC); their prevalence, role in pathogenesis, and possible use as diagnostic markers are outlined individually later, but it is essential to remember the multifactorial nature of disease processes, and the probability that effective probes will be those that seek combinations of virulence determinants.

A. Adhesins

The major nonspecific defense mechanism of the urinary tract is frequent unidirectional flushing during micturition; bacteria that are not tightly adherent to urinary epithelia are likely to be expelled from the body at such times. Ultrastructural studies of most pathogenic strains of *E. coli* reveal the presence of fine fibrils, 1–2 μm long and up to ~9 nm across, radiating from the bacterial surface. These structures, called fimbriae, are protein oligomers that interact with carbohydrate moieties of components of mammalian cell surfaces to promote bacterial adherence. Indeed the high degree of specificity of such adhesion–receptor interactions in bacterial adherence means that most pathogens infect only a restricted range of animal species, and within any host they tend preferentially to colonize particular body tissues, sometimes only of a particular age. In terms of the "special pathogenicity" theory, therefore, adhesins can be regarded as the most specialized of characteristics. Thus a fecal strain unable to adhere to any nonintestinal tissues will be unlikely to be able to colonize extraintestinal sites in the uncompromised host.

Bacteria of several species from a wide range of sources adhere to a variety of animal cell types by a mechanism that is inhibited by low concentrations of D-mannose or certain D-mannose-containing compounds [hence the designation mannose-sensitive (MS) adherence]. Electron-microscopic examination of MS-adherent bacteria indicates that so-called type 1, or "common," fimbriae are associated with the phenotype (30). The widespread distribution of type 1 fimbriae among pathogenic and benign bacteria and of mannose-containing receptors on mammalian cell surfaces, however, suggests that, while MS adherence may be selectively advantageous in some general physiological function, it has no obvious or specific role in bacterial virulence. Indeed, although *E. coli* expressing type 1 fimbriae will bind to urinary mucus because it contains the mannose-rich Tamm–Horsfall protein, any advantage that this ability provides for colonization of the lower urinary tract is likely to be somewhat counteracted by bacterial removal as mucus is shed during micturition (33).

Uropathogenic *E. coli* may also express mannose-resistant adhesins, of which the most common and best studied are the P fimbriae (43), so called because they bind to a carbohydrate component of the human P blood group antigen (23). They mediate adherence to uroepithelial cells, and it is thought that they aid translocation of bacteria from the bladder to the upper urinary tract. The incidence of P fimbriae among pyelonephritis strains is very high, while among strains isolated from lower UTI and from the feces of healthy individuals they are less common. At least eight serologically distinguishable P-specific adhesins have been recognized

(29), designated F7 (two variants) to F13, with considerable similarities at the genetic level between several of the gene clusters (41). In addition, a few P fimbriae have not yet been assigned F numbers, and some UPEC express P-independent adhesins with other specificities; these are named according to their receptor structures where known [e.g., M fimbriae recognize part of the M blood group system (22), and S fimbriae adhere to sialyl moieties of sialyl glycoproteins (24)], or else are simply designated "X-adhesins" (42).

Fimbrial structures are strongly antigenic; evolutionary pressures for divergence of major epitopes (within the limits imposed by maintenance of functional integrity) are therefore very strong on the genes encoding subunit proteins. It may be that other sequences in fimbrial gene clusters are more conserved, but until such common sequences are identified, a suitable general-purpose probe is unlikely to be available.

B. Capsules

Fresh bacterial isolates from patients with invasive infections are usually encapsulated by acidic polysaccharide matrices (21,30) ranging in structure from simple homopolymers (e.g., the K1 capsule of *E. coli* and the group b meningococcal capsule are both composed of polymerized sialic acid) to more complex mixed polymers, which can be serologically distinguished because they are antigenic to varying degrees. Capsules cause mucoid colonial morphology on solid media and cell clumping in liquid media, but capsular synthetic ability is easily lost on subculture in the laboratory. The role of capsules as virulence determinants is believed to be defensive. Some capsules (e.g., K12 of *E. coli*) are very poorly immunogenic and may confer a level of serum resistance by masking potentially strong antigens from the immune system. Others (e.g., *E. coli* K1 and K5 capsules) resemble host structures to such an extent that they are recognized as "self" by the immune system (13,21). Moreover, although the number of capsular types in *E. coli* is large, only a few are particularly associated with individual diseases (K1 with neonatal meningitis, for example, and K1, K2, and K5 with pyelonephritis) and with independent isolates from different disease outbreaks or geographic areas (21). They are therefore excellent candidates as targets for DNA probes.

C. Iron Uptake

The average human body contains 4–4.5 g of iron, virtually all of it present in a variety of complexed forms, in metalloproteins such as hemoglobin, with the iron-binding glycoproteins lactoferrin and transferrin, or in insoluble aggregates such as ferritin (46). About 1–1.5 mg is lost daily in feces, urine, and sloughed cells in body secretions, although women lose significantly greater amounts during menstruation. Losses are made up from the diet; the average human daily intake is ~10 mg, of which only ~1.5 mg is in a form capable of being absorbed.

The tissues and body fluids of a healthy individual are normally sterile, in part at least because the level of available ferric iron is too low to support the growth of microorganisms (46). For instance, transferrin is normally only ~20–30% saturated with iron, and the level of "free" ferric ions in solution in the serum is in the attomolar (10^{-18}) range; bacterial growth requires micromolar amounts. Moreover, the natural response of an otherwise healthy body to microbial invasion is the so-called hypoferremic response in which relocation of circulating iron to hepatic storage tends to reduce its availability still further (46,47). Conversely, patients with conditions in which free serum iron levels are abnormally high, such as anemia, are particularly susceptible to opportunistic bacterial infections.

Most bacterial species acquire iron by secreting low molecular weight (<1000) high-affinity iron-specific chelators called siderophores (19). Uptake of iron is initiated by the interaction of ferrisiderophore complexes with cognate receptor proteins (usually in the range 70,000–90,000 Da) in the bacterial cell envelope. It is characteristic of siderophore systems that their expression is iron-regulated; that is, they are switched off when iron is readily available, and on in conditions of iron stress (deprivation or nonavailability). Most enterobacterial isolates from systemic and extraintestinal infections of humans use the siderophore aerobactin, a hydroxamate derivative of lysine coupled to citric acid (45). This molecule effectively competes with transferrin and lactoferrin for bound iron and shuttles it to the cell surface receptor, even at very low cell densities and siderophore concentrations (50). Isolates that make aerobactin, however, usually have the capacity to make enterochelin (enterobactin), a phenolic siderophore (36) with the highest affinity for iron of any known natural compound. But despite its power as a chelating agent, enterochelin is (for a variety of reasons) much less effective as a siderophore than aerobactin in the stringent iron stress conditions that prevail at extraintestinal sites of infection. Apart from various structural considerations, such as the propensity of enterochelin to bind to serum proteins, the genetic determinants of the aerobactin system seem to be preferentially expressed, particularly at very low cell densities (50), and indeed in some enterobacterial species (e.g., *Shigella flexneri*) the enterochelin genes seem to be permanently repressed.

While enterochelin is a virtually universal feature of *E. coli* strains, whether pathogenic or harmless, aerobactin is significantly correlated with extraintestinal diseases in several independent epidemiological studies

(6,26,31,37), and is therefore a suitable diagnostic marker for probe technology.

D. Hemolysin

A significant proportion of *E. coli* isolates from extraintestinal infections secrete a protein with hemolytic activity (10)—up to 50% of pyelonephritis strains, for example, compared with only 5–20% among fecal isolates. Although it has been proposed that hemolysin is a virulence determinant in UTI (48), the mechanism of action *in vivo* is unclear. Release of erythrocyte-bound iron has been suggested as a possible role (25), but as well as erythrocyte lysis at high concentrations, hemolysin also causes the inhibition of neutrophil chemotaxis and other cytotoxic effects (44). Genetic determinants of hemolysin synthesis and secretion from several strains of animal and human origin have been cloned and studied in detail, but the biological determination of activity on blood agar plates is so quick and easy that the application of probe technology would be a redundant complication.

E. Serum Resistance

Strains of enteric bacteria capable of causing bacteremia and other extraintestinal infections are often said to be significantly resistant to the bactericidal effects of serum (40). The effect is probably multifactorial, the overall result being to block the activity of the so-called membrane attack complex (MAC) of terminal proteins of the complement cascade. The MAC forms normally at the surface of serum-resistant organisms but is released before it can penetrate the outer membrane to destroy its integrity and kill the organism. It has been suggested that structural features that tend to reduce membrane fluidity may block integration of MAC into the membrane and so confer resistance to killing (40). For instance, a high degree of substitution of the lipopolysaccharide (LPS) core with O-antigenic side chains may cause lowered membrane fluidity and so enhance refractoriness to MAC activity. However, antigenically indistinguishable smooth isolates may vary dramatically in their resistance, so that not only is the correlation between serum resistance and virulence suspect, but our ability to make rational choices of gene probes is at present somewhat doubtful.

As mentioned before, acid polysaccharide capsules may provide some level of resistance either by masking more powerful antigens or by mimicking host structures. Also certain outer-membrane proteins are said to confer added serum resistance, but the mechanism of action is unknown; all examples are abundant proteins whose presence in the outer membrane may reduce membrane fluidity sufficiently to affect the efficiency of MAC penetration, at least of very susceptible strains such as *E. coli* K-12. At most they are probably only marginal components of a complex phenomenon, and there is certainly no obvious correlation between their expression and virulence (51).

III. RESULTS AND DISCUSSION

A. Choice of Screening Targets

The genetic determinants of several P-fimbrial systems of human UTI isolates have been cloned and subjected to physical and functional analyses (41). There is considerable similarity between restriction maps of the various cloned systems, and genes of one system frequently complement mutational defects in other gene clusters. However, no clone or fragment of a clone has so far been suggested as a hybridization probe that will unequivocally recognize a range of serologically distinct P fimbriae. Of the other potential virulence determinants of extraintestinal *E. coli* described earlier, hemolysin is so easily detected biologically and serum resistance is so ill-defined a phenomenon that the use of DNA probes to screen for either is not a reasonable option. Probes for capsules and the aerobactin system do, however, offer distinct advantages as quick and reproducible diagnostic tests.

B. Detection of Capsules

The genetic determinants of some of the commonest capsular types have been cloned (5,34). Physical and functional mapping suggests that the organization of genes for the synthesis and surface assembly of acidic polysaccharides is similar in all systems so far examined, comprising three functional regions as indicated in Fig. 1. Region 2 encodes enzymes involved in sugar biosynthesis and polymerization, region 3 may be important in various postpolymerization modifications of the polysaccharide chain, and region 1 specifies the process of translocation of the polysaccharide to the cell surface. To demonstrate the effectiveness of colony hybridization (17) for identifying encapsulated organisms and to illustrate its potential to discriminate between individual capsular types, various probes were used to screen isolates of *E. coli* whose K serotype had been determined by standard diagnostic methods (35). The probes used were derived from region 1 of a clone of the K1 genetic determinants, and from region 2 of each of the cloned K1, K5, and K12 determinants (Fig. 1).

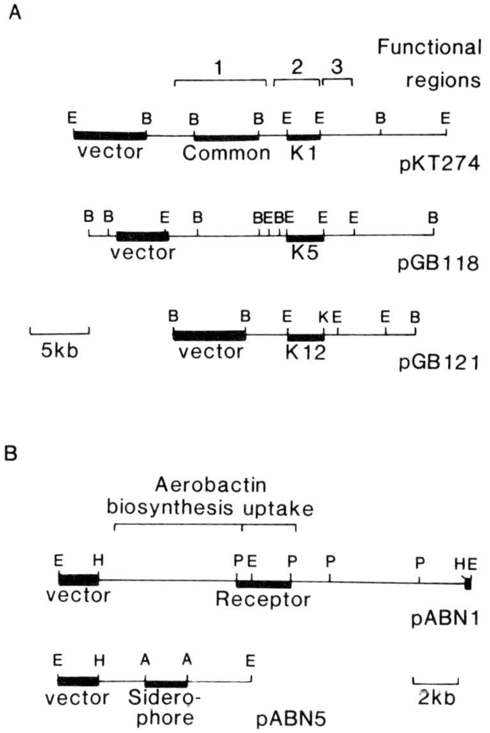

Fig. 1. Probes for detection of acidic polysaccharide capsules and the aerobactin iron uptake system among clinical isolates of *Escherichia coli*. (A) Restriction maps of plasmids carrying the K1 (pKT274), K5 (pGB118), and K12 (pGB121) determinants (5,3,35) showing the approximate extents of functional regions 1 (membrane transport), 2 (sugar biosynthesis and polymerization), and 3 (postpolymerization modification). Two probes were derived from pKT274; the "common" probe, a 5.5-kb *Bam*HI fragment representing the K1 region 1, and the K1-specific probe, 2.8-kb *Eco*RI fragment comprising region 2. K5 and K12 region 2 probes were a 3.1-kb *Eco*RI fragment of pGB118 and a 3-kb *Eco*RI–*Kpn*I fragment of pBG121, respectively. (B) Restriction maps of plasmid pABN1, which carries the entire aerobactin system, and the derived plasmid pABN5, which has determinants for aerobactin biosynthesis but not for uptake (3.7). The receptor probe was a 2.3-kb *Pvu*II fragment of pABN1 and the siderophore probe a 2-kb *Ava*I fragment from within the biosynthesis genes carried on pABN5. Only *Ava*I (A), *Bam*HI (B), *Eco*RI (E), *Hin*dIII (H), *Kpn*I (K), and *Pvu*II (P) cleavage sites relevant to the preparation of probes are shown; detailed restriction maps of the plasmids are published elsewhere. All vectors (pHC79 for pKT274, cos4 for pGB plasmids, and pPlac for pABN plasmids) confer ampicillin resistance.

The K1 region 1 appeared to be strongly homologous with sequences in the genome of all K^+ strains tested (Table I), whatever their associated O serotype, indicating that the genes for membrane transport functions are well conserved among capsule gene clusters. This "common" probe did not, however, hybridize with the DNA of any strains determined by other means to lack a capsule; thus, while it is obviously conceivable that mutations in capsule genes would render a strain phenotypically K^-, among fresh clinical isolates at least failure to express a capsule is indicative of loss of part or all of the capsule gene cluster. The common probe may in fact also be species-specific; so far no other bacterial species commonly isolated from human UTI and other extraintestinal infections, or indeed from intestinal diseases, has been shown to carry sequences

TABLE I

Colony Hybridization Screening with Capsule Probes of Strains of *Escherichia coli* Serotyped by Conventional Methods

K serotype	Associated O serotype[a]	Hybridization with indicated probe: Common	K1 (region 2)	K5 (region 2)	K12 (region 2)
1	1,2,6,7,12 16,22,R	+	+	−	−
2	2,6	+	−	−	−
5	2,6,18,25, 75,R	+	−	+	−
12	4,18	+	−	−	+
13	6,22	+	−	−	−
14	21	+	−	−	−
25	8	+	−	−	−
34	9,55	+	−	−	−
52	88	+	−	−	−
53	6	+	−	−	−
92	73	+	−	−	−
93	73	+	−	−	−
95	75,83	+	−	−	−
96	77	+	−	−	−
99	77	+	−	−	−
100	75	+	−	−	−
K^-	1,2,4,8,9 12,18,22,26 50,55,77, 83,111,119	−	−	−	−

[a] R, rough.

homologous with any probe representing part of an *E. coli* capsule gene cluster.

In the case of the three region 2 probes tested there was absolute coincidence of hybridization data with the results of conventional tests. Thus, polysaccharide biosynthesis is K antigen-specific at the DNA sequence level even among capsules that show some biological cross-reactivity. K5 and K95, for example, are sufficiently similar that they act as receptors for the same phage; nevertheless the K5 region 2 probe showed no homology with the DNA of an O83:K95 organism. Similarly, although K12 and K82 cross-react serologically, the K12 region 2 probe gave no hybridization signal with the DNA of an O139:K82 strain isolated from an animal infection. When posthybridization washes were performed at low stringency (1× SSC instead of the normal 0.1× SSC; see Section VIII,D) there was weak binding of the K1 region 2 probe to the DNA of K92 strains, but this serotype is sufficiently rare that potential mix-ups are unlikely to be a problem in diagnostic tests on clinical isolates. The implication of these data is that each encapsulated strain carries genes for only a single capsular antigen, rather than a battery of determinants of which only one is selectively expressed.

C. Detection of the Aerobactin System

The genetic determinants of the aerobactin iron uptake system have been cloned from an *E. coli* plasmid as a multicopy recombinant plasmid designated pABN1 (3). The system comprises a cluster of five contiguous genes, four of which encode enzymes for the biosynthesis of the siderophore and the fifth specifies the outer membrane ferriaerobactin receptor protein (7). Two probes have been extensively used (6,26,49): an *Ava*I fragment containing part of the biosynthesis genes and a *Pvu*II fragment representing virtually the entire receptor gene (Fig. 1). While the *Pvu*II fragment is easily isolated from pABN1, the *Ava*I probe is most conveniently obtained by digestion of a smaller derivative of pABN1 designated pABN5 (3). Isolates from human extraintestinal infections fall into a restricted number of groups on the basis of serotype, and indeed some serotypes such as O1:K1 and O4:K12 are particularly well represented among UTI isolates. The aerobactin system is a feature of the majority of isolates of all but two of the most common serotypes screened by colony hybridization in my laboratory (Table II). Among the strains tested, discrepant results with the two probes were very rare (<1%), confirming the truism that strains that make aerobactin also require a receptor for its uptake. Most discrepancies revealed a class of strains that did not make aerobactin, but could utilize it if, for instance, it was secreted by co-

TABLE II

Incidence of the Aerobactin System among the Major Clonal Groups of *Escherichia coli* Associated with Human Extraintestinal Diseases

Serotype (arranged by K type)	Main diagnostic group[a]	Aerobactin production by >50% of isolates
O1:K1:H7	Pyelonephritis	+
O1:K1:H⁻	ABU	+ (−)[b]
R:K1:H33	Sepsis	+
O2:K1:H4	Pyelonephritis	+
O7:K1:H⁻	Pyelonephritis, sepsis	+
O16:K1:H6	Pyelonephritis	+
O16:K1:H⁻	Pyelonephritis	+
O18:K1:H7	UTI, sepsis	+
O2:K2:H1	Lower UTI	+
O6:K2:H1	UTI, Sepsis	+
O2:K5:H4	Pyelonephritis	+
O6:K5:H1	UTI	+ (−)
O18:K5:H7	Pyelonephritis	+
O18:K5:H⁻	UTI	+
O25:K5:H1	Lower UTI	+
O75:K5:H⁻	UTI	+
O4:K12:H1	Pyelonephritis	+
O4:K12:H5	Pyelonephritis	+
O6:K13:H1	Cystitis	−
O75:K95:H⁻	Cystitis	−
O75:K100:H5	Cystitis	+ (−)
O4:K⁻:H?	UTI in boys	+

[a] Disease from which the serotype was most frequently (but not exclusively) recovered. ABU, Asymptomatic bacteriuria; UTI, any urinary tract infection.

[b] + (−) Indicates that strains showing positive hybridization signals all gave negative results in bioassays.

infecting organisms; while such a phenotype was rare among *E. coli* isolates, it was relatively common among strains of *Klebsiella pneumoniae*. In the other type of discrepancy, lack of hybridization only with the receptor probe, posthybridization washing at low stringency usually allowed weak binding, perhaps indicating a markedly diverged receptor gene in these rare strains.

Hybridization of the aerobactin probes with DNA sequences of clinical isolates was an effective indicator of the presence of the aerobactin system (Table III) and therefore of its potential involvement in the pathogenesis of these strains. It was more effective than biological assays mainly because the latter were susceptible to trivial cultural variations that may affect the

TABLE III

Comparison of Colony Hybridization and Bioassay Methods of Screening for the Aerobactin System

Source of isolates[a] (references)	Total number tested	Number (%) aerobactin positive	Number (%) of positive strains in: Both tests	Colony hybridization only	Bioassay only
1. Human extraintestinal infections (8)	516	313 (60.7)	258 (82.4)	43 (13.7)	12 (3.8)
2. Animal extraintestinal infections (23)	576	155 (26.9)	139 (89.7)	13 (8.4)	3 (1.9)
3. Human UTI (32)	466	261 (56.0)	232 (88.9)	23 (8.8)	6 (2.3)

[a] Each study included fecal isolates from healthy individuals (human or animal) as ''control'' material for comparison with infection isolates from various diagnostic groups. The total numbers of fecal isolates were 67, 201, and 97, of which 23 (34.3%), 15 (7.5%), and 42 (43.3%) were aerobactin-positive for studies 1, 2, and 3, respectively.

efficiency of induction of iron-regulated systems (although there may also be inherent strain differences in the regulation of the aerobactin system; see Table II and Section V,B). A small but significant percentage of isolates from virtually any source, however, displayed the intriguing paradox of an aerobactin-positive phenotype [i.e., they made a derivative of lysine that was electrophoretically and chromatographically indistinguishable from authentic aerobactin (26)], but an apparently negative genotype with the probes shown in Fig. 1. Thus there is a class of extraintestinal *E. coli* isolate that carries aerobactin genetic determinants sufficiently diverged from the ones cloned as pABN1 that hybridization even at low stringency was not detectable. It is recommended that bioassays and the probe test should both be carried out if possible, so that these strains (perhaps up to 2% of all isolates) do not go undetected.

IV. CONCLUSIONS

It has been known for many years that certain surface features are often associated with bacterial isolates from human or animal disease. The correlation of the *E. coli* K1 capsule with human neonatal infections such as septicemia and meningitis, and with pyelonephritis in older children has already been noted. Moreover, of the almost 200 different *E. coli* O serotypes, <50 are significantly associated with disease isolates and different modes of pathogenesis (38). Recent detailed studies of a variety of properties of pathogenic *E. coli* have led to the suggestion that, rather than there being a causal relationship between all these characters and disease, they may simply be independently conserved in clonally derived bacteria (1). That is, the prevalence of a particular phenotype in a pathogen does not necessarily mean that it is a virulence determinant, but rather that it is a conserved feature of genetically related organisms, or clones.

In this context, clones are defined as bacterial isolates from geographically and temporally independent sources with so many similar genotypic and phenotypic traits that they are likely to be derived from a common ancestor. This implies genetic stability during linear descent, due either to direct selection for several characters that identify the clone or to selection for a single (virulence) determinant to which other identifying characters are genetically closely linked. Specific clonal groups are often associated with particular diseases (e.g., O1:K1 and O18:K5 with pyelonephritis, and O18:K1 with neonatal meningitis), but no clonal groupings have yet been detected that distinguish between isolates from healthy and diseased individuals.

Analysis of genetic variation among natural populations of *E. coli*, however, suggests that serotypes do not necessarily define genetically homogeneous or even related groups of isolates (9). Thus, while full O:K:H serotyping is usually complete enough to identify related organisms (although cases have been noted of strains with the same three-antigen serotype that were genetically diverse at several multiallele enzyme loci), isolates with merely the same O or O:K determinants are generally quite diverse in genotype. Thus, surface structures, either individually or in combination, do not by themselves adequately reflect overall genetic relatedness; extensive analysis of enzyme variants would also be required (37).

In effect, then, the identification of a phenotype as a virulence determinant requires different criteria from those needed to assert confidently that a particular marker would be useful for diagnosis. In the former case it is essential to know that the proposed virulence determinant exists among populations of clonally unrelated pathogens; in the latter case, however, it is simply necessary to know that the screening procedure being used will identify most, if not all, of the isolates obtained at a particular time and place, whether in an epidemic or as sporadic outbreaks. Since capsular typing is part of the serotyping scheme, and aerobactin is so widespread among isolates of a wide range of serotypes, it is clear that a composite probe that detects the two genotypes will be very effective in identifying pathogenic isolates.

V. GENE PROBES COMPARED WITH OTHER METHODS OF DETECTION

A. Capsules

Routine identification of specific K types is usually performed by serological techniques; countercurrent immunoelectrophoresis is the preferred method (20). Of the most frequent K types among *E. coli* isolates from extraintestinal infections, K12 antiserum is readily available, but K1 and K5 are such poor immunogens, probably because they resemble host structures, that only low-titer polyclonal antisera are obtainable. Recently, however, monoclonal antibodies against *E. coli* antigens have been generated (4,14,37a).

Bacteriophages that specifically lyse strains possessing K1 or K5 capsules are also used in typing (16,18). Indeed it is often necessary to use a combination of serological and phage typing methods to obtain unequivocal identification of K1, since certain isolates may be positive in one or other test but not both. In addition, it is possible that other surface structures may interfere with capsule determination by phage typing, perhaps by shielding phage attachment sites. A further complication is that serologically different K antigens, such as K5 and K95, may act as receptors for the same phage, although in most cases this is unlikely to cause practical problems if one of the receptor types is rare.

Colony hybridization has several advantages over traditional methods of antigen typing. It obviates the requirement for high-titer antisera or phage preparations, and is particularly useful if large numbers of clinical isolates are to be screened. It is also independent of the vagaries of variable environmental effects on capsule gene expression.

B. Aerobactin

Detection of siderophore secretion requires either chemical or biological tests. Aerobactin, being a hydroxamic acid, can be quantified by the so-called Csáky method described over 40 years ago (11); however, the test is complicated and dangerous (it involves the potential carcinogen 1-naphthylamine), and it may not in fact detect the levels of aerobactin secreted by most *E. coli* strains without prior concentration of the culture supernatant by passage over an anion exchange resin such as Dowex-1. A method involving acidic ferric chlorate, described a mere 20 years ago (2), is quicker and safer, but less sensitive. However, both tests suffer from not being specific for any particular hydroxamate compound, unlike a bioassay developed in my laboratory (8), which specifically detects aerobactin. The main disadvantage of the bioassay is that it may be susceptible to trivial variations in bacterial physiology. For instance, bacteria grown in iron-rich conditions (such as storage media) accumulate iron, which represses siderophore systems; thus, while growth on bioassay test plates is usually adequate to deplete iron pools sufficiently that induction of the aerobactin system will occur, in some strains repression may not be lifted sufficiently to give a positive reaction in the bioassay without repeated passaging in iron-restricted conditions. Moreover, it has been observed (Table II) that isolates of certain serotypes (e.g., O1:K1:H$^-$, O6:K5:H1, and O75:K100:H5), which carry sequences homologous with the aerobactin probes, do not express the aerobactin system at all in the conditions of the bioassay (31,32), but may, of course, in the very different environment of the infected body.

Alternative biological tests for the aerobactin system are based on the presence of an outer-membrane protein that acts as the aerobactin receptor. This protein is also the receptor for cloacin DF13 (7,8), a bacteriocin of *Enterobacter cloacae*, and for a virulent bacterophage recently isolated in my laboratory. However, while sensitivity to either of these agents is a convenient indicator of the presence of a functional receptor in laboratory strains of *E. coli,* their usefulness for screening clinical isolates is limited by the presence of surface structures, primarily LPS, which act as barriers to cloacin or phage adsorption.

Hybridization assays test directly for the presence of particular genes rather than their expression in the conditions of a laboratory assay. In the case of established virulence phenotypes, the presence of genetic determinants in a pathogenic isolate strongly suggests that they must be selectively advantageous in the complex dynamic environment of the body of an infected animal; a negative result in a bioassay, however, may indicate only that the simple controlled test environment is not adequate to induce expression of the appropriate genetic determinants.

VI. PROSPECTS FOR THE FUTURE

Urinary tract infection is diagnosed by detecting bacteria in the urine. However, normal urine from uninfected individuals may often be slightly contaminated with fecal organisms, and so a statistical definition of UTI must be applied for accurate diagnosis. If in a milliliter specimen of urine there are $>10^5$ organisms of a single type, there is a high probability (>80%) that the bladder is infected. Two or three such results on independent urine samples would virtually rule out fecal contamination as the source of the bacteria present (38).

Any proposed new method for the detection of a particular pathogen must have distinct advantages (e.g., of speed, cheapness, or enhanced predictive value) over existing technology if it is to win general acceptance in diagnostic and research laboratories. Hybridization with the probes described in this chapter has such advantages; it detects the presence of bacteria that have two key virulence determinants of extraintestinal disease, both of which are rare among "normal" fecal isolates that might simply be contaminants. Moreover, DNA probe technology allows more rapid identification of smaller numbers of infectious bacteria [at least an order of magnitude less than the 10^5 organisms/ml of significant infection (35)], especially those with the potential to invade the upper urinary tract, so permitting earlier diagnosis and consequent prompt treatment.

VII. SUMMARY

Escherichia coli is the major bacterial species isolated from human UTI and a significant cause of neonatal meningitis and septicemia. Isolates from such extraintestinal diseases express a number of virulence determinants. Our understanding of their contribution to pathogenesis has in large part developed from recombinant DNA methodology, and the consequent existence of well-defined cloned genes provides the opportunity for the application of useful hybridization probes to the epidemiology and diagnosis of extraintestinal infections. Particularly appropriate are the genetic determinants of capsule biosynthesis and of the aerobactin iron uptake system. Encapsulated organisms are better able to survive the host's immune system and progress to the upper urinary tract where their effects are more severe and long-lasting, and from which systemic infections may begin. Cloned capsule genes comprise both common and serotype-specific regions, the former allowing detection of any encapsulated organism in a urine sample, the latter permitting precise identification if required. Aerobactin promotes bacterial proliferation in the iron-restricted microenvironments of the urinary mucosa and serum. Isolates that utilize aerobactin for growth are identified by probes comprising parts of the aerobactin gene cluster. A composite probe would allow earlier diagnosis of UTI.

VIII. MATERIALS AND METHODS

A. Preparation of Filters

Bacterial isolates to be screened by colony hybridization (7) were inoculated as short (1-cm) streaks in a grid array on Amersham Hybond-N nylon filter disks laid on the surface of nutrient agar plates and grown overnight at 37°C. Filters were carefully transferred, growth side upward, to a pad of Whatman 3MM paper dampened with 0.5 *M* NaOH–1.5 *M* NaCl for 7–10 min at room temperature to lyse bacteria (until bacterial streaks became "glassy" in appearance) and denature the DNA. Filters were then moved to a 3MM pad dampened with a neutralizing solution of 1.5 *M* NaCl–0.5 *M* Tris-HCl pH 7.2–1 m*M* EDTA for two periods of 3 min each, rinsed with 2× SSC (1× SSC is 0.15 *M* NaCl–0.15 *M* sodium citrate) to remove colony debris and residual agar, and air-dried. DNA was immobilized on filters by wrapping them in a single layer of plastic film (Dow Saran Wrap) and exposing them to short-wave UV light (e.g., on a transilluminator) for 2–5 min. Filters were stored dry at room temperature until required.

B. Preparation of Probe DNA Fragments

Plasmid DNA was cleaved with appropriate restriction enzymes, and restriction fragments were separated by electrophoresis in 0.8% low-melting-temperature agarose (FMC Sea Plaque LGT) gels. After electrophoresis, gels were stained with ethidium bromide by immersion in a 0.5 μg/ml solution for ~20 min and the DNA visualized on a short-wavelength UV transilluminator. The required bands were cut from gels with a scalpel, extraneous agarose was removed, and the gel slices were placed in preweighed 1.5-ml Eppendorf tubes and reweighed to determine the total weight of agarose. Distilled water was added (1.5 ml/g agarose), and tubes were placed in a boiling-water bath for 7 min to melt the agarose and denature the DNA.

C. Radiolabeling Probe DNA Fragments

Probe DNA was radioactively labeled to high specific activity by the random primer technique (12). "Oligo-labeling buffer" (OLB) comprised two parts of 0.24 *M* Tris-HCl pH 8.0–0.24 *M* $MgCl_2$–0.25 *M* 2-mercaptoethanol, and 0.48 m*M* each of dATP, dGTP, and dTTP : five parts 2 *M* HEPES pH 6.6 : three parts hexadeoxynucleotides (Pharmacia, 27-2166-01, 90 OD units/ml) in 10 m*M* Tris-HCl pH 7.6–1 m*M* EDTA. Boiled DNA fragments (5 ng in 15 μl H_2O) were incubated at 37°C for 1 hr or at room temperature overnight in a 25-μl reaction mixture containing 6 μl OLB with hexadeoxynucleotides, 1 μl bovine serum albumin (BSA; Pharmacia, enzyme grade, 10 mg/μl), 2 μl [α^{32}P]dCTP (Amersham, 0.37 MBq/μl, 110 TBq/mmol), and 1 μl *E. coli* DNA polymerase I Klenow fragment (Pharmacia, 1 unit/μl).

When radiolabeled probe DNA was used immediately for hybridization it was found to be unnecessary either to stop the labeling reaction or to remove unincorporated dCTP. On occasions where the efficiency of incorporation of label was assessed, the reaction was stopped by the addition of 75 μl 20 m*M* NaCl–20 m*M* Tris-HCl pH 7.5–2 m*M* EDTA–0.25% SDS, and the proportion of precipitable to total counts in a given volume of stopped reaction mixture determined (27). Incorporation levels of 40% (1 hr incubation) to 80% (overnight incubation) were typically obtained.

D. Hybridization

It was found to be generally unnecessary to use complex hybridization solutions when the probe DNA and the target DNA in clinical isolates were highly conserved. We used a simple solution of 1 m*M* EDTA and 7% (w/v) SDS in 0.5 *M* sodium phosphate buffer (made by dilution of a stock of 1 *M* Na_2HPO_4 solution adjusted to pH 7.2 with phosphoric acid). It has advantages, both practical and financial, over other protocols (27) that involve Denhardt's solution, competitor DNA, and dextran sulfate or polyethylene glycol, and gives clean interpretable results in the type of analyses described in this chapter.

Filters with bound DNA were washed in excess hybridization solution at 65°C for 30 min prior to addition of probe DNA. The wash solution was then replaced by sufficient fresh hybridization solution, to which radioactively labeled probe DNA had been added (after denaturation by boiling for 5 min), just to cover the filters. As many as six filters were processed in the same hybridization chamber without obvious loss of sensitivity or discrimination.

Incubation was continued at 65°C for 4 hr, or overnight if more convenient, after which the solution was carefully discarded, and the filters washed extensively at 65°C with three changes of 0.1 *M* sodium phosphate buffer (pH 7.2) containing 1% SDS for 15 min each, and finally with 0.1× SSC containing 0.1% SDS for a further 15 min. Filters were dried, wrapped in plastic film, and autoradiographed to detect specific hybridization of probe DNA to the total DNA of test isolates. Nonspecific binding of the probe was discounted by comparison with the signal from a known negative control strain on each filter.

ACKNOWLEDGMENTS

I am very grateful to Nick Carbonetti, Mark Roberts, Hilary Gavine, Tom Baldwin, and Karl Woolridge, past and present members of my laboratory, who developed and simplified the tests described in this chapter. I also acknowledge the many colleagues and collaborators who provided strains, expertise, hospitality, and encouragement. In particular I thank Stephen Parry and Margaret Linggood of Unilever Research, Graham Boulnois and Ian Roberts of the Microbiology Department at Leicester, Timo Korhonen of the University of Helsinki, and Ida and Frits Ørskov of the International Escherichia and Klebsiella Center (WHO) in Copenhagen.

REFERENCES

1. Achtman, M. (1985). Clonal groups and virulence factors among *Escherichia coli* strains. *In* "Enterobacterial Surface Antigens: Methods for Molecular Characterization" (T. K. Korhonen, E. A. Dawes, and P. H. Mäkelä, eds.), pp. 65–74. Elsevier, Amsterdam.
2. Atkin, C. L., and Neilands, J. B. (1968). Rhodotorulic acid, a diketopiperazine dihydroxamic acid with growth-factor activity. I. Isolation and characterization. *Biochemistry* **7,** 3734–3839.
3. Bindereif, A., and Neilands, J. B. (1983). Cloning of the aerobactin-mediated iron assimilation system of plasmid ColV. *J. Bacteriol.* **153,** 1111–1113.
4. Bitter-Suermann, D., Görgen, I., and Frosch, M. (1986). Monoclonal antibodies to weak immunogenic *Escherichia coli* and meningococcal capsular polysaccharides. *In* "Protein–Carbohydrate Interactions in Biological Systems" (S. Normark, ed.), pp. 395–396. Academic Press, London.
5. Boulnois, G. J., Roberts, I. S., Hodge, R., Hardy, K. R., Jann, K. B., and Timmis, K. N. (1987). Analysis of the K1 capsule biosynthesis genes of *Escherichia coli;* Definition of three functional regions for capsule production. *Mol. Gen. Genet.* **208,** 243–246.

6. Carbonetti, N. H., Boonchai, S., Parry, S. H., Väisänen-Rhen, V., Korhonen, T. K., and Williams, P. H. (1986). Aerobactin-mediated iron uptake by *Escherichia coli* isolates from human extraintestinal infections. *Infect. Immun.* **51,** 966–968.
7. Carbonetti, N. H., and Williams, P. H. (1984). A cluster of five genes specifying the aerobactin iron uptake system of plasmid ColV-K30. *Infect. Immun.* **46,** 7–12.
8. Carbonetti, N. H., and Williams, P. H. (1985). Detection of synthesis of the hydroxamate siderophore aerobactin by pathogenic isolates of *Escherichia coli. In* "The Virulence of *Escherichia coli*" (M. Sussman, ed.), pp. 419–424. Academic Press, London.
9. Caugant, D. A., Levin, B., and Ørskov, F. (1985). Genetic diversity in relation to serotype in *Escherichia coli. Infect. Immun.* **49,** 407–413.
10. Cavalieri, S. J., Bohach, J. A., and Snyder, I. S. (1984). *Escherichia coli* alpha hemolysin: Characteristics and probable role in pathogenicity. *Microbiol. Rev.* **48,** 326–344.
11. Csáky, T. A. (1948). On the estimation of bound hydroxylamine in biological materials. *Acta Chem. Scand.* **2,** 450–454.
12. Feinberg, A. P., and Vogelstein, B. (1984). A technique for radiolabeling DNA restriction endonuclease fragments to high specific activity. Addendum. *Anal. Biochem.* **137,** 266–267.
13. Finne, J., Leinonen, M., and Mäkelä, P. H. (1983). Antigenic similarities between brain components and bacteria causing meningitis. *Lancet* **2,** 355–357.
14. Frosch, M., Görgen, I., Boulnois, G. J., Timmis, K. N., and Bitter-Suermann, D. (1985). NZB mouse system for production of monoclonal antibodies to weak bacterial antigens: Isolation of an IgG antibody to the polysaccharide capsules of *Escherichia coli* K1 and group B meningococci. *Proc. Natl. Acad. Sci. U.S.A.* **82,** 1194–1198.
15. Ginsburg, C. M., and McCracken, G. H. (1982). Urinary tract infections in young infants. *Pediatrics* **69,** 409–412.
16. Gross, R. J., Cheusty, T., and Rowe, B. (1977). Isolation of bacteriophage specific for the K1 polysaccharide capsule antigen of *Escherichia coli. J. Clin. Microbiol.* **6,** 548–550.
17. Grunstein, M., and Hogness, D. S. (1975). Colony hybridization, a method for the isolation of cloned DNAs that contain a specific gene. *Proc. Natl. Acad. Sci. U.S.A.* **72,** 3961–3965.
18. Gupta, D. S., Jann, B., Schmidt, G., Golecki, J. R., Ørskov, I., Ørskov, F., and Jann, K. (1982). Coliphage K5, specific for *E. coli* exhibiting the capsular K5 antigen. *FEMS Microbiol. Lett.* **14,** 75–78.
19. Hider, R. C. (1984). Siderophore-mediated absorption of iron. *Struct. Bonding* (*Berlin*) **58,** 25–87.
20. Jann, K. (1985). Isolation and characterization of capsular polysaccharides (K antigens) from *Escherichia coli. In* "The Virulence of *Escherichia coli*" (M. Sussman, ed.), pp. 375–379. Academic Press, London.
21. Jann, K., and Jann, B. (1983). The K antigens of *Escherichia coli. Prog. Allergy* **33,** 53–79.
22. Jokinen, M., Ehnholm, C., Väisänen-Rhen, V., Korhonen, T. K., Pipkorn, R., Kalkkinen, N., and Gahmberg, C. G. (1985). Identification of the major sialoglycoprotein from red cells, glycophorin A^M, as the receptor for *Escherichia coli* IH11165 and characterization of the receptor site. *Eur. J. Biochem.* **147,** 47–52.
23. Källenius, G., Möllby, R., Svenson, S. B., Winberg, J., Lundblad, A., Svensson, S., and Cedergren, B. (1980). The P^K antigen as receptor for the haemagglutination of pyelonephritic *Escherichia coli. FEMS Microbiol. Lett.* **7,** 297–302.
24. Korhonen, T. K., Valtonen, M. V., Parkkinen, J., Väisänen-Rhen, V., Finne, J., Ørskov, F., Ørskov, I., Svenson, S. B., and Mäkelä, P. H. (1985). *Escherichia coli* strains associated with neonatal sepsis and meningitis: Serotypes, hemolysin production and receptor recognition. *Infect. Immun.* **48,** 486–491.

25. Linggood, M. A., and Ingram, P. L. (1982). The role of alpha haemolysin in the virulence of *Escherichia coli* for mice. *J. Med. Microbiol.* **15,** 23–30.
26. Linggood, M. A., Roberts, M., Ford, S., Parry, S. H., and Williams, P. H. (1987). Incidence of the aerobactin iron uptake system among *Escherichia coli* isolates from infections of farm animals. *J. Gen Microbiol.* **133,** 835–842.
27. Maniatis, E., Fritsch, E. F., and Sambrook, J. (1982). "Molecular Cloning: A Laboratory Manual." Cold Spring Harbor Lab., Cold Spring Harbor, New York.
28. Morris, J. A., and Sojka, W. J. (1985). *Escherichia coli* as a pathogen in animals. *In* "The Virulence of *Escherichia coli*" (M. Sussman, ed.), pp. 47–77. Academic Press, London.
29. Ørskov, I., and Ørskov, F. (1983). Serology of *Escherichia coli* fimbriae. *Prog. Allergy* **33,** 80–105.
30. Ørskov, I., and Ørskov, F. (1985). *Escherichia coli* in extraintestinal infections. *J. Hyg.* **95,** 551–575.
31. Ørskov, I., Svanborg Edén, C., and Ørskov, F. (1988). Aerobactin production of serotyped *Escherichia coli* from urinary tract infections. *Med. Microbiol. Immunol.* **177,** 9–14.
32. Ørskov, I., Williams, P. H., Svanborg Edén, C., and Ørskov, F. (1989). Assessment of biological and colony hybridization assays for detection of the aerobactin system in *Escherichia coli* from urinary tract infections. *Med. Microbiol. Immunol.* **178,** 143–148.
33. Parry, S. H., Abraham, S. N., and Sussman, M. (1982). The biological and serological properties of adhesion determinants of *Escherichia coli* isolated from urinary tract infection. *In* "Clinical, Bacteriological and Immunological Aspects of Urinary Tract Infection in Children" (H. Schulte-Wisserman, ed.), pp. 113–126. Thieme, Stuttgart.
34. Roberts, I., Mountford, N., High, N., Bitter-Suermann, D., Jann, K., Timmis, K. N., and Boulnois, G. J. (1986). Molecular cloning and analysis of genes for production of K5, K7, K12, and K92 capsular polysaccharides in *Escherichia coli. J. Bacteriol.* **168,** 1228–1233.
35. Roberts, M., Roberts, I., Korhonen, T. K., Jann, K., Bitter-Suermann, D., Boulnois, G., and Williams, P. H. (1988). DNA probes for K-antigen (capsule) typing of *Escherichia coli. J. Clin. Microbiol.* **26,** 385–387.
36. Rosenberg, H., and Young, I. G. (1974). Iron transport in the enteric bacteria. *In* "Microbial Iron Metabolism" (J. B. Neilands, ed.), pp. 67–82. Academic Press, New York.
37. Selander, R. K., Korhonen, T. K., Väisänen-Rhen, V., Williams, P. H., Pattison, P. E., and Caugant, D. A. (1986). Genetic relationships and clonal structure of strains of *Escherichia coli* causing neonatal septicemia and meningitis. *Infect. Immun.* **52,** 213–222.
37a. Söderström, T. (1985). *Escherichia coli* capsules and pili: Serological, functional, protective and immunoregulatory studies with monoclonal antibodies. *In* "Monoclonal Antibodies against Bacteria" (A. J. L. Macario and E. Conway de Macario, eds.), Vol. 2, pp. 185–212. Academic Press, Orlando, Florida.
38. Sussman, M. (1985). *Escherichia coli* in human and animal disease. *In* "The Virulence of *Escherichia coli*" (M. Sussman, ed.), pp. 7–45. Academic Press, London.
39. Sussman, M., and Asscher, A. W. (1979). Urinary tract infection. *In* "Renal Disease" (D. Black and N. F. Jones, eds.), pp. 400–436. Blackwell, Oxford.
40. Taylor, P. W. (1983). Bactericidal and bacteriolytic activity of serum against Gram-negative bacteria. *Microbiol. Rev.* **47,** 46–83.
41. Uhlin, B. E., Båga, M., Goransson, M., Lindberg, F. P., Lund, B., Norgren, M., and Normark, S. (1985). Genes determining adhesin formation in uropathogenic *Escherichia coli. Curr. Top. Microbiol. Immunol.* **118,** 163–178.

42. Väisänen, V., Elo, J., Tallgren, L. G., Siitonen, A., Mäkelä, P. H., Svanborg Edén, C., Källenius, G., Svenson, S. B., Hultberg, H., and Korhonen, T. K. (1981). Mannose-resistant haemagglutination and P-antigen-recognition are characteristic of *Escherichia coli* causing primary pyelonephritis. *Lancet* **2,** 1366–1369.
43. Väisänen-Rhen, V., Elo, J., Väisänen, E., Siitonen, A., Ørskov, I., Ørskov, F., Svenson, S. B., Mäkelä, P. H., and Korhonen, T. K. (1984). P-fimbriated clones among uropathogenic *Escherichia coli* strains. *Infect. Immun.* **43,** 149–155.
44. Waalwijk, C., McLaren, D. M., and de Graaf, J. (1983). In vivo function of hemolysin in the nephropathogenicity of *Escherichia coli. Infect. Immun.* **42,** 245–249.
45. Warner, P. J., Williams, P. H., Bindereif, A., and Neilands, J. B. (1981). ColV plasmid-specified aerobactin synthesis by invasive strains of *Escherichia coli. Infect. Immun.* **33,** 540–545.
46. Weinberg, E. D. (1978). Iron and Infection. *Microbiol. Rev.* **42,** 45–66.
47. Weinberg, E. D. (1984). Iron withholding: A defence against infection and neoplasia. *Physiol. Rev.* **64,** 65–102.
48. Welch, R., Patchen Dellinger, E., Minshew, B., and Falkow, S. (1981). Haemolysin contributes to the virulence of extraintestinal *Escherichia coli* infections. *Nature (London)* **294,** 665–667.
49. Westerlund, B., Siitonen, A., Elo, J., Williams, P. H., Korhonen, T. K., and Mäkelä, P. H. (1988). Properties of *Escherichia coli* strains isolated from urinary tract infections in boys. *J. Infect. Dis.* **158,** 996–1002.
50. Williams, P. H., and Carbonetti, N. H. (1986). Iron, siderophores, and the pursuit of virulence: Independence of the aerobactin and enterochelin iron uptake systems in *Escherichia coli. Infect. Immun.* **51,** 942–947.
51. Williams, P. H., Roberts, M., and Hinson, G. (1988). Stages in bacterial invasion. *J. Appl. Bacteriol. Symp., Suppl.,* pp. 131S–147S.
52. Wiswell, T. E., Smith, F. R., and Bass, J. W. (1985). Decreased incidence of urinary tract infections in circumcised male infants. *Pediatrics* **75,** 901–903.

6

Identification of Enterotoxigenic *Escherichia coli* by Colony Hybridization Using Biotinylated LTIh, STIa, and STIb Enterotoxin Probes

HIROFUMI DANBARA
Department of Bacteriology
The Kitasato Institute
Tokyo, Japan

I. INTRODUCTION

Enterotoxigenic *Escherichia coli* (ETEC) produces heat-labile (LT) or heat-stable (ST) enterotoxin or both. Laboratory tests for enterotoxin production in ETEC strains include animal models, tissue culture methods, immunological tests, and DNA–DNA hybridization tests. Animal models for the detection of LT are the ligated rabbit intestinal loops test (11) and the vascular permeability factor test (2). The widely used method to detect STI is the suckling mouse test (4,8,25). STII is detectable in ligated pig intestinal loops (1), but this test is used only in specialized laboratories. Cell lines of Chinese hamster ovary (8,9) and Y1 mouse adrenal (5) are used for the detection of LT. A wide range of immunological tests has been developed for the detection of LT, including passive immune hemolysis (6,27), GM_1-enzyme-linked immunosorbent assay (ELISA) (21), and a precipitin test, the Biken test (10). An ELISA using monoclonal antibodies (MAb) has been developed for the detection of STI (17,24,26). A colony hybridization test has been developed for the detection of LTI- or STI-producing ETEC strains using radioisotope-labeled enterotoxin probe DNA (19,20). Biotinylated polynucleotides have been synthesized (14) and incorporated into a DNA probe by an enzymatic reaction. Colony hybridization using biotinylated enterotoxin probes is easily applicable to clinical examination because no difficulty in handling radioisotopes is involved and the biotinylated probes can be stored for a longer period than the radioisotope-labeled ones. Identification of ETEC strains by colony hybridization using biotinylated enterotoxin probes has been reported (12). The successful use of the biotinylated probes for colony hybridization has also been reported for *Shigella* and enteroinvasive *E. coli* (25).

In this chapter a method of colony hybridization with ETEC strains using biotinylated LTIh, STIa, and STIb enterotoxin probes is reported.

II. BACKGROUND

Enterotoxigenic *E. coli* causes diarrhea in humans and animals. Human diarrhea due to ETEC appears to be prevalent mainly in tropical and developing countries and affects those traveling to these places. They produce LT or ST or both. Two classes of LT protein, LTI and LTII, are found, and both are further subdivided into two types: LTIh (human origin) and LTIp (porcine origin) for LTI and LTIIa and LTIIb for LTII. LTI is encoded by plasmids and LTII is mediated by a chromosome. The LT operon consists of genes for A and B subunits. The organization of LT operons of this family is closely homologous. The A subunit with ADP-ribosyltransferase activity is responsible for diarrhea as a result of its action on small-intestine fluid transport processes. Genes encoding the A subunit of LTIh and LTIp are extensively homologous, but they have restricted homology with LTII (57%). The B subunit binds to specific gangliosides with high affinity. Genes encoding the B subunit of LTI and CT (choleratoxin) are homologous, but they have no significant homology with those of LTII. Gangliosides, to which toxins bind, are GM_1 for LTI, GD_{1b} for LTIIa, and GD_{1a} for LTIIb.

Two classes of ST peptides are found. One is the methanol-soluble STI produced by human and animal ETEC isolates. The other is the methanol-insoluble STII, which is found in porcine ETEC strains. STI is further subdivided into two types: STIa and STIb originally isolated from porcine and human ETEC isolates, respectively. Both STI and STII are mediated by plasmids. The gene for both STI and STII is formed by a single operon encoding the structural gene for ST peptide and the signal peptide involved in the transmembrane transport of ST. The matured STIa and STIb consist of 18 and 19 amino acid residues, respectively, and differ in four amino acids. The matured STII seems to consist of 48 amino acid residues. STI and STII are not homologous. STI stimulates guanylate cyclase activity, and is responsible for diarrhea as a result of its action on small-intestine fluid transport processes.

Fimbrial adhesins are found among most ETEC strains. The adhesive property of the fimbriae to the mucosal surface would help overcome the mechanical clearance mechanism of peristalsis and would therefore promote colonization of the small intestine by ETEC. CFA/I (15,058 molecular weight) and CFA/II (consisting of three subtypes: MW 16,300 for CS1, 15,300 for CS2, and 14,700 for CS3) are adhesive fimbriae frequently found in ETEC isolates from human diarrheal disease. K88 (consisting of three subtypes: MW 27,540 for K88ab, 25,000 for K88ac, and 26,000 for K88ad), K99 (MW 18,500–19,500) 987P (MW 18,900), and F41 (MW 29,500) are common elements of animal ETEC adhesive fimbriae. CFA/I, CFA/II, K88, and K99 are plasmid-encoded whereas 987P and F41 are mediated by a chromosome.

Many epidemiological studies have demonstrated that O serotypes of ETEC strains isolated from diarrheal humans and animals are restricted to certain types. Strains ETEC with O serotypes such as O6, O8, O20, O25, O27, O78, O114, O115, O126, O128, O148, and O159 are frequently isolated from diarrheal patients from a wide range of countries. The strains producing both LTI and STI are frequently O6, O8, or O78. However, the strains producing only LTI belong to various O serotypes. Major constituents of isolates from diarrheal pigs are O8, O9, O45, O101, O138, O141, and O157 strains. O Serotypes of human and animal ETEC strains differ.

III. RESULTS AND DISCUSSION

A. Preparation of Enterotoxin Probes

Three enterotoxin probes, LTIh, STIa, and STIb, were prepared as shown in Fig.1. For the LTIh probe, a 660-bp *Xba*I–*Hinc*II region of 1032H-19 plasmid was used. This region contains a part of the structural gene for subunit A of LTIh. 1032H-19 plasmid produces LTI, detected by the Biken test (10), and is a recombinant between pBR322 and two *Hin*dIII fragments (1680 bp and 670 bp) of pMY1190 plasmid encoding LTIh and STIa (12). For the STIa probe, a 157-bp *Hin*f I region of pTE5014 plasmid was used. This region contains a part of the structural gene of STIa and 16-bp sequences flanked on 3′ end of the STIa gene. pTE5014

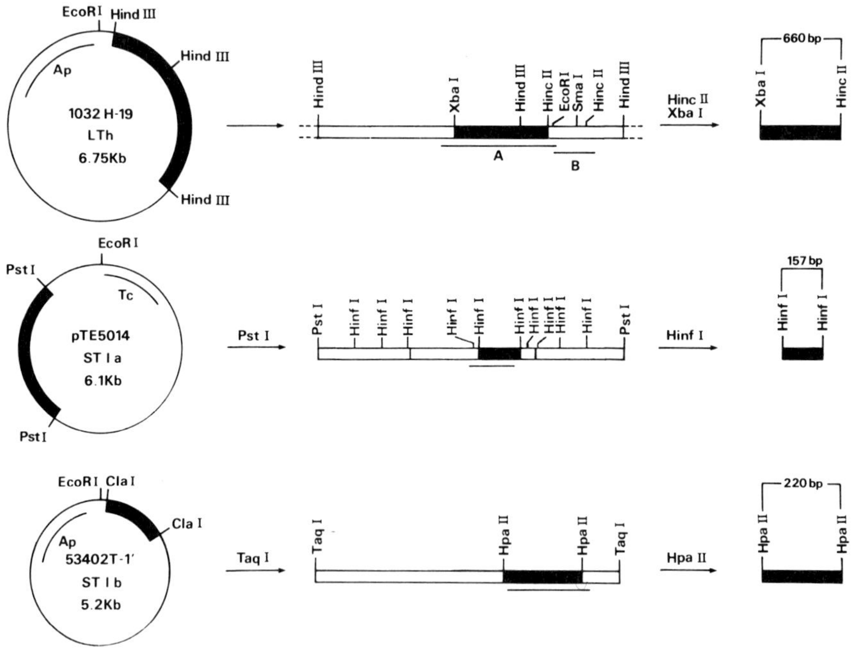

Fig. 1. Recombinant plasmids consisting of pBR322 and the structural gene for LTIh (LTh), STIa, or STIb enterotoxin, and preparation of the enterotoxin gene probes. Closed boxes in circular maps indicate the cloned restriction endonuclease fragments that encode the enterotoxins. Ap, Ampicillin; Tc, tetracycline. Closed boxes on linear maps indicate the location of the probe fragments, and thin lines beneath the maps indicate the ranges of the structural gene for the enterotoxins. A and B beneath the thin lines for the LTIh structural gene correspond to subunits A and B, respectively. The size of probes used for colony hybridization is shown above the closed boxes by the number of base pairs (bp).

produces STI, detected by the suckling mouse test (25), and is a recombinant between pBR322 and a 1700-bp *Pst*I fragment of pTE501 plasmid encoding STIa (12). For the STIb probe, a 220-pb *Hpa*II region of 53402T-l′ plasmid was used. This region contains a part of the structural gene of STIb and 10-bp sequences flanked on 5′ end of the signal peptide of STIb. 53402T-l′ plasmid produces STI, detected by the suckling mouse test, and is a recombinant between pBR322 and an 810-bp *Taq*I fragment of pMY1191 encoding STIb (12).

B. Labeling of Enterotoxin Probes

The LTIh probe was labeled with Bio-11-dUTP (2′-deoxyuridine triphosphate 5-allylamine-biotin) or [^{32}P]dATP by nick translation (16) with DNA polymerase I. The STIa and STIb probes were labeled with Bio-11-dUTP, [^{32}P]dATP and Bio-11-dCTP, or [^{32}P]dCTP, respectively, by 3′-end labeling (16) with the large fragment of DNA polymerase I. The biotinylated deoxyribonucleotide triphosphates, Bio-11-dUTP and Bio-11-dCTP, were purchased from Enzo Biochemicals (New York, New York). The radioisotope-labeled deoxyribonucleotide triphosphates, [^{32}P]dATP and [^{32}P]dCTP, were obtained from ICN Pharmaceuticals Inc., (Irvine, California). DNA polymerase I and the large fragment of the DNA polymerase I were purchased from Toyobo Co., Inc. (Osaka, Japan).

C. Detection of Enterotoxigenic *Escherichia coli* by Colony Hybridization with Enterotoxin Probes

Sterile nitrocellulose (NC) filters (60 × 60 mm; B85) were placed on nutrient agar plates. *Escherichia coli* strains to be tested were inoculated onto the filters. After overnight incubation at 37°C, the filters were removed from the agar plates, placed on Whatman no. 3 filter papers saturated with 0.5 *N* NaOH, and kept there for 10 min. The filters were then placed on Whatman no. 3 filters saturated with 1 *M* Tris-HCl (pH 7.5) and kept there for 5 min. This procedure was repeated three times. After the filters were kept on Whatman no. 3 filter paper saturated with 1.5 *M* NaCl–0.5 *M* Tris-HCl (pH 7.5) for 5 min, they were immersed in 0.05 *M* Tris-HCl (pH 7.5) containing 2 mg of proteinase K per milliliter and 0.5% Triton X-100 for 15 min at room temperature. The filters were air-dried and baked at 80°C for 2 hr. They were then incubated for 3 hr in a solution containing 10× Denhardt's solution, 5× SSC solution, 0.1% sodium dodecyl sulfate (SDS), and 50 μg of heat-denatured calf thymus DNA per milliliter. The filters were then incubated for 24 hr at 37°C in a solution containing 50% formamide, 1× Denhardt's solution, 5× SSC, 0.1% SDS, 10% dextran sulfate, 50 μg of heat-denatured calf thymus DNA per milliliter, and 10^5 cpm of ^{32}P-labeled DNA probe or 50 ng of biotinylated DNA probe per milliliter. The filters were washed in a solution containing 0.1× SSC and 0.1% SDS for 45 min at 65°C, rinsed twice in 2× SSC at room temperature for 5 min, and air-dried.

To detect hybridization between ^{32}P-labeled probes and the ETEC strain, the filters were exposed to X-ray film for 24 hr at −70°C. To detect hybridization between biotinylated probes and the ETEC strain, streptavidin- and biotin-conjugated acid phosphatase were used according to the instructions of the manufacturer (Detek I-acp; Enzo Biochemicals). The air-dried filters were incubated with blocking buffer containing 1× phosphate-buffered saline (PBS: 0.13 *M* NaCl–7 m*M* Na_2HPO_4–3 m*M* NaH_2PO_4), 2% bovine serum albumin (BSA)–0.05% Triton X-100–5 m*M* EDTA for 30 min at room temperature. The filters were incubated with streptavidin and biotin-conjugated acid phosphatase complex solution for 1 hr in a plastic bag and then placed in a washing buffer containing 10 m*M* KH_2PO_4 (pH 6.5)–0.5 *M* NaCl–0.5% Triton X-100–1 m*M* EDTA–2% BSA in a petri dish. After 25 min, the filters were placed in a clean dish containing 3-hydroxy-2-naphthoic acid 2,4-dimethylanilide phosphate, and 40 μg of fast violet B salt per milliliter and were allowed to stand at 37°C until color developed.

Nitrocellulose filters (Schleicher and Schuell Co., Dassel, Federal Republic of Germany), proteinase K (Boeringer-Mannheim, Federal Republic of Germany), and streptavidin and biotin-conjugated acid phosphatase complex solution (Detek I-acp; Enzo Biochemicals, New York, New York) were commercially purchased. The composition of the reagents used was as follows: 1× PBS (0.13 *M* NaCl–7 m*M* Na_2HPO_4–3 m*M* NaH_2PO_4), Denhardt's solution (0.02% Ficoll MW 400,000–0.02% polyvinylpyrrolidone MW 360,000–0.02% BSA) and SSC solution (0.15 *M* NaCl–0.015 *M* sodium citrate).

D. Comparison of Enterotoxin Production and Colony Hybridization with Enterotoxin Probes

A total of 200 *E. coli* strains isolated from patients with "travelers diarrhea" at the Osaka International Airport Quarantine (3) were tested. These were standard strains that have been authenticated by the World Health Organization (Geneva, Switzerland) and provided by Y. Takeda, Kyoto University, Kyoto, Japan.

Biological enterotoxin assays of the 200 strains were performed using the Biken test for LTI (10) and the suckling mouse test for STI (25). Colony hybridization using ^{32}P-labeled enterotoxin probes, previously established by Moseley *et al.* (18), was done as a control experiment.

Colony hybridization with the 200 *E. coli* strains was done using biotinylated LTIh probes (Table I). All strains produced both LTI and STI, and only LTI were positive on hybridization with biotinylated probes. None of the STI and nonenterotoxigenic strains was hybridized with biotinylated LTIh probes. Complete agreement was obtained between colony hybridization with ^{32}P-labeled and biotinylated LTIh probes.

A total of 36 *E. coli* strains of the 200 isolates were randomly chosen, and colony hybridization was done using STIa and STIb probes labeled with biotin (Table II). Eleven strains of 13 STIa producers hybridized with the biotinylated STIa probe. The remaining two STIa producers, however, were negative on hybridization with biotinylated STIa. All 13 STIa producers were negative on hybridization with biotinylated STIb probe. All 10 STIb producers hybridized with the biotinylated STIb probe but not with the biotinylated STIa probe. None of the 6 LTI-producing strains and none of the 7 nonenterotoxigenic strains hybridized with either STIa or STIb probes labeled with biotin.

The treatment of colonies fixed on NC filter with proteinase K (2 mg/ml) and Triton X-100 (0.5%) was an essential step for successful colony hybridization with ETEC strains using biotinylated enterotoxin probes. Biotinylated LTIh probe nonspecifically hybridized with *E. coli* C600 strain when this step was omitted (data not shown). Both biotinylated STIa and STIb probes might give similar nonspecific hybridization, but they were not tested. The basis for the nonspecific hybridization seems to be the binding of the streptavidin–biotinylated acid phosphatase complex to the endogenous biotin in bacterial cells.

TABLE I

Colony Hybridization Using ^{32}P- or Biotin-Labeled LTIh Probe with *Escherichia coli* Strains Isolated from Travelers' Diarrhea

		Positive hybridization with LTIh probes	
Enterotoxigenicity	Number of strains[a]	^{32}P-labeled	Biotin-labeled
LTI + STI	39	39	39
LTI	47	47	47
STI	53	0	0
Nontoxigenic	61	0	0

[a] A total of 200 *E. coli* strains were used. These were the standard strains authorized by WHO. The enterotoxigenicity of these strains was tested before colony hybridization experiments by the Biken test for LTI (10) and the suckling mouse assay for STI (25).

TABLE II

Colony Hybridization Using Biotin-Labeled STIa and STIb Probes with *Escherichia coli* Strains Chosen as the STIa or STIb Standard Strain

			Hybridization of biotin-labeled probes			
			STIa		STIb	
Enterotoxigenicity	Type of STI toxins	Number of strains[a]	Positive	Negative	Positive	Negative
LTI + STI	STIa	1	1	0	0	1
STI	STIa	12	10	2	0	12
STI	STIb	10	0	10	10	0
LTI	—	6	0	6	0	6
Nontoxigenic	—	7	0	7	0	7

[a] A total of 36 strains were randomly chosen from the 200 *E. coli* standard strains. The STI-toxin types were subclassified into STIa and STIb by colony hybridization using ^{32}P-labeled STIa and STIb probes prepared in this study. The classification obtained was the same as that authorized by WHO.

Two strains producing STIa detected by the ^{32}P-labeled STIa probe were not detected by the biotinylated STIa probe. Southern blots prepared from purified DNA from each of the two strains hybridized with both the biotinylated and radiolabeled STIa probes (data not shown). One possibility for this difference might be that some non-nucleic acid factor characteristic of these two strains interfered with the biotin–avidin interaction or detection.

IV. CONCLUSIONS

Colony hybridization with biotin-labeled enterotoxin probes was used for the detection of ETEC among *E. coli* isolates from patients with "travelers diarrhea." The biotinylated LTIh and STIb probes specifically hybridized with 86 LTI and 10 STIb producers tested, respectively. The biotinylated STIa probe was almost specific for STIa producers; however, two strains of 13 STIa producers failed to hybridize with the probe.

V. GENE PROBES VERSUS ANTISERA AND MONOCLONAL ANTIBODIES

Immunological methods of the Biken test (10) and GM_1-ELISA (23) have been proposed as the practical methods for the detection of LTIh-producing *E. coli* strains. Colony hybridization using gene probes for LTIh described in this study is also a practical method for the specific identification of LTI producers. Gene probe for LTII reported by Pickett *et al.* (20) also specifically hybridizes with LTII producers. The gene probe method is more valuable than the immunological method for testing large numbers of samples. The gene probe method is applicable to fecal samples. Neither the immunological nor gene probe method can be used to differentiate between strains of LTIh and LTIp producers.

The ELISA method described in Section I has been proved to be applicable to the diagnostic identification of STI producers. Monoclonal antibodies against STIa and STIb react with both types of STI, STIa and STIb. However, STIa and STIb gene probes described in this study specifically hybridize with STI producers by Southern hybridization or dot hybridization. Therefore, DNA–DNA hybridizations are particularly useful for differentiating between *E. coli* strains of STIa and STIb producers. Colony hybridization using biotinylated STI probes could become a practical diagnostic method if the hybridization signal were enhanced and the sensitivity increased. The detection of STII in ligated pig intestinal loops seems to be impractical. An STII gene probe (14) is useful in detecting STII-producing *E. coli* strains by hybridization.

VI. PROSPECTS FOR THE FUTURE

A. Improvements for a More Reliable Hybridization

A weaker hybridization signal was obtained in colony hybridization using a biotinylated STIa probe. This could be why two STIa producers could not be detected by the STIa probe in this study. Although the hybridization signal of the biotinylated STIb probe was strong enough to detect all the STIb-producing strains, it was weaker than that of the biotinylated LTIh probe. A stronger hybridization signal will give a more reliable colony hybridization by biotinylated enterotoxin probes. The following procedures might be useful for this purpose. (1) The oligo-labeling method (7) may incorporate biotinylated polynucleotides into the enterotoxin probe DNA with a higher specific activity. This method will be particularly useful for labeling a shorter probe DNA such as that for STIa or STIb. (2) It has been reported that treatment of the colonies with steam fixed the target DNA more tightly on a filter paper (15,22). These methods may enhance the hybridization signal and increase the sensitivity for the detection of ETEC strains by colony hybridization using biotinylated enterotoxin probes.

B. Improvements for the Practical Application

This study demonstrated that most of the STI-producing strains were specifically detected using STIa and STIb enterotoxin probes. These probes could be used for the diagnosis of ETEC strains by colony hybridization. However, preparation of the probes was time-consuming. Two steps of a digestion by restriction endonucleases and elution from agarose gel procedures were required for the preparation of STIa and STIb probes. A "cassette-probe plasmid," which carries the STIa or STIb probe flanked by an *Eco*RI recognition site, is under construction in our laboratory. Another "cassette-probe plasmid," which carries the LTIh, STIa, and STIb probe DNA on a high copy number plasmid, is also under construction in our laboratory. Only a single step of digestion and elution procedure will be enough to prepare the three types of probe (LTIh, STIa, or STIb probe) from the latter type of "cassette-probe plasmids."

The direct application of the colony hybridization technique to fecal samples will be required for a rapid diagnosis of ETEC strains. This technique has already been established by using radioisotope-labeled enterotoxin probes (18). However, successful results using biotin-labeled enterotoxin probes have not been reported.

VII. SUMMARY

A method of colony hybridization using biotinylated enterotoxin DNA probes was established to detect enterotoxigenic *Escherichia coli* (ETEC) strains. The enterotoxin probes used in this study were heat-labile (LTIh, 660 bp) and heat-stable enterotoxins (STIa, 157 bp and STIb, 220 bp). The biological enterotoxin assays used were the Biken test for LTIh and the suckling mouse test for STIa. A total of 200 *E. coli* strains isolated from travelers with diarrhea were tested for colony hybridization using the LTIh probe. All strains (86) that produced LTI, but none of the non-LTI producers, hybridized with ^{32}P-labeled and biotinylated LTIh probes. A total of 36 strains chosen randomly from the 200 isolates were tested for colony hybridization using STIa and STIb probes. All but two strains (11,13) that hybridized with the ^{32}P-labeled STIa probe also hybridized with the biotinylated STIa probe. All strains (10) that hybridized with the ^{32}P-labeled STIb probe also hybridized with the biotinylated STIb probe. It was found that the treatment of colonies on nitrocellulose filters with proteinase K and Triton X-100 was essential for the specific hybridization using the biotinylated enterotoxin probes. The results obtained in this study demonstrated that almost all *E. coli* strains tested were judged to be the same by the biological enterotoxin assays and colony hybridization using biotinylated or ^{32}P-labeled enterotoxin probes. Thus, the biotinylated enterotoxin probes might be useful in the diagnosis of ETEC strains by colony hybridization.

ACKNOWLEDGMENTS

This work was done at the Institute of Medical Science, University of Tokyo, Tokyo, Japan, in consultation with M. Yoshikawa. I am indebted to Yasuyuki Kirii, Katsuhiro Komase, and Hiroshi Arita, with whom most of this work was done. I am grateful to Tae Takeda, National Children's Medical Research Center, for critical reading of this manuscript.

REFERENCES

1. Burgess, M.N., Bywater, R.J., Cowley, C.M., Mullan, N.A., and Newsome, P.M. (1978). Biological evaluation of a methanol-soluble, heat-stable *Escherichia coli* enterotoxin in infant mice, pigs, rabbit, and calves. *Infect. Immun.* **21,** 526–531.
2. Craig, J.P. (1965). A permeability factor (toxin) found in cholera stools and culture filtrates and its neutralization by convalescent cholera sera. *Nature (London)* **207,** 614–616.
3. Danbara, H., Komase, K., Arita, H., Abe, S., and Yoshikawa, M. (1988). Molecular analysis of enterotoxin plasmids of enterotoxigenic *Escherichia coli* of 14 different O serotypes. *Infect. Immun.* **56,** 1513–1517.
4. Dean, A.G., Ching, Y.-C., Williams, R.G., and Harden, L.B. (1972). Test for *Escherichia coli* enterotoxin using infant mice: Application in a study of diarrhea in children in Honolulu. *J. Infect. Dis.* **125,** 407–411.
5. Donta, S.T., Moon, H.W., and Whipp, S.C. (1974). Detection of heat-labile *Escherichia coli* enterotoxin with the use of adrenal cells in tissue culture. *Science* **183,** 334–336.
6. Evans, D.J., Jr., and Evans, D.G. (1977). Direct serological assay for the heat-labile enterotoxin of *Escherichia coli,* using passive immune hemolysis. *Infect. Immun.* **16,** 604–609.
7. Feinberg, A.P., and Vogelstein, B. (1983). A technique for radiolabeling DNA restriction endonuclease fragments to high specific activity. *Anal. Biochem.* **132,** 6–13.
8. Guerrant, R.L., Bruton, L.L., Schnaitman, T.C., Rebhun, L.I., and Gilman, A.G. (1974). Cyclic adenosine monophosphate and alteration of Chinese hamster ovary cell morphology: A rapid, sensitive in vitro assay for the enterotoxins of *Vibrio cholerae* and *Escherichia coli. Infect. Immun.* **10,** 320–327.
9. Honda, T., Shimizu, M., Takeda, Y., and Miwatani, T. (1976). Isolation of a factor causing morphological changes of Chinese hamster ovary cells from the culture filtrate of *Vibrio parahaemolyticus. Infect. Immun.* **14,** 1028–1033.
10. Honda, T., Taga, S., Takeda, Y., and Miwatani, T. (1981). A modified Elek test for detection of heat-labile enterotoxin of enterotoxigenic *Escherichia coli. J. Clin. Microbiol.* **13,** 1–5.
11. Kasai, G.J., and Burrows, W. (1966). The titration of cholera toxin and antitoxin in the rabbit ileal loop. *J. Infect. Dis.* **116,** 606–612.

12. Kirii, Y., Danbara, H., Komase, K., Arita, H., and Yoshikawa, M. (1987). Detection of enterotoxigenic *Escherichia coli* by colony hybridization with biotinylated enterotoxin probes. *J. Clin. Microbiol.* **25,** 1962–1965.
13. Langer, P.R., Waldrop, A.A., and Ward, D.C. (1981). Enzymatic synthesis of biotin-labeled polynucleotides: Novel nucleic acid affinity probes. *Proc. Natl. Acad. Sci. U.S.A.* **78,** 6633–6637.
14. Lee, C.H., Moseley, S.L., Moon, H.W., Whipp, S.C., Gyles, C.L., and So, M. (1983). Characterization of the gene encoding heat-stable toxin II and preliminary molecular epidemiological studies of enterotoxigenic *Escherichia coli* heat-stable toxin II producers. *Infect. Immun.* **42,** 264–268.
15. Maas, R. (1983). An improved colony hybridization method with significantly increased sensitivity for detection of signal genes. *Plasmid* **10,** 296–298.
16. Maniatis, T., Fritsch, E.F., and Sambrook, J. (1982). ''Molecular Cloning: A Laboratory Manual.'' Cold Spring Harbor Lab., Cold Spring Habor, New York.
17. Mol, P.D., Hemelhof, W., Petoré, P., Takeda, T., Miwatani, T., Takeda, Y., and Butzler, J.P. (1985). A competitive immunosorbant assay for the detection of heat-stable enterotoxin of *Escherichia coli. J. Med. Microbiol,* **20,** 69–74.
18. Moseley, S.L., Echeverria, P., Seriwatana, J., Tirapat, C., Chaicumpa, W., Sakuldaipeara, T., and Falkow, S. (1982). Identification of enterotoxigenic *Escherichia coli* by colony hybridization using three enterotoxin probes. *J. Infect. Dis.* **145,** 863–869.
19. Moseley, S.L., Huq, I., Alim, A.R.M.A., So, M., Samadpour-Motalebi, M., and Falkow, S. (1980). Detection of enterotoxigenic *Escherichia coli* by DNA colony hybridization. *J. Infect. Dis.* **142,** 892–898.
20. Pickett, C.L., Twiddy, E.M., Belisle, B.W., and Holmes, R.K. (1986). Cloning of genes that encode a new heat-labile enterotoxin of *Escherichia coli. J. Bacteriol.* **165,** 348–352.
21. Sack, D.A., Huda, S., Neogi, P.K.B., Daniel, R.R., and Spira, W.M. (1980). Microtiter ganglioside enzyme-linked immunosorbent assay for *Vibrio* and *Escherichia coli* heat-labile enterotoxins and antitoxin. *J. Gen. Microbiol.* **11,** 35–40.
22. Seriwatana, J., Eschverria, P., Taylor, D.N., Sakuldaipeara, T., Changchawalit, S., and Chivoratanond, O. (1987). Identification of enterotoxigenic *Escherichia coli* with synthetic alkaline phosphatase-conjugated oligonucleotide DNA probes. *J. Clin. Microbiol.* **25,** 1438–1441.
23. Sethabutr, O., Hanchalay, H., Echeverria, P., Taylor, D.N., and Leksoboon, U. (1985). A non-radioactive DNA probe to identify *Shigella* and enteroinvasive *Escherichia coli* stools of children with diarrhea. *Lancet* **2,** 1095–1097.
24. Svennerholm, A.-M., Wikström, M., Lindbald, M., and Holmgren, J. (1986). Monoclonal antibodies against *Escherichia coli* heat-stable toxin (STa) and their use in a diagnostic ST ganglioside GM1-enzyme-linked immunosorbent assay. *J. Clin. Microbiol.* **24,** 585–590.
25. Takeda, Y., Takeda, T., Yano, T., Yamamoto, K., and Miwatani, T. (1979). Purification and partial characterization of heat-stable enterotoxin of enterotoxigenic *Escherichia coli. Infect. Immun.* **25,** 978–985.
26. Thompson, M.R., Brandwein, H., LaBine-Racke, M., and Giannella, R.A. (1984). Simple and reliable enzyme-linked immunosorbent assay with monoclonal antibodies for detection of *Escherichia coli* heat-stable enterotoxins. *J. Clin. Microbiol.* **20,** 59–64.
27. Tsukamoto, T., Kinoshita, Y., Taga, S., Takeda, Y., and Miwatani, T. (1980). Value of passive hemolysis for detection of heat-labile enterotoxin produced by enterotoxigenic *Escherichia coli. J. Clin. Microbiol.* **12,** 768–771.

7

Early Days in the Use of DNA Probes for *Mycobacterium tuberculosis* and *Mycobacterium avium* Complexes

PETER W. ANDREW AND GRAHAM J. BOULNOIS
Department of Microbiology
University of Leicester
Medical Sciences Building
Leicester, England

I. INTRODUCTION

Recombinant DNA techniques are having a major impact in biological and medical sciences with applications ranging from production of novel compounds to the use of DNA probes for diagnosis of inherited disorders and microbial infections. The concept that DNA probes can be used for the rapid detection and simultaneous identification of microbes is well established. Clearly, in the first instance, this type of application will find ready use for those microbial infections for which laboratory diagnosis is either currently impossible or time-consuming. Nevertheless, the adoption of DNA probes for the diagnosis of microbial infections for which present systems are convenient will surely follow. This is especially so where the use of DNA probes can provide information on, for example, likely virulence, as well as providing a rapid means of detection and identification.

Having said this, both the isolation and characterization of suitable probes and their application to the clinical situation are still in the early phases of research. Mycobacteria provide an ideal testing ground for DNA probes, since mycobacterial infections pose some difficult problems for DNA probe use. These include difficulties in the handling and processing of clinical material and the refractory nature of the mycobacterial cell wall to lytic agents. In addition, sensitivity of detection is a crucial issue along with the complex taxonomy of these microorganisms, which produces difficulties in terms of the desired specificity of DNA probes.

In this article we describe progress, to date, in the production and use of mycobacterial DNA probes. The experience so far is cause for cautious optimism and the lessons learned with mycobacteria will considerably facilitate the development and exploitation of DNA probes for diagnosis of other microbial infections.

II. BACKGROUND

The genus *Mycobacterium* contains at least 54 recognized species (74). The majority of these are saprophytes but some do cause disease in humans. The most important are *Mycobacterium tuberculosis* and *Mycobacterium leprae*; the cause of tuberculosis and leprosy, respectively. Others cause disease in other animals. For example, *Mycobacterium bovis* and *Mycobacterium paratuberculosis* are the cause of serious disease in cattle (23). For humans, *M. tuberculosis* is the most important mycobacterial pathogen, both in terms of morbidity and mortality. Tuberculosis is one of the most important infectious diseases worldwide, with an estimated 10 million new cases developing annually, resulting in 3 million deaths per year (71,78). Although the incidence and death rate from this disease have shown a steady decline in Europe and the United States, recent evidence suggests that the incidence may be beginning to increase, possibly as a result of the increase in patients with AIDS (70). *Mycobacterium tuberculosis* is very closely related to *M. bovis* and *Mycobacterium africanum*, which also cause tuberculosis in humans (74). Because of their very close taxonomic relationship these three species together with *Myco-*

bacterium microti, the cause of tuberculosis in the vole, are referred to as tubercle bacilli, tubercle complex, or the *M. tuberculosis* complex. Other mycobacterial species (the majority) are, perhaps oddly, known as "atypical" mycobacteria, or maybe more accurately as "nontuberculous" mycobacteria. While the majority of cases of mycobacterial disease in humans are due to *M. tuberculosis* or *M. leprae*, nontuberculous mycobacteria cannot be ignored. Particularly in tropical regions, saprophytic species may contaminate clinical samples.

Several of these nontuberculous mycobacteria are able to colonize the nasopharynx and intestinal mucosa of individuals without causing disease. However, sometimes infections with nontuberculous mycobacteria result in disease that is often indistinguishable from tuberculosis (50, 51). The nontuberculous mycobacteria most commonly causing disease in humans in the United Kingdom are *Mycobacterium kansasii, Mycobacterium avium,* and *Mycobacterium intracellulare*.

There is an increasing incidence of disease caused by nontuberculous mycobacteria, probably a consequence of increased numbers of immunocompromised patients undergoing either transplantation surgery or therapy for neoplasms, or, more recently, in patients with AIDS. The situation is highlighted in AIDS where up to 40% of patients in some studies were infected with members of the *M. avium–intracellulare* complex (MAI) (25,35,43). The taxonomy of this complex is unresolved.

There has been debate as to whether *M. avium* and *M. intracellulare* are truly separate species. For this reason, the term *M. avium–intracellulare* has arisen. To add further confusion, at least two serotypes have been identified (77) as belonging to either *M. avium* or *M. intracellulare*.

Although members of the MAI cause a tuberculous-type disease, they can also produce a systemic bacteremia. Also closely related to *M. avium–intracellulare* are *M. paratuberculosis*, the cause of Johne's disease in cattle, and *Mycobacterium lepraemurium*, the cause of leprosy in rats (24). Other mycobacteria also produce patterns of disease that differ from classical pulmonary tuberculosis in humans. These range from diseases of the skin caused by, for example, *Mycobacterium ulcerans* or *Mycobacterium marinum*, to chronic granulomatous diseases where mycobacteria are the suspected, but not proven etiological agents—Crohn's disease being an example.

Rapid and accurate diagnosis of mycobacterial infection are of obvious importance both for the infected individual and for community infection control programs. The principal methods for reducing the incidence of tuberculosis in a community are bacillus Calmette-Guèrin (BCG) vaccination plus case-finding combined with treatment. However, as demonstrated by the recent trial in South India, BCG vaccination does not always prove as effective as desired (72). The importance of diagnosis of new cases of pulmonary tuberculosis and speedy application of chemotherapy, particularly to heavily infected individuals who have tubercle bacilli in their sputum and are excreting the bacteria, are well recognized as part of any control program. Speed of diagnosis also must be of value to the patient, yet bacteriological examination can be time-consuming to such a point that it only confirms a physician's diagnosis long after treatment has begun.

The diagnosis of mycobacterial infections is symbolic of the potential advantages of DNA probes. They offer the potential of increased speed and accuracy and possibly sensitivity, in comparison with the often time-consuming nature of classical procedures and the sometimes confused taxonomy.

Bacteriological examination can be by microscopy or culture. Microscopy for acid-fast bacteria is rapid but can only detect the most heavily infected, though the most infectious, individuals. Microscopy can not determine which mycobacterial species is present. Given that conventional antituberculous chemotherapy is often ineffective against nontuberculous mycobacteria (35,59), this is not an unimportant consideration. Culture is an important diagnostic procedure. It is a more sensitive technique than microscopy and permits speciation. However, it is time-consuming, requiring incubation of cultures for as long as 8 weeks, at which stage further tests may be necessary to speciate nontuberculous mycobacteria (32). Even with the radiometric Bactec system, detection time can be as long as 45 days (37).

The development of serological methods for diagnosis of active mycobacterial disease has been an aspiration for very many years. However, in spite of the advent of many promising monoclonal antibodies (MAb), the solution to this quest remains elusive.

DNA probes and associated detection systems offer the possibilities of improvement in diagnostic methods by combining (or even improving?) the specificity and sensitivity of culture with the speed of microscopy.

III. RESULTS AND DISCUSSION

Probes for mycobacteria range from total genomic DNA through random clones in plasmids or phage to specifically designed or isolated DNA fragments. Almost without exception, these probes have been used in conjunction with labeling with radioisotopes. To date, nonradioactive labeling has found little favor. Currently, two products for detection of mycobacteria have been produced commercially. These are the Rapid

Diagnostic Systems for *M. tuberculosis* complex and for *M. avium* complex (contains probes for *M. avium* and *M. intracellulare*) produced by Gen-Probe Inc. (9880 Campus Point Drive, San Diego, California 92121). They are marketed in the United Kingdom by Laboratory Impex Ltd. (111 Waldegrave Road, Teddington, Middlesex). These products, which use ^{125}I-labeled DNA homologous to rRNA, have been approved by the U.S. Food and Drug Administration for use in culture confirmation assays. The assays have not been tested directly on clinical specimens. At the time of writing, however, only the product for detection of *M. avium* complex is available in the United States and United Kingdom. The *M. tuberculosis* kit has been withdrawn temporarily by Gen-Probe for further evaluation.

Use of these probes is the most extensively reported of all mycobacterial probes described to date. The manufacturer claims the probes can detect 10^5–10^6 bacteria per assay within 2 hr. Gen-Probe has produced clinical studies summaries (16,17). In Gen-Probe's own evaluation (16), the probe for *M. tuberculosis* complex correctly identified 21 isolates of *M. tuberculosis* complex including *M. bovis* BCG and *M. microti*. The average percentage hybridization value was 52.7 ± 6.8 (SD), where $<10\%$ hybridization was considered as a negative result. The average hybridization with 104 other species including 42 mycobacteria was $1.2 \pm 0.5\%$, with the highest being 3.4% with *Mycobacterium asiaticum*. The manufacturers also have reported a study done by the Center for Disease Control (CDC) in Atlanta (16) of 128 isolates of *M. tuberculosis* and 105 nontuberculous mycobacteria, which produced no false positives or negatives. Finally is the description of a premarket trial done in 23 institutions in the United States in which 593 isolates of *M. tuberculosis* complex and 305 nontuberculous isolates including 290 mycobacteria were tested. Three false negatives were described and one isolate of *M. asiaticum* gave 14% hybridization. Some results from workers involved in the premarket trial have been published independently. Gonzalez and Hanna (22) report that the one *M. tuberculosis* culture to give a false negative was subsequently found to have a nonmycobacterial contaminant. They speculated that the contribution of contaminants to the turbidity of the bacterial suspension during sample preparation could have led to fewer mycobacteria being present in the test than was required and hence led to a negative result. It appears that contaminants per se do not result in false negatives (12,19). Musial and co-workers (55) have reported that the one false negative found in their study was positive on retesting.

A similar series of trials has been reported for the kit for detection of *M. avium* complex (17). This kit contains probes for *M. avium* and *M. intracellulare*. The trials indicated that use of these probes may be more problematic than the kit for *M. tuberculosis*, with 39 of 1480 giving discrepant results. Perhaps it is rather surprising that it was not this product that was withdrawn! However, it seems reasonable that part of this discrepancy could be due to problems in the taxonomy of this group of mycobacteria and in conventional identification methods rather than reflecting a lack of specificity in the probes.

The in-house trial by Gen-Probe (17) included 98 isolates of *M. avium* complex and 110 other isolates covering 41 genera and 92 species, including 49 other mycobacterial species. No false positives were reported, while the two false negatives were described as being subsequently identified as *M. asiaticum* and *Mycobacterium simiae* on further analysis of mycolic acids. After use of the *M. avium* complex, Drake and colleagues reported (13) that all isolates of *M. avium* complex tested were positive while no false positives were identified with 66 other mycobacterial isolates, which include 22 *M. tuberculosis* and 8 *M. kansasii* cultures. The average percentage hybridization (±SD) of *M. avium* isolates was 48.0 ± 8.8% and for non-*M. avium* complex isolates it was in the range 1.0–4.2%, except for one isolate of *M. kansasii*, which was 9.7%. A report of a premarket evaluation trial in 24 laboratories in United States (17) indicated that from 609 *M. avium* complex isolates and 338 non-*M. avium* cultures there were 8 false negatives and 30 false positives. Some of those involved in the premarket trial have independently reported their results (15,22,34,55). All report similar levels of specificity and sensitivity. For example, in a study (34) of 56 *M. avium* complex cultures on agar or in Bactec 12B bottles, all 52 were positive with the *M. avium* complex probe. The four negatives were direct from Bactec bottles and tested positive after subculture on Lowenstein–Jensen medium. All 56 isolates were negative with the *M. tuberculosis* probe.

Results so far with these kits indicate that they will prove useful to the clinical laboratory. The possibility of a 2-hr test after culture is a particular boon. However among the factors that need to be considered are the need for specialized equipment (γ counter and sonic waterbath), the use of the isotope ^{125}I, and the considerable expense. In the United Kingdom, at the time of writing, the *M. avium* complex kit was £179 for 20 tests. These factors suggest that at the present time these kits are more suited to larger, well-resourced laboratories and will be inappropriate for use in developing countries, where most cases of mycobacterial infection are found.

Total chromosomal DNA has been used as a probe to identify *M. tuberculosis* and *M. avium* complexes (3,62,64). In dot blots with purified DNA and using ^{32}P to label probes, 100 pg of mycobacterial DNA were detectable. This corresponds to $1–2 \times 10^4$ bacteria, a value that compares favorably with the 10^4 acid-fast bacilli/ml detection limit of light microscopy (80).

Roberts and co-workers (62) used total genomic DNA from *M. bovis* BCG as a probe for *M. tuberculosis* complex. Not surprisingly, given the extremely high DNA homology between *M. bovis* and *M. tuberculosis* (2,5), this probe could not distinguish members of the *M. tuberculosis* complex. These workers also prepared a *M. avium* complex probe using a combination of total genomic DNA from *M. avium* serotype 1, *M. intracellulare* serotype 14, and *Mycobacterium scrofulaceum*. DNA from *M. bovis* BCG did not react with DNA from *M. avium* complex, and vice versa. This is probably not surprising given that *M. tuberculosis* has been shown to have only ~25% DNA homology with members of *M. avium* complex (2). None of the probes hybridized to DNA from 18 other mycobacterial species and 11 nonmycobacterial species identified as common contaminants of the Bactec system. When these probes were used in dot-blot assays of 61 cultures of clinical isolates, 57 were correctly identified as *M. tuberculosis* or *M. avium*. When the four false negatives (3 *M. tuberculosis* and 1 *M. avium*) were regrown in the presence of ampicillin and *D*-cycloserine to aid cell lysis, they were correctly identified.

Genomic DNA from *M. tuberculosis* H37Ra has been shown to hybridize to DNA from *M. tuberculosis* and *M. bovis* and not to *M. avium* or *M. smegmatis* DNA when incubated at ≥55°C. At lower stringency (<55°C), binding to non-*M. tuberculosis* complex DNA was seen. Shoemaker and co-workers have reported (64) similar observations but were more circumspect in drawing conclusions. Under high-stringency conditions, total genomic DNA from *M. tuberculosis* did not hybridize to any DNA sample from 18 nonmycobacterial species commonly isolated from the respiratory tract. However, they found that *M. tuberculosis* did hybridize to DNA from *M. avium* complex. Even though they could detect as few as 100 times less *M. tuberculosis* DNA than *M. avium*, they concluded that total *M. tuberculosis* DNA cannot be used to differentiate between *M. tuberculosis* and other mycobacterial species. They were probably correct in drawing this conclusion.

Mycobacteria have been implicated in inflammatory diseases such as Crohn's disease and sarcoidosis for a long time without actual proof. DNA probes are offering the opportunity of a fresh appraisal of this question. A mycobacterial isolate (strain Linda) from a patient with Crohn's disease was shown by S1 nuclease-resistant hybridization to resemble *M. paratuberculosis* more than any other mycobacterial species studied (82). Total genomic DNA from Linda was used to probe DNA prepared from human intestinal tissue. Sequences that reacted with strain Linda DNA were detected in 10 of 19 Crohn's disease samples, 2 of 6 ulcerative colitis samples, and 1 of 6 controls. Further analysis showed these sequences to be present in DNA from muscle of the intestinal wall rather than surface mucosa. This probably eliminates the possibility of surface contaminants contributing the homologous sequences. Even after considering possible limitations in the sensitivity of the technique used, the authors considered that the failure to demonstrate an exclusive association with Crohn's disease meant that the nature of the etiology of this disease remained an open question.

Cloned random restriction enzyme fragments have been used to explore the etiology of Crohn's disease. By means of determination of S1 nuclease-resistant hybrids, McFadden and co-workers were unable to differentiate a mycobacterial isolate (strain Ben) from a Crohn's disease patient from *M. paratuberculosis* and *M. avium–intracellulare* serotypes 2 and 5 (46). However, they found that a cloned *Bam*HI fragment of strain Ben in combination with Southern blotting (67) identified restriction fragment length polymorphisms (RFLP) between *M. paratuberculosis* and both *M. avium–intracellulare* serotypes 2 and 5. No RFLP were detected between three isolates from Crohn's disease and *M. paratuberculosis*.

These workers (47) have reported the use of the cloned fragment from strain Ben to examine if RFLP correlated with 21 serotypes of *M. avium–intracellulare* (77). Using a method of estimating the value of base substitution between strains (73), McFadden (47) found that the *M. avium* complex split into distinct groups; a very homologous group similar to serotype 2 (*M. avium*), which included *M. paratuberculosis* and *M. lepraemurium*, and a distinct but heterogeneous group that included serotype 16 (*M. intracellulare*-type strain). Imaeda and Kaminski (29) reported that, using total genomic DNA as a probe, 18 strains of *M. avium–intracellulare* could also be divided into two distinct groups.

Random restriction fragments have also been used to study strain differences in the *M. tuberculosis* complex. Bhattacharya and colleagues (3) have described a random clone from *M. tuberculosis* that identified a RFLP between *M. tuberculosis* and *M. bovis* digested with *Pst*I. We have also found that a random clone of sonicated *M. bovis* DNA could distinguish between *Hae*III digests of *M. tuberculosis* and *M. bovis* (unpublished data). A random *Bam*HI fragment of *M. tuberculosis* H37Rv cloned into λ1059 detected two *Bam*HI fragments present in five isolates of *M. tuberculosis*, but these were absent from two strains of *M. bovis* and from *M. bovis* BCG (14).

Random restriction fragments can be of use in the detection of members of the *M. tuberculosis* complex and their differentiation from nontuberculous mycobacteria. We found that a random fragment of *M. bovis* DNA generated by sonication and cloned in M13 hybridized to 27 isolates of *M.*

bovis and *M. tuberculosis*, at high stringency (0.1× SSC, 65°C), but not to DNA from *M. avium, Mycobacterium fortuitum*, or *Mycobacterium phlei*, nor to DNA from six nonmycobacterial species. At low stringency (0.1× SSC, 42°C) the probe bound to DNA from nontuberculous mycobacteria but not to DNA from nonmycobacterial species (unpublished data). These data suggest that certain DNA fragments may be used both as *M. tuberculosis* complex-specific probes or as panmycobacterial probes, depending on the hybridization and washing conditions.

Pao and co-workers (57) have described the use of two random clones of *M. tuberculosis* DNA in pUC8 that, labeled with ^{32}P, could detect 50 pg of DNA. These probes did not hybridize to DNA from 19 nonmycobacterial species. Hybridization to mycobacterial DNA other than *M. tuberculosis* was low, in the range 0.8–18.3% of hybridization to *M. tuberculosis*, except from *Mycobacterium xenopi*, which was 78%. Potentially the most exciting part of this report was that percentage hybridization to *M. bovis* was only 4.1%. The sensitivity of these probes was examined by assay of 441 uncultured clinical samples. Given the reported specificity of these probes, the results of this study were disappointing. Of 84 culture-positive samples, 8 were negative with the probes, and 58 of 357 culture-negative samples were positive with the probes. Given the detection limit of the method used (10^4 mycobacteria), failure to culture bacteria seems an unlikely explanation for the large number of false positives. However, especially in view of the reported ability of these probes to differentiate *M. tuberculosis* and *M. bovis* by simply dot-blotting, rather than Southern blotting, these probes warrant further analysis.

Analysis of restriction fragments has proved useful in differentiating strains of *M. tuberculosis* and *M. bovis*. Shoemaker (65) found that more discriminatory patterns were obtained if ^{32}P-labeled total *M. tuberculosis* DNA was used to visualize fragments rather than conventional staining with ethidium bromide.

IV. CONCLUSIONS

The limited progress to date in the development of DNA probes for the diagnosis of mycobacterial infections indicates substantial potential for the future. However, a consensus has not yet emerged on the optimal design and use of these DNA probes. There remains a requirement for the more sophisticated probes useful for the differentiation of species and strains within the *Mycobacterium tuberculosis* complex.

V. GENE PROBES VERSUS ANTISERA AND MONOCLONAL ANTIBODIES

The current status of immunodiagnosis of tuberculosis has been recently reviewed (31). To date there are no immunological tests that are recommended for the routine diagnosis of active tuberculosis. This is so in spite of extensive searches over many decades for effective serological and skin tests for tuberculosis. These searches have been driven by two problems that hinder rapid diagnosis, namely the very slow growth rate of mycobacteria, which delays culture results, and the difficulty in obtaining suitable material for diagnosis of extrapulmonary tuberculosis. DNA probes are more likely to solve the problem of delay in cultural analysis while effective immunological procedures offer the best chance of rapid diagnosis of extrapulmonary disease.

Although levels of antibodies against *M. tuberculosis* antigens do rise following infection, previous attempts at developing serological tests have failed to distinguish patients from controls. The main reason for this lack of specificity is that *M. tuberculosis* contains antigens and epitopes that are present in other mycobacterial species (69). Individuals are continuously exposed to nontuberculous mycobacteria in the environment and may have been vaccinated with *M. bovis* BCG. Development of specific immunological tests depends on preparations that do not have cross-reacting epitopes, yet even purified antigens from *M. tuberculosis* contain mixtures of specific and cross-reactive epitopes (31).

Recent attempts to improve specificity have focused on the use of MAb and T-cell clones and their corresponding epitopes. These approaches have so far failed to identify epitopes specific to *M. tuberculosis*. Only two proteins with epitopes specific for the *M. tuberculosis* complex have been found (79). Yet there has been progress. Monoclonal antibodies against a 38 kDa protein have given very good results with patients smear-positive for pulmonary disease (31). However, results with smear-negative or extrapulmonary disease have been disappointing. Results with these patients indicate that these patients are responding to epitopes other than those defined by existing MAb (6,31,61).

An alternative approach to detection of antibody is detection of antigen in situations in which detection of acid-fast bacteria in body fluids is difficult. These are situations in which either serological or DNA probes may find a use. There has been success in use of serology to detect mycobacterial antigen in cerebrospinal fluid from patients with tuberculous meningitis (6,33,38). However, cross-reactivity remains a problem. For example, in one study 21% of patients with pyogenic meningitis and up to 8% of controls were positive (6).

Delayed-type hypersensitivity reactions to mycobacterial antigens are seen soon after infection. Crude mixtures of mycobacterial proteins, such as tuberculin, have been used for decades as skin test reagents. Here again, cross-reactivity has limited the usefulness of these reagents. Even when purified antigens of *M. tuberculosis* have been used, they are no more specific as skin test reagents than tuberculin because of the antigens also found in *M. bovis* (11,39,52). The full potential of skin testing will only be exploited when preparations are found that allow the individual with active disease from *M. tuberculosis* to be distinguished from those who have been BCG-vaccinated or exposed to environmental mycobacteria. This requires identification of species-specific T-cell epitopes.

Attempts to define T-cell clones that can discriminate between *M. tuberculosis* and *M. bovis* BCG have met with only limited success (36, 56). Perhaps a recently described method of stimulating T-cells with mycobacterial antigen separated by electrophoresis then blotting onto nitrocellulose will be more successful (40). This approach has identified antigen that was not recognized by existing MAb. An inherent problem of existing approaches of isolating MAb and T-cell clones is that neither is designed to enhance identification of species-specific epitopes. Also approaches that rely on screening antigen preparation with cell lines or clones from whole blood will tend to identify immunodominant epitopes rather than necessarily species-specific epitopes. We would argue that DNA probes may offer a route out of this dilemma. We propose that once pieces of DNA from *M. tuberculosis* are found that have sequences not found in *M. bovis* (i.e., a *M. tuberculosis*-specific probe), there will be a possibility that these sequences may also code for *M. tuberculosis*-specific B- or T-cell epitopes. Although a species-specific DNA sequence does not necessarily lead to species-specific protein epitopes, it seems a good place to search for such epitopes.

DNA probes may have something new to offer in the study of taxonomy of mycobacteria. As mentioned previously in this chapter, analysis of RFLP offers the possibility for better definition of strains of *M. tuberculosis* and *M. avium*.

As might be expected from the foregoing discussion, serology has been of little value in the study of the taxonomy of the *M. tuberculosis* complex (69). Monoclonal antibodies had failed to resolve this situation (30). Analysis of soluble antigens in immunodiffusion studies failed to differentiate members of the *M. tuberculosis* complex (69). Analysis of insoluble cell wall antigens by agglutination is not possible because of the tendency of *M. tuberculosis* to self-agglutinate (24).

Phage typing has proved more useful in identifying strains of *M. tuberculosis* complex, members of the complex falling into one of three phage groups, with *M. bovis* BCG strains being in a separate fourth group (24). Shoemaker and co-workers (65) have suggested that RFLP analysis may complement phage typing. They found that in some cases isolates of *M. tuberculosis* had the same restriction enzyme fragment pattern but were different phage type. Elsewhere the same phage type had different restriction patterns.

Serology has been employed quite extensively in the analysis of the taxonomy of the *M. avium* complex but the classification of strains within this group remains confused, divisions seemingly being dependent on the method of analysis. On the basis of agglutination Wolinsky and Schaefer (77) described 21 serotypes. Serotypes 1–3 were considered *M. avium* and the remainder *M. intracellulare*. After immunodiffusion analysis (48,49,68), four groups (subspecies) were defined. Group A contained serotypes 1–3, group B serotypes 4–12 (considered by some to be intermediate between *M. avium* and *M. intracellulare*), 20, and 21, and group C serotypes 13–19. The fourth group contained *M. lepraemurium*. Strains of *M. paratuberculosis* and other mycobactin-dependent strains fell into groups A or B. Conventional taxonomy has failed to provide a solution (68). Attempts have been made to clarify this picture by analysis of DNA.

Following measurement of DNA homology by solution hybridization, Baess (2) concluded that two groups with <56% homology between them could be defined. Serotypes 1–3 constituted one group (*M. avium*) together with serotypes 4–6 and 8 (previously intermediate of *M. intracellulare*). Serotypes 7, 12, 14, 16, and 18 constituted a second group (*M. intracellulare*). DNA–DNA hybridization in solution has shown the mycobactin-dependent mycobacteria, including *M. paratuberculosis*, belong to the same hybridization group and to be closely related to serotypes 2, 5, 8, and 9 (46,81). As previously mentioned, analysis of banding patterns after Southern blotting also defines two groups, with serotypes 2, 5, 8, 12, and *M. paratuberculosis* together in one group but RFLP could be seen that differentiate the members of each group (47).

VI. PROSPECTS FOR THE FUTURE

Apart from the obvious requirement of production of more probes covering more of the genus, especially those most often found in association with disease or as contaminants of clinical samples (58), three general areas suggest themselves as areas where progress would be desirable. These are (1) the production of probes that can discriminate below the complex or species level, (2) better systems for detection, and (3) the elimination of the need to culture samples before application of probes. Of course, the desire for progress in these areas is not confined to the mycobacterial field.

It is considered that there is probably no clinical value in producing probes that distinguish between species or strains within the *M. tuberculosis* complex, since all isolates are considered as pathogens and the treatment regimen is the same whatever the isolate. However, there are advantages to be gained from probes that are more discriminatory. These advantages are principally in epidemiology and in taxonomy. More defined knowledge of the distribution of strains leads to greater understanding of the reservoirs of infection, routes of transmission, efficacy of control programs, and maybe of the basic mechanisms of the disease process.

This need is illustrated by *M. bovis*. It is highly likely, for example, that the number of cases of human disease caused by *M. bovis* is being greatly underestimated in industrial countries at a time when the number of recognized cases of infection by *M. bovis* is rising (7). The true incidence may not be known until rapid, simple methods of identification are available. The principal measures for control of *M. bovis* infections in humans center on elimination of transmission of infection from cattle by pasteurization of milk and culling of infected animals. However, complete eradication has been difficult to achieve. Two of the unanswered questions are the following: Does transmission from infected humans to animals or other humans occur? Do domestic and wild animals act as a reservoir from which cattle become infected? Two animals have been widely suspected, the badger (*Meles meles*) (7) in the United Kingdom and the possum (*Trichosuris vulpecula*) in New Zealand (9).

There is also confusion in the taxonomy of *M. bovis*. Some consider there may be as many as six variants of *M. bovis*, with much of the confusion surrounding *M. africanum* which resembles *M. bovis* in several ways (7,74). The true status of *M. africanum* may be important in obtaining a true epidemiological picture. All of the issues discussed here have been well reviewed elsewhere (7). However, the reason they remain issues is in part due to the absence of discriminatory typing systems.

Better typing systems for *M. tuberculosis* would also be of value. Among the unanswered questions, where better typing systems may be of value, are the following: Why did the recent trial of BCG vaccination in South India fail (72)? Is disease in an individual patient due to exogenous or endogenous infection? There are two variants of *M. tuberculosis* defined by virulence in the guinea pig, the classical type and South Indian type (53). The South Indian type, of low virulence in the guinea pig, predominates in the BCG vaccination trial area (60), and it has been suggested that its presence in part explains the failure of the trial (60,66). Clearly, fuller understanding of the epidemiology of these strains is required, but widespread epidemiological studies will require simpler effective typing systems.

Prabhaker and co-workers (60) have proposed some important hypotheses regarding patterns of infection and disease in tuberculosis. They predicted that in cases of reactivation of dormant infections, mixtures of biovars would be found in sputum. They also proposed that tubercle bacilli isolated from extrapulmonary sites will be biovars of higher virulence in the guinea pig. Again, it is clear that answers to these important questions will be easier to obtain if effective typing systems, including typing for virulence, are available. It is in the provision of effective typing systems that DNA probes offer exciting prospects. None of the probes described in this chapter are suitable for typing purposes. Indeed, apart from one case, they cannot even distinguish *M. bovis* and *M. tuberculosis*. While the technique of identifying potentially unique sequence within rRNA would appear to be effective in producing species-specific probes, it is unlikely to be the route to strain or even "intracomplex" probes. New approaches are required, involving either the definition of specific genes associated with the character of interest or the development of new methods for isolation of fragments of DNA, regardless of function, that are uniquely associated with the pertinent character. Refinement of "subtractive hybridization" (75) techniques may be an answer to the latter problem.

Improvements in detection are required in three areas: (1) the improvement of sensitivity; (2) the effective use of nonradioactive probes; and (3) the need to eliminate culture are other factors of importance. These facets are linked in many ways. Above all, there is the question of cost. If only "high-tech," high-cost solutions are found to these questions, mycobacterial probes will not find their place into laboratories in developing countries. It is in these countries that mycobacterial disease is most prevalent, yet it is here that technical and financial resources are at their weakest. Often centers for diagnosis of tuberculosis will only have facilities for simple microscopy without the equipment for bacteriology (1).

The reported sensitivity limit of existing radioisotope-labeled probes is $\sim 10^4$ bacilli per assay, about the same level as light microscopy. These sensitivity limits are a consequence of the activity of the label, the size of the probe, and the amount of the probe bound. Radioisotope labeling has been used, to date, to maximize sensitivity. However, the disadvantages of radioisotopes, to manufacturers and users, in terms of shelf-life, safe handling, and equipment requirements are well known. Alternative labels include enzymes whose activities can produce colored or fluorescent products (e.g., peroxidase) or fluorescent compounds such as fluorescein. However, these suffer the disadvantage of being less sensitive than radiolabels (41). Methods of amplification of the signals from these markers are required.

Probably the method of improving sensitivity that is receiving the most attention at the moment is amplification of target DNA sequence via the polymerase chain reaction (PCR) (41), in which synthetic oligonucleotides (the probe) are used to direct the synthesis of DNA complementary to the target sequence. It offers the ultimate sensitivity of detection of a single copy of target sequence. Use of heat-stable polymerase (Taq polymerase) means that cycles of synthesis and denaturing can occur, without the need for continuous addition of enzyme, resulting in an exponential increase in the amount of target sequence. Instrumentation exists to do these reactions. The disadvantages of this approach in terms of worldwide application include the high cost of reagents and instrumentation.

Other improvements in the use of mycobacterial probes would eliminate the need for culture, the test being done directly in the clinical sample or after fixation of bacteria to a suitable support, followed by *in situ* hybridization. *In situ* hybridization has been done successfully with tissue specimens containing *Chlamydia trachomatis* (27) but may prove more problematic with mycobacteria given the nature of their cell walls.

The ideal (utopian?) test would be one of low cost and rapidity that could be done directly on the specimen and a positive reaction detected by a color change or alternatively on a sample of the specimen on a glass side with the positive reaction being observed by microscopy.

VII. SUMMARY

A variety of approaches have been adopted for the isolation of DNA probes for the detection and identification of mycobacterial species. These range from total mycobacterial genomic DNA, randomly cloned fragments of DNA, and specifically designed, synthetic oligonucleotides representative of species-specific segments of rRNA. These probes have been used in a variety of hybridization protocols and several ways adopted to detect and quantitate hybrids. On the basis of the available evidence it is not possible to make a rational decision as to the optimal route to take in probe design and use. An open mind is crucial if the promise of automated, rapid, and sensitive detection and identification of mycobacteria using DNA probes is to be realized. Progress to date gives cause for guarded optimism.

VIII. MATERIALS AND METHODS

Essentially all of the techniques employed in the production and use of probes, to date, for mycobacteria are collections of standard and very well-documented methodology. We can recommend the DNA cloning manuals of

Maniatis and colleagues (44) and of Boulnois (4). It is only the breakage of mycobacterial cells that presents unique problems.

A. Isolation of Mycobacterial DNA

Mycobacteria are not among the easiest of bacteria to lyse. Mycobacterial DNA has been generally obtained from cells lysed by a combination of detergent and enzymatic action, namely, sodium dodecyl sulfate (SDS), following treatment of cells with lysozyme (8,14,57). Experience has shown that addition of *D*-cycloserine (1 mg/ml) (10) or glycine (0.2 *M*) (54) to mycobacterial cultures 24 hr prior to lysozyme treatment enhances subsequent lysis. Various proteases, but usually proteinase K (100 μg/ml), have been included with SDS to inactivate endogenous DNase (14,64,82). DNA has then been recovered from cell lysates by the conventional means of phenol–chloroform extraction combined with ethanol precipitation. These techniques have also proved suitable for extraction of mycobacterial DNA from tissue homogenates (47,82).

Some workers (45,82) have combined lysozyme–SDS treatment with mechanical methods of breaking cells, typically French press or Hughes press. One of these groups subsequently found mechanical shearing unnecessary, including for extraction from tissue homogenates. It is interesting therefore that the Gen-Probe rapid diagnostic kits for mycobacteria utilize mechanical breakage in the presence of a "lysing agent" (18,19). Gen-Probe reports that breakage of mycobacteria is enhanced by sonication in a sonic waterbath in the presence of small beads (21). The beads could be of various materials such as glass, sand, plastic, or metal and of any size from 0.05 to 1.0 mm diameter.

Mechanical methods could reduce the time required for lysis by obviating the need for prolonged incubation in *D*-cycloserine or detergent–lysozyme. This could be important for preparation of DNA from clinical samples but is not essential for the preparation of probes. However, it is noteworthy that some (3) have reported that mycobacteria can be lysed directly on filters in 15 min with 0.5 *M* NaOH–1.5 *M* NaCl.

We have found a more simplified procedure based on the method of Winder and MacNaughton (76) results in very satisfactory cell breakage and yields of DNA. Where large amounts of DNA are required (e.g., for preparation of genomic libraries), mycobacteria were grown as a pellicle in Sauton's medium (63) containing 0.5% (v/v) pentane to aid flotation. For more rapid growth or where smaller amounts of DNA are required as, for example in dot blots, bacteria were grown in Middlebrook 7H9 medium with ADC enrichment (available from Difco). Cells were harvested by centrifugation, washed once in 0.01% (v/v) Tween-80 in H_2O, then resuspended in 100 ml of 0.15 *M* NaCl–0.1 *M* EDTA–1 *M* sucrose containing 0.5 mg lysozyme. The cell suspension was incubated for 16 hr at 36°C. Cells were collected by centrifugation and resuspended in 10 ml of 0.15 *M* NaCl–0.1 *M* EDTA–1% SDS, incubated at 65°C for 15 min, then allowed to cool to room temperature before the addition of 1.6 g sodium perchlorate. This concoction was then mixed for 5 min at room temperature. The lysate was then extracted twice with chloroform–isoamyl alcohol (26.5 : 1). DNA was precipitated by addition of 20 ml of ethanol followed by incubation at −20°C for 16 hr. The precipitate was collected by centrifugation, then resuspended in 5 ml H_2O and extracted with phenol–chloroform (100 g phenol and 0.2 g hydroxyquinoline in 100 ml of chloroform and saturated 0.5 *M* Tris-HCl and saturated with pH 8.0). Phenol–chloroform was extracted from DNA with diethylether. All procedures until this point were done under conditions of microbiological containment.

B. Preparation of Probes

The majority of attempts have been to examine total genomic DNA as a species-specific hybridization probe. Others have described potentially useful probes identified by screening random clones for species-specific patterns of hybridization. Such clones have been isolated from genomic libraries in phage (14) or plasmid vectors (45,57).

There are well-documented procedures for preparation of genomic libraries in phage λ or M13 or in plasmids. See, for example, DNA cloning manuals by Maniatis and colleagues (44) or Boulnois (4).

There have been few reports of successful attempts to use or devise nonrandom approaches for the isolation of species-specific probes. That is, use of methods for isolating or enriching species-specific DNA fragments or in which a probe is synthesized following analysis of the properties of the mycobacterial genome. Yet, while use of total genomic DNA or random clones may be successful for discrimination of distantly related mycobacteria, they may be inappropriate for more closely related species or strains.

Given the paucity of descriptions of such approaches, it is ironic that probably the most widely reported probes generated, to date, have relied on rational design. These are DNA probes for the *M. tuberculosis* complex and *M. avium* complex produced by Gen-Probe Inc. (16,17). These probes are oligonucleotides complementary to sequences in rRNA of *M. tuberculosis, M. avium,* and *M. intracellulare*. This approach relies on the observation that within the generally highly conserved 16 S and 23 S rRNA molecules, there are regions of high variation between species. RNA sequences that are potential targets for hybridization probes are identified by comparison of the sequences of the variable regions of rRNA from species of interest with known rRNA sequences of related microbes (20,26,28). The sequence of the variable region of rRNA can be readily obtained by direct sequencing of rRNA using extension of primers to the highly conserved regions by reverse transcriptase in the presence of dideoxynucleotides. Using this approach, Gen-Probe (20) described a primer derived from *E. coli* 16 S rRNA to sequence a region of the rRNA of *M. avium, M. intracellulare,* and *M. tuberculosis*. This region corresponded to bases 185–225 of *E. coli* 16 S rRNA. The sequence of the primer was as follows:

5° GGCCGTTACCCCACCTACTAGCTAAT-3°

Among the potential rRNA-specific probes identified (20) were

M. avium	ACCGCAAAAGCTTTCCACCAGAAGACATGCGTCTTGAG
M. intracellulare	ACCGCAAAAGCTTTCCACCTAAAGACATGCGCCTAAAG
M. tuberculosis	TAAAGCGCTTTCCACCACAAGACATGCATCCCGTG

C. Labeling of Probes

Probes have been labeled with either radioactive or nonradioactive markers. ^{3}H, ^{32}P, and ^{125}I have all been used to label mycobacterial probes, but ^{32}P has been the isotope of choice for most workers. The commercially available probes from Gen-Probe Inc. are ^{125}I-labeled. Nucleotides labeled with any of these isotopes are available from Amersham International.

Most reports of mycobacterial probes use the technique of nick translation to incorporate labeled nucleotides into probes (3,14,62,64). Others (45), including ourselves, have incorporated isotope with the Klenow fragment of *E. coli* DNA polymerase I by oligo-labeling. Again, both nick translation and oligo-labeling are standard procedures well described elsewhere (4,44). Using ^{32}P, probes with a specific activity of $1–5 \times 10^{8}$ cpm/μg of DNA can be expected (57,64,81).

Use of nonradioactive probes in conjunction with mycobacterial probes has been limited, with biotin the nonradioactive marker of choice (29). Biotinylated nucleotides, which are available from Amersham International, are incorporated by nick translation or oligo-labeling. Biotin is detected by its ability to bind avidin or streptavidin covalently linked to enzymes such as peroxidase or phosphatase. Biotinylated DNA is ultimately detected by the colored product produced during the assay of the enzymes. Again these techniques are in no way uniquely applicable to mycobacterial probes and are well described elsewhere; see, for example, Leary *et al.* (42).

D. Use of Hybridization Probes

Hybridization analysis can be done either in solution or on filters after dot blotting or Southern (67) blotting. Probes for the detection of mycobacteria have been used in both techniques.

1. *Filter Hybridization*

To date, most studies have used purified target DNA for hybridization studies. However, as mentioned earlier, in one report (3) it was found that satisfactory results were obtained with mycobacterial cells applied to nitrocellulose (NC) filters, which were then placed for 15 min on Whatman 3 MM paper soaked in 0.5 *M* NaOH–1.5 *M* NaCl to lyse bacteria. Methods for dot blotting and Southern hybridization have been extensively described in the literature (4,44). Nitrocellulose (Hybond C, Amersham International) has generally been the filter material used, but nylon filters (Gene Screen, New England Nuclear, or Hybond N, Amersham International) have also been used. Conditions for hybridization have typically been those for hybrids with a high degree of homology. This involves the use of Denhardt's solution (0.1% polyvinylpyrolidine–0.1% Ficoll–0.1% bovine serum albumin) plus 5× SSC (1× SSC is 0.15 *M* NaCl–15 m*M* sodium citrate, pH 7.0) at 65°–68°C, or at 42°C in Denhardt's–5× SSC–50% formamide. Filters were then washed to remove unreacted probe, first under moderately stringent conditions (2× SSC–0.1% SDS at 65°C for 30–60 min) followed by a high-stringency wash (0.1× SSC–0.1% SDS at 65°C) and bound probe detected by autoradiography.

2. *Solution Hybridization*

Filter hybridization, particularly dot blotting, is ideally suited to the qualitative analysis of a large number of samples. Although filter hybridization can be a semiquantitative technique, it is solution hybridization that is generally used for quantitative work. However, solution hybridization is less well suited for analysis of large numbers of samples. In spite of this, solution hybridization is the method recommended when using the Gen-Probe rRNA probes (18,19).

Conditions for solution hybridization and methods of measuring extent of hybridization have been well described elsewhere (83).

Typical conditions used for hybridization studies with mycobacterial probes are 10 m*M* HEPES pH 7.0 containing 0.6 *M* NaCl and 2 m*M* EDTA at 68°–72°C for 1 hr. Various methods had been used for detection of hybridization of mycobacterial probes. Absorption of hybrids with hydroxyapatite is used as part of the Gen-Probe Rapid Diagnostic System, and the methodology is clearly described in the manufacturer's instructions (18,19). As described elsewhere by Gen-Probe (20), 4 ml of hydroxyapatite suspension (2% hydroxyapatite in 0.14 *M* phosphate buffer pH 6.8–0.02% SDS) were added to 1 ml of hybridization reaction and incubated at 72°C for 5 min. Hydroxyapatite was then pelleted by centrifugation (3000 *g*, 2 min), washed once in 0.14 *M* phosphate buffer, pH 6.8, and radioactivity bound to hydroxyapatite measured by scintillation counting or γ counting, depending on the isotope used.

An alternative method of enumerating hybrids is digestion of nonhybridized molecules with single strand-specific S1 nuclease followed by precipitation of undigested hybrids with trichloroacetic acid (TCA). As described by Yoshimura and co-workers (81,82), hybridization samples are brought to 0.2 *M* acetate pH 4.5–0.4 *M* NaCl–2.5 m*M* zinc acetate. S1 nuclease was then added to a minimum concentration previously shown empirically to degrade totally the amount of single-stranded probe used. After 1 hr at 43°C, 0.2 ml of bovine serum albumin (1 mg/ml) was added. Macromolecules were then precipitated in 25% TCA, the precipitate recovered by centrifugation, washed twice in 5% TCA containing 5% sodium pyrophosphate, then solubilized in 0.5 ml of NCS tissue solubilizer (Amersham International) and radioactivity measured by scintillation counting. The percentage of probe hybridized was calculated relative to samples incubated without S1 nuclease.

ACKNOWLEDGMENTS

Work in the authors' laboratories is funded by grants from the World Health Organization, The Trent Regional Health Authority, and the Leicestershire Area Health Authority. GJB is a Lister Institute–Jenner Research Fellow. We would like to thank Judy Ravenhill for effort beyond the call of duty in the typing of the manuscript.

REFERENCES

1. Allen, B. W. (1984). Tuberculosis bacteriology in developing countries. *Med. Lab. Sci.* **41,** 400–409.
2. Baess, I. (1979). Deoxyribonucleic acid relatedness among species of slowly-growing mycobacteria. *Acta Path. Microbiol. Scand., Sect. B* **87,** 221–226.
3. Bhattacharya, S., Ranadive, S. N., and Bhattacharya, A. (1988). Distinction of *Mycobacterium tuberculosis* from other mycobacteria through DNA hybridisation. *Indian J. Med. Res.* **87,** 144–150.
4. Boulnois, G. J. (1987). "Gene Cloning and Analysis: A Laboratory Guide." Blackwell, Oxford.
5. Bradley, S. G. (1973). Relationships among mycobacteria and norcardiae based upon deoxyribonucleic acid reassociation. *J. Bacteriol.* **113,** 645–651.
6. Chandramuki, A., Allen, P. R. J., Keen, M., and Ivanyi, J. (1985). Detection of mycobac-

terial antigen and antibodies in the cerebrospinal fluid of patients with tuberculous meningitis. *J. Med. Microbiol.* **20,** 239–247.
7. Collins, C. H., and Grange, J. M. (1983). The bovine tubercle bacillus. *J. Appl. Bacteriol.* **55,** 13–29.
8. Collins, D. M, and DeLisle, G. W. (1985). DNA restriction endonuclease analysis of *Mycobacterium bovis* and other members of the tuberculosis complex. *J. Clin. Microbiol.* **21,** 562–564.
9. Collins, D. M., DeLisle, G. W., and Gabric, D. M. (1986). Geographic distribution of restriction types of *Mycobacterium bovis* isolates from brush-tailed possums. (*Trichosurus vulpecula*) in New Zealand. *J. Hyg.* **96,** 431–438.
10. Crawford, J. T., and Bates, J. H. (1979). Isolation of plasmids from mycobacteria. *Infect. Immun.* **24,** 979–981.
11. Daniel, T. M., Balestrino, E. A., Balestrino, O. C., Davidson, P. T., Debaune, S. M., Katarina, S., Kataria, Y. P., and Scocozya, J. B. (1982). The tuberculin specificity in humans of *Mycobacterium tuberculosis* to antigen 5. *Am. Rev. Respir. Dis.* **126,** 600–606.
12. Dinuzzo, A., Bannister, E., Gatson, A., and Lucia, H. (1988). Detection of *Mycobacterium tuberculosis* and *Mycobacterium avium* dual infection by Gen-Probe rapid diagnostic system. *Am. Soc. Microbiol., Abstr. Annu. Meet.* Abstr. U-35.
13. Drake, T. A., Hindler, J. A., Berlin, O. G. W., and Bruckner, D. A. (1987). Rapid identification of *Mycobacterium avium* complex in culture using DNA probes. *J. Clin. Microbiol.* **25,** 1442–1445.
14. Eisenbach, K. D., Crawford, J. T., and Bates, J. H. (1986). Genetic relatedness among strains of the *Mycobacterium tuberculosis* complex. *Am. Rev. Respir. Dis.* **133,** 1065–1068.
15. Ellner, P. D., Kiehn, T. E., Cammarata, R., and Hosmer, M. (1988). Rapid detection and identification of pathogenic mycobacteria by combining radiometric and nucleic acid probe methods. *J. Clin. Microbiol.* **26,** 1349–1352.
16. Enns, R. K. (1987). "Clinical Studies Summary Report: The Gen-Probe Rapid Diagnostic System for the Mycobacterium TB Complex." Gen-Probe Inc.
17. Enns, R. K. (1987). "Clinical Studies Summary Report: The Gen-Probe Rapid Diagnostic System for the *Mycobacterium avium* Complex. Gen-Probe Inc.
18. Gen-Probe Incorporated (1986). "*Mycobacterium avium* Complex Rapid Diagnostic System," Package insert. Gen-Probe Inc.
19. Gen-Probe Incorporated (1986). "*Mycobacterium tuberculosis* Complex Rapid Diagnostic System," Package insert. Gen-Probe Inc.
20. Gen-Probe Incorporated (1987). "Nucleic Acid Probes for Detection and/or Quantitation of Nonviral Organisms," Patent No. W088/03957. Gen-Probe Inc.
21. Gen-Probe Incorporated (1987). "Method for Releasing RNA and DNA from Cells," Patent No. AU70404-87. Gen-Probe Inc.
22. Gonzalez, R., and Hanna, B. A. (1987). Evaluation of the Gen-Probe DNA hybridisation systems for the identification of *Mycobacterium avium–intracellulare. Diagn. Microbiol. Infect. Dis.* **8,** 69–77.
23. Grange, J. M. (1983). The mycobacteria. *In* "Topley & Wilsons Principles of Bacteriology, Virology and Immunity" (M.T. Parker, ed.), 7th ed., pp. 60–93. Edward Arnold, London.
24. Grange, J. M. (1988). "Mycobacteria and Human Disease." Edward Arnold, London.
25. Greene, J. B., Sidlin, G. S., Lewin, S., Levine, J. F., Masur, H., Sinnerhoff, M. S., Nicholas, P., Good, R. C., Zolla-Paznor, S. B., Pollock, A. A., Tapper, M. L., and Holzman, R. S. (1982). *Mycobacterium avium–intracellulare*: A cause of disseminated

life-threatening infection in homosexuals and drug abusers. *Ann. Intern. Med.* **97,** 539–546.
26. Haun, G., and Gobel, V. (1987). Oligonucleotide probes for genus, species and subspecies-specific identification of representatives of the genus *Proteus*. *FEMS Microbiol. Lett.* **43,** 187–193.
27. Horn, J. E., Kappus, E. W., Galkow, S., and Quinn, T. C. (1988). Diagnosis of *Chlamydia trachomatis* in biopsied tissue specimens by using in situ DNA hybridisation. *J. Infect. Dis.* **157,** 1249–1253.
28. Huysmans, E., and De Wachter, R. (1986). Compilation of small ribosomal subunit RNA sequences. *Nucleic Acids Res.* **14,** Suppl., 73–118.
29. Imaeda, T., and Kaminski, Z. (1987). Identification of *Mycobacterium avium–intracellulare* complex including those isolated from AIDS patients by characterisation of their genomic DNAs. *Am. Soc. Mircobiol., Abstr. Annu. Meet.* Abstr. U-33.
30. Ivanyi, J., Morris, J. A., and Keen, M. (1985). Studies with monoclonal antibodies to mycobacteria. *In* "Monoclonal Antibodies against Bacteria" (A. J. L. Macario and E. Conway, eds.), Vol. 1, pp. 59–90, Academic Press, London.
31. Ivanyi, J., Bothamley, G. H., and Jackett, P. S. (1988). Immunodiagnostic assays for tuberculosis and leprosy. *Br. Med. Bull.* **44,** 635–649.
32. Jenkins, P. A., Pattyn, S. R., and Portaels, F. (1982). Diagnostic bacteriology. *In* "The Biology of the Mycobacteria" (C. Ratledge and J.L. Stanford, eds.), Vol. 1, pp. 441–470. Academic Press, New York.
33. Kadivel, G. V., Samuel, A. M., Mazardo, T. M. B. S., and Chaparas, S. D. (1987). Radioimmunoassay for detecting *M. tuberculosis* antigen in cerebrospinal fluids of patients with tuberculous meningitis. *J. Infect. Dis.* **155,** 608–611.
34. Kielin, T. E., and Edwards, F. F. (1987). Rapid identification using a specific DNA probe of *Mycobacterium avium* complex from patients with acquired immunodeficiency syndrome. *J. Clin. Microbiol.* **25,** 1551–1552.
35. Kielin, T. E., Edwards, F. F., Brannon, P., Tsang, A. Y., Maio, M., Gold, J. W. M., Whimby, E., Wong, B., McClatchy, J. K., and Armstrong, D. (1985). Infections caused by *Mycobacterium avium* complex in immunocompromised patients: Diagnosis by blood culture and fecal examination, antimicrobial susceptibility tests and morphological and seroagglutination characteristics. *J. Clin. Microbiol.* **21,** 168–173.
36. Kingston, A. E., Salquane, P. R., Mitchison, N. A., and Colston, M. J. (1987). Immunological activity of a 14-kilodalton recombinant protein of *Mycobacterium tuberculosis* H37Rv. *Infect. Immun.* **55,** 3149–3154.
37. Kirihara, J. M., Hillier, S. L., and Coyle, M. B. (1985). Improved detection times for *Mycobacterium avium* complex and *Mycobacterium tuberculosis* with the Bactec radiometric system. *J. Clin. Microbiol.* **22,** 841–845.
38. Krambovitis, E., Lock, P. E., McIllmurray, M. B., Hendruckse, W., and Hetzel, H. (1984). Rapid diagnosis of tuberculous meningitis by latex particle agglutination. *Lancet* **2,** 1229–1231.
39. Kuwabara, S. (1975). Purification and properties of tuberculin-active proteins from *Mycobacterium tuberculosis*. *J. Biol. Chem.* **250,** 2556–2562.
40. Lamb, J. R., and Young, D. B. (1987). A novel approach to the identification of T cell epitopes in *Mycobacterium tuberculosis* using human T lymphocyte clones. *Immunology* **60,** 1–5.
41. Landegren, V., Kaiser, R., Caskey, C. T., and Hood, L. (1988). DNA diagnostics—molecular techniques and automation. *Science* **242,** 229–237.
42. Leary, J. J, Brigati, D. J., and Ward, D. C. (1983). Rapid and sensitive colorimetric method for visualising biotin-labelled DNA probes hybridised to DNA or RNA immobilised on nitrocellulose: bioblots. *Proc. Natl. Acad. Sci. U.S.A.* **80,** 4045–4049.

43. Macher, A. M., Kovacs, J. A., Gill, V., Roberts, G. D., Ames, J., Parke, C. H., Strans, S., Lane, H. C., Parrillo, J. E., Fanci, A. S., and Masur, H. (1983). Bacteremia due to *Mycobacterium avium–intracellulare* in the acquired immunodeficiency syndrome. *Ann. Intern. Med.* **99,** 782–785.
44. Maniatis, T., Fritsch, E. F., and Sambrook, J. (1982). "Molecular Cloning: A Laboratory Manual." Cold Spring Harbor Lab., Cold Spring Harbor, New York.
45. McFadden, J. J., Butcher, P. D., Chiodini, R., and Hermon-Taylor, J. (1987). Crohn's disease-isolated mycobacteria are identified to *Mycobacterium paratuberculosis* as determined by DNA probes that distinguish between mycobacterial species. *J. Clin. Microbiol.* **25,** 796–801.
46. McFadden, J. J., Butcher, P. D., Chiodini, R. J., and Hermon-Taylor, J. (1987). Determination of genome size and DNA homology between an unclassified *Mycobacterium* species, isolated from patients with Crohn's disease and other mycobacteria. *J. Gen. Microbiol.* **133,** 211–214.
47. McFadden, J. J., Butcher, P. D., Thompson, J., Chiodini, R. J., and Hermon-Taylor, J. (1987). The use of DNA probes identifying restriction-fragment-length polymorphisms to examine the *Mycobacterium avium* complex. *Mol. Microbiol.* **1,** 283–291.
48. McIntyre, G., and Stanford, J. L. (1986). The relationship between immunodiffusion and agglutination serotypes of *Mycobacterium avium* and *Mycobacterium intracellulare. Eur. J. Respir. Dis.* **69,** 135–141.
49. McIntyre, G., and Stanford, J. L. (1986). Immunodiffusion analysis shows that *Mycobacterium paratuberculosis* and other mycobactin-dependent mycobacteria are variants of *Mycobacterium avium. J. Appl. Bacteriol.* **61,** 295–298.
50. Meissner, G., and Anz, W. (1977). Sources of *Mycobacterium avium* complex infection resulting in human disease. *Am. Rev. Respir. Dis.* **116,** 1057–1064.
51. Merchx, J. J., Soule, E. H., and Karlson, A. G. (1964). The histopathology of lesions caused by infection with unclassified acid-fast bacteria in man. *Am. J. Clin. Pathol.* **41,** 244–255.
52. Minden, P., Kelleher, P. J., Freed, J. H., Nielsen, L. D., Brennan, B., McPheron, L., and McClatchy, J. K. (1984). Immunological evaluation of a component isolated from *Mycobacterium bovis* BCG with a monoclonal antibody to *M. bovis* BCG. *Infect. Immun.* **46,** 519–525.
53. Mitchison, D. A., Wallace, J. G., Bhatia, A. L., Selkon, J. B., Subbaiah, T. V., and Lancaster, M. C. (1960). Comparison of the virulence in guinea pigs of South Indian and British tubercle bacilli. *Tubercle* **41,** 1–22.
54. Mizuguchi, Y., and Tokunaga T. (1970). Method of isolation of deoxyribonucleic acid from mycobacteria. *J. Bacteriol.* **104,** 1020–1021.
55. Musial, C. E., Tice, L. S., Stockman, L., and Roberts, G. D. (1988). Identification of mycobacteria from cultures by using the Gen-Probe rapid diagnostic system for *Mycobacterium avium* complex and *Mycobacterium tuberculosis* complex. *J. Clin. Microbiol.* **26,** 2120–2123.
56. Mustafa, A. S., Kralheim, G., Degre, M., and Godal, T. (1986). *Mycobacterium bovis* BCG-induced human T cell clones from BCG-vaccinated healthy subjects: Antigen specificity and lymphokine production. *Infect. Immun.* **53,** 491–497.
57. Pao, C. C., Lin, S. -S., Wu, S. Y., Juang, W. -M., Chang, C. -H., and Lin, J. -Y. (1988). The detection of mycobacterial DNA sequences in uncultured clinical specimens with cloned *Mycobacterium tuberculosis* DNA as probes. *Tubercle* **69,** 27–36.
58. Parmasivan, C. N., Govindan, D., Prabhakar, R., Somasundaram, P. R., Subbammal, S., and Tripathy, S.P. (1985). Species level identification of non-tuberculous mycobacteria from South Indian BCG Trial Area during 1981. *Tubercle* **66,** 9–15

59. Pinching, A. J. (1987). The acquired immune deficiency syndrome: With specific reference to tuberculosis. *Tubercle* **68,** 65–69.
60. Prabhaker, R., Venkataraman, P., Vallishayee, R. S., Reeser, P., Musa, S., Hashin, R., Kim, Y., Dimmer, C., Wiegeshans, E., Edwards, M. L., and Smith, D. W. (1987). Virulence for guinea pigs of tubercle bacilli isolated from the sputum of participants in the BCG trial. Chingleput district, South India. *Tubercle* **68,** 3–17.
61. Ranadive, S. N., Battacharya, S., Kale, M. K., and Battacharya, A., (1986). Humoral immune response in tuberculosis—initial characterisation by immunoprecipitation of ^{125}iodine-labelled antigens and sodium dodecyl sulphate–polyacrylamide gel electrophoresis. *Clin. Exp. Immunol.* **64,** 277–284.
62. Roberts, M. C., McMillan, C., and Coyle, M. B. (1987). Whole chromosomal DNA probes for rapid identification of *Mycobacterium tuberculosis* and *Mycobacterium avium* complex. *J. Clin. Microbiol.* **25,** 1239–1243.
63. Sauton, B. (1912). Sur la mitrition minerale du bacille tuberculeux. *C. R. Hebd. Sean Acad. Sci.,* **155,** 860.
64. Shoemaker, S. A., Fisher, J. H., and Scoggin, C. H. (1985). Techniques of DNA hybridisation detect small numbers of mycobacteria with no cross-hybridisation with non-mycobacterial respiratory organisms. *Am. Rev. Respir. Dis.* **131,** 760–763.
65. Shoemaker, S. A., Fisher, J. H., Jones, W. D., and Scoggin, C. H. (1986). Restriction fragment analysis of chromosomal DNA defines different strains of *Mycobacterium tuberculosis. Am. Rev. Respir. Dis.* **134,** 210–213.
66. Smith, D. W., Wiegehaus, E. H., and Edwards, M. L. (1988). The protective effects of BCG vaccination against tuberculosis. *In* "*Mycobacterium tuberculosis:* Interactions with the Immune System" (M. Bendinelli and H. Friedman, eds.), pp. 341–370. Plenum, New York.
67. Southern, E. M. (1975). Detection of specific sequences among DNA fragments separated by gel electrophoresis. *J. Mol. Biol.***98,** 503–517.
68. Stanford, J. L. (1983). Immunologically important constituents of mycobacteria: Antigens. *In* "The Biology of the Mycobacteria" (C. Ratledge and D. L. Stanford, eds.), Vol. 2, pp. 85–127. Academic Press, New York.
69. Stanford, J. L., and Grange, J. M. (1974). The meaning and structure of species as applied to mycobacteria. *Tubercle* **55,** 143–152.
70. Stead, W. W., and Dutt, A. K. (1988). Changing faces of clinical tuberculosis. *In* "*Mycobacterium tuberculosis*. Interactions with the Immune System" (M. Bendinelli and H. Friedman, eds.), pp. 371–388. Plenum, New York.
71. Stylbo, K. (1983). Tuberculosis and its control: Lessons to be learned from past experience and its implications for leprosy control programmes. *Ethiop. Med. J.* **21,** 101–122.
72. Tuberculosis Prevention Trial (1980). Trial of BCG vaccines in South India for tuberculosis prevention. *Indian J. Med. Res.* **72,** Suppl., 1–74.
73. Upholt, W. B. (1977). Estimation of DNA sequence divergence from comparison of restriction endonuclease digests. *Nucleic Acid. Res.* **4,** 1257–1265.
74. Wayne, L. G., and Kubica, G. P. (1986). Genus *Mycobacterium*. *In* "Bergey's Manual of Systematic Bacteriology" (P. H. A. Sneath, N. Mair, and M. E. Sharpe, eds.), Vol. 2, pp. 1436–1457. Williams & Wilkins, Baltimore, Maryland.
75. Welcher, A. A., Torres, A. R., and Ward, D. C. (1986). Selectivbe enrichment of specific DNA, cDNA and RNA sequences using biotylated probes, avidin and copper-chelate agarose. *Nucleic Acids Res.* **14,** 10027–10044.
76. Winder, F. G., and MacNaughton, A. W. (1978). A relatively rapid procedure for the preparation of lysis-susceptible forms of *Mycobacterium smegmatis*. *J. Gen. Microbiol.* **109,** 177–180.

77. Wolinsky, E., and Schaefer, W. B. (1973). Proposed numbering scheme for mycobacterial serotypes by agglutination. *Int. J. Systm. Bacteriol.* **23,** 182–183.
78. World Health Organization (1982). Tuberculosis control: Report of a joint IUAT/WHO study group. *W.H.O. Tech. Rep. Ser.* **671.**
79. World Health Organization (1986). Results of a World Health Organisation-sponsored workshop to characterize antigens recognized by *Mycobacterium*-specific monoclonal antibodies. *Infect. Immun.* **51,** 718–720.
80. Yeager, H., Lacy, J., Smith,, L. R., and Lemaistre, C. A. (1964). Quantitative studies of mycobacterial populations in sputum and saliva. *Am. Rev. Respir. Dis.* **95,** 998–1004.
81. Yoshimura, H. H., and Graham, D. Y. (1988). Nucleic acid hybridisation studies of mycobacteria-dependent mycobacteria. *J. Clin. Microbiol.* **25,** 1309–1312.
82. Yoshimura, H. H., Graham, D. Y., Estes, M. K., and Merkal, R. S. (1987). Investigation of association of mycobacteria with inflammatory bowel disease by nucleic acid hybridization. *J. Clin. Microbiol.* **25,** 45–51.
83. Young, B. D., and Anderson, M. L. M. (1985). Quantitative analysis of solution hybridisation. *In* "Nucleic Acid Hybridisation: A Practical Approach (B. D. Hames and S. J. Higgins, eds.), pp. 47–71. IRL Press, Washington, D.C.

8

Att Sites, *Tox* Gene, and Insertion Elements as Tools for the Diagnosis and Molecular Epidemiology of *Corynebacterium Diphtheriae*

RINO RAPPUOLI AND ROY GROSS
Sclavo Research Center
Siena, Italy

GENE PROBES
FOR BACTERIA

I. INTRODUCTION

A. The Disease

Diphtheria is a severe disease caused by the noninvasive infection of the upper respiratory tract by a toxinogenic strain of *Corynebacterium diphtheriae*. The fatality of the disease, which was often close to 50% before the introduction of serotherapy, is still high: on the order of 5–15%. The serious consequences of diphtherial infection are entirely due to the production and systemic dissemination of diphtheria toxin, one of the most potent bacterial protein toxins. The infection, often spread by droplets from a healthy carrier to a susceptible host, starts with the multiplication of the bacteria in the throat where they elaborate the toxin, which causes necrosis of the surrounding tissues and host inflammatory response resulting in the typical diphtheritic pseudomembranes composed of fibrin, bacteria, and leukocytes. The spread of the toxin through the bloodstream to other organs may cause neurological and cardiac complications. Death may occur by suffocation due to the obstruction of the respiratory tract or by the systemic complications. Early neutralization of the toxin with antitoxin antibodies is the only way to prevent the severe complications of diphtheria. Typically occurring in large epidemics before the introduction of mass immunization, diphtheria is now a rare disease that occurs in small, localized outbreaks.

B. *Corynebacterium diphtheriae*

Corynebacterium diphtheriae is a gram-positive, rodlike bacillus belonging to the genus *Corynebacterium*, a genus related to mycobacteria and nocardia in the serological relatedness of the principal cell wall antigens (13). The genus *Corynebacterium* comprises *C. diphtheriae, C. ulcerans, C. pseudotuberculosis, C. hofmannii, C. xerosis, C. equi, C. kutscheri, C. minutissimum, C. flavidum, C. bovis, C. hoagii, C. glutamicum, C. equi, C. betae, C. renale,* and other species (1). Most of these species, including the nontoxinogenic *C. diphtheriae* strains, are harmless commensals. Following lysogenization by a phage of the β family, which carries the structural gene for diphtheria toxin, the harmless *C. diphtheriae* strains acquire the capability of producing the toxin and become virulent (16). Although the phages carrying the *tox* gene can occasionally also lysogenize *C. pseudotuberculosis* and *C. ulcerans,* very rarely are these associated with disease.

C. Diphtheria Toxin

The diphtheria toxin is a protein with MW 58,350 (28,39) that is released into the supernatant by toxinogenic *C. diphtheriae*. The toxin is synthesized as a single polypeptide chain, which can be divided into two functionally different moieties following mild trypsin treatment and reduction of a disulfide bond. The fragment A has MW 21,150 and is an NAD^+-binding enzyme that transfers the ADP-ribosyl groups to a posttranslationally modified histidine residue (diphthamide) of the cytoplasmic elongation factor 2 (EF-2) of eukaryotic cells (4). This modification leads to the inhibition of protein synthesis and cell death. The toxin is one of the most powerful bacterial toxins, with a minimal lethal dose of $<0.1\ \mu g/kg$ body weight. Fragment B has MW 37,200. It is required for the recognition of specific surface receptors present on sensitive eukaryotic cells and for the translocation of the enzymatically active fragment A across the cell membrane into the cytoplasm where fragment A elaborates its toxic activity. Structure–function analyses of the toxin and its fragments have been extensively reviewed (51).

D. History of Diphtheria

While the history of diphtheria dates back to Hippocrates, who described its typical symptoms in the fourth century BC, its scientific history started in 1884, when Loeffler described the bacillus and proved that *C. diphtheriae* was the etiological agent of diphtheria (22). A few years later (1888) Roux and Yersin (41) were able to produce diphtheria in laboratory animals by inoculating cell-free filtrates of cultures of *C. diphtheriae*, showing that the disease was caused solely by an exogenous poison produced by the bacteria. This observation provided the basis for the treatment of the disease by von Behring and Kitasako in 1890 (3) with an antiserum against the toxin, and for the active prevention of the disease through immunization with a toxin inactivated by antitoxic antibodies (2) and by formaldehyde (31,32). The introduction of mass immunization using the formaldehyde-detoxified toxin, which is given to infants with tetanus and pertussis (DPT) vaccination, has allowed the eradication of the disease in countries in which vaccination is practiced. Diphtheria is now confined to a few areas of the world, such as Indonesia, India, and some parts of western Africa. In Western countries, vaccination resulted not only in the disappearance of the disease but also of the toxinogenic strains among healthy carriers. As a consequence, the immunity against diphtheria is no longer boosted by asymptomatic infections, and a large proportion of the adult population of Western countries is again susceptible to diphtheria. In these countries, diphtheria now occurs sporadically when a toxinogenic bacterium from another country in which diphtheria is still endemic finds a susceptible population (9,27,30). Many times, a phage deriving from an imported strain can convert the nontoxinogenic strains already present in the population to toxinogenicity and thus lead to the spread of the disease.

E. Diagnosis of Diphtheria

Rapid diagnosis of diphtheria is mandatory, since the serotherapy is effective only if practiced at the very early stage of the disease. When diphtheria was prevalent or in epidemic situations, experienced personnel were highly skilled in recognizing the disease from the clinical symptoms and from the methylene blue- or toluidine blue-stained smears of throat swabbings containing *C. diphtheriae*. Present-day competence requires the cultural isolation and biochemical identification of *C. diphtheriae,* including laboratory proof of toxinogenicity.

Therefore, the diagnosis of diphtheria requires three main tests: identification of *C. diphtheriae,* determination of its toxinogenicity, and epidemiological correlation of different isolates. So far the identification of *C. diphtheriae* is determined by classical microbiological and biochemical methods. The toxinogenicity is determined *in vitro* by the immunoprecipitation of the toxin produced by the bacteria on agar plates containing strips of filter paper embedded in antitoxic antiserum (Elek test) (15), or *in vivo* by observing the necrosis produced by the toxin in the skin of rabbits or guinea pigs (20). The epidemiological studies rely mostly on three classification schemes, one based on the subdivision of the strains of the biotypes *mitis, intermedius,* and *gravis,* and the other two based on the classification of the strains according to their sensitivity to bacteriophages and bacteriocins. In this chapter we describe four DNA probes that allow (1) the identification of corynebacteria, (2) the identification of *C. diphtheriae*, (3) the determination of the toxinogenicity, and (4) epidemiological correlation between different isolates.

II. BACKGROUND

A. Biotypes of *Corynebacterium diphtheriae*

The classification of *Corynebacterium* in different species and biotypes is based on their morphological and biochemical properties, which rely on such key reactions as nitrate reduction, urease activity, and carbohydrate fermentation, in addition to hemolysin production and catalase reaction (49). With some experience the different biotypes of *C. diphtheriae,* which have been described by McLeod (24), can be distinguished on CTBA solid medium (44) after 48 hr of incubation.

The biotype *gravis* forms large gray or black colonies that are granular on the surface and have a characteristic colony morphology with radial striations and crenated margins. The mitis biotype grows to middle-sized colonies that are intensely black and have regular margins. The *interme-*

dius biotype usually grows in small colonies with regular margins and a granular surface; the colonies are flat or convex with a black center. The species *C. ulcerans* forms black, dense, and smooth colonies. Only the *C. diphtheriae* subsp. *mitis* biotype is hemolytic, whereas *gravis* is only occasionally hemolytic. The *intermedius* biotype and *C. ulcerans* are not hemolytic.

The initial idea of a correlation between the colony type and the severity of the disease was later found to be incorrect. All three *C. diphtheriae* biotypes, *C. ulcerans,* and *C. pseudotuberculosis* can be toxinogenic and can cause disease in humans. Different strains of the various biotypes can differ enormously in their pathogenic properties. They can be normal commensals in the throat or skin. They can also cause disease depending on the presence of the *tox* gene encoded on lysogenic phages, on the expression of other still uncharacterized virulence factors, and on the immune status of the host. Because of these strong differences of strains, even within one biotype, there is a need for a more specific classification for the evaluation of their pathogenic properties and for epidemiological studies.

An early attempt for a more detailed classification scheme was made using serological techniques, which have been very useful in other examples such as *Salmonella* (21). Although these methods allowed some success in specific cases, the serological approach was not satisfactory as a general and standardized classification system for epidemiological studies. Some reasons for this are as follows: (1) Although there is a clear-cut specific agglutination reaction for the *gravis* biotype, the *mitis* and *intermedius* biotypes show cross-reactions. (2) Single serotypes can dominate huge geographic regions for long periods. (3) Most of the non-toxinogenic strains are not agglutinable and therefore cannot be classified.

An important result of the early epidemiological studies is, however, that one biotype can dominate for long periods in certain areas: the *gravis* biotype was usually found in the last big epidemics in Europe in the 1940s (47); the *mitis* biotype has been dominant in more recent years in the United States (14); and the *intermedius* biotype is predominantly found in outbreaks in Madagascar and India (43).

B. Bacteriophage Typing

A big step forward in the classification of strains of the genus *Corynebacterium* was the typing of bacteria according to their sensitivity or resistance against bacteriophages. The first systematic use of this approach was described by Saragea and Maximesco (42). They used in their "provisorial phage-typing scheme for *C. diphtheriae*" a set of 24 lysogenic reference phages for the screening of the resistance pattern of the test strains. With this scheme they succeeded in dividing the three classical *C. diphtheriae* biotypes *gravis, mitis,* and *intermedius* into 21 "phage types" consisting of 14 *gravis,* 4 *mitis,* and 3 *intermedius* biotypes. The large proportion of the *gravis* biotype in the scheme is explained by the dominance of this biotype during the epidemics in Europe. These phage types are stable and correlate specifically with the biotypes and with the toxinogenicity. The method has been used on several occasions in various countries. For example, in Romania, from 1956 to 1969, nearly 100% of the toxinogenic *gravis* strains and about 75% of all tested *C. diphtheriae* strains could be typed (44).

This original phage-typing system was the first method that allowed a reproducible and stable subdivision of the *C. diphtheriae* biotypes in a standardized manner. Once established, the technique was simple enough to be used by different groups in various parts of the world. Because of the standardized typing set of phages the results obtained in different geographic areas at different times could be compared. Therefore, using this typing scheme it was possible for the first time to monitor the origin and spread of epidemics and to clarify the role of healthy carriers in the epidemiology of the disease.

A limitation of this scheme is that it is of value only for the typing of toxinogenic *gravis* strains, of which nearly 100% of the tested strains could be classified. For the nontoxinogenic strains, especially for *mitis* and *intermedius* biotypes, only 20% and 40%, respectively, of the tested strains could be classified. This became a greater problem because it was found in many countries that as a consequence of the large-scale immunization programs the distribution of the *C. diphtheriae* biotypes changed significantly. For example, for Europe it was found that the dominant *gravis* biotype was gradually replaced by nontoxinogenic *gravis, mitis,* and *intermedius* strains. Nontoxinogenic *mitis* and *intermedius* strains are often found in healthy carriers or in nonepidemic areas or periods. This resulted in a decrease of the typing efficiency from ~75% to ~47% of the tested strains. Maximesco (44) tried to respond to this increase in nontypable strains by the introduction of the "additional phage-typing scheme." This approach considers, in addition to the lysogenic properties of the test strains, their restriction properties against a highly virulent phage. A set of host range descendants of the virulent phage 951 (11) was defined, and the lysis properties of these phages in the test strains were analyzed. The phage types of these strains obtained by this additional typing scheme are stable and can therefore be used in manner similar to that in the former typing scheme. The number of phages (between 19 and 33) will probably grow in the future because of the isolation of new strains not typable with the established set.

In summary, while the original phage-typing scheme is useful for epidemiological studies of epidemic occurrence of toxinogenic *C. diphtheriae* subsp. *gravis*, the additional scheme should be helpful for endemic and postepidemic situations in which nontoxinogenic strains are dominant. Undoubtedly, this approach provided for the first time a tool for analyzing in a standardized manner the complicated epidemiology of diphtheria outbreaks. On the other hand, the change in the distribution of the biotypes and toxinogenic and nontoxinogenic strains after the introduction of mass immunization limited the value of this method. This led to an increase in the number of phages necessary for the typing and therefore to more complex typing schemes. Furthermore, typing efficiency became quite low.

C. Bacteriocin Typing

Another approach to the classification of strains of the genus *Corynebacterium* arose after the finding of Thibaut and Fredericq (48) that a *C. diphtheriae* subsp. *gravis* strain produced a bacteriocin. Bacteriocins are bacterial protein toxins that usually are only active on related strains or species. Such bacteriocinogenic strains have then been observed in all corynebacterial species. Bacteriocinogenic strains are immune against their own bacteriocin (25), and the coexistence of lysogeny and bacteriocinogeny is observed. Meitert and Bica-Popii (25) proposed a typing scheme based on the determination of the sensitivity of the test strains against the bacteriocins produced by a set of 20 different defined bacteriocin-producing strains (44). This method has been applied successfully in the classification of some phage-untypable *C. diphtheriae* strains (50,52). The set of the 20 bacteriocin producers consists of strains of different species of *Corynebacterium: C. diphtheriae, C. ulcerans, C. hofmannii,* and *C. xerosis*. This broad spectrum of species already implies that there is no correlation between the bacteriocin type of a strain and its biotype. Besides the more complex microbiological techniques required for the method, there is no standardization possible. The bacteriocin type of a strain is too variable depending on the growth conditions, agar concentrations, growth factors, and storage conditions for the bacteriocinogenic assay strains. Therefore, this method can only be useful in special cases. The results should be interpreted carefully in a more local sense and over a short time period (44).

D. DNA Probes

In 1983 Pappenheimer and Murphy (30) introduced the gene probe technology in the epidemiological analysis of diphtheria outbreaks. They compared the pattern of chromosomal DNA of clinical *C. diphtheriae* isolates after digestion with restriction enzymes and separation on agarose gels, and they used radioactively labeled DNA probes of the β phage for hybridization analysis of chromosomal DNA preparations. With this technology they were able to confirm the results of the phage-typing analysis of small diphtheria outbreaks and to show that diphtheria can be spread not only by toxinogenic *C. diphtheriae* strains but also by phages that can convert *in situ* nontoxinogenic diphtheria strains to toxinogenicity. This approach, which opened the field to the use of restriction enzymes and DNA probes for the diagnosis of diphtheria, is however, not generally applicable. DNA probes are described in the following section.

III. RESULTS AND DISCUSSION

A. Description of the DNA Probes

1. Probes specific for the genus Corynebacterium *and for* Corynebacterium diphtheriae

Corynebacterium diphtheriae is a nonpathogenic bacterium that can become virulent following lysogenization by a bacteriophage of the β family, which carries the structural gene coding for diphtheria toxin. Phage integration into the bacterial chromosome occurs by the mechanism described by Campbell for the bacteriophage λ of *E. coli* (7): site-specific recombination between a site of the phage DNA (phage attachment site or *attP*) and a site of the bacterial chromosome (bacterial attachment site or *attB*). *attP* and *attB* share some sequence homology, which consists of 15 bp in the case of λ and of 96 bp in the case of the β phage. *Corynebacterium diphtheriae* contains two *attB* sites (*attB1* and *attB2*) where the corynephages can integrate with similar frequency (34,35,38). This property has been used for the construction of double lysogens that hyperproduce diphtheria toxin or toxin-related molecules (33). The chromosomal region containing the bacterial attachment sites in *C. diphtheriae* C7(−) has been cloned and partially sequenced. *attB1* and *attB2* are contained in a 3.5-kb *Eco*RI fragment, the map of which is shown in Fig. 1. *attB1* and *attB2* share a sequence homology of ~120 bp, 96 of which are also homologous to *attP*. Plasmid pA634, the map of which is reported in Fig. 1, contains sequences that are common to all corynebacteria (the *attB* sites) and sequences that are unique to *C. diphtheriae* (the 0.7-kb *Hin*cII fragment internal to the *attB1–attB2* region). By using the 3.5-kb *Eco*RI fragment derived from plasmid pA634 as a probe, it has been shown that all species of the genus *Corynebacterium* contain sequences homologous to the *attB* region (10). Southern blots of *Bam*HI-digested chromosomal

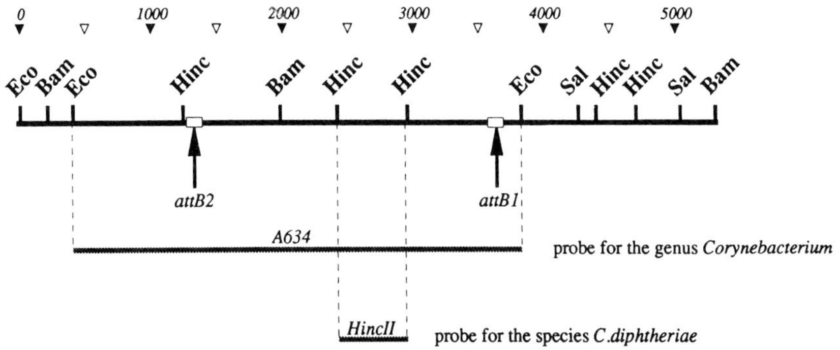

Fig. 1. Restriction map of the chromosomal region of *Corynebacterium diphtheriae* C7(−) containing the loci *attB1* and *attB2*, where the phages of the β family integrate. The 3.5-kb *Eco*RI fragment contained in the clone pA634, which can be used as a probe specific for the genus *Corynebacterium*, is shown. The 0.7-kb *Hinc*II fragment, which can be used as a probe specific for *C. diphtheriae,* is also shown.

DNA from *C. diphtheriae, C. pseudotuberculosis, C. minutissimum, C. xerosis, C. flavidum, C. bovis,* and *C. glutamicum* showed two hybridizing fragments, suggesting that these strains contain two *attB* sites. *Corynebacterium ulcerans, C. renale, C. hofmanii, C. hoagii, C. betae,* and *C. equi* had only one hybridizing fragment, suggesting the presence of only one *attB* site. Although the intensity of hybridization with DNA of *C. diphtheriae* strains was significantly stronger than with the DNA from the other strains, it can be concluded that the 3.5-kb *Eco*RI fragment derived from plasmid pA634 can be used as a probe specific for the species of the genus *Corynebacterium* (Fig. 2a). The hybridization described previously can also be reproduced using as a probe only the 96-bp fragment common to *attB* and *attP*, suggesting that the DNA region necessary for integration of β phage is common to all corynebacteria. While the 3.5-kb *Eco*RI fragment of plasmid pA634 can be used as a probe specific for the genus *Corynebacterium*, the 0.7-kb *Hinc*II fragment internal to the clone A634 can be used as a probe specific for the species *C. diphtheriae*. When this fragment is used as a probe on Southern blots of *Bam*HI-digested chromosomal DNA from *C. diphtheriae* or other corynebacteria, only the band containing the *attB1* site of *C. diphtheriae* will hybridize (Fig. 2b). Therefore, the 0.7-kb DNA fragment can be used for the positive identification of *C. diphtheriae*.

2. The Diphtheria tox *Gene as a Probe for Toxinogenic Strains*

One of the major problems associated with the diagnosis of *C. diphtheriae* strains is whether they are toxinogenic. So far two methods have been

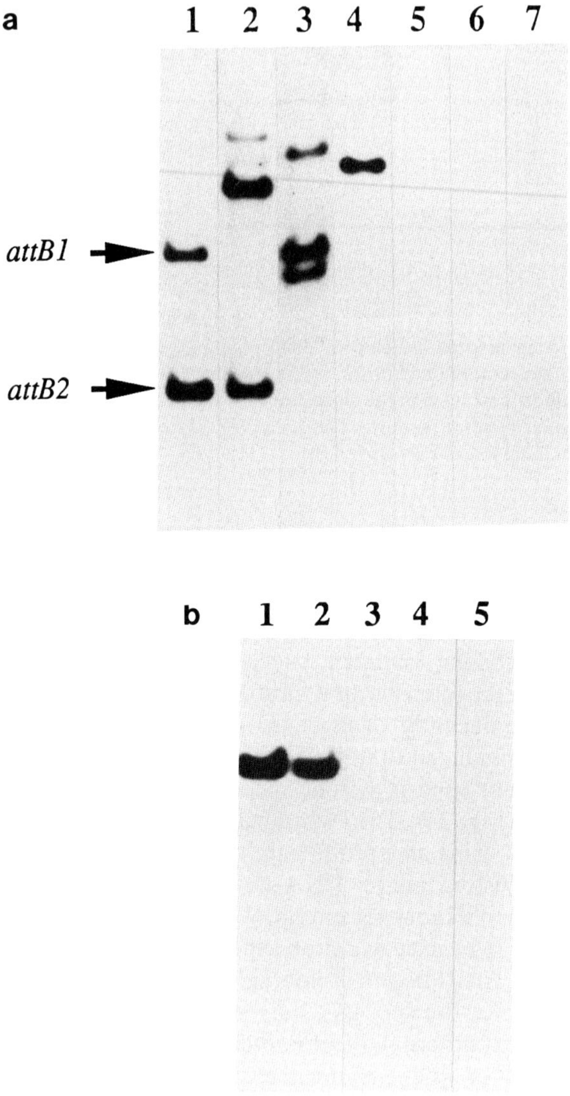

Fig. 2. (a) Detection of *Corynebacterium* species by hybridizing Southern blots of *Bam*HI-digested chromosomal DNA with the 3.5-kb *Eco*RI fragment of plasmid pA634. (1) *Corynebacterium diphtheriae* C7(−); the two empty *attB* sites hybridize. (2) *Corynebacterium diphtheriae* C7 (β). The β phage integrated into the *attB1* site splits the *attB1* band into two bands of higher molecular weights. (3) *Corynebacterium diphtheriae* C7(β). The β phage integrated in the *attB2* site splits the *attB2* band into two bands of higher molecular

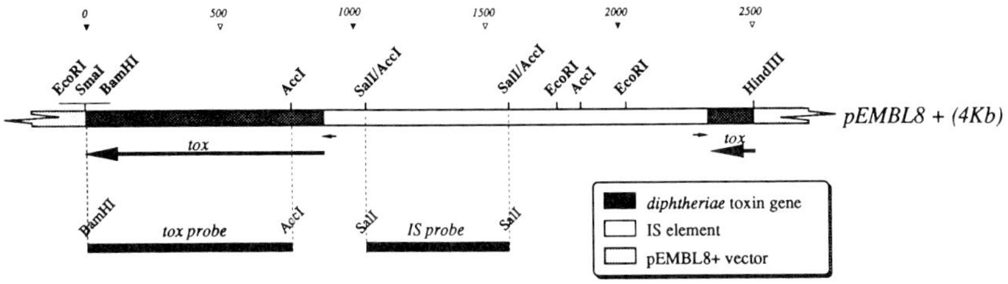

Fig. 3. Restriction map of the plasmid pDγ2 (ATCC 67011), which contains a probe specific for the gene of diphtheria toxin (*tox* probe) and a probe specific for the IS element of *Corynebacterium diphtheriae* (IS probe). The two small arrows at the end of the IS element indicate the 40-bp inverted repeats.

used for this purpose: the Elek test (15) and the rabbit skin test (20). Both methods are satisfactory, but each of them will miss 10–20% of the toxinogenic strains. For this reason it is important to check the negative strains by hybridization with a probe specific for the diphtheria toxin gene. The gene encoding diphtheria toxin has been cloned and sequenced from a variety of different sources (12,17,18,39). The *tox* gene is contained in an 1880-bp *Hin*dIII–*Eco*RI fragment of the phage DNA (the start codon is in position 122 and the stop codon in position 1804). Since the cloning of the full-length gene is not allowed by the NIH guidelines for recombinant DNA, several clones are available that contain only part of the *tox* gene. The plasmid pDγ2 (ATCC 67011), which contains a fragment internal to the *tox* gene, is shown in Fig. 3. A 0.77-kb *Bam*HI–*Acc*I fragment spanning nucleotides 297–1066 of the *tox* gene can be obtained from this plasmid and used as a probe (Fig. 4). We and others have extensively used a *tox* gene as a probe for the toxinogenicity of *C. diphtheriae* strains (19). The common finding is that often strains that were nontoxinogenic by the Elek test or the rabbit skin test can be positive to the hybridization with the *tox* gene. When these strains are then further analyzed by classical methods, most of the time they are found to be toxinogenic. In rare instances same of these strains will still be nontoxinogenic. In these cases they can either be very low toxin producers or contain an incomplete *tox* gene. In no case did strains that were positive by classical methods give a negative result by hybridization.

---

weights. (4) *Corynebacterium ulcerans*. (5) *Escherichia coli* JM101. (6) *Salmonella typhimurium*. (7) *Pseudomonas aeruginosa*. (b) Detection of *C. diphtheriae* strains by hybridizing Southern blots of *Bam*HI-digested chromosomal DNA with the 0.7-kb *Hin*cII fragment of plasmid pA634. (1) *Corynebacterium diphtheriae* C7(−); (2) *C. diphtheriae* 17093A (see Table I); (3) *C.. bovis*; (4) *C. betae;* (5) *C. glutamicum*.

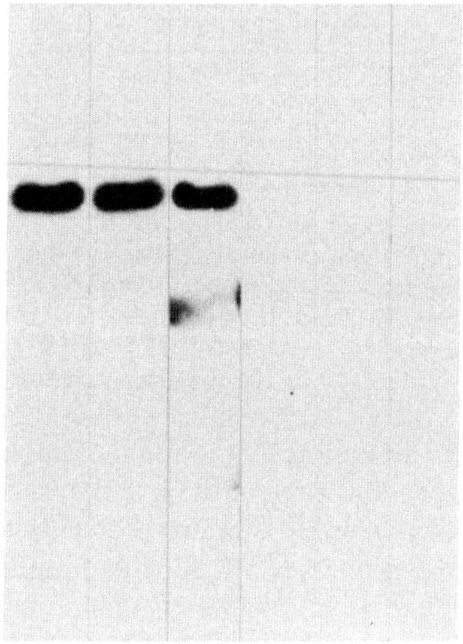

Fig. 4. Detection of toxinogenic *Corynebacterium diphtheriae* strains by hybridizing Southern blots of *Bam*HI-digested chromosomal DNA with the 0.77-kb *Bam*HI–*Acc*I probe specific for the *tox* gene deriving from the plasmid pDγ2. (Lanes 1–3) Three toxinogenic strains of *C. diphtheriae*. (Lanes 4-6) Two nontoxinogenic *C. diphtheriae* strains.

3. Insertion Elements as a Tool for the Molecular Epidemiology of Corynebacterium diphtheriae

Insertion (IS) elements are transposable DNA segments that can insert themselves into several sites of a bacterial genome. They have been described in most of the bacterial species where they can be located on the chromosomes, plasmids, or bacteriophages (6). They have a size ranging from 800 to 2200 bp and a typical structure that includes one or more internal open reading frames (ORF) flanked by a short terminal sequence that is invested at the two ends. The ORF usually codes for a transposase, while the terminal inverted repeats are required for the transposition process. The transposition events of the IS elements are sufficiently rare so that strains that are closely related have the IS elements inserted into the same sites of the chromosome, while the strains that diverged long ago usually have the IS elements located in different loci of the chromosome.

Two typical examples of the distribution of IS elements are *C. diphtheriae* and *Bordetella pertussis*: the former is a very old bacterium and its insertion sequence is found in a different chromosomal location in isolates deriving from different countries or different outbreaks. Whooping cough, on the other hand, is a disease caused by a bacterium of recent origin, and all the isolates derive from the same clone. In this case the distribution of the known insertion elements is almost identical in all isolates (23a,24a). Therefore, while the chromosomal location of the IS element can be used for the epidemiology of diphtheria, in the case of pertussis, the IS element can only be used for discriminating *Bordetella pertussis* from other *Bordetella* species.

The presence of an IS element in *C. diphtheriae* was first detected in bacteriophage γ (5). This phage was known to contain the *tox* gene and to be unable to convert nontoxinogenic *C. diphtheriae* strains to toxinogenicity (26). The analysis of the structure of the *tox* gene in this bacteriophage revealed that it contained an insert of ~1.5 kb, which interrupted the *tox* gene. A 2.5-kb *Hin*dIII–*Cla*I fragment containing the first 1060 bases of the *tox* gene and the insertion sequence was cloned in the plasmid vector pEMBL8, and its nucleotide sequence determined. Figure 3 shows the map of the plasmid pDγ2. The *tox* gene is interrupted within the region coding for the leader peptide of diphtheria toxin, 176 bp after the *Hin*dIII site, by an insert of 1450 bp that has the typical features of an insertion sequence: (1) two ORF spanning most of the insert in the direction opposite to the *tox* gene; (2) a sequence of 40 bp inverted and repeated at the two ends; and (3) the presence of a 9-bp direct repeat at the site of insertion (37; G. Ratti and R. Rappuoli, unpublished data).

To verify whether the IS element present in bacteriophage γ was also present in the chromosome of *C. diphtheriae* strains, the 0.6-kb DNA fragment internal to the IS element was used as a probe in a Southern blot of chromosomal DNA from the strain *C. diphtheriae* Belfanti 1030. Several hybridizing bands were detected (Fig. 5), indicating that this strain contained many (~30) copies of the IS element integrated into the chromosome. When the same probe was used on Southern blots of chromosomal DNA deriving from other strains, we found that the IS element was present in 88% of the *C. diphtheriae* strains tested and that the number and the size of the hybridizing bands varied among strains of different origin, although it was identical in strains isolated from the same outbreak. These properties suggested that the presence or absence and the hybridization pattern of this IS element could be used to study the epidemiology of diphtheria. The usefulness of this IS element was verified in a number of examples, which are described in the following paragraphs.

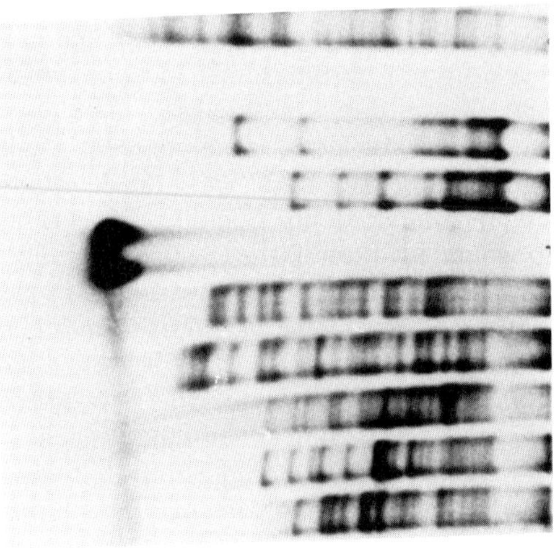

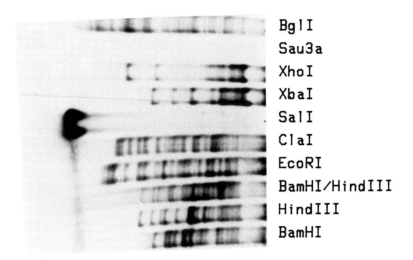

Fig. 5. Southern blot of chromosomal DNA from *Corynebacterium diphtheriae* Belfanti 1030 cut with different restriction enzymes and hybridized with the 0.55-kb *Sal*I fragment of plasmid pDγ2. Reproduced from the *Journal of Bacteriology* (37).

B. Examples for Epidemiological Use of DNA Probes

1. The Strains from Toronto

Three different strains (A, B, and C) of *C. diphtheriae* subsp. *mitis* were isolated from the throat of one healthy woman in Toronto (8). During the first analysis, they were of the Belfanti type (unable to reduce nitrates). Strains A and C were toxinogenic but their colony morphology and bacteriophage type were different. Strain B was nontoxinogenic but otherwise indistinguishable from A. The same strains were then analyzed by Pappenheimer and Murphy using restriction patterns of chromosomal DNA and hybridization with phage DNA (30). They concluded that strains A and C carried identical phages and that strain B was converted to A *in situ* by the phage coming from strain C. The identity of strains A and B and their difference from strain C were even more evident using the IS element as a probe (Fig. 6a): A and B had many comigrating bands which hybridized to the probe, while strain C had none. In this case, the use of the IS element gave the same result obtained by classical methods such as colony morphology, nitrate reduction, bacteriophage typing, or analysis of the restric-

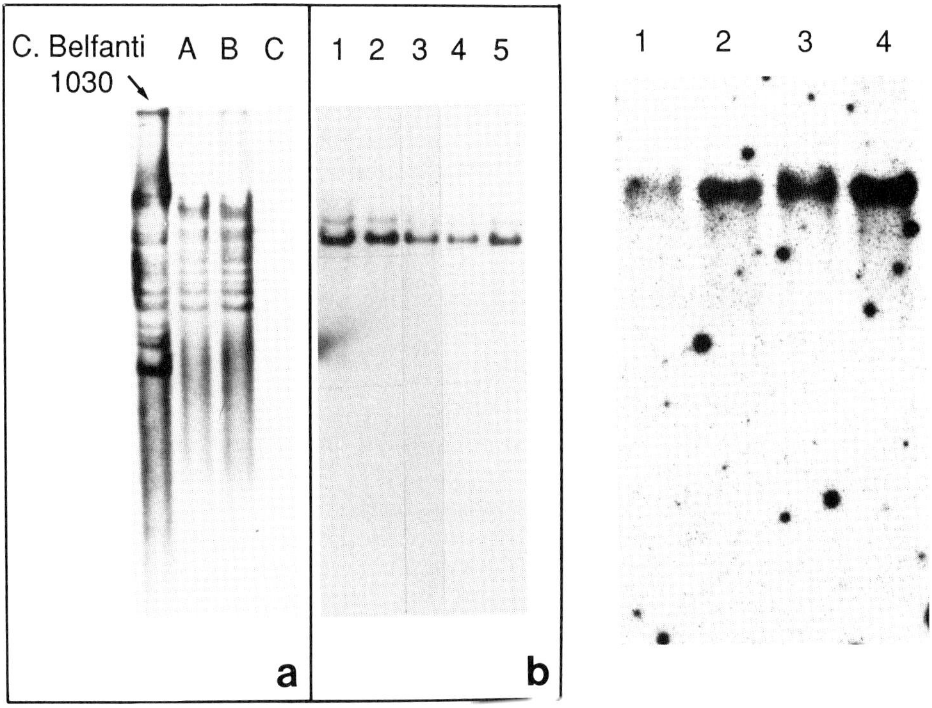

Fig. 6. Southern blot of *Bam*HI-digested chromosomal DNA from different isolates of *Corynebacterium diphtheriae* hybridized with the 0.55-kb *Sal*I fragment of plasmid pDγ2. (a) Strains A, B, and C isolated in Toronto. The first lane contains chromosomal DNA from Belfanti 1030 to show that most of the bands of strains A and B comigrate with those of Belfanti 1030. (b) Five different strains from Manchester. (c) Strains isolated in Indonesia. Lanes 1 and 2 contain tetracycline-resistant strains and lanes 3 and 4 tetracycline-sensitive strains. Partially reproduced from the *Journal of Bacteriology* (37).

tion pattern of the chromosomal DNA. However, the interpretation of the results using the IS probe was absolutely unequivocal and easy to derive. In addition to the results obtained by the classical methods, we were able to show that strains A and B (isolated in Toronto in 1977) were somehow related to the strain *C. diphtheriae* Belfanti 1030 (isolated in Austria in 1953), because many of the bands hybridizing to the IS element in strains A and B were also present in Belfanti 1030 (Fig. 6a). These latter results could not easily be obtained using the other methods.

2. *The Manchester Case*

In 1977, after the isolation of a toxinogenic strain of *C. diphtheriae* from the throat of a 10-week-old baby in Manchester (England), an epidemiological survey was carried out for a period of 6 months. Several toxinogenic and nontoxinogenic strains were isolated from children attending primary schools (46). By analyzing the restriction pattern of their chromosomal DNA, Pappenheimer and Murphy (30) concluded that all the toxinogenic and nontoxinogenic strains isolated in Manchester were indistinguishable, but could be differentiated from strains isolated during previous outbreaks of diphtheria that occurred elsewhere (30). Using the IS element as a probe, we could unequivocally come to the same conclusion; in fact all the stains isolated from Manchester had an identical pattern of hybridization (Fig. 6b) which differed from that of strains isolated elsewhere (Fig. 6a–c).

3. *Tetracycline-Sensitive and Tetracycline-Resistant Strains from Indonesia*

Diphtheria is still a major cause of morbidity and mortality in Indonesia where mass vaccination is not yet practiced, although the situation is slowly changing with the introduction of the Expanded Program of Immunization of the World Health Organization. In 1982 it was calculated that for every 100,000 children born in this country, ~600 developed the faucial form of the disease and 100 died. Antibiotics were commonly used in addition to serotherapy to treat the disease, erythromycin, tetracycline, clindamycin, and penicillin being the drugs of choice. Tetracycline was also empirically used to treat various bacterial infections in adults and children. Under these conditions, many tetracycline-resistant strains were isolated. In 1980, 86% of the 133 *C. diphtheriae* isolates were resistant to tetracycline; however, it was not clear whether the antibiotic marker was acquired by the strains endemic in Indonesia or whether the antibiotic treatment had favored the spread of a tetracycline-resistant strain of different origin (40). We obtained some of the tetyracycline-resistant and tetracycline-sensitive strains from Indonesia and tested them with the IS probe. Figure 6c shows that all of them had the same hybridization pattern, indicating that the tetracycline-resistant marker was acquired by the tetracycline-sensitive strains endemic in that area.

4. *The 1984–1986 Outbreak of Diphtheria in Sweden*

Following the introduction of mass vaccination against diphtheria, the disease and the toxinogenic strains have disappeared from most of the Western countries. With the disappearance of the toxinogenic strains, the natural infections that were necessary to boost the immunity acquired with vaccination have also disappeared, and, as a result, a large proportion of the adult population lack protective immunity against diphtheria. The high proportion of people without immunity, which can be as high as 50–70%, has allowed the occurrence of many small outbreaks of diphtheria in several European countries (9,27,30). During the period 1984–1986 a major outbreak occurred in Göteborg and Stockholm (Sweden). During the outbreak, which occurred largely among alcohol and drug abusers, 17 cases of diphtheria and 65 carriers were identified. Among the cases, 3 patients died and 6 had reversible paralysis. The carriers were mostly people with a high titer of antitoxin antibodies who did not develop symptoms and were probably the source of *C. diphtheriae* from 1984 to 1986. The presence of the Culture Collection of the University of Göteborg, in the same city where the outbreak occurred, allowed the collection of strains during the duration of the outbreak. The strains collected plus other strains isolated in Sweden or Denmark during the same period were analyzed using classical methods, including bacteriophage typing, and also with more recent methods such as the analysis of the restriction pattern of the chromosomal DNA or sodium dodecyl sulfate–polyacrylamide gel electrophoresis (SDS–PAGE) of the total bacterial proteins. None of these methods resulted in a satisfactory explanation for the epidemiology of the Swedish outbreak. When we used the IS probe on a selected sample of 36 strains provided by Enevold Falsen from the Culture Collection, University of Göteborg, we found that 31 of the 36 strains (86%) contained bands hybridizing to the IS probe (36). On the basis of the hybridization pattern, they could easily be classified in 17 different groups (from A to Q in Fig. 7 and Table I). The correlation between the hybridization pattern and the epidemiology was striking (Fig. 7). All strains isolated from alcohol and drug abusers and from patients with clinical diphtheria had the same hybridization pattern (group A in Table I), suggesting that the outbreak had been caused by a single strain. Strains isolated from carriers who said they had no contact with alcohol or drug abusers and those imported in leg wounds from tropical regions (e.g., India, Nepal and Africa) had a hybridization pattern different from those of group A and different from one another. Furthermore, it was possible to determine that the strain that caused the outbreak in Sweden had probably been imported from Denmark where it had been isolated as early as 1983 from a fatal case (strain 1 in Fig. 7 and Table I). In conclusion, in this case, the use of the IS element as a probe was superior to all other methods so far described, being the only one that allowed the epidemiological features of the Swedish outbreak of diphtheria to be seen. The study showed that a single strain probably imported from Denmark was responsible for all the cases of diphtheria that occurred in Göteborg and Stockholm during the period 1984–1986. Since during the same period at least six other different toxinogenic strains (strains 19,21,30,31,33, and 36 of Table I) were isolated in Sweden to which, in theory, a large proportion of the population was susceptible, these strains also should have

TABLE I

Strains of *Corynebacterium diphtheriae*[a]

Strain number[b]	Group	*C. diphtheriae* isolates[c]	Toxigenicity[d]	Copies of the insertion element	Source and year isolated
1		17233, var *mitis*	+		Copenhagen, Denmark, 1983; clinical diphtheria
2		17024A, var. *mitis*	+		Göteborg, Sweden, 1984; pharynx
3		17093A, var. *mitis*	−		Göteborg, 1984; pharynx
4		16574, var. *mitis*	+		Göteborg, 1984; throat; fatal diphtheria
5		15935, var. *mitis*	+		Göteborg, 1984; throat; clinical diphtheria
6		17156, var. *mitis*	+		Stockholm, Sweden, 1984; throat, fatal diphtheria
7	A	17269, var. *mitis*	+		Stockholm, 1984; throat
8		17074A, var. *mitis*	+	2	Boras, Sweden, 1984; pharynx
9		17945, var. *mitis*	+		Göteborg, 1985; trachea
10		17093, var. *mitis*	+		Göteborg, 1985; throat; clinical diphtheria
11		17995, var. *mitis*	+		Copenhagen, Jan. 1985; fatal diphtheria
12		18321, var. *mitis*	+		Stockholm, 1986
13		18673, var. *mitis*	+		Stockholm, 1986
14		18957, var. *mitis*	+		Stockholm, 1986; throat; fatal diphtheria
15	B	9740, var. *mitis*	−	3	Karlstad, Sweden, 1980; tropical wound
16		18642, var. *mitis*	−		Karlstad, 1980; leg wound
17	C	17083A, var. *mitis*	−	Many	Göteborg, 1984; pharynx
18		17133, var. *mitis*	−		Göteborg, 1984; throat/pharynx

19	D	17398, var. *gravis*	+	5	Växjö/Lund, Sweden, 1985
20	E	17274, var. *gravis*	–	5	Stockholm, 1984; throat
21	F	18645, var. *mitis*	+	5	Sweden, 1981
22	G	18637, var. *mitis*	–	4	Sweden, 1979; right leg wound
23	H	10091, var. *mitis*	–	Many	Sundsvall, Sweden, 1985; sore throat
24	I	17907, var. *mitis*	–	Many	Göteborg, 1985; trachea
25	J	17158, var. *mitis*	–	Many	Göteborg, 1984; pharynx
26	K	17052, var. *mitis*	–	Many	Göteborg, 1984; throat
27	L	17890, var. *mitis*	–	Many	Göteborg, 1985
28	M	17674, var. *gravis*	–	3	Uppsala, Sweden; wound received in tropical region
29	N	18918, var. *gravis*	–	2	Kristianstad, Sweden, 1986; toe wound
30	O	17141, var. *gravis*	+	2	Stockholm, 1976; imported from Nepal and India
31	P	18639, var. *mitis*	+	2	Sweden, 1979; leg wound in India
32		17920, var. *mitis*	–		Göteborg, 1985; trachea
33		18644, var. *mitis*	+		Sweden, 1981; imported from India
34	Q	17265, var. *mitis*	–	None	Stockholm, 1984; throat
35		18912, var. *gravis*	–		Kristianstad, 1986; toe wound
36		18646, *C. ulcerans*	+		Stockholm, 1985; healthy carrier

[a] Reproduced from the *New England Journal of Medicine* (36).
[b] The strains were divided into groups A to Q according to their hybridization pattern with the insertion element.
[c] All isolates were collected at the Culture Collection of the University of Göteborg, Sweden.
[d] Toxigenicity was determined by the Elek test and by hybridization with a genetic probe specific for the gene coding for diptheria toxin.

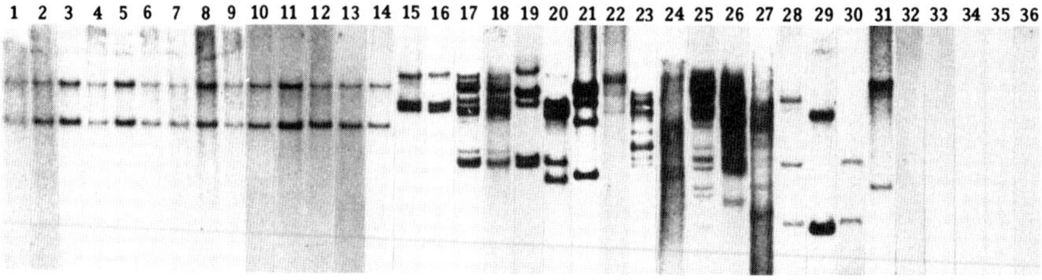

Fig. 7. Southern blots of chromosomal DNA for *Corynebacterium diphtheriae* isolates hybridized with the 0.55-kb *Sal*I fragment of plasmid pDγ2. Strains with identical patterns were assigned to the same group (see Table I). Reproduced from the *New England Journal of Medicine* (36).

caused diphtheria. The observation that only the strain of group A was able to do so suggests that is more virulent than the others (36).

IV. CONCLUSIONS AND PROSPECTS FOR THE FUTURE

The DNA probes described in this chapter are useful for the identification and the epidemiology of *C. diphtheriae*. Three of them, which allow the identification and the determination of the toxinogenicity of *C. diphtheriae* strains, are alternatives to the classical technologies, which rely chiefly on biochemical and microbiological methods. The fourth probe, based on a DNA insertion (IS) element specific for *C. diphtheriae*, is superior to all the methods previously available for the epidemiology of *C. diphtheriae* and suggests that the use of IS elements as DNA probes may be useful for the epidemiology of many other bacteria.

V. SUMMARY

The diagnosis of *Corynebacterium diphtheriae* involves three main steps: identification of *C. diphtheriae*, determination of toxinogenicity, and the epidemiological correlation of different isolates. In this chapter we described two plasmids containing a full set of probes, which allow the complete diagnosis of *C. diphtheriae*.

Restriction fragments of plasmid pA634 contain a probe specific for the genus *Corynebacterium* as well as a probe specific only for *C. diphtheriae*. Plasmid pDγ2 contains part of the diphtheria *tox* gene interrupted by a DNA insertion (IS) element. Restriction fragments of this plasmid provide (1) a *tox* gene-specific probe that allows differentiation between toxinogenic and nontoxinogenic strains of *C. diphtheriae* and (2) a probe specific for the transposable (IS) element. This IS element is present in the same chromosomal location in strains that are closely related, while unrelated strains contain different numbers of copies of the IS element integrated into different sites of the chromosome. The use of this probe allows one to distinguish between strains that are closely related by observing their identical pattern on Southern blots of chromosomal DNA. The usefulness of this probe to solve the epidemiology of a number of cases of diphtheria is discussed.

VI. MATERIALS AND METHODS

A. Growth Conditions

Media commonly used for the isolation and growth of *C. diphtheriae* have been widely described (44). We describe the CY medium, which we have routinely used for the laboratory growth of *C. diphtheriae* on agar plates or in liquid medium.

CY medium

20 g Yeast extract
10 g Casamino acids
5 ml 1% Tryptophan
5 g KH_2 PO_4
2 ml 50% $CaCl_2$
1000 ml Distilled H_2O

Adjust the pH to 7.4 and boil the solution. While boiling, filter the solution through a Whatman no. 40 filter paper. Then add 2 ml of solution I and 1 ml of solution II. Autoclave for 20 min at 121°C. Before the use of the medium add 3 ml of solution III per 100 ml of medium.

Solution I

22.5 g $MgSO_4 \cdot 7H_2O$
115 mg β-Alanine
115 mg Nicotinic acid
7.5 mg Pimelinic acid
5 ml 1% $CuSO_4 \cdot 5H_2O$

4 ml 1% $ZnSO_4 \cdot 5H_2O$
1.5 ml 1% $MnCl_2 \cdot 4H_2O$
3 ml HCl
Add distilled H_2O to a final volume of 100 ml.

Solution II
20 g L-Cystine
20 ml HCl
Add distilled H_2O to a final volume of 100 ml.

Solution III
50 g Maltose
2 ml 50% $CaCl_2$
1 g KH_2PO_4
Add distilled H_2O to a final volume of 100 ml and adjust the pH to 7.4. Then filter the solution through a Whatman no. 40 filter and autoclave.

For the preparation of solid medium add 15 g of agar per 1000 ml of medium.

B. DNA Probes

1. DNA Probe Specific for the Genus Corynebacterium

By digesting the plasmid pA634 (Fig. 1) with *Eco*RI, two DNA fragments are generated: the 3.5-kb insert and the 2.7-kb pUC8 vector. The 3.5-kb fragment is purified by electroelution or any other suitable method, labeled with ^{32}P, and then used as a probe on Southern blots of *Bam*HI-digested chromosomal DNA. After hybridization under stringent conditions and autoradiography, the presence of any hybridizing fragment will indicate that the bacterium is a *Corynebacterium*; if no hybridization is obtained, the bacterium does not belong to the genus *Corynebacterium*. In the case that a positive hybridization is obtained, one, two, or even three or four hybridizing bands can be observed, depending on whether the bacterium has one or two *attB* sites and whether one or both the *attB* sites are split into two fragments by the integration of a β-related phage (Fig. 2a).

2. DNA Probe Specific for Corynebacterium diphtheriae

By digesting the plasmid pA634 (Fig. 1) with the restriction enzyme *Hin*cII, four fragments are generated: a 3.5-kb fragment containing the vector and part of the insert, and three low molecular weight fragments of 1.2, 0.85, and 0.7 kb, respectively. The 1.2-kb fragment contains *attB2*, the 0.85-kb fragment contains *attB1*, and the 0.7-kb fragment contains the region between *attB1*, and *attB2*, which is present only in *C. diphtheriae*. The 0.7-kb fragment is purified by electroelution, labeled with ^{32}P, and then used as a probe in a Southern blot of *Bam*HI-digested chromosomal DNA. After hybridization under stringent conditions and autoradiography, the presence of any hybridizing fragment will indicate that the bacterium is a *C. diphtheriae* (Fig. 2b). After very long exposures (days after obtaining a strong signal with *C. diphtheriae*), a very weak hybridization can be detected in *C. ulcerans* and *C. pseudotuberculosis* but not in the other corynebacteria; however, this signal is so weak that it cannot be confused with the one deriving from *C. diphtheriae* strains.

3. DNA Probe Specific for the **tox** *Gene*

The plasmid pDγ2 (ATCC 67011) contains an insert of 2.5 kb including the *tox* gene of bacteriophage γ from the *Hin*dIII site to the *Cla*I site in position 1066, interrupted in position 176 by a 1450-bp-long *C. diphtheriae* IS element. By cutting this plasmid with *Bam*HI and *Acc*I, five fragments are obtained of 3.1, 0.77, 0.55, 0.28, and 0.27 kb, respectively. The 3.1-kb fragment contains the vector, the 0.77-kb fragment contains nucleotides 297–1066 of the *tox* gene, which include the region coding for most of fragment A, and the amino-terminal part of fragment B of diphtheria toxin. The 0.55- and 0.27-kb fragments are internal to the IS element and can be used as probes specific for it. To obtain a probe specific for the diphtheria toxin gene, the 0.77-kb fragment is purified by electroelution, ^{32}P-labeled, and then used to hybridize Southern blots of *Bam*HI- or *Eco*RI-digested chromosomal DNA. After hybridization under stringent conditions and autoradiography, any hybridizing fragment will indicate the presence of sequences homologous to the *tox* gene and therefore the strain should be considered toxinogenic or potentially toxinogenic (Fig. 4).

4. A Probe Internal to the IS Element for the Epidemiology of Diphtheria

The plasmid pDγ2 (ATCC 67011) contains a 2.5-kb insert, including the first 1066 bp of the *tox* gene of bacteriophage γ interrupted in position 176 by the 1450-bp long IS element of *C. diphtheriae* (Fig. 3). By cutting this plasmid with *Sal*I, only one fragment of 0.6 kb is obtained. This fragment is totally internal to the IS element and, after purification by electroelution and labeling with ^{32}P, should be used to hybridize under stringent conditions Southern blots of *Bam*HI-digested chromosomal DNA from *C. diphtheriae* isolates. Following autoradiography, ~85% of the *C. diphtheriae* strains are expected to hybridize to the probe. The IS element has never been detected in strains other than *C. diphtheriae* and, therefore, the presence of any hybridizing band will indicate that the strain is indeed a *C.*

diphtheriae. The relatedness between the strains will be easily determined by comparing the pattern of hybridization: closely related strains will have identical patterns (see Figs. 6,7).

C. Preparation of Chromosomal DNA of *Corynebacterium diphtheriae*

Two methods are described; a fast method using a small quantity of bacteria, which is useful for the screening of many isolates, and a preparative method, which results in very pure chromosomal DNA preparations.

DNA prepared with both methods can be used after digestion with restriction enzymes in Southern blot analysis and also in simple dot-blot hybridizations using standard methods described by Maniatis *et al.* (23).

1. Fast-Isolation Method

The bacteria are grown overnight at 35°C on a CY plate. Then some bacteria are taken with an inoculation loop and resuspended in an Eppendorf tube containing 0.5 ml 10 m*M* Tris-HCl–10 m*M* EDTA–10 mg/ml lysozyme, pH 8.0. The tube is incubated at 37°C for 2–4 hr. Then 50 μl of a 10% SDS solution are added and the tube is agitated. For bacterial lysis the cells are incubated at 65°C until the solution becomes viscous. After the addition of 50 μl of 5 *M* potassium acetate (pH 4.8), the solution is placed in ice for 15–30 min and then centrifuged for 15 min. The supernatant is transferred to a new tube and ethanol is added slowly at room temperature until the precipitation of the DNA is observed. The solution is then centrifuged again for 15 sec. The pellet is washed with 100% ethanol and dried with air. Then the DNA is resuspended in 100 μl of distilled H_2O. Usually 20 μl are used for the digestion with restriction enzymes. For preservation, 20 μl of choroform are added and the DNA in stored at 4°C.

2. Preparative Isolation Method

A preculture is prepared in 10 ml of CY liquid medium and incubated overnight at 35°C. The next morning a new flask containing 10 ml of CY medium is inoculated with 0.3 ml of the overnight culture and grown with good aeration at 35°C for 3 hr. Then 10 μl of a 1 mg/ml solution of penicillin G are added. The bacteria are then grown for another 2 hr. The cells are centrifuged in Corex tubes and washed once with 3 ml of 10 m*M* Tris-HCl, pH 8.2. The cells are then resuspended in 2 ml of 10 m*M* Tris-HCl–0.5 *M* sucrose–5 mg/ml lysozyme, pH 8.2. After an incubation of 10 min at 37°C the bacteria are centrifuged and resuspended in 0.5 ml 10 m*M* Tris-HCl–10 m*M* EDTA, pH 8.2. The bacteria are then transferred to an Eppendorf tube, 40 μl of 20% SDS are added, and the tube is agitated until the solution becomes viscous. Then the test tube is incubated for 20 min at 65°C. After cooling to 37°C, 50 μl of a self-digested pronase solution (50 mg/ml) are added. The tube is incubated at least for 1 hr (better overnight) at 37°C. The next day 0.5 ml of phenol are added and the solution is agitated vigorously. Then the tube is centrifuged for 15 min. With a Pasteur pipette the aqueous phase is transferred to a new tube and extracted a second time with phenol and two times with chloroform. The DNA is then precipitated by adding 50 μl of 5 *M* NaCl–0.6 ml of isopropanol with slight shaking. After 30 min the tube is centrifuged for 2 min and the sediment is resuspended in 0.5 ml 10 m*M* Tris-HCl, pH 8.0. Then 10 μl of RNase (1 mg/ml) are added and the solution is incubated at 37°C for 30 min. Finally, the DNA is precipitated by the addition of 50 μl 5 *M* NaCl–0.6 ml of isopropanol and incubated for 10 min at −80°C. After centrifugation the DNA is resuspended in 200 μl of distilled H_2O. Usually 5 μl of this preparation are enough for restriction enzyme digestions. For preservation of the DNA, 20 μl of chloroform are added and stored at 4°C.

ACKNOWLEDGMENTS

The authors would like to thank Maria Perugini for excellent technical help, Lucia Filippeschi for typing and editing the manuscript, and Giorgio Corsi for the graphic work.

REFERENCES

1. Barksdale, L. (1970). *Corynebacterium diphtheriae* and its relatives. *Bacteriol. Rev.* **34,** 378–422.
2. Behring, E. von (1913). Über ein neues Diphtherieschutzmittel. *Dtsch. Med. Wochenschr.* **39,** 873.
3. Behring, E. von, and Kitasako, S. (1890). Über das Zustandekommen der Diphtherie-Immunität bei Tieren. *Dtsch. Med. Wochenschr.* **16,** 1113.
4. Brown, B. A., and Bodley, J. W. (1979). Primary structure at the site in beef and wheat elongation factor 2 of ADP-ribosylation by diphtheria toxin. *FEBS Lett.* **103,** 253–255.
5. Buck, G., and Groman, N. B. (1981). Genetic elements novel for *Corynebacterium diphtheriae*: Specialized transducing elements and transposons. *J. Bacteriol.* **148,** 143–152.
6. Bukhari, A. I., Shapiro, J. A., and Adhya, S. L. (1977). "DNA: Insertion Elements, Plasmids, and Episomes." Cold Spring Harbor Lab., Cold Spring Harbor, New York.
7. Campbell, A. M. (1962). Episomes, *Adv. Genet.* **11,** 101–145.
8. Chang, D. N, Laughren, G. S., and Chalvardjian. (1978). Three variants of *Corynebacterium diphtheriae* subsp. *minis* (Belfanti) isolated from a throat specimen. *J. Clin. Microbiol.* **8,** 767–768.
9. Christenson, B., and Böttiger, M. (1986). Serological immunity to diphtheria in Sweden in 1978 and 1984. *Scand. J. Infect. Dis.* **18,** 227–233.

10. Cianciotto, N., Rappuoli, R., and Groman, N. (1986). Detection of homology to the beta bacteriophage integration site in a wide variety of *Corynebacterium* spp. *J. Bacteriol.* **168,** 103–108.
11. Ciucà, M., Calalb, G., Saragea, A., and Maximesco, P. (1960). Particularities des "eco-systèmes phage-bactérie" des souches de *Corynebacterium diphtheriae* type *gravis*, isolées d'un foyer epidémique. *Arch. Roum. Pathol. Exp. Microbiol.* **19,** 1–8.
12. Costa J., Michel, J. L., Rappuoli, R., and Murphy J. (1981). Restriction map of Corynebacteriophages β_c and β_{vir} and physical localization of diphtheria *tox* operon. *J. Bacteriol.* **148,** 124–130.
13. Cummins, C. S. (1965). Chemical and antigenic studies on cell walls of mycobacteria, corynebacteria and nocardias. *Am. Rev. Respir. Dis.* **92,** 63–72.
14. Dixon, J. M. S. (1984). Diphtheria in North America. *J. Hyg.* **93,** 419–432.
15. Elek, S. D. (1949). The plate virulence test for diphtheria. *J. Clin. Pathol.* **2,** 250–258.
16. Freeman, V. J. (1951). Studies on the virulence of bacteriophage-infected strains of *Corynebacterium diphtheriae. J. Bacteriol.* **61,** 675–688.
17. Giannini, G., Rappuoli, R., and Ratti, G. (1984). The amino acid sequence of two non-toxic mutants of diphtheria toxin: CRM45 and CRM197. *Nucleic Acids Res.* **12,** 4063–4069.
18. Greenfield, L., Bjorn, M. J., Horn, G., Fond, D., Buck, G. A., Collier, R. J., and Kaplan, D. A. (1983). Nucleotide sequence of the structural gene for diphtheria toxin carried by corynephage β. *Proc. Natl. Acad. Sci. U.S.A.* **80,** 6853°857.
19. Groman, N., Cianciotto, N., Bjorn, M., and Rabin, M. (1983). Detection and expression of DNA homologous to the *tox* gene in nontoxinogenic isolates of *Corynebacterium diphtheriae. Infect. Immun.* **42,** 48–56.
20. Jerne, N. K. A. (1951). Study of avidity based on rabbit skin responses to diphtheria toxin–antitoxin mixtures. *Acta Pathol. Microbiol. Scand, Suppl.* **87.**
21. Kauffmann, F. (1966). *In* "The Bacteriology of the Enterobacteriaceae," p. 25. Williams & Wilkins, Baltimore, Maryland.
22. Loeffler, F. (1884). Untersuchungen über die Bedeutung der Mikroorganismen für die Entstehung der Diphtherie beim Menschen, bei der Taube und beim Kalbe. *Mitt. Klin. Gesund.* **2,** 451–499.
23. Maniatis, T., Fritsch, E. F., and Sambrook, J. (1982). "Molecular Cloning: A Laboratory Manual." Cold Spring Harbor Lab., Cold Spring Harbor, New York.
23a. McLafferty, M. A., Harcus, D. R., and Hewlett, E. L. (1988). Nucleotide sequence and characterization of a repetitive DNA element from the genome of *Bordetella pertussis* with characteristics of an insertion sequence. *J. Gen. Microbiol.* **134,** 4726–4735.
24. McLeod, J. W. (1943). The types *mitis, intermedius* and *gravis* of *Corynebacterium diphtheriae. Bacteriol. Rev.* **7,** 1–41.
24a. McPheat, W. L., and McNally, T. (1986). Isolation of a repeated DNA sequence from *Bordetella pertussis. J. Gen. Microbiol.* **133,** 323–330.
25. Meitert, E., and Bica-Popii, V. (1972). Etude des relations antigeniques entre les phages anti-corynebacterium hofmanni. *Arch. Roum. Pathol. Exp. Microbiol.* **31,** 475–480.
26. Michel, J. L., Rappuoli, R., Murphy, J. R., and Pappenheimer, A. M., Jr. (1982). Restriction endonuclease map of the nontoxinogenic corynephage γc and its relationship to the toxigenic corynephage βc. *J. Virol.* **42,** 510–518.
27. Naumann, P., Krech, T., Maximesco, P. *et al.* (1986). Phagenlysotypie und Epidemiologie der Diphtherie—Erkrankungen 1975 bis 1984. *Dtsch. Med. Wochenschr.* **111,** 288–292.
28. Pappenheimer, A. M., Jr. (1977). Diphtheria toxin. *Annu. Rev. Biochem.* **46,** 69–94.
29. Pappenheimer, A. M., Jr. (1984). The diphtheria bacillus and its toxin: A model system. *J. Hyg.* **93,** 397 404.

30. Pappenheimer, A. M., Jr., and Murphy, J. R. (1983). Studies on the molecular epidemiology of diphtheria. *Lancet.* **2,** 923–926.
31. Ramon, G. (1922). Flocculation dans un mélange neutre de toxine–antitoxine diphtherique. *C. R. Seances Soc. Biol. Ses Fil.* **86,** 661–771.
32. Ramon, G. (1924). Sur la toxine et sur l'anatoxine diphtériques. Pouvoir floculant et propriétés immunisantes. *Ann. Inst. Pasteur* **38,** 1.
33. Rappuoli, R. (1983). Isolation and characterization of *Corynebacterium diphtheriae* non tandem double lysogens hyper-producing CRM197. *Appl. Environ. Microbiol.* **45,** 560–564.
34. Rappuoli, R., Michel, J. L., and Murphy, J. R. (1983). Integration of corynephages βtox−, ωtox− and γtox− into two attachment sites on the *Corynebacterium diphtheriae* chromosome. *J. Bacteriol.* **153,** 1202–1210.
35. Rappuoli, R., Michel, J. L., and Murphy, J. R. (1983). Restriction endonuclease map of corynebacteriophage w_c^{tox+} isolated from the Park Williams no. 8 strain of *Corynebacterium diphtheria. J. Virol.* **45,** 524–530.
36. Rappuoli, R., Perugini, M., and Falsen, E. (1988). Molecular epidemiology of the 1984–1986 outbreak of diphtheria in Sweden. N. Engl. J. Med. **318,** 12–14.
37. Rappuoli, R., Perugini, M., and Ratti, G. (1987). DNA element of *Corynebacterium diphtheriae* with properties of an insertion sequence and usefulness for epidemiological studies. *J. Bacteriol.* **169,** 308–312.
38. Rappuoli, R., and Ratti, G. (1984). Physical map of the chromosomal region of *C. diphtheriae* containing corynephage attachment sites *attB1* and *attB2*. *J. Bacteriol.* **158,** 325–330.
39. Ratti, G., Rappuoli, R., and Giannini, G. (1983). The complete nucleotide sequence of the gene coding for diphtheria toxin in the corynephage $\omega(tox^+)$ genome. *Nucleic Acids Res.* **11,** 6589.
40. Rockhill, R. C., Hadiputranto, S. H., Siregar, S. P., and Muslihun, B. (1982). Tetracycline resistance of *Corynebacterium diphtheriae* isolated from diphtheria patients in Jakarta, Indonesia. *Antimicrob. Agents. Chemother.* **21,** 842–843.
41. Roux, E., and Yersin, A. (1888). Contribution á l'ètude de la diphthérie. *Ann Inst. Pasteur* **2,** 629–661.
42. Saragea, A., and Maximesco, P. (1964). Schema provisoire de lysotypie pour *Corynebacterium diphtheriae. Arch. Roum. Pathol. Exp. Microbiol.* **23,** 817–838.
43. Saragea, A., and Maximesco, P. (1966). Phage typing of *Corynebacterium diphtheriae.* Incidence of *C. diphtheriae* phage types in different countries. *Bull. W.H.O.* **35,** 681–689.
44. Saragea, A., Maximesco, P., and Meitert E. (1979). *Corynebacterium diphtheriae:* Microbiological methods used in clinical and epidemiological investigations. *In* "Methods in Microbiology" (T. Bergan, and J. R. Norris, eds.), Vol. 13, pp. 61–176. Academic Press, New York.
45. Schick, B. (1908). Kutanreaktion bei Impfung mit Diphtherietoxin. *Münch. Med. Wochenschr.* **55,** 504.
46. Simmons, L. E., Abbott, J. D., Macauley, M. E., Jones, A. E., Ironside, A. G, Mandal, B. K., Stambridge, T. N., and Maximesco, P. (1980). Diphtheria carriers in Manchester: Simultaneous infection with toxigenic and nontoxigenic *mitis* strains. *Lancet:*, 304–305.
47. Tasman, A., and Lansberg, H. P. (1957). Problems concerning the prophylaxis, pathogenesis and therapy of diphtheria. *Bull. W.H.O.* **16,** 939–973.
48. Thibaut, J., and Fredericq, P. (1956). Actions antibiotiques reciproques chez *Corynebacterium diphtheriae. C. R. Seances Soc. Biol. Ses. Fil.* **150,** 1513–1514.
49. Thompson, J. S., Gates-Davis, D. R., and Yong, C. T. D. (1983). Rapid microbiochemical identification of *Corynebacterium diphtheriae* and other medically important corynebacteria. *J. Clin. Microbiol.* **18,** 926–929.

50. Toshach, S., Valentine, A., and Sigurdson, S. (1977). Bacteriophage typing of *Corynebacterium diphtheriae*. *J. Infect. Dis.* **136,** 655–659.
51. Yoshimori, T., and Uchida, T. (1986). Monoclonal antibodies against diphtheria toxin: Their use in analysis of the function and structure of the toxin and their application to cell biology. *In* "Monoclonal Antibodies against Bacteria" (A. J. L. Macario and E. Conway de Macario, eds.), Vol. 3, pp. 229–248. Academic Press, Orlando, Florida.
52. Zamiri, I., and McEntegart, M. G. (1972). Diphtheria in Iran. *J. Hyg. C.* **70,** 619–625.

9

Nucleic Acid Probes for *Bacteroides* Species

DAVID J. GROVES
Department of Pathology
McMaster University, and
Department of Laboratory Medicine
St. Joseph's Hospital
Hamilton, Ontario, Canada

I. INTRODUCTION

Members of the *Bacteroides* genus make up a large part of the normal human flora and are significant agents of human disease. *Bacteroides* species are also involved in colonization, infection, and metabolic activities of agriculturally significant animals. Anaerobic bacteria, including the genus *Bacteroides*, have proved difficult to isolate and identify using conventional culture methods. The application of techniques for direct detection of specific genetic material by hybridization with nucleic acid (NA) probes holds promise for better understanding of these organisms.

II. BACKGROUND

It is important to be able to detect and identify specific members of the genus *Bacteroides*. Anaerobic bacteria, and in particular the members of the genus *Bacteroides*, are the predominating organisms of the normal human bacterial flora. *Bacteroides fragilis*, though only a relatively small component of intestinal flora, is the most common isolate from infections involving this flora. Resistance to commonly used antibiotics makes the detection of *B. fragilis* essential. Other *Bacteroides* spp., including relatives of *B. melaninogenicus*, are common members of the oral flora of humans. As such, they are common opportunistic agents in periodontal disease and infections involving aspiration of oral secretions (41). *Bacteroides* spp. are major members of rumen flora, and the response of such flora to shifts in the feed supply have been studied (1). *Bacteroides nodosus* is a major cause of foot rot disease in sheep, and the genetic determinants of fimbriae involved in pathogenesis have been of interest (6).

The normal bacterial flora of humans and other mammals consists of a complex of many organisms, with varying characteristics. Anaerobes often grow relatively slowly because of the low energy yield of fermentative metabolism, and can be overgrown by facultative organisms under appropriate conditions. Special techniques to exclude toxic oxygen must be used during collection and transport of specimens, as well as during isolation and identification of bacterial isolates. The genus *Bacteroides* is complex, with at least 40 recognized species, and additional new species being identified (19). *Bacteroides* spp. of the bile-sensitive group also have specific nutritional requirements. All of these characteristics have combined to make conventional culture of *Bacteroides* strains from studies of normal flora or infection exacting and expensive (12,41). This has led to attempts to develop techniques whereby significant anaerobes can be directly detected in clinical and environmental specimens or in which identification can be achieved without requiring further growth (41).

Because of the clinical significance of the *B. fragilis* group, the development of antibiotic resistance in these organisms has been of great interest (46). Resistance to tetracycline has developed so extensively that two-thirds of *B. fragilis* isolates are no longer susceptible to this drug, and alternative drugs such as clindamycin and chloramphenicol have been used. Development of resistance to these antibiotics in turn led to their being replaced by less toxic but more expensive antimicrobials, and to the hypothesis that transferable antibiotic resistance was involved, as had been demonstrated for aerobic bacteria. Initial studies of antibiotic resistance in the *B. fragilis* group focused on studies of transfer of resistance between strains. Phenotypic expression of changes in resistance profiles following transfer experiments have resulted in the characterization of resistance transfer mechanisms for β-lactamases as well as resistance to clindamycin–erythromycin, tetracycline, and chloramphenicol (46).

With the recognition of these transfer mechanisms and identification of the plasmid nature of several of the resistance determinants, it has become possible to develop genetic systems for more detailed molecular analysis of these determinants [for a review, see Salyers *et al.* (35)]. A major contribution to this analysis has been the identification of transposons coding for resistance to these antibiotics and the use of probes specific for the transposons to survey resistant strains and to dissect the mechanisms involved. The development of cloning and shuttle vectors (27,37,46), as well as a polyethylene glycol-mediated transformation system (39), allow transfer of specific genes into *Escherichia coli* strains, detailed genetic analysis, and reintroduction of engineered genes into *B. fragilis*. The development of specific probes for antibiotic resistance and other characteristics has become feasible.

III. RESULTS AND DISCUSSION

A. Whole Chromosomal DNA Probes

Rapid and specific identification of the members of the *B. fragilis* group is a major potential use of NA probes. Clear differentiation of *B. fragilis, B. vulgatus, B. uniformis,* and two strains of *B. ovatus* (3) was possible when whole chromosomal DNA from different species of the *B. fragilis* group was isolated and subjected to restriction endonuclease analysis (REA). This could be coupled with elution of the REA fragments onto nitrocellulose (NC) by Southern blotting, followed by hybridization with whole chromosomal DNA. Total DNA was isolated from an appropriate reference culture. This DNA was labeled with ^{32}P by nick translation for use as a probe against the separated and blotted endonuclease fragments. Because the DNA homology between the members of the *B. fragilis* group range from <4% to >80%, the usefulness of whole chromosomal probes was limited to investigations where the strains compared demonstrated little homology (e.g., <24% between *B. fragilis* and *B. ovatus*). Comparisons between strains with higher homology (e.g., different strains of *B. ovatus* with >75% homology) resulted in smeared tracks. Though useful for investigation of common regions of DNA in strains with low overall homology, the hybridization technique as described had only limited applicability to clinical strains because of its complex procedures.

Dot-blot hybridization, with whole cells as targets and ^{32}P-labeled whole chromosomal DNA from type strains as probes, has been developed to monitor carcinogen-producing species (30). Bacterial cells were applied to nylon membranes by vacuum and lysed in place. This system identified 95% of 62 members of the *B. fragilis* group isolated from human fecal specimens. Two strains identified by hybridization as *B. ovatus* and one strain of *B. caccae* could not be identified conventionally. As described, the system promised to be adequate for the identification of isolates isolated during enumeration of the members of the fecal flora of populations at different risks for colon cancer.

Members of the genus *Bacteroides* are important members of the normal flora of the oral cavity. *Bacteroides intermedius* and *B. gingivalis* have been associated with adult periodontitis, and early diagnosis and treatment can be facilitated by early identification and quantitation of specific organisms (41). Whole chromosomal probes for five of the oral *Bacteroides* species (*B. asaccharolyticus, B. intermedius, B. gingivalis, B. loeschii,* and *B. melaninogenicus*) were used to screen 243 clinical strains identified as oral *Bacteroides* (31). These probes identified 94% of the clinical strains in agreement with conventional identifications. The majority of the discrepancies in identification were with *B. intermedius* strains that did not react with any of the five probes. The dot-blot assay was much faster than culture, requiring only 36 hr to complete, as compared to 7–21 days for conventional identification.

Whole chromosomal probes have also been investigated for three oral gram-negative organisms; *B. intermedius, B. gingivalis,* and *Haemophilus actinomycetemcomitans* (36). Whole cells of reference cultures were applied to filters and lysed. Under these conditions, 10^3 cells were required for detection. Some cross-reaction was found with 10^5 cells per slot. This cross-reactivity, which is consistent with previous hybridization studies (14,18,45), requires that the other probes must be used as controls, and the amount of hybridization with the two probes must be compared. These probes were also used to evaluate their ability to detect the three species in subgingival plaque samples (36). In a comparison of conventional culture and probe assay of duplicate paper points from 60 diseased patients as well as healthy controls, the probes were significantly more sensitive for the three pathogens than culture when at least 10^3 cells were present on the paper point. *Bacteroides gingivalis* and *B. intermedius* were found in 74 and 77% of the adult periodontitis sampled and analyzed by probe technology, compared to 21 and 26% by culture analysis ($p < .001$ level). DNA probing can detect nonviable cells after sampling and transport. Further, culture may not be able to detect all cells physiologically and nutritionally adapted to the highly specialized niche occupied by the flora involved in periodontitis. Use of healthy patients, with low detection rates by hybridization, ruled out false positive reactions by the probes used.

Because of overlapping sequence homology (15,16,18,19) and differences in the quantitative makeup of intestinal (or oral) flora by *Bacteroides*, it is often necessary to isolate the strains before identification by whole chromosomal probe (30). However, the combination of isolation and use of whole chromosomal probes is accurate and sensitive and has several advantages over the use of species-specific probes without culture (31). The whole chromosomal probe is more likely to react with all members of the species and can be up to 100 times as sensitive as probes for single-copy genes.

B. Cloned Random Fragment DNA Probes

Other workers have attempted to eliminate the culture step by development of species-specific probes that could be applied directly to clinical specimens or mixed cultures. Random fragments of chromosomal DNA from *B. thetaiotaomicron* cloned into plasmids of *E. coli* were screened for their specificity in binding to *B. thetaiotaomicron* DNA but not to DNA from other *Bacteroides* or members of the intestinal flora (33). Although the probe for *B. thetaiotaomicron* was used only in pure culture, the techniques developed were further used to develop a probe specific for *B. vulgatus* (21), which could be used for enumeration of *B. vulgatus* in feces. Briefly, random fragments of chromosomal DNA from *B. vulgatus* were cloned into the plasmid pBR322 in *E. coli*. They were tested for specificity in filter hybridization against DNA from a variety of *Bacteroides* spp. Plasmid pBV-1 bound specifically to DNA from *B. vulgatus* and not to other colon *Bacteroides* or *Fusobacterium prausnitzii*. The specific fragment was labeled with ^{32}P and allowed estimation of the concentration of *B. vulgatus* in human feces ($2–3 \times 10^{10}$ per gram of dry weight). This estimate agreed well with results obtained by quantitative culture ($3–6 \times 10^{10}$ per gram of dry weight) (13,29).

In a study to determine the role of the ability to degrade polysaccharide in determining the bacterial content of human feces, a series of specific DNA random fragment probes was used to enumerate *B. thetaiotaomicron, B. uniformis, B. distasonis,* "*Bacteroides* group 3452-A," and *B. ovatus* (22). Prior studies had indicated that *Bacteroides* spp. were the predominant polysaccharide-degrading organisms. *Bacteroides vulgatus* and *B. uniformis* could ferment many plant polysaccharides but not the mucopolysaccharides from human cells. *Bacteroides distasonis* and *B. eggerthii* could only ferment a limited range of plant polysaccharides. *Bacteroides thetaiotaomicron, B. ovatus,* and "*Bacteroides* group 3452-A" could ferment hyaluronic acid and chondroitin sulfate from sloughed epithelial cells as well as a wide variety of polysaccharides of plant origin. Use of the probes allowed direct investigation without relying on prior culture or conventional identification techniques. The results of the previous study on *B. vulgatus* (21) and this enumeration are summarized in Table I. The species capable of degrading a wider range of polysaccharides (*B. thetaiotaomicron* and "*Bacteroides* group 3452-A") are present in high concentrations, but not more so than the species with a relatively narrow range of polysaccharide-degradative capabilities. Thus other factors than ability to compete for a wide spectrum of energy sources must also be important determinants of the composition of human feces. The system was incapable of detecting organisms that were present in numbers similar to *B. ovatus* (e.g., *B. fragilis*).

The major limitation of the enumeration system was identified as sensitivity (21,22). Because of the single-copy genes targeted and because the strain may make up a small part of the total species present in a complex culture, only those strains representing at least 2% of the mixture could be detected and enumerated (21). Further limitations of the procedure included the necessity partially to purify the DNA from the feces sample prior to application to the filters, the requirement to adjust the amount of sample added to the filter to allow linear response of amount hybridized to numbers of cells present, the possibility of cross-reaction with labeled vector, and the difficulties with using short-lived radioactive ^{32}P to label the probes (21).

Specific DNA probes were also developed to identify clinically important *Bacteroides* spp. (20). These probes included three that specifically bound to *B. fragilis* DNA. Five additional probes were able to differentiate specifically between the two homology groups of *B. fragilis*. An additional probe bound DNA from all members of the *B. fragilis*, while a further probe hybridized to DNA from all *Bacteroides* spp. as well as *Fusobacterium nucleatum* and *Fusobacterium necrophorum*. These probes could only detect 10^6 bacteria, even with 48-hr exposure in autoradiography. Specific hybridization took place in blood culture media and in mixtures with other gram-negative bacteria. Preextraction of the target DNA before application to the filter was required. Attempts to replace the radioactive labeling system with biotin labeling were unsuccessful because of nonspecific binding of the streptavidin-conjugated enzyme (20).

TABLE I

Comparison of Concentrations of *Bacteroides* spp. in Human Feces Obtained by Different Methods[a]

Former species designation	Concentration of bacteria by conventional methods[b] (× 10^9 per g dry wt)	Current species designation[c]	Concentration of bacteria by DNA probe method (× 10^9 per g dry wt)
B. fragilis subsp. *vulgatus*	30–60	*B. vulgatus*	20–30
B. fragilis subsp. *thetaiotaomicron*	20–30	*B. thetaiotaomicron*	8–12
		B. uniformis	12–20
B. fragilis subsp. *distasonis*	10–20	*B. distasonis*	5.8–8.4
		"*Bacteroides* group 3452-A"	3.0–6.0
B. fragilis subsp. *ovatus*	0.3–3.0	*B. ovatus*	Not detectable

[a] Adapted by permission of the publisher (22). Copyright 1986 by American Society of Microbiology.

[b] Concentrations determined previously by Moore and Holdeman (13,29) by using conventional plating techniques and biochemical tests for identification of colonic *Bacteroides* spp.

[c] New species designations based on DNA–DNA homology studies (15).

C. Cloned Random Fragment RNA Probes

Ribonucleic acid probes developed by cloning specific DNA sequences and transcribing multiple RNA copies have potential advantages over using the DNA sequence directly as a probe (28). If a vector is used with an appropriate promoter, it is relatively convenient to purify specific probes with high specific activity and without having to isolate the DNA insert. Under appropriate conditions there is no cross-reaction of the probe with the vector sequences. The relative stability of RNA–DNA duplexes compared to the DNA–DNA duplex allows the use of hybridization and washing conditions of higher stringency. Nonspecific background binding of probe can be reduced by using RNase to degrade the probe specifically after hybridization has taken place (28). These potential advantages have led to the development of probes specific for the members of the *B. fragilis* group (8). Following the schema outlined in Fig. 1, random fragments of chromosomal DNA from *B. fragilis, B. thetaiotaomicron, B. vulgatus, B. ovatus,* and *B. distasonis* generated by *Hin*dII and *Eco*RI endonuclease cleavage have been force-cloned into the plasmid pGEM-1. This vector

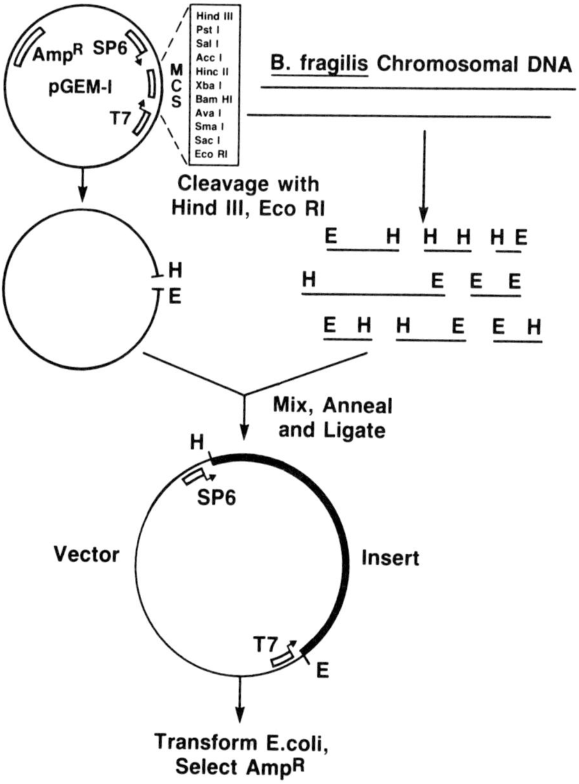

Fig. 1. Schema for forced cloning of random fragments of *Bacteroides fragilis* DNA into pGEM-1 vector. Reprinted with permission of the publisher (8). Copyright 1987 by Elsevier Science Publishing Co., Inc.

contains promoters for SP6 and T7 RNA polymerase at opposite ends and on opposite strands of a multiple cloning site (Riboprobe, Promega Biotec). For *B. fragilis*, 18 strains with inserts of various sizes were identified, and four inserts were further characterized. Plasmid pDJG6 contained an insert of 800 bp and pDJG13 an insert of 2.4 kbp. Plasmid pDJG5 contains an extra *Eco*RI site, so cleavage with *Eco*RI results in a 400-base transcript while cleavage with *Hin*dIII produces a 700-base transcript. Similarly, pDJG18 produced a full 7-kb probe when cleaved with *Eco*RI, while *Hin*dIII produced a 1.3-kb probe. Purified target DNA was fixed on NC filters by slot filtration or colony lysis.

RNA probes were labeled with [^{32}P]-UTP using standard conditions (Promega Biotec) and hybridized to test DNA using aqueous conditions without prehybridization. Probe prepared from plasmid pDJG13 bound only to DNA from type and clinical strains of *B. fragilis*. No homology was found to strains of *B. thetaiotaomicron, B. ovatus, B. vulgatus,* or *B. distasonis*. Identical patterns of hybridization were found for colony or slot blotting. Under these conditions the pDJG13 probe was able to detect 10 ng of DNA (10^6 cells), a result comparable to that found with DNA probes for single-copy genes (20).

A similar series of random fragment probes were developed for *B. thetaiotaomicron* and partially characterized (unpublished results). Plasmids pDJG103, pDJG109, pDJG112, and pDJG118 were ~0.9, 2.1, 2.9, and 4.0 kbp, respectively. No cross-homology was found for these four probes with each other or with another 14 *B. thetaiotaomicron* probes. No reaction was found for any of the type strains other than the *B. thetaiotaomicron* ATCC 29741 from which they were derived. Four reaction patterns were found for the 4 probes against 7 clinical strains identified as *B. thetaiotaomicron* (Table II). Two of the strains (numbers 4 and 5) had equivocal identifications due to intermediate biochemical reactions. All of the probes reacted with the parent strain. Only 3 of the 7 strains reacted with pDJG103. Five had homology with pDJG112/118. Two strains demonstrated homology with both pDJG103 and pDJG112/118. Probe prepared from pDJG109 did not hybridize with any of the clinical isolates. No homology was demonstrated for any of the *B. thetaiotaomicron* probes with the *B. fragilis, B. vulgatus, B. ovatus,* or *B. distasonis* clinical isolates.

On the basis of these results, as well as others (21,33), it is clear that even well-controlled biochemical identifications cannot unequivocally separate the members of the *B. fragilis* group. Further, there is considerable hetero-

TABLE II

Hybridization of *Bacteroides thetaiotaomicron* Probes to Reference and Clinical Strains of *B. thetaiotaomicron*

	Strain[a]							
Probe	1	2	3	4	5	6	7	C
pDJG103	−	−	+	+	−	−	+	+
pDJG109	−	−	−	−	−	−	−	+
pDJG112	+	+	+	−	−	+	+	+
pDJG118	+	+	+	−	−	+	+	+

[a] Strain 4 had an equivocal biochemical identification as *B. thetaiotaomicron/B. uniformis,* and strain 5 as *B. thethaiotaomicron/B. ovatus*. Strain C is *B. thetaiotaomicron* ATCC 29741, source of DNA for the probes.

geneity in the genomes of those strains currently described as *B. thetaiotaomicron, B. ovatus, B. distasonis,* and *B. uniformis*. Similar heterogeneity exists within *B. fragilis* as well (20). Application of the probes also prompted reassessment of conventional identifications and resulted in clearer interpretations. Further studies to use such specific probes to clarify species boundaries and to serve as epidemiological and pathological markers are under way.

D. DNA Probes for Specific Genes of Known Function

With the development of more sophisticated genetic techniques, it has become possible to clone genes coding for specific characteristics in *B. fragilis* (9,34,35,39,46) and to use these sequences as probes.

1. Polysaccharide Metabolism Genes

Following on their studies in enumerating the *Bacteroides* spp. involved in degrading polysaccharides in human feces, Salyers's group cloned the gene for *B. thetaiotaomicron* chondroitin lyase II (CSaseII) into an *E. coli* vector (11). With the subcloning of this gene and characterization of the enzyme produced in an *E. coli* expression system, it became possible to use targeted insertional mutagenesis to construct mutants deficient in CSaseII (10). Using an 0.8-kb fragment from the gene sequence for the enzyme, a suicide vector was constructed that also carried a replication origin for *E. coli* and transposable genes for macrolide resistance. This vector could be mobilized into *B. thetaiotaomicron* but could not replicate in that host. Erythromycin-resistant clones were screened for integration of the vector into the CSaseII gene by REA of whole chromosomal DNA from the transconjugants and controls, followed by Southern hybridization with the whole CSaseII gene. Using the cloned gene as a probe in this way showed that the vector had inserted into the CSaseII gene and resulted in an increase in the number of bands which would hybridize with the probe. These mutants were still able to metabolize chondroitin sulfate, but were shown to produce no CSaseII activity, indicating that CSaseII was not essential, at least under conditions of substrate excess.

To determine the significance of mucopolysaccharides as growth substrates *in vivo,* several mutant strains were developed by both chemical and insertional mutagenesis (34). One strain derived by chemical mutagenesis could not grow on *N*-acetylgalactosamine, a component sugar of chondroitin sulfate and several related polysaccharides. An additional two mutants could not grow on chondroitin sulfate, but could metabolize the component sugars as well as other unrelated polysaccharides. These strains were used in mixed culture with wild type to colonize previously germ-free mice, and the competition between the mutants and wild-type strains were monitored. In order to quantitate the ratio it was necessary to pick 80 colonies to both chondroitin sulfate-containing defined medium and glucose-containing defined medium. Similar techniques were necessary to detect potential reversion of the mutants.

Construction of a mutant with a 0.5-kb deletion in the CSaseII gene provided more convenient data about the role of this enzyme in the survival of *B. thetaiotaomicron in vivo* (32). The gene for CSaseII previously cloned into *E. coli* (see above) was subcloned by cleavage with *Eco*RV resulting in two sequences, one with a deletion of the 0.5-kb fragment flanked by the two *Eco*RV sites (pEG1003) and the other consisting of the *Eco*RV segment alone (pARV1). Insertion of pEG1003 into the wild-type *B. thetaiotaomicron* CSaseII gene followed by reversion to erythromycin susceptibility resulted in one mutant. This strain did not produce CSaseII, was extremely stable to reversion, and did not contain the *Eco*RV sequence when probed using the pARV1 plasmid. During experiments to monitor competition between the strain containing the deletion and CSaseII-producing parent organisms, the pARV1 probe was used to enumerate the proportion of strains containing the deletion. Colony blots of a selection of strains from broth competition indicated that the wild-type organisms would eventually overgrow the CSaseII-deficient strains (Table III). However, the failure of CSaseII-producing strains to outgrow the

TABLE III

Competition of the Deletion Mutant with the Wild Type in Defined Laboratory Medium Containing Glucose or Chondroitin Sulfate as the Sole Source of Carbohydrate[a]

Carbohydrate source	Organisms in mixture	Percentage wild type[b] after generations[c]		
		0	24	50
Glucose	Deletion/ wild type	64	58	68
Chondroitin sulfate	Deletion/ wild type	50	72	>99
Chondroitin sulfate	Reisolated deletion/ wild type	59	87	>99

[a] Adapted by permission of the publisher (32). Copyright 1988 by American Society for Microbiology.

[b] Numbers are the means of at least duplicate measurements. The range of values was within 10% of mean.

[c] Percentage wild type in inoculum.

TABLE IV

Percentage Wild Type in Ceca and Feces of Previously Germ-Free Mice at Various Times after Colonization with a Mixture of 16% Wild Type and 84% Deletion Mutant[a]

	Percentage wild type[b] after (days)			
Type of sample	5	15	25	35
Cecum	14,14	12,17	32,71	40,46
Feces	16,17	11,27	27,80	46,40

[a] Adapted by permission of the publisher (32). Copyright 1988 by American Society for Microbiology.

[b] The two numbers are the values obtained for the two mice sacrificed at each time point. The values for the feces are given in the same order as the values for cecal contents.

deletion mutant strains in trials using germ previously free mice (Table IV) provided convincing evidence that CSaseII is indeed not essential to survival of *B. thetaiotaomicron* in the mouse intestine. Other factors must determine the ratio of *B. thetaiotaomicron* to the more plentiful *Bacteroides* spp. in the gut. This is an elegant use of a cloned gene both to derive truly isogenic strains and to enumerate, by hybridization with the specific probe, populations containing the mutant strains.

2. *Ribosomal RNA Sequences*

With the description of the nucleotide sequence of 16 S rRNA (17), complementary species- and group-specific oligonucleotide probes (44) were developed. These probes were used to enumerate various strains of *B. succinogenes* and *Lachnospira multiparus*-like organisms during antibiotic disturbance of bovine rumen bacterial populations. Three different 21-base probes were prepared to enumerate the *B. succinogenes* group. One probe bound to the DNA of all of the *B. succinogenes* strains except one. A second probe bound to strains of *B. succinogenes* found in the rumen, and a third probe bound only to DNA from extrarumenal (cecum) strains. To reduce variability due to uncertainties in NA recovery, hybridization of specific sequences was compared as a fraction of the total ribosome population detected by hybridization to a universal probe. This probe binds to corresponding sequences of rRNA from all tested species. Total NA, mostly rRNA, was extracted and spotted to nylon membranes by dot blotting for hybridization with the oligonucleotide probes.

After treatment with the antibiotic monensin, a sodium ionophore, the proportion of the rumen-type strains increased ~5-fold, then fell transiently below the premonensin level. During this period, the total *B. succinogenes* level, as detected by the "signature" probe, was larger than the total of the two subtypes, indicating the presence of a third group of *B. succinogenes*-like organisms not accounted for by the cecal- or rumen-type probes. The amplification of the target NA, by selection of rRNA, present in many copies per cell, as the target for hybridization greatly increases the theoretical sensitivity of such probes. The relative ease and sensitivity of the probing technique argues strongly for its use in place of conventional culture and identification.

3. Antibiotic Resistance Genes

There are a limited number of antibiotic resistance determinants in the *Bacteroides* genus that have been cloned and are available for use as probes. The clindamycin resistance determinant is the most intensively studied. Three resistance plasmids, pBFTM10, pBF4, and pBI136, have been characterized as containing a common sequence coding for resistance to erythromycin and clindamycin (40,46). The whole plasmid pBFTM10 as well as a cloned *Eco*RI fragment pMJS100 containing the clindamycin resistance sequence were used to survey the mechanisms of resistance in clindamycin-resistant members of the *B. fragilis* group. Total DNA isolated from the resistant strains was separated electrophoretically, Southern blotted, and probed with pMJS100 and pBFTM10. The resistance sequence was found in 15 of 16 resistant strains, either on the chromosome or on plasmids (24). In another study (9), whole-cell DNA from 13 clinical strains resistant to clindamycin was digested with *Eco*RI, separated electrophoretically, and blotted to NC. DNA from five strains highly resistant to clindamycin hybridized with the complete pBF4 as probe, as well as with a labeled *Eco*RI-derived fragment. No homology was found with DNA from three strains resistant to low levels of clindamycin. It is clear that the majority of clinically significant resistance to clindamycin could be probed for by use of this specific sequence common to the plasmids characterized rigorously so far. More information is necessary about the range and distribution of other sequences coding for this phenotype.

Other antibiotic resistance determinants are less well characterized and their use as probes has not been described. These determinants have not been further exploited because of an inability to express them as cloned genes in *E. coli* (35) and their relative lack of clinical significance. Two types of tetracycline resistance have been described, one of which can be expressed only in *E. coli* under aerobic conditions (43). Transferable determinants for β-lactamase and resistance to chloramphenicol have been described (25,46) but not extensively characterized or used as probes.

4. Miscellaneous Known Genes

A selection of other genes have been cloned and are potentially available for use as probes. By using a *rec*A-deficient mutant of *E. coli* as a host for cloning, it has been possible to clone for and express a *rec*A-like gene from *B. fragilis* (7). There was no homology detected between the *B. fragilis* sequence and *E. coli* chromosomal DNA, although specificity with regard to other genera and species was not tested. Using a similar strategy, the gene coding for glutamine synthetase in *B. fragilis* was cloned in *E. coli* (42). There was no detectable homology between the cloned sequence and *E. coli* chromosomal DNA, and no immunological cross-reaction between the cloned protein and antibodies to *E. coli* glutamine synthetase by Ouchterlony gel diffusion. However, there were weak cross-reactions detected by Western blot analysis. A similar strategy was employed to clone the xylanase gene sequence from *B. succinogenes* (38). Production of xylanase is a major bacterial activity in breaking down the hemicellulose of plant material in ruminants. As *E. coli* does not produce this enzyme, cloned xylanase genes were detected by assaying for expression of the enzyme activity. Although these specific genes have not been tested as diagnostic or investigative probes, the strategy for cloning and characterization of similar genes that can be expressed in *E. coli* and detected by complementation with an appropriately deficient *E. coli* strain holds potential for future specific probe development.

Expression of the gene coding for the fimbrial subunit of *B. nodosus* was detected by screening a gene bank with a colony immunoassay, using antisera to purified fimbrae. This cloned gene was further used in an *E. coli* expression vector to produce a potential vaccine for use against ovine foot rot caused by *B. nodosus* (6).

Using the previously described amino acid sequence of the fimbrial subunit protein of *B. gingivalis,* oligonucleotide probes were prepared and used to screen gene banks for the presence of the sequence for fimbrilin (5). By this technique it was possible to clone the fimbrilin gene. Although this study concerned the use of this probe to investigate structural similarities and control of fimbrillation in *B. gingivalis* and related organisms, the sequence could potentially be used as a NA probe in studies of this organism in clinical periodontal disease.

IV. CONCLUSIONS

A variety of NA probes are available for detection and quantitation of the *Bacteroides* spp., as well as for fundamental investigations of their role in normal flora and disease. Currently, the most useful probes would appear to be whole chromosomal DNA preparations, with their higher sensitivity. This is countercd by the relative lack of specificity of the whole chromosomal probes due to shared sequences. Substantial effort is required to develop improved labeling and hybridization conditions, as well as procedures to amplify the target sequences, in order to increase sensitivity while taking advantage of the much greater specificity possible when using individual gene sequences as probes. With continued development of appropriate genetic and cloning systems, a wider variety of specific probes is feasible.

V. GENE PROBES VERSUS MONOCLONAL ANTIBODIES

Although a number of specific antisera, either polyclonal or monoclonal, have been used in the studies of *Bacteroides* spp. (2,4,6a,47), no comparisons of the effectiveness of monoclonal antibodies (MAb) with NA probes have been reported. Such studies are potentially possible only in a limited number of situations. Specific gene sequences that could be used as probes (6) and MAb (2) are available for the pilin gene in *B. nodosus*. Though highly specific, the NA probe has been developed in order to produce pilin antigen in *E. coli* for vaccine use, and there has been no evaluation of its use as a diagnostic or investigative probe. MAb has only been used in immunocytochemical investigations of pilus structure in *B. nodosus*, although the potential use of this immune probe to screen for pilin antigen has been recognized (2).

Monoclonal antibodies specific for *B. gingivalis* have been described, but the antigenic component with which the MAb bind has not been defined (4). When used in an immunofluorescence microscopic technique, the MAb was found have 91–100% correlation with anaerobic culture of periodontal specimens. As previously noted, the gene for the fimbrial protein subunit of *B. gingivalis* has been cloned and used in comparisons with fimbriae production in other related strains. This probe has not been used to quantitate *B. gingivalis* in clinical specimens. However, as previously discussed, the whole chromosomal probe, which can detect clinically significant numbers of *B. gingivalis*, has been shown to be significantly more sensitive than culture (36). No direct comparisons of these immunological and hybridization detection systems have been published, but the aforementioned results demonstrate that the potential for the use of NA probes shows at least as much promise as the use of specific MAb, both for clinical and investigational purposes.

VI. PROSPECTS FOR THE FUTURE

The application of gene probes to diagnostic bacteriology has not proceeded as rapidly as might have been predicted. This is largely due to problems of convenience and sensitivity in using the probes in routine clinical laboratories. Introduction of nonradioactive detection systems has resulted in reduced sensitivity and high backgrounds, as compared to radioactively labeled probes, in situations where mixed organisms are encountered. Further development of convenient, sensitive, and specific labeling systems for detection of hybridization will undoubtedly result in much more widespread use of NA probe technology.

Another major potential improvement is the amplification of specific target sequences through use of the polymerase chain reaction (PCR) system (26). This allows the synthesis of multiple copies of specific target genes if they are present in the specimen, thus greatly increasing the sensitivity of hybridization and reducing the problems with using specific gene sequences, single-copy genes, and nonradioactive probes. Polymerase chain reaction technology is commercially available but is still under evaluation.

Utilization of highly specific gene sequences will provide much more convenient and reliable speciation of the members of the genus *Bacteroides*, as well as providing a genotyping system through using arrays of probes that can define groups within the individual species. This type of rapid speciation, combined with the potential for rapid and specific strain typing and development of specific probes for genes involved in pathogenicity, will result in greatly increased knowledge about the role of *Bacteroides* in health and disease.

VII. SUMMARY

A variety of NA probes for members of the *Bacteroides* genus have been described. These include whole chromosomal DNA probes used for rapid and specific identification of the colonic *B. fragilis* group, as well as for the oral *Bacteroides*. Random fragments of chromosomal DNA have also been cloned for use as DNA probes or as templates for production of complementary RNA transcripts. These probes have proved specific, if less sensitive than whole chromosomal DNA, and useful in both identification and enumeration of members of the *B. fragilis* group in clinical and investigational settings. Sequences coding for specific known genes have also been cloned and used as probes. Probes specific for chondroitin lyase II in *Bacteroides* have been used both in the construction of true isogenic mutants for the enzyme as well as for monitoring the relative numbers of mutant and wild-type strains in competition experiments. A probe specific for 16 S rRNA sequences has been used to monitor *B. succinogenes* in rumen flora experiments. Probes specific for clindamycin resistance from *B. fragilis* and related organisms have been used to compare for homology and distribution of this clinically significant gene.

VIII. MATERIALS AND METHODS

All manipulations should be carried out using sterile reagents and other materials, and wearing gloves to prevent endonuclease contamination of NA preparations. Standard methods of endonuclease manipulation have been previously described (23).

A. Forced Cloning of *Bacteroides* Chromosomal DNA

Cloning of random chromosomal DNA fragments into a transcriptional vector is outlined in Fig. 1.

1. High molecular weight chromosomal DNA is isolated from 1-liter cultures of *Bacteroides* sp. after harvesting and washing by centrifugation, by standard sodium dodecyl sulfate (SDS)/proteinase K1 lysis, phenol– chloroform extraction, and dialysis (23).
2. Chromosomal DNA (5 μg) or pGEM-1 DNA (2 μg) is digested simultaneously with *Eco*RI (10 units) and *Hin*dIII (10 units) restriction endonucleases (BRL, Bethesda Research Laboratories; Gaithersburg, Maryland) in buffer (100 m*M* Tris-HCl pH 7.5–50 m*M* NaCl–10 m*M* MgCl). Double cleavage prevents reannealing of completely restricted plasmid molecules.
3. Linear *Hin*dIII/*Eco*RI-treated pGEM-1 (300 μg) is mixed with similarly treated chromosomal DNA (75 μg), heated to 68°C for 10 min, and chilled on ice to anneal.
4. Ligase buffer (66 m*M* Tris pH 7.6–6.6. m*M* $MgCl_2$–10 m*M* dithiothreitol (DTT)–0.4 m*M* ATP) and 2 units of T4 DNA ligase (BRL) are used to ligate the annealed DNA fragments during incubation for 18 hr at 4°C. The ligated DNA molecules are used to transform *E. coli* JM83 made competent by standard CaCl treatment (23). Transformed cells are plated on MacConkey agar plates containing ampicillin, 30 μg/ml. Because of the double cutting, only those cells transformed by a recombinant plasmid containing the pGEM-1 ampicillin resistance and an insert of the *Bacteroides* DNA should be able to grow.

5. A selection of colonies putatively containing recombinant plasmids are screened by rapid boiling, plasmid isolation, and gel electrophoresis (23). Clones with plasmids of increased size (slower electrophoretic migration) are selected for further characterization.

B. Preparation of Labeled RNA Probes from Recombinant Plasmids

Transcription of inserts in recombinant pGEM-1/*E. coil* JM83 plasmids is carried out in the Riboprobe system (Promega Biotec, Madison, Wisconsin) according to the manufacturer's recommendations.

1. Plasmid DNA isolated by standard CsCl procedures is linearized by digestion to completion with appropriate enzyme (e.g., *Hin*dIII in 1× CORE buffer (BRL)), then purified and concentrated as necessary by extracting with phenol–chloroform, chloroform, and sodium acetate–ethanol precipitation (23). The linear plasmid DNA is resuspended to 0.5 mg per microliter TE buffer (10 m*M* Tris-HCl pH 8.0–1 m*M* EDTA).
2. Remove Riboprobe kit from freezer; make sure reagents thaw and store on ice.
3. Add to microfuge tube, in order, at room temperature:
   4.0 μl 5× transcription buffer
   2.0 μl 100 m*M* DTT
   0.8 μl RNAsin
   1.0 μl each of 10 m*M* ATP, GTP, and CTP
   1.0 μl sterile H_2O
   2.4 μl of 100 μ*M* UTP (1 μl 10 m*M* UTP–99 μl H_2O)
   1.0 μl linearized DNA (0.5 μg)
   5.0 μl 10 mCi/ml (50 μCi) [$\alpha-{}^{32}$P]UTP
   0.5 μl of T7 RNA polymerase
   Incubate 1 hr at 37°C.
4. Separate unincorporated [^{32}P]UTP using Sephadex G50-80 spin column. Apply reaction mixture followed by 100 μl TE 7.6 and spin (23). Add 3 μl RNAsin; use in hybridization.

C. Preparation of DNA Membrane Filters

1. Slot Blotting

DNA from reference strains or clinical isolates is applied to NC filters (0.45 μm, S+S) using a slot blotter (Hybri-Slot, BRL).

a. For reference strains, purified chromosomal DNA is diluted to 6 μg in 0.4 ml TE pH 7.6 buffer.
b. For clinical isolates, 2 × 10^8 cells are suspended in 350 μl TE and

lysed with an equal volume of lysing buffer (0.4 *M* Tris–0.1 *M* EDTA–1.0% SDS–200 μg/ml proteinase K1 pH 8.0) by incubation at 55°C for 2 hr, followed by extraction with phenol–chloroform (1 : 1) and chloroform. The aqueous layer is removed to a clean tube for further processing.

c. Both purified DNA and the rapid extracts are denatured by adding a one-tenth volume of 3 *M* NaOH and heating at 67°C for 45 min. The denatured DNA is neutralized by addition of an equal volume of 2 *M* ammonium acetate and ~30 ng of DNA (60 μl) deposited onto NC premoistened with 1 *M* ammonium acetate using the slot-blot apparatus under vacuum.

2. *Colony Blotting*

A modified colony hybridization is used for the clinical isolates.

a. A swab moistened with brain heart infusion broth is used to swab lightly a *Brucella* agar plate of the culture to be tested.
b. The swab is rotated on the surface of a 0.45-μm NC disk at the appropriate grid location. The NC disk is processed by a standard colony lysis procedure (23).

All NC filters are baked at 80°C for 2 hr under vacuum and stored in airtight bags at room temperature.

D. Hybridization

1. The radioactively labeled RNA probe is hybridized to reference DNA on NC using standard aqueous hybridization conditions (23). No prehybridization is necessary, and hybridization is carried out at 68°C overnight with shaking.
2. Filters are washed three times at room temperature in 2× SSPE (20× SSPE is 20 m*M* Na_2EDTA–0.16 *M* NaOH–0.2 *M* $NaH_2PO4 \cdot H_2O$–3.6 *M* NaCl pH 7.0) with 0.1% SDS, followed by a wash in 0.1× SSPE with 0.1% SDS at 50°C for 15 min with agitation.
3. The filters can also be treated for 1 hr in 2× SSPE containing 40 μg RNase/ml at 37°C, followed by four washes in 2× SSPE. Air-dried filters are used to expose Kodak X-Omat film at −70°C in film cassettes with intensifier screens.

ACKNOWLEDGMENTS

I gratefully acknowledge the guidance and support of Virginia Clark during the investigations described. This work was supported in part by St. Joseph's Hospital Foundation, Hamilton, Ontario, Canada.

REFERENCES

1. Attwood, G. T., Lockington, R. A., Xue, G. P., and Brooker, J. D. (1988). Use of a unique gene sequence as a probe to enumerate a strain of *Bacteroides ruminicola* introduced into the rumen. *Appl. Environ. Microbiol.* **54,** 534–539.
2. Beesley, J. E., Day, S. E. J., Betts, M. P., and Thorley, C. M. (1984). Immunocytochemical labelling of *Bacteroides nodosus* pili using an immunogold technique. *J. Gen. Microbiol.* **130,** 1481–1487.
3. Bradbury, W. C., Murray, R. G. E., Mancini, C., and Morris, V. L. (1985). Bacterial chromosomal restriction endonuclease analysis of the homology of *Bacteroides* species. *J. Clin. Microbiol.* **21,** 24–28.
4. Chen, P., Bochacki, V., Reynolds, H. S., Beanan, J., Tatakis, D. N., Zambon, J. J., and Genco, R. J. (1986). The use of monoclonal antibodies to detect *Bacteroides gingivalis* in biological samples. *Infect. Immun.* **54,** 798–803.
5. Dickinson, D. P., Kubiniec, M. A., Yoshimura, F., and Genco, R. J. (1988). Molecular cloning and sequencing of the gene encoding the fimbrial subunit protein of *Bacteroides gingivalis*. *J. Bacteriol.* **170,** 1658–1665.
6. Elleman, T. C., Hoyne, P. A., Emery, D. L., Stewart, D. J., and Clark, B. L. (1986). Expression of the pilin gene from *Bacteroides nodosus* in *Escherichia coli*. *Infect. Immun.* **51,** 187–192.
6a. Gmür, R., and Wyss, C. (1985). Monoclonal antibodies to characterize the antigenic heterogeneity of *Bacteroides intermedius*. *In* "Monoclonal Antibodies against Bacteria" (A.J.L. Macario and F. Conway de Macario, eds), Vol. I, pp. 91–119. Academic Press, Orlando, Florida.
7. Goodman, H. J. K., Parker, J. R., Southern, J. A., and Woods, D. R. (1987). Cloning and expression in *Escherichia coli* of a *rec*A-like gene from *Bacteroides fragilis*. *Gene* **58,** 265–271.
8. Groves, D. J., and Clark, V. (1987). Preparation of ribonucleic acid probes specific for *Bacteroides fragilis*. *Diagn. Microbiol. Infect. Dis.* **7,** 273–278.
9. Guiney, D. G., Hasegawa, P., Stalker, D., and Davis, C. E. (1983). Genetic analysis of clindamycin resistance in *Bacteroides* species. *J. Infect. Dis.* **147,** 551–558.
10. Guthrie, E. P., and Salyers, A. A. (1986). Use of targeted insertional mutagenesis to determine whether chondroitin lyase II is essential for chondroitin sulfate utilization by *Bacteroides thetaiotaomicron*. *J. Bacteriol.* **166,** 966–971.
11. Guthrie, E. P., Shoemaker, N. B., and Salyers, A. A. (1985). Cloning and expression in *Escherichia coli* of a gene coding for a chondroitin lyase from *Bacteroides thetaiotaomicron*. *J. Bacteriol.* **164,** 510–515.
12. Holdeman, L. V., Cato, E. P., and Moore, W. E. C. (1984). "Anaerobe Laboratory Manual," 4th ed. Virginia Polytechnic Institute and State University, Blacksburg.
13. Holdeman, L. V., Good, I. J., and Moore, W. E. C. (1976). Human fecal flora: Variation in bacterial composition within individuals and a possible effect of emotional stress. *Appl. Environ. Microbiol.* **31,** 359–375.
14. Holdeman, L. V., and Moore, W. E. C. (1982). Description of *Bacteroides loeschii* sp. nov. and emendation of the descriptions of *Bacteroides melaninogenicus* (Oliver and Wherry) Roy and Kelly 1939 and *Bacteroides denticola* Shah and Collins 1981. *Int. J. Syst. Bacteriol.* **32,** 399–409.
15. Johnson, J. L. (1978). Taxonomy of the bacteroides. I. Deoxyribonucleic acid homologies among *Bacteroides fragilis* and other saccharolytic *Bacteroides* species. *Int. J. Syst. Bacteriol.* **28,** 245–256.

16. Johnson, J. L., and Ault, D. A. (1978). Taxonomy of the *Bacteroides*. II. Correlation of phenotypic characteristics with deoxyribonucleic acid homology grouping for *Bacteroides fragilis* and other saccharolytic *Bacteroides* species. *Int. J. Syst. Bacteriol.* **28,** 257–268.
17. Johnson, J. L., and Harich, B. (1986). Ribosomal RNA homology among species of the genus *Bacteroides*. *Int. J. Syst. Bacteriol.* **36,** 71–79.
18. Johnson, J. L., and Holdeman, L. V. (1983). *Bacteroides intermedius* comb. nov. and description of *Bacteroides levii* sp. nov. *Int. J. Syst. Bacteriol.* **33,** 15–25.
19. Johnson, J. L., Moore, W. E. C., and Moore, L. V. H. (1986). *Bacteroides caccae* sp.nov., *Bacteroides merdae* sp.nov., and *Bacteroides stercoris* sp.nov. isolated from human feces. *Int. J. Syst. Bacteriol.* **36,** 499–501.
20. Kuritza, A. P., Getty, C. E., Shaughnessy, P., Hesse, R., and Salyers, A. A. (1986). DNA probes for identification of clinically important *Bacteroides* species. *J. Clin. Microbiol.* **23,** 343–349.
21. Kuritza, A. P., and Salyers, A. A. (1985). Use of a species-specific DNA hybridization probe for enumerating *Bacteroides vulgatus* in human feces. *Appl. Environ. Microbiol.* **50,** 958–964.
22. Kuritza, A. P., Shaughnessy, P., and Salyers, A. A. (1986). Enumeration of polysaccharide-degrading *Bacteroides* species in human feces by using species-specific DNA probes. *Appl. Environ. Microbiol.* **51,** 385–390.
23. Maniatis, T., Fritsch, E. F., and Sambrook, J. (1982). "Molecular Cloning: A Laboratory Manual." Cold Spring Harbor Lab., Cold Spring Harbor, New York.
24. Marsh, P. K., Malamy, M. H., Shimell, M. J., and Tally, F. P. (1983). Sequence homology of clindamycin resistance determinants in clinical isolates of *Bacteroides* spp. *Antimicrob. Agents Chemother.* **23,** 726–730.
25. Martinez-Suarez, J. V., Baquero, F., Reig, M., and Perez-Diaz, J. C. (1985). Transferable plasmid-linked chloramphenicol acetyltransferase conferring high-level resistance in *Bacteroides uniformis*. *Antimicrob. Agents Chemother.* **28,** 113–117.
26. Marx, J. L. (1988). Multiplying genes by leaps and bounds. *Science* **240,** 1408–1410.
27. Matthews, B. G., and Guiney, D. G. (1986). Characterization and mapping of regions encoding clindamycin resistance, tetracycline resistance, and a replication function on the *Bacteroides* R plasmid pCP1. *J. Bacteriol.* **167,** 517–521.
28. Melton, D. A., Krieg, P. A., Rebagliati, M. R., Maniatis, T., Zinn, K., and Green, M.R. (1984). Efficient *in vitro* synthesis of biologically active RNA and RNA hybridization probes from plasmids containing a bacteriophage SP6 promoter. *Nucleic Acids Res.* **12,** 7035–7056.
29. Moore, W. E. C., and Holdeman, L. V. (1974). Human fecal flora: The normal flora of 20 Japanese-Hawaiians. *Appl. Microbiol.* **27,** 961–979.
30. Morotomi, M., Ohno, T., and Mutai, M. (1988). Rapid and correct identification of intestinal *Bacteroides* spp. with chromosomal DNA probes by whole-cell dot-blot hybridization. *Appl. Environ. Microbiol.* **54,** 1158–1162.
31. Roberts, M. C., Moncla, B., and Kenny, G. E. (1987). Chromosomal DNA probes for the identification of *Bacteroides* species. *J. Gen. Microbiol.* **133,** 1423–1430.
32. Salyers, A. A., and Guthrie, E. P. (1988). A deletion in the chromosome of *Bacteroides thetaiotaomicron* that abolishes production of chondroitinase II does not affect survival of the organism in gastrointestinal tracts of exgermfree mice. *Appl. Environ. Microbiol.* **54,** 1964–1969.
33. Salyers, A. A., Lynn, S. P., and Gardner, J. F. (1983). Use of randomly cloned DNA fragments for identification of *Bacteroides thetaiotaomicron*. *J. Bacteriol.* **154,** 287–293.

34. Salyers, A. A., Pajeau, M., and McCarthy, R. E. (1988). Importance of mucopolysaccharides as substrates for *Bacteroides thetaiotaomicron* growing in intestinal tracts of exgermfree mice. *Appl. Environ. Microbiol.* **54,** 1970–1976.
35. Salyers, A. A., Shoemaker, N. B., and Guthrie, E. P. (1987). Recent advances in *Bacteroides* genetics. *CRC Crit. Rev. Microbiol.* **14,** 49–71.
36. Savitt, E. D., Strzempki, M. N., Vaccaro, K. K., Peros, W. J., and French, C. K. (1987). Comparison of cultural methods and DNA probe analyses for the detection of *Actinobacillus actinomycetemcomitans, Bacteroides gingivalis,* and *Bacteroides intermedius* in subgingival plaque samples. *J. Periodontol.* **59,** 431–438.
37. Shoemaker, N. B., Getty, C., Gardner, J. F., and Salyers, A. A. (1986). Tn*4351* transposes in *Bacteroides* spp. and mediates the integration of plasmid R751 into the *Bacteroides* chromosome. *J. Bacteriol.* **165,** 929–936.
38. Sipat, A., Taylor, K. A., Lo, R. Y. C., Forsberg, C. W., and Krell, P. J. (1987). Molecular cloning of a xylanase gene from *Bacteroides succinogenes* and its expression in *Escherichia coli. Appl. Environ. Microbiol.* **53,** 477–481.
39. Smith, C. J. (1985). Polyethylene glycol-facilitated transformation of *Bacteroides fragilis* with plasmid DNA. *J. Bacteriol.* **164,** 466–469.
40. Smith, C. J., and Gonda, M. A. (1985). Comparison of the transposon-like structures encoding clindamycin resistance in *Bacteroides* R-plasmids. *Plasmid* **13,** 182–192.
41. Socranksy, S. S., Haffajee, A. D., Smith, G. L., and Dzink, J. L. (1987). Difficulties encountered in the search for the etiologic agents of destructive periodontal diseases. *J. Clin. Periodontol.* **14,** 588–593.
42. Southern, J. A., Parker, J. R., and Woods, D. R. (1986). Expression and purification of glutamine synthetase cloned from *Bacteroides fragilis. J. Gen. Microbiol.* **132,** 2827–2835.
43. Speer, B. S., and Salyers, A. A. (1988). Characterization of a novel tetracycline resistance that functions only in aerobically grown *Escherichia coli. J. Bacteriol.* **170,** 1423–1429.
44. Stahl, D. A., Flesher, B., Mansfield, H. R., and Montgomery, L. (1988). Use of phylogenetically based hybridization probes for studies of ruminal microbial ecology. *Appl. Environ. Microbiol.* **54,** 1079–1084.
45. Strzempko, M. N., Simon, S. L., French, C. K., Lippke, J. A., Raia, F. F., Savitt, E. D., and Vaccaro, K. K. (1987). A cross-reactivity study of whole genomic DNA probes for *Haemophilus actinomycetemcomitans, Bacteroides intermedius,* and *Bacteroides gingivalis. J. Dent. Res.* **66,** 1543–1546.
46. Tally, F. P., and Malamy, M. H. (1986). Resistance factors in anaerobic bacteria. *Scand. J. Infect. Dis., Suppl.* **49,** 56–63.
47. Viljanen, M. K., Linko, L., Arstila, P., and Lehtonen, O. P. (1986). Monoclonal antibodies to the lipopolysaccharide and capsular polysaccharide of *Bacteroides fragilis. In* "Monoclonal Antibodies against Bacteria" (A.J.L. Macario and E. Conway de Macario, eds.), Vol. 3, pp. 119–142. Academic Press, Orlando, Florida.

10

Nucleic Acid Probes for *Campylobacter* Species

BRUCE L. WETHERALL* AND ALAN M. JOHNSON†
**Department of Clinical Microbiology*
Flinders Medical Center
†Department of Clinical Microbiology
School of Medicine
Flinders University and Flinders Medical Center
Bedford Park, South Australia
Australia

I. INTRODUCTION

Although the importance of campylobacters in veterinary medicine has been recognized for many years, their role as significant human pathogens has only relatively recently been realized. This realization has led microbiologists to seek numerous improved methods for the isolation, cultivation, and identification of *Campylobacter* species from various sources. As a result, several new species have been described, and their ubiquitous nature in the environment, animals, and humans is becoming obvious. The true significance of *Campylobacter* species, including their epidemiology and role in infections, will only be revealed by continuing to develop ways of detecting and identifying unusual campylobacters. The traditional techniques for the isolation of *Campylobacter* species involving the use of selective agars and a selective temperature of incubation (43°C) have been important in establishing the role of *Campylobacter jejuni* and *Campylobacter coli* in enteric infections of humans (115,117). However, these methods may be responsible for the lack of isolation, and therefore recognition of the importance of other campylobacters in enteric disease. Such species include "*Campylobacter cinaedi,*" "*Campylobacter fennelliae*" (15,29), and atypical strains of *C. jejuni* (121) that favor a temperature of 37°C for growth. Other species such as "*Campylobacter hyointestinalis*" are sensitive to cephalothin, which is commonly included in selective media (28). These examples illustrate but a few of the diverse growth characteristics of *Campylobacter* species and highlight just some of the reasons why they can be difficult to isolate. Further conditions that are important for optimal recovery of the many species include incubation under various atmospheric conditions, the time of incubation, and the use of appropriate media (8,84,119). In addition, failure to isolate campylobacters may be due to competition from the normal microbial flora or the presence of viable but nonculturable organisms. Thus several different isolation techniques would be required for the complete detection of all members of this genus. A review by Penner (97) on the various species of *Campylobacter* provides an insight into the many methods required for the isolation of the various strains.

Once isolated, campylobacters are generally difficult to identify at the species level because of their inactivity in most biochemical tests (84,119). Thus very few practical tests are available to distinguish between species and subspecies. Identification is often dependent on only a few phenotypic characters. These can be difficult to interpret or may be unreliable because they usually involve susceptibility to antibiotics, ability to grow at various temperatures, and growth in the presence of inhibitors (119). Because of these problems with the biochemical characterization of *Campylobacter* species, considerable confusion has arisen over their taxonomy and epidemiology.

In an endeavor to improve the identification of campylobacters to species level, alternative methods to the conventional biochemical tests have been sought. Examples of these include serological methods (45), enzymatic methods (82), analysis of cellular fatty acid compositions (64), electrophoretic protein patterns (16), and DNA–DNA hybridization (131). This latter method appeared to be particularly successful and was subsequently used by others (126). Such nucleic acid (NA) probes would also theoretically be suitable for direct testing of clinical specimens. There exist, however, several problems and limitations associated with the use of these probes, many of which are equally applicable to NA probe assays for other bacteria (127).

This chapter describes the various types of NA probes that have been used for *Campylobacter* species, different labels that have been used, and also the different formats. Emphasis will be placed on those techniques that provide, or have the potential to provide, rapid, reliable, stable, and hardy systems for use in a diagnostic laboratory for both the detection in clinical specimens and identification of these nonsaccharolytic pathogens.

II. BACKGROUND

A. Epidemiology and Clinical Significance of *Campylobacter* Infections

In 1957 King (57) recognized the possible importance of campylobacters (then thought to be "related vibrios") in human gastroenteritis, but because of the difficulties associated with the isolation of this organism from feces, little attention was paid to them until Butzler (13) described a selective coproculture technique and Skirrow (115) reported the successful use of an antibiotic-selective medium. The latter author's technique was designed to isolate the thermophilic campylobacters, *C. jejuni* and *C. coli,* the names given to the "related vibrios" of King (57) by Veron and Chatelain (133). These species were recovered from the feces of 7.1% of 803 unselected patients with diarrhea and none of the controls. This suggested that they may be one of the more common identifiable causes of bacterial diarrhea. Subsequent studies have confirmed these findings, and *C. jejuni* and *C. coli* are now implicated as being major causes of gastroenteritis throughout the world (6,116). The true frequency of infection with each of these species is not known because most clinical laboratories do not distinguish between them. The reason for this is the lack of definitive differentiating characteristics (see Section II,C).

Of the other species of *Campylobacter* implicated in human infections, *Campylobacter fetus* subsp. *fetus* is probably the best known. It is generally regarded as being an opportunist species, with systemic infections of debilitated or immunosuppressed patients the usual presentation. Several studies, however, indicate that this organism may also cause human gastroenteritis (23,42). In contrast, *C. fetus* subsp. *venerealis* has not been associated with human infections, and only infects the bovine genital tract (84).

Increasing awareness of the importance of *Campylobacter* species in human infections has resulted in the description of several new species involved in gastroenteritis, some of which were originally isolated from only animals or birds. *Campylobacter laridis,* another thermophilic species that was primarily isolated from animals and sea gulls (3), has now been implicated in human gastroenteritis (124). A urease-positive variant of *C. laridis* has been isolated for the first time from the feces of two patients with diarrhea and from the appendix of another patient with appendicitis (83). By a whole DNA hybridization technique, this *C. laridis* variant was found to be closely related to the urease-positive thermophilic campylobacters (UPTC) described by Bolton *et al.* (7) and to the type strain of *C. laridis*. The UPTC strains were previously thought to be exclusively environmental organisms, as they were isolated from river water and seawater, as well as mussels and cockles.

Campylobacter hyointestinalis, a new species described in 1983, was originally isolated from pigs with proliferative enteritis (33). Subsequent reports indicate that this species may now also be associated with human disease. Fennell *et al.* (28) isolated *C. hyointestinalis* from the rectal culture of a homosexual man with proctitis, and Edmonds *et al.* (24) reported the isolation of this organism from the feces of four patients with watery diarrhea.

Thermotolerant catalase-negative or weakly reacting campylobacters are yet another example of the original isolation from animals, with subsequent recovery from humans. This group of campylobacters, for which the name "*Campylobacter upsaliensis*" has been proposed [see review by Penner (97)], was initially isolated from dog feces in 1983 (113), but has since been grown from human feces (121,126). Although the two strains isolated by Tee *et al.* (126) were from patients with acute gastroenteritis, the exact role of this species of *Campylobacter* in human infections remains to be defined.

In a study of the cause of gastrointestinal symptoms in homosexual men, Quinn *et al.* (100) reported the recovery of *C. jejuni, C. fetus* subsp. *fetus,* and a group of campylobacterlike organisms (CLO) from rectal swabs. These CLO have been phenotypically and genetically characterized, and have been tentatively classified into two groups as the new species, "*C. cinaedi*" and "*C. fennelliae*" (131). Both species are capable of causing systemic infections indicated by their isolation from blood cultures of a bisexual male (92) and a homosexual male (15).

Two of the remaining species of *Campylobacter* isolated from humans have not been associated with infection. *Campylobacter sputorum* biovar sputorum is thought to be part of the normal flora of the gingival crevice of humans, and has not been associated with disease (84). *Campylobacter concisus* has been isolated from patients with gingivitis, periodontitis, and periodontosis, but its pathogenicity is unknown (123).

The newly described species *Campylobacter pylori,* which was originally isolated by Marshall *et al.* (74) from human gastric mucosa, has been associated with active chronic gastritis, gastric ulcer, duodenal ulcer, and nonulcer dyspepsia in adults (11,53,75). In addition, several studies indicate that this organism is associated with primary antral gastritis and antral lymphoid hyperplasia in children (20,22), and may also play a role in protein-losing enteropathy in this age group (47). Although a gastric CLO that appears to be identical to *C. pylori* has been isolated from the stomach of rhesus monkeys (89), it has never been isolated from any other source, including human saliva and feces.

Clearly, our knowledge of infections caused by *Campylobacter* species is incomplete. In many instances this is due to a lack of epidemiological data, because studies have not been done, because we have simply not searched for campylobacters in the right circumstances, or because of shortcomings in our ability to isolate and identify them in clinical specimens.

B. Problems with Conventional Isolation and Detection Techniques

The most widely used method for the isolation of *Campylobacter* species known to cause enteritis is the use of the selective medium developed by Skirrow (115) in conjunction with incubation at 42°–43°C in a microaerophilic atmosphere. The antibiotics contained in this medium (vancomycin, polymyxin B, and trimethoprim) and the raised incubation temperature are important for the recovery of the thermophilic *Campylobacter* species *C. jejuni, C. coli,* and *C. laridis.* In order to provide greater selectivity and the ability to suppress normal bacterial flora, numerous modifications to the antibiotic content and the base medium have been made [see review by Penner (97)]. In particular, the antibiotic supplement of a medium developed by Blaser *et al.* (5) consists of vancomycin, trimethoprim, polymyxin B, amphotericin, and cephalothin. However, *C.*

fetus subsp. *fetus* is susceptible to cephalothin and thus could not be isolated on this medium, even if incubated at 37°C. Although this species is thought not to grow at 42°C, fecal strains associated with enteritis have been reported to be capable of growth at this temperature (23,42). Therefore the combination of these two factors may have inadvertently resulted in an underestimation of the true incidence of gastrointestinal infection with *C. fetus* subsp. *fetus*. In addition, *C. hyointestinalis* and "*C. upsaliensis*," which are capable of growth at 42°C, are susceptible to cephalothin and could not be isolated on this type of selective medium (32,121). Although antibiotic-containing media are important for the recovery of fecal campylobacters, they should not be excessively inhibitory. Ng *et al.* (90) found that thermophilic campylobacters, in particular *C. coli*, may be inhibited by the combination of antibiotics used in primary isolation media, and thus may account for the lower incidence of reported infections with *C. coli* compared with *C. jejuni*.

Increasing reports of the isolation of "*C. cinaedi*" and "*C. fennelliae*" are largely due to the efforts of Fennell and co-workers (29), who recognized the importance of incubating primary isolation media at 37°C for longer than the usual 2 days that are used for the recovery of thermophilic campylobacters. Laughon *et al.* (67), in a study on the recovery of *Campylobacter* species from homosexual men, found that the type of CLO isolation medium was important for the recovery of "*C. cinaedi*," and reported a requirement for freshly prepared media for optimum recovery of *C. jejuni*. Perhaps more important than the isolation of CLO from the homosexual male population by incubation at 37°C is the necessity to incubate isolation media at this temperature for the recovery of atypical *C. jejuni*-like strains (121,126).

Consequently, for the optimum recovery of all strains of *Campylobacter* associated with enteric infections, several different media are required, two different incubation temperatures should be used, and the time of incubation may need to be extended for up to 5 days. In addition, thermophilic *Campylobacter* species do not survive well in specimens left at room temperature (38) and in buffered glycerol–saline, a transport medium commonly used for other enteric pathogens (84), and thus may not be isolated if inappropriately transported. Furthermore, the optimum recovery of "*C. cinaedi*" and "*C. fennelliae*" appears to require rectal swabs in conjunction with immediate inoculation of isolation media (67). The earliest a presumptive identification of *C. jejuni* and *C. coli* can be made is 24 hr, but more often than not is 48 hr. This places constraints on early appropriate antibiotic treatment, which is important for the clearance of these organisms from the stools of patients with severe infections (5).

The isolation and detection of the newly described species *C. pylori* may also present problems. The presence of this organism is usually determined by histological and microbiological examination of endoscopic biopsy samples of gastric mucosa. The organism has fastidious growth requirements and can only be grown in microaerophilic conditions in an atmosphere of high humidity (36). In addition, optimum recovery of this organism from antral biopsy specimens requires the inoculation of special solid media and incubation for at least 5 days (62). These fastidious growth requirements may, in part, be responsible for the failure to isolate *C. pylori* from any organ other than the stomach in humans nor from the environment (35). Even in specimens of gastric antrum, *C. pylori* may not be isolated because of poor technique either during the biopsy or later in the laboratory (35).

A characteristic feature of *C. pylori* is the production of a large quantity of extracellular urease (65). This property has been used as the basis for the rapid detection of this organism by testing for the presence of preformed urease in gastric antral biopsy specimens (81). However, although this rapid urease test appears to give satisfactory results in the initial screening of patients, Borromeo *et al.* (9) reported that negative results may be found in specimens culture-positive for *C. pylori,* and they concluded that the reliability of the test in patients being reevaluated after treatment remains to be established. In addition, Das *et al.* (18) found that the urease test was not reliable when results from esophageal and duodenal biopsy specimens were included with results from gastric antrum. False negative rapid urease test results may be due to a low initial inoculum, and can be expected to be negative at $<10^4$ colony-forming units (CFU)/ml (40). Alternatively, false negative results may be due to the presence of urease-negative strains of *C. pylori,* although such strains are only rarely isolated (134).

C. Problems with Conventional Identification Techniques

The genus *Campylobacter* is composed of gram-negative microaerophilic, slender, spirally curved bacilli. Most species exhibit corkscrewlike motility by means of a single polar unsheathed flagellum at one or both ends of the cell. The exception to this is *C. pylori,* which has four to six unipolar sheathed flagella (37). They are chemoorganotrophs and do not ferment or oxidize carbohydrates. All *Campylobacter* species have respiratory metabolism. There has been considerable conflict in classification schemes and nomenclature (55), resulting in much taxonomic and epidemiological confusion. In part, this situation has arisen because of the lack of convenient differential tests. Campylobacters are generally difficult to identify at the species level because of their asaccharolytic nature and inactivity in most biochemical tests (84). In addition, some strains have atypical or inconclusive reactions with conventional tests (121,132). Moreover, for some biochemical tests, different media formulations and different interpretations have been used by various investigators. For instance, a variety of media have been used for the detection of hydrogen sulfide production by enteropathogenic campylobacters and CLO (3,7,24,44,113,118). Different results can be obtained, depending on the medium used (55,111). The use of different concentrations of sodium chloride in growth inhibition studies may also give rise to conflicting results (3,41,87). Other tests that are likely to provide inconsistent results include the DNA hydrolysis test (44,68), tolerance to 1% glycine (41), and hippurate hydrolysis test. The latter is the only test, apart from genetic methods, capable of distinguishing *C. jejuni* from *C. coli,* and is largely species-specific. However, a small percentage of strains have been reported to give aberrant results (43,132).

Campylobacter pylori and other CLO are even less reactive than the enteropathogenic campylobacters in biochemical tests. *Campylobacter pylori* is frequently identified on the basis of growth characteristics, Gram stain morphology, oxidase and catalase reactions, and a positive urease test (48,103,135). However, for accurate results, isolates should not be identified solely on the basis of a positive urease test because other urease-producing campylobacters have been described: UPTC (7) (which are now thought to be identical to a *C. laridis* variant; see above) and *Campylobacter nitrofigilis* (76). In addition, strongly urease-positive CLO have been found in the stomach of laboratory-raised ferrets (31) and have been isolated from the gastric antrum of rhesus monkeys (89).

Despite these problems of identification, the increased interest in campylobacters as human pathogens has resulted in several new species being described. This emphasizes the need to establish consistent ways of differentiating species of *Campylobacter* and to ascertain the identity of new isolates of CLO in order to provide meaningful epidemiological data. The aforementioned developments have seen the 1985 grouping of Smibert (119) for *Campylobacter* species changed and expanded to include 14 species, as reviewed by Penner (97). The classification and nomenclature outlined by Penner is used in this chapter. Thus, *C. sputorum* subsp. *mucosalis* is regarded as being a separate species, *C. mucosalis,* and *C. sputorum* is separated into three biovars: sputorum, bubulus, and fecalis. In addition, the names of some of the newly described species have now been validated: *C. hyointestinalis, C. laridis, C. pylori, C. cryaerophila,* and *C. nitrofigilis*. The names of others, however, have only been proposed: "*C. upsaliensis,*" "*C. cinaedi,*" and "*C. fennelliae.*"

D. Alternative Detection and Identification Techniques

Because of the time taken to isolate enteropathogenic campylobacters and the need to provide a rapid presumptive diagnosis in cases of clinical urgency, direct microscopic techniques have been tried. Using phase-contrast microscopy, Karmali and Fleming (54) detected *C. jejuni* in 75% of patients with culture-proven *Campylobacter* enteritis. However, specimens must be no more than a few hours old, and the recognition of the characteristic morphology and motility of these bacteria requires a skilled microscopist. This same technique was used to examine rectal swabs from homosexual men, but did not prove a success (67). The value of dark-field microscopy of fresh fecal specimens for the detection of organisms exhibiting typical *Campylobacter* motility was assessed by Paisley *et al.* (95). Only 36% of specimens that were culture-positive for *C. jejuni* demonstrated this microscopic morphology. A direct fluorescent-antibody test for the detection of *C. jejuni* and *C. coli* in fecal specimens has been compared with detection by culturing (49), but the sensitivity of the antibody test was found to be too low to make it useful in a laboratory setting (see Section V,A). Clearly, there is a wide scope for further improvements and additional techniques for the direct detection of campylobacters in clinical specimens.

The paucity of reliable and definitive biochemical tests for characterizing campylobacters (see Section II,C) has led investigators to seek other means of classifying these bacteria. Megraud *et al.* (82) performed an enzyme profile characterization of strains of *C. pylori* and reported the presence of γ-glutamyltranspeptidase in all strains. Although this enzyme was found in some strains of *C. jejuni,* it was not detected in any of the other campylobacters or CLO tested. Tests for the detection of preformed enzymes in several *Campylobacter* species have since been reported to be of value for the rapid identification of *C. pylori* (80). Analysis of the isoprenoid quinone content and cellular fatty acid composition of 36 campylobacters was done by Moss *et al.* (85). Because the results were promising, Lambert *et al.* (64) examined 368 strains of *Campylobacter* species and CLO for their cellular fatty acid composition, using gas–liquid chromatography. Although only five of eight *Campylobacter* species could be differentiated, it proved particularly useful for identifying the newly described species "*C. cinaedi,*" "*C. fennelliae,*" *C. cryaerophila,* and *C. pylori*. Sodium dodecyl sulfate–polyacrylamide gel electrophoresis (SDS–PAGE) and subsequent protein profile analysis have also been found to be of value for differentiating among species (16), and have recently been used as an aid in the identification of an unclassified CLO associated with gastroenteritis (2).

III. RESULTS AND DISCUSSION

However, for ease of performance, rapidity, and accuracy, the technique of DNA–DNA hybridization using specific probes is likely to be the most successful approach for identifying and classifying campylobacters. In a study on the differentiation of *Campylobacter* species isolated from homosexual men with intestinal symptoms, Totten *et al.* (131) described the use of a taxonomic spot-blot test, which used whole bacteria rather than purified DNA. This technique enabled rapid confirmation of the phenotypic grouping of CLO, but more importantly would be suitable for screening clinical isolates for homology with a standard probe. Others have also reported success using this technique (91) and with a similar method using target blots of purified DNA (126). In these studies, and indeed in many other investigations using NA probes, the DNA has been labeled with ^{32}P. However, these have the inherent problems of potential hazard to the user and a requirement for specially equipped laboratories. In addition, because of their short half-lives, radioactive probes need to be repeatedly and freshly prepared.

Considerable recent research has been conducted into obtaining suitable nonradioactive methods for labeling nucleic acids. Most experience has been gained with biotinylated probes. The most widely used method for the preparation of these probes involves the labeling of nucleotides with biotin by nick translation. Another technique that has been recently used incorporates linking photobiotin, a photoactivatable analog of biotin, to nucleic acids (30). However, these have been reported to be less sensitive than comparative radioactive probes for the detection of both purified DNA and target sequences in clinical material (4,52,56,127). In addition, problems with the nonspecific reaction of intracellular biotin or biotin analog with streptavidin in the detection reagent may be encountered (4,63). Other approaches have been to modify DNA chemically by the introduction of sulfone moieties into cytosine residues (51), or by using *N*-acetoxy-2-acetylaminofluorene (AAF) (125), and detecting hybridized probe with monoclonal antibodies (MAb) to the respective modified DNA and enzyme-labeled anti-species immunoglobulin. Also, enzymes have been covalently linked directly to DNA (104). However, only few investigations have been reported on the use of sulfonated probes and AAF–DNA probes (21,83), and reports on direct alkaline phosphatase-labeled probes have yielded conflicting results (71,114).

A. *Campylobacter jejuni* Probes

Tompkins and colleagues were the first to develop a gene probe detection system for *C. jejuni* (129,130). They prepared a probe consisting of whole genomic DNA and another probe consisting of a mixture of eight unique chromosomal fragments ranging in size from 7 to 11 kb cloned into pBR322. This latter probe gave no advantage over the whole chromosomal probe, as neither could detect $<2 \times 10^6$ organisms/ml of seeded stool sample. Both probes were labeled with ^{32}P. In fact, genomic (or chromosomal) probes have advantages over several other types of probes such as repetitive DNA or ribosomal RNA (rRNA) in that cloning or sequencing is not required and the genetic organization of an organism need not be known. In addition, in principle, total genomic DNA probes should be the most sensitive, with the capacity to hybridize with all target sequences present (79,107). Furthermore, because cloning is not required, the vector homology problem will not occur during hybridization with clinical specimens. This problem may occur with techniques that use cloned probes and has been shown to be due to the hybridization of material, homologous to the plasmid vector pBR322 present in many clinical samples, containing segments of DNA from the cloning vector (1).

Compared with culture, the whole genomic probe of Tompkins *et al.* (129,130) gave a sensitivity of 44% and a specificity of 94%. None of the samples containing other important fecal pathogens (*Salmonella, Shigella, Yersinia*) were probe-positive.

The NA probe produced by Korolik *et al.* (60) also showed exquisite specificity. It consisted of a 6.1-kb piece of DNA isolated by immunoscreening a pBR322 expression library of *Sau*3A-partially cut *C. jejuni* chromosomal DNA with polyclonal antibody to whole cells and purified outer-membrane proteins. On Southern blotting, this insert did not hybridize to a range of other fecal pathogens (see Table I) nor other campylobacters (*C. laridis, C. fetus* subsp. *fetus, C. fetus* subsp. *venerealis, C. fecalis, C. pylori*). It did, however, weakly cross-hybridize with *C. coli*.

In an attempt to remove this type of cross-hybridization, Picken *et al.* (99) screened a *Sau*3A-partially cut library of *C. jejuni* in EMBL 4, with both *C. jejuni* and *C. coli* DNA. The majority of the clones positive with *C. jejuni* DNA did cross-hybridize with the *C. coli* DNA (in spite of the fact that these DNA share only a 30% homology), but 30 putative *C. jejuni*-specific clones were chosen for further screening. However, some of these also hybridized with *C. coli,* some hybridized with *C. laridis,* and one hybridized with all four species (*C. jejuni, C. coli, C. laridis, C. fetus*) tested. Consequently, only four were chosen for detailed study. Fragments of these four clones were tested individually against a range of campylobacters and other organisms. A 4.6-kb fragment and a 3.6-kb fragment from one clone and a 2.3-kb fragment of another clone showed less cross-reactivity than the others, with a differential hybridization ratio greater than 100 : 1 of *C. jejuni* to *C. coli* and *C. laridis*. The other organisms, the DNA of which were found not to hybridize, are listed in Table I.

TABLE I

Hybridization of Nucleic Acid from Various Species with Putative *Campylobacter jejuni*- Specific Probes

Reaction	Organism[a]	Organism[b]	Organism[c]
Strong	*C. jejuni* (11 isolates)	*C. jejuni*	*C. jejuni* (9 isolates)
		Campylobacter coli	
		Campylobacter laridis	
Weak	*C. coli*		
None	*Salmonella enteritidis*	*Salmonella enteritidis*	*Neisseria lactamica*
	Salmonella typhimurium	*Salmonella typhimurium*	*Neisseria meningitidis*
	Escherichia coli	*E. coli*	*Neisseria gonorrhoeae*
	Shigella sonnei	*Shigella sonnei*	*Lactobacillus casei*
	Enterobacter cloacae	*E. cloacae*	*Legionella pneumophila*
	Citrobacter freundii	*Citrobacter freundii*	*E. aerogenes*
	Klebsiella pneumoniae	*K. pneumoniae*	*K. pneumoniae*
	Proteus vulgaris	*Proteus vulgaris*	*C. fetus*
	Vibrio parahemolyticus	*V. parahemolyticus*	*Corynebacterium diphtheriae*
	Serratia marcescens	*Serratia marcescens*	*C. coli*
	Pseudomonas aeruginosa	*Pseudomonas aeruginosa*	*Pseudomonas aeruginosa*
	C. laridis	*Salmonella arizonae*	*C. laridis*[d]
	Campylobacter fetus subsp. *fetus*	*Streptococcus* (groups B,C)	*Streptococcus* (groups A,C,E,G)

	C. fetus subsp. *venerealis*	*Staphylococcus aureus*	*Staphylococcus aureus*
	Campylobacter fecalis	*Shigella dysenteriae*	
	Campylobacter pylori	*C. pylori*	
		Shigella flexneri	
		Shigella boydii	
		Enterobacter aerogenes	
		Yersinia enterocolitica	
		Bacteroides fragilis	
		Clostridium difficile	
		Candida albicans	
		Vibrio cholerae	
		Aeromonas hydrophila	
		Aeromonas sobria	
		Pleisiomonas shigelloides	
		Clostridium perfringens	
		"*Campylobacter fennelliae*"	
		"*Campylobacter cinaedi*"	

[a] Korolik *et al*. (60).
[b] Rashtchian *et al*. (102).
[c] Picken *et al*. (99).
[d] See text for details.

Another way to try and ensure the species specificity of the NA probe is to use a synthetic oligonucleotide. The shortness of this type of probe would make it extremely unlikely that it would hybridize to the DNA of organisms other than the species from which the sequence was obtained. Bryan *et al.* (10) produced a 26-mer specific for a unique sequence in the *C. jejuni* genome. Under stringent hybridization conditions, the ^{32}P-labeled probe reacted with DNA from *C. jejuni* but not with DNA from *C. coli, C. laridis, C. fetus,* or 30 control enteric organisms. Unfortunately, the sequence of the 26-mer was not given and the 30 control organisms were not listed. The species specificity of this probe is in contrast to that developed by Rashtchian *et al.* (101,102,122).

These authors screened a partial cut *Hin*dIII genomic library of *C. jejuni* with ^{32}P-end-labeled rRNA from *C. jejuni* and cloned a 1.7-kb fragment that contained the major part of the coding region for the 16 S rRNA (101). They then synthesized three oligonucleotide probes, each ~50 bases long, complementary to sections of this 16 S rRNA, and labeled them with biotin (102). The biotinylated rRNA probes were then used in an assay to detect the corresponding rRNA sequences of *C. jejuni* in preparations of organisms isolated and cultured from feces. Biotinylated synthetic oligomer is hybridized with the test organism after sodium hydroxide denaturation. Polyclonal antibody produced against the DNA–RNA hybrid is bound to a plastic tube and immobilizes any such hybrids formed. Theoretically, high stringency ensures that only *C. jejuni* rRNA binds to the complementary oligonucleotides. The biotin binds to added streptavidin–peroxidase conjugate, which causes a color change in an added chromogen–substrate mix. The assay was optimized to detect as few as 7×10^4 *Campylobacter* cells. Compared with culture, the assay had a sensitivity of 98.7% and a specificity of 98.2% for a test population ($n = 1448$) that had a 5.2% prevalance of *Campylobacter* infection. In addition, the assay is nonisotopic, takes only 2.5 hr to complete, and is similar in format to radioimmunoassays (RIA) and enzyme-linked immunosorbent assays (ELISA) already commercially available (122). However, as can be seen in Table I, this assay is still not species-specific. Although it does not react with a wide range of non-*Campylobacter* species, nor *C. pylori,* "*C. fennelliae,*" or "*C. cinaedi,*" it does not have strong cross-hybridization with *C. coli* and *C. laridis*.

B. Genus-Specific Probes

Romaniuk and Trust (109) also used an oligonucleotide complementary to 16 S rRNA as a probe. However, they chose their sequence, TCATCCTCCACGC(G,T)GCGT, because it was identified as being

Campylobacter-specific from a phylogenetic comparison of partial *Campylobacter* 16 S rRNA sequences with a data base of similar sequences of other organisms (110). The ^{32}P-labeled oligonucleotide hybridized with Southern blots of *Eco*RV and *Rsa*I-restricted DNA from four isolates of *C. jejuni,* two isolates of *C. coli,* one isolate of *C. fetus,* and two isolates of *C. laridis,* but not with DNA from *Salmonella typhimurium* or *Escherichia coli.*

Although they did not look for a genus-specific probe as such, Chevrier *et al.* (14) labeled the sonicated whole genomic DNA of seven *Campylobacter* strains (*C. jejuni, C. coli, C. laridis, C. fetus, C. hyointestinalis,* "*C. upsaliensis,*" *C. sputorum*) with AAF, and used them to investigate the cross-hybridization among *Campylobacter* species. They observed 30–50% homology between *C. jejuni* and *C. coli,* and 20–30% homology between *C. hyointestinalis* and *C. fetus,* and <20% homology among the DNA of the remaining strains. The AAF-labeled probes were then used to identify 60 *Campylobacter* strains that had been characterized by conventional tests after isolation from human feces. Using a rapid DNA extraction procedure, the bank of seven AAF-labeled probes led to the correct identification of 28 *C. jejuni* isolates, 16 *C. coli* isolates, 3 *C. fetus* isolates, 7 "*C. upsaliensis*" isolates, 4 *C. laridis* isolates, and 1 isolate each of *C. hyointestinalis* and *C. sputorum.* The test could be performed in <18 hr. However, AAF is potentially carcinogenic and requires careful precautions during handling.

C. *Campylobacter pylori* Probes

The feasibility of using NA probes for the rapid definitive identification and detection of *C. pylori* is currently under investigation (137). As part of this study we have used DNA from *C. pylori* to compare three nonradioactive probes with a ^{32}P-labeled probe. The nonradioactive probes and detection systems used were a biotin-labeled DNA (biotin–DNA) probe together with the BluGene detection system and a photobiotin-labeled DNA (photobiotin–DNA) probe and detection system (see Section VIII). The sensitivity of detection of these total genomic DNA probes for homologous purified DNA was found to be 100 pg with the ^{32}P, biotin–DNA, and sulfonated probes, and 10 ng with the photobiotin–DNA probe. Equivalent sensitivity was obtained for nylon and nitrocellulose (NC) membranes, but because the latter became extremely brittle after treatment with denaturing solutions, they were not used thereafter. The degree of sensitivity of the ^{32}P, biotin–DNA, and sulfonated probes was consistent with previously reported results of the detection sensitivity of these types of probes for homologous DNA (51,52,78). The somewhat surprising low sensitivity of the photobiotin–DNA probe and detection system, combined with its propensity to give high backgrounds when tested with the nylon membrane used, meant that it was not suited to our requirements. Others have also reported a lack of sensitivity of photobiotin-labeled probes in comparison with ^{32}P-labeled probes (56). However, in a comparison of nonradioactive probes for the detection of human immunodeficiency virus (HIV), Donovan *et al.* (19) found that a photobiotinylated probe was only two times less sensitive than a radioactive (^{32}P) probe. The reason for these differences is unclear, but it may reflect a dependency of the photobiotin system on the nature of the probe DNA that is to be labeled.

From a practical point of view, more meaningful results can be obtained by testing for the minimum number of organisms detectable. This was done by using a similar technique to the differential spot-blot test of Totten *et al.* (131). Serial dilutions of homologous whole bacteria were dotted onto nylon membranes. The dot blots were denatured and hybridized with the different types of *C. pylori* LL probe. The lowest detectable number of organisms was 5×10^4 per dot for the ^{32}P, biotin–DNA, and sulfonated probes and 5×10^6 per dot with the photobiotin–DNA probe (Fig. 1). Few studies on the detection sensitivity for absolute numbers of bacteria by NA probes are available for comparison. In addition, this sensitivity is likely to vary depending on the nature of the probe, the label, and the hybridization format. However, others who have used genomic probes labeled with ^{32}P and the dot-blot format have reported detection sensitivities of $\sim 10^4$ *Mobiluncus* species and 10^4 *Mycobacterium* species (105,106). In the study of Roberts *et al.* (105) on the detection of *Mobiluncus* species, biotinylated DNA probes were found to be 100- to 1000-fold less sensitive than ^{32}P-labeled probes and required 10^5–10^6 bacteria for a positive signal. In contrast, in our study the biotin–DNA probe gave the same sensitivity as the ^{32}P-labeled probe, although the intensity of the reaction with the former probe was rather weak. This may reflect improvements that have occurred in nonradioactive probe and hybridization methodologies. The *C. pylori* sulfonated probe was clearly equal in detection sensitivity to the ^{32}P-labeled probe. Although to our knowledge no other studies have been done on the detection of whole organisms with sulfonated DNA systems, our results for purified DNA are consistent with those of Hyman *et al.* (51) in which the lower limit of detectability for *Mycoplasma pneumoniae* DNA was 100 pg. An advantage of sulfonated-probe systems over biotinylated systems is the apparent lack of nonspecific backgrounds, which allows probes to be used at high concentrations (21,51). In addition, unlike the biotinylation system, nonspecific reactions with cellular structural analogs of biotin with sulfonated probes and its detection systems are

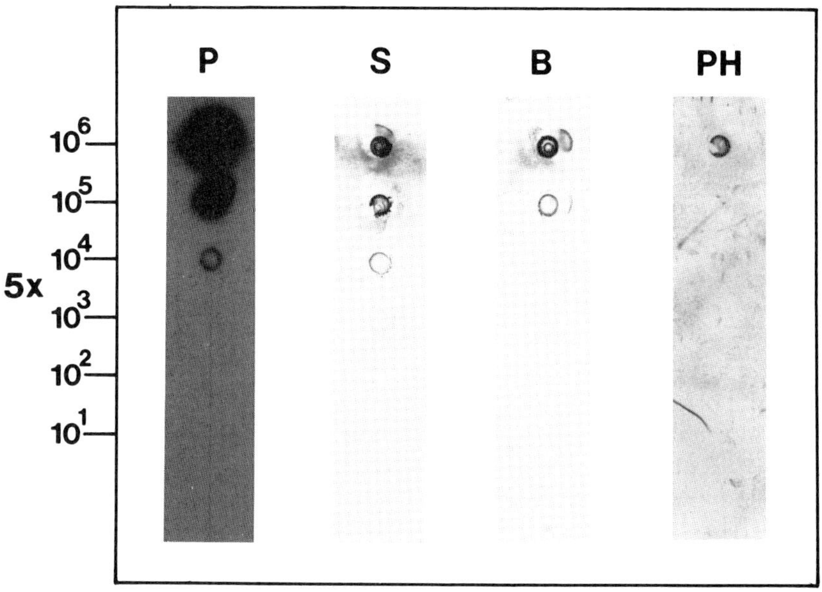

Fig. 1. Dot-blot hybridization of *Campylobacter pylori* LL genomic probes with whole-cell preparations. Homologous bacterial cells in 10-fold dilutions were spotted onto nylon membranes to give concentrations ranging from 5×10^6 to 5×10^1 cells. After denaturing, treating, and baking, the dots were hybridized with various types of probes: P, ^{32}P; S, sulfonated DNA; B, biotin DNA, PH, photobiotin–DNA.

unlikely to occur (51,137). The 100-fold lower detection sensitivity of the photobiotin–DNA probe for whole bacteria is consistent with its lack of sensitivity for purified DNA, as outlined previously.

We have used genomic *C. pylori* probes because they have several advantages. As mentioned previously, cloning is not required. They have the capacity to hybridize with all target sequences present, resulting in a detection sensitivity reported to be 10- to 100-fold higher than cloned single-copy gene probes (52,77). In addition, because cloning is not required, the vector homology problem will not occur during hybridization with clinical specimens (1). However, a potential deficiency of genomic probes is that they might cross-hybridize with phylogenetically related organisms, as has been demonstrated by McLaughlin *et al.* (78), who found that a *Plasmodium falciparum* genomic probe hybridized with heterologous *Plasmodium* DNA. It was therefore important to establish the specificity of our *C. pylori* genomic probe.

Initial cross-hybridization results using a whole-cell dot-blot method and *C. pylori* LL probes had revealed no reaction with *C. jejuni* NCTC

11351, *C. fetus* subsp. *fetus* NCTC 10842, gastric CLO2 NCTC 11847, and *E. coli* ATCC 25922 (137). Because of the success obtained with sulfonated probes, we have used this sytem to examine further bacteria for reactions with the *C. pylori* genomic probe, and thus determine its suitability for differentiating *C. pylori* from other *Campylobacter* species. Whole-cell dot blots of a range of bacteria were hybridized with a sulfonated *C. pylori* LL probe (Fig. 2). Strong hybridization occurred with the homologous strain of *C. pylori,* while the intensity of the reaction with *C. pylori* NCTC 11637 was less, but clearly positive. The explanation for this may lie in the considerable genomic or subspecies variation that appears to exist in *C. pylori* (73). Apart from slight hybridization signals with *C. coli* NCTC 11366 and *C. laridis* NCTC 11352, no other reactions were evident. Visual estimates of the intensity of these two dots compared with the homologous dots indicated that they were of the order of ≤5%, and thus would not present a problem of misidentification as *C. pylori*. These results are consistent with reports that suggest that *C. pylori* occupies a unique taxo-

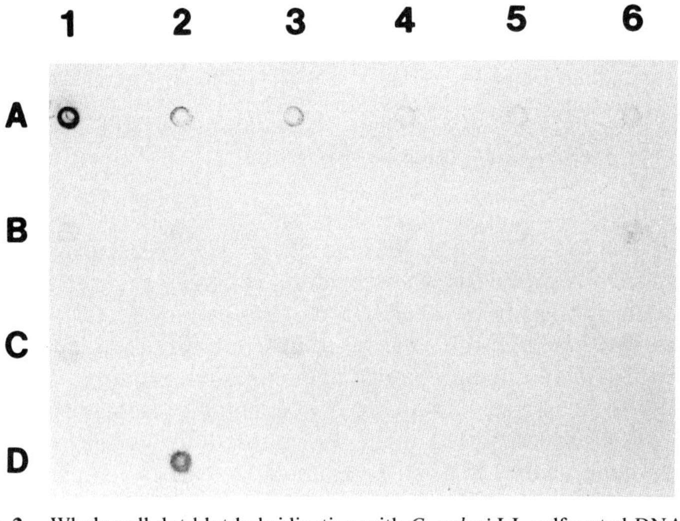

Fig. 2. Whole-cell dot-blot hybridization with *C. pylori* LL sulfonated-DNA probe and 1–2 × 10^7 cells of the following organisms, indicated by row and column numbers: A1, *C. pylori* LL; A2, *C. coli* NCTC 11366; A3, *C. laridis* NCTC 11352; A4, "*C. upsaliensis*" NCTC 11540; A5, *C. sputorum* biovar fecalis NCTC 11415; A6, atypical *C. jejuni* NCTC 11924; B1, *C. concisus* NCTC 11485; B2, *C. cryaerophila* NCTC 11885; B3, *C. hyointestinalis* NCTC 11608; B4, *C. fetus* subsp. *venerealis* NCTC 10354; B5, UPTC 7/85; B6, *C. sputorum* biovar sputorum NCTC 11528; C1, *C. sputorum* biovar bubulus NCTC 11367; C2, *C. mucosalis* NCTC 11000; C3, *W. succinogenes* NCTC 11488; C4, *Serratia marcescens* ATCC 8100; C5, *Pseudomonas aeruginosa* ATCC 27853; C6, *Klebsiella pneumoniae* ATCC 13883; D1, *Haemophilus influenzae* ATCC 19418; D2, *C. pylori* NCTC 11637.

monic position relative to most other campylobacters, based on 16 S rRNA sequence comparisons (110,128). However, *Wolinella succinogenes* NCTC 11488, an organism that has been shown to be phylogenetically closely related to *C. pylori,* showed no sign of cross-hybridization. Our results are in agreement with others who have reported no cross-hybridization between purified *C. pylori* DNA and *C. jejuni* probes (126), and between *C. jejuni* DNA and *C. pylori* probes (134). Definitive, unambiguous identification of bacterial species using the dot-blot hybridization procedure usually requires the use of different stringency conditions or the use of several dilutions of target cells (91,131). In this study, *C. pylori* could be differentiated from other campylobacters and *W. succinogenes* using a single concentration of organisms and single hybridization and washing temperatures. This is a further indication of the low degree of relatedness between *C. pylori* and the other organisms examined.

The combination of using the whole-cell dot-blot procedure and *C. pylori* sulfonated DNA probes should prove valuable for the identification and differentiation of this organism from other *Campylobacter* species and CLO. The technique is more accurate and less time-consuming than conventional biochemical tests and does not require the isolation of pure DNA or the handling of radioisotopes. The performance of this probe in the detection of *C. pylori* in clinical specimens of gastric biopsies is presently being assessed. In addition, it may prove useful for the presumptive identification of *C. pylori* in sites from which the organism has not been able to be cultured (112).

D. *Campylobacter*-Like Organism Probes

Totten and colleagues were the first to use the dot-blot hybridization assay to investigate CLO (29). They extracted genomic DNA from representatives of three groups of CLO, and used these to probe CLO isolated from rectal cultures of homosexual men. The DNA hybridized with DNA from other strains in the same group, but not with strains in other groups or with reference strains of catalase-positive *Campylobacter* species (*C. coli, C. jejuni, C. fetus* subsp. *fetus, C. fetus* subsp. *venerealis, C. fecalis*). Based on these results, they tentatively classified two of the CLO groups into species, "*C. cinaedi*" and "*C. fennelliae*" (131).

Steele *et al.* (121) also used this dot blotting to assist in the identification of campylobacters in feces. They isolated CLO other than *C. sputorum* biovar sputorum from 6% of 1672 fecal specimens. Thirty of the isolates were from 217 specimens obtained from children aged ≤3 with gastroenteritis. Ng *et al.* (91) used dot-blot hybridization to classify atypical strains of *Campylobacter* species and CLO. Three CLO did not hybridize with probes from *C. jejuni, C. coli, C. laridis, C. hyointestinalis,* or *C. fetus* subsp. *fetus,* but there was a high degree of homology among the three CLO.

Tee *et al.* (126) used conventional DNA hybridization to assist in the identification of 9 atypical campylobacters isolated along with 148 typical *Campylobacter* species from the feces of 1005 patients. They found that 3 of the strains were highly related to *C. jejuni*-type strains and 3 had characteristics similar to those of "*C. cinaedi.*"

These studies illustrate the utility of gene probe assays for the identification of CLO and thus may have prevented their misidentification by the use of conventional biochemical tests. However, in some of these investigations purified DNA was used as the target, whereas others used whole cells. In addition, various concentrations of organisms and hybridization conditions were used, and ^{32}P was used for the probe label. It would be helpful for future studies and for interlaboratory comparisons if a standardized technique could be adopted, preferably using nonradioactively labeled probes. Furthermore, it may be more appropriate to use an in-solution hybridization format rather than immobilization of target on solid supports, because of the former method's many advantages (127).

To our knowledge, CLO probes have not been used for the direct detection of these bacteria in clinical specimens.

IV. CONCLUSIONS

Campylobacter gene probes have been instrumental in resolving some of the problems of classification and nomenclature of species belonging to this genus. In addition, they have aided the characterization of newly described species where conventional techniques were limited. The whole-cell dot-blot hybridization technique has the potential to replace routine biochemical tests and may prove to be as accurate as quantitative DNA hybridization (132). However, it will be important to standardize such differential dot-blot tests for routine use in taxonomic studies (126). Most studies have used radioactively labeled probes, but will need to be replaced with nonradioactive methods before such identification techniques can be routinely incorporated into diagnostic laboratories.

We have found that a sulfonated-DNA probe for *C. pylori* has the same sensitivity as a radioactive (^{32}P) probe for homologous purified DNA and whole cells, using dot-blot techniques. In addition, it was found to be very specific, and gave little nonspecific background staining. This nonradioactive label may be equally applicable to probes for other *Campylobacter* species. Clinical studies using the sulfonated *C. pylori* probe remain to be pursued, a situation that similarly applies to gene probe studies for other *Campylobacter* species.

Several probes for the enteropathogenic campylobacters have been described, particularly those for *C. jejuni*. Although these have been useful for the identification and differentiation of species, the direct detection of *C. jejuni* in stool has not met with success. The presence of interfering substances that contribute to nonspecific background, inadequate sensitivity in clinical material, and difficulties with extraction procedures are problem areas that remain to be surmounted.

Promising studies by Rashtchian *et al.* (102), using a nonradioactive oligonucleotide probe to *C. jejuni*, indicate that these problems are being addressed. The assay, however, did not preclude the need for primary enrichment culturing before hybridization. Future prospective studies on the direct detection of *Campylobacter* in patients' samples, together with refinements and advances in methodologies, will ultimately determine the role of these probes in the diagnosis of infections caused by *Campylobacter* species.

V. GENE PROBES VERSUS ANTISERA AND MONOCLONAL ANTIBODIES

Although we are unaware of published work describing detailed direct comparison between the use of gene probes and immunological detection systems for *Campylobacter*, we can gain some insight into the future roles of each by comparing and contrasting the respective advantages and disadvantages of the technologies. Tompkins (129) has suggested that the two techniques have equal sensitivity and specificity but did not give results.

A. Immunological Detection Systems

Direct immunofluorescence with fluorescein isothiocyanate (FITC)-labeled IgG antibodies to *C. jejuni*, *C. coli*, and *C. fetus*, have been used to define serogroups of the three species (45). The polyclonal antibodies were raised in rabbits by inoculation of whole, formalinized organisms, and used to screen 316 strains of *C. jejuni*, 121 strains of *C. fetus*, 7 strains of *C. sputorum*, 4 strains of *C. fecalis*, 29 CLO and 256 other organisms representing 39 species from 21 other genera. The 14 conjugates produced defined 10 serogroups of *C. jejuni*, 2 serogroups of *C. coli*, and 2 serogroups of *C. fetus*, but were negative with the 256 bacteria from other genera. The morphology of the reactive antigen shared by the *Campylobacter* species showed much variety. The conjugates used to define serogroups of *C. jejuni, C. coli,* and *C. fetus* reacted with the flagella, but not the cells of the other *Campylobacter* species, suggesting genus-common flagella antigens. This fluorescence serotyping system was compared with the passive hemagglutination system (98) for typing strains of *C. jejuni* and *C. coli* (46). The results obtained suggested that the two test systems were measuring completely different antigenic fractions. The two methods were complementary, their combined use discriminating among strains more effectively than did either passive hemagglutination or direct immunofluorescence alone. Hodge *et al.* (49) used the direct-immunofluorescence method of Hebert for the detection of *C. jejuni* and *C. coli* in human fecal specimens. This test was almost (99.7%) as specific, but much less (40%) sensitive than culture of the stool specimens. As well as this low sensitivity, factors such as the subjectivity of the results and technical expertise needed to perform the direct-immunofluorescence assay make it not suited for screening large numbers of samples such as would be obtained in epidemiological or diagnostic surveys.

A latex slide agglutination test has been developed for *C. jejuni, C. coli, C. laridis,* and *C. fetus* subsp. *fetus* (50), but it still requires the organisms to be grown on primary plates of *Campylobacter*-selective media before testing. In addition, it has been shown that flagella protein is the key antigenic determinant involved in slide agglutination (136), whereas exposed outer-membrane protein or lipopolysaccharide is likely to be the major antigenic component involved in direct immunofluorescence. This is probably why there has been little agreement in results obtained by agglutination tests and direct immunofluorescence (49).

One way to overcome the problem of different antigenic fractions reacting in different tests, is to select for a defined antigenic fraction. Klipstein and colleagues (58,59) developed an ELISA using the Ig fraction of rabbit antiserum to formalinized whole cells of an invasive strain of *C. jejuni* absorbed with a noninvasive strain, as the first antibody. This has two advantages. The ELISA technology is sensitive and specific, and can be used to test large numbers of specimens, so it is an improvement over both agglutination and immunofluorescence tests. In addition, this specific ELISA could be used to differentiate invasive from noninvasive *C. jejuni*. However, it still suffered from the main disadvantage of having to culture the organisms before testing, and it still used polyclonal antisera.

The problem with the use of polyclonal antisera is that different animals of the same species may produce different antibodies to the same antigen (17), and this has led to the widespread use of monoclonal antibodies (MAb) in diagnosis (96). Several groups have produced MAb to *Campylobacter* species.

Kosunen *et al.* (61,61a) used MAb to *C. jejuni* in an ELISA to devise a serotyping scheme for *Campylobacter* species. No cross-reactions with unrelated bacteria were detected, but several MAb reacted in the ELISA with many of the 24 *Campylobacter* strains (*C. coli, C. jejuni, C. fetus*) studied, and one MAb reacted with them all. Nachamkin and Hart (86) produced MAb to *C. jejuni* flagella and found two distinct epitopes that define a common genus-specific determinant and a determinant restricted to *C. jejuni* and *C. coli*. These epitopes were additional to the serotype-specific epitope detected on the flagellin found by Wenman *et al.* (136). Newell (88) also produced MAb to *C. jejuni* flagella. One of these was found to react with all other *Campylobacter* species (*C. jejuni, C. coli, C. laridis, C. fetus,* and *C. pylori*) tested, except *C. sputorum* biovar bubulus. This MAb was then used to develop an experimental antigen-capture ELISA for flagella antigen in an emulsion of human feces. The use of this test in an epidemiological or diagnostic survey should establish the value, or otherwise, of the use of MAb and ELISA for the rapid detection of enteric *Campylobacter* species.

Monoclonal antibodies have also been produced against *C. pylori* (25,26,39). These MAb did not react in a direct-immunofluorescence test against *C. jejuni, C. coli,* or a range of unrelated bacteria. They could be used for the immunohistochemical localization of *C. pylori* in paraffin-embedded, cryostat, or smeared gastric biopsy specimens, using indirect immunofluorescence, the peroxidase–antiperoxidase technique, and the alkaline phosphatase–anti-alkaline phosphatase method. These new MAb tests gave good correlation with culture and were quicker than both culture and the urea hydrolysis test.

Clearly, antibody probes for *Campylobacter* have evolved from polyclonal specificity to monoclonal specificity, just as they have for hundreds of other organisms. However, MAb do have disadvantages (139), and gene probe technology is rapidly providing alternatives or replacing many of the immunological detection systems (138).

B. Gene Probes

Gene probes have many advantages. The genome of an organism is unlikely to change and give a false negative result in a similar manner to that likely if the antigenic structure of the organism changes and a MAb to the lost epitope is used to probe for it.

Nucleic acid probes are robust and are therefore less likely to be destroyed than the proteins they encode and that form the target of immunological detection systems. This is particularly important in the design of diagnostic probe systems, where interfering substances may have to be extracted from clinical specimens by pH changes or temperature increases, and where a specific antigen, but not a specific NA sequence, would also be destroyed. In fact, this is an added advantage with gene probes, because often the NA is denatured before dot blotting onto a membrane, by using low pH or heat.

The chemical bonding involved in nucleotide base-pairing reactions is strong and precise, and is therefore more likely to be much more specific than the reaction between an antigen and an antibody (138).

The hybridization between two single-stranded NA chains is rapid and under appropriate test conditions is effectively irreversible, promising shorter reaction times and ease of handling.

There are also likely to be cost savings in the use of gene probe technology (127), although this will obviously depend on the type of diagnostic laboratory, the number of tests requested, and the workflow practices used.

Perhaps the biggest disadvantage of gene probe detection systems in the past has been the requirement for the use of a radioactive label to increase their sensitivity. However, the recent use of nonradioactive diagnostic detection systems such as that described here reduces the reliance on radioactive labels. Some nonradioactive labels give results comparable in sensitivity to those obtained with radioactive gene probes and will hasten the already increasing use of gene probes into diagnostic laboratories. We certainly still appear to be on the crest of a wave of enthusiasm for this new gene probe technology. However, the enthusiasm for gene probe detection systems will probably follow the life cycle of any new diagnostic assay system, as described by Yolken (138), where high levels of enthusiasm are followed by disappointment, with reality eventually ending up somewhere between these two earlier levels.

Gene probe detection systems are likely to have their biggest impact on organisms that are difficult or slow to grow. *Mycobacteria, Chlamydia,* and *Legionella* are obvious examples, and we believe that in the future gene probes for *Campylobacter* will also have wide diagnostic usage.

VI. PROSPECTS FOR THE FUTURE

The current upsurge in interest in *Campylobacter* species, CLO, and similar spiral-shaped bacteria is likely to continue for some time, because to a large extent they have been a neglected group of bacteria with unrealized importance in human infections. Methods will be required for their detection, identification, and characterization. We have seen the success of dot-blot hybridization techniques for the differentiation of some species of *Campylobacter*. These methods are eminently suited to the characterization of unusual, putative campylobacters such as the unclassified microaerophilic bacterium associated with human gastroenteritis, reported by Archer *et al.* (2).

It can be confidently predicted that nonradioactive labels without the deficiencies of nonspecific background reactions (biotin), low sensitivity (photobiotin), or carcinogenicity (AAF) will have the same sensitivity as comparatively labeled radioactive probes brought into routine use. The sulfonated-DNA probe and immunological detection system may be the forerunner of such methods; undoubtedly many improvements will be forthcoming. Such probes, used in conjunction with a standardized dot-blot hybridization format, could provide a relatively simple, rapid, and safe means of identifying species of bacteria that are difficult to characterize biochemically, such as the campylobacters.

In this chapter we have described some of the NA probes that have been used for the detection and identification of species and subspecies of *Campylobacter*. However, perhaps probes directed against putative virulence determinants will be even more useful. The importance of identifying potentially pathogenic strains rather than all strains of these species is conjectural at present. Three potentially pathogenic properties have been described for *C. jejuni:* invasiveness and the production of enterotoxin and cytotoxin. Although the pathogenic role of the cytotoxin is uncertain, Klipstein *et al.* (50) reported that enterotoxin production or invasiveness or both could be found in many *C. jejuni* fecal isolates, and that there was a relationship between the clinical symptoms and the infecting strain exhibiting pathogenic properties. In addition, their findings suggested that some isolates may not be pathogenic.

Similarly, recent evidence suggests that not all isolates of *C. pylori* are equally virulent (12), that only a proportion of strains may produce a cytotoxin (70), and that they differ in their ability to cause hemolysis of erythrocytes (137a). To what extent this situation occurs in other strains of *Campylobacter* and CLO implicated in human infections is unknown, and awaits further studies. Perhaps it will be necessary to develop *Campylobacter* gene probes to the above virulence determinants, analogous to the variety of probes for the pathogenic properties of diarrheogenic strains of *E. coli*. In the immediate future such developments will be the province of research laboratories and will be directed toward the characterization of strains after they have been cultured.

However, developments on the practical application of *Campylobacter* gene probes should be directed toward the elimination of routine cultures for reasons previously outlined in this chapter. Detection of bacterial nucleic acid in complex clinical specimens such as feces and sputum is notoriously difficult because of the presence of interfering substances and low sensitivity. One of the first research priorities should be the development of better ways to treat such clinical specimens and at the same time eliminate the complex extraction procedures currently in use for the preparation of bacterial DNA. Given that this development will be successful, it is likely that improvements in labels, labeling techniques, and hybridization formats will result in rapid, easy to perform, and more sensitive assays. It is evident that this is already occurring. The in-solution hybridization assay using DNA probes complementary to 16 S rRNA of *Campylobacter* and the immunological capture of NA hybrids, described by Rashtchian *et al.* (102) (see Section III,A), was reported to be simple to perform and rapid (2.5 hr). This approach is well worth exploring. It should be possible to construct genus-specific oligonucleotide probes complementary to 16 S rRNA sequences common to all species of *Campylobacter* and CLO given that much of this information is already available (128). Such probes could be used as the "first-line" test on clinical specimens to detect the presence of any strains of this group of organisms. However, to provide accurate information for clinical management and epidemiological studies, secondary probes will need to be constructed from unique sequences in the genome of each species. Whether this will be possible and whether the specimen should be probed further for virulence determinants of the infecting strain using defined DNA sequence probes remain to be determined.

Phylogenetically based hybridization probes have proved valuable for the monitoring of groups of bacteria in their natural setting without culturing (120). In addition, Rollins and Colwell (108) have described a viable but nonculturable stage of *C. jejuni* capable of survival in the environment. The detection of this form by using an appropriate gene probe assay may provide valuable epidemiological information during investigations of outbreaks of *Campylobacter* infection in which organisms cannot be cultured from the suspected source. It is likely that similar forms of other campylobacters and CLO (including *C. pylori*) also exist, and it will be interesting to see if probes to rRNA can be used for their detection.

The problem of low detection sensitivity in clinical specimens may, in part, be solved by future developments of recently described techniques. Examples of such techniques include the polymerase chain reaction (PCR) to multiply the number of copies of target DNA (93), exponential amplification of probe sequences (72), and methods to amplify the signal-generating capacity of the probe system by concentrating more label at the site of the target molecule or by designing each label to produce a stronger signal (27). We envisage that these developments will revolutionize infectious disease diagnosis.

VII. SUMMARY

The recent description of several new species of *Campylobacter* associated with human disease has stimulated interest in this group of organisms. Conventional techniques for the isolation and detection of both the established species (e.g., *C. jejuni, C. coli, C. fetus* subsp. *fetus*) and new species (e.g., "*C. cinaedi,*" "*C. fennelliae,*" "*C. upsaliensis,*" *C. pylori*) have several disadvantages. In addition, even after isolation these organisms are difficult to identify to species level using conventional biochemical techniques. Gene probe assays have proved invaluable for the identification and differentiation of campylobacters and CLO, and have been of assistance in clarifying difficulties associated with their nomenclature and classification. In contrast, probe assays for the direct detection of enteropathogenic campylobacters in clinical material have not been successful and have not been attempted for CLO detection. However, gene probe technology is evolving rapidly, and it can be predicted that many of the problems presently encountered with the detection of campylobacters will be overcome. *Campylobacter* gene probes will always be valuable in the research laboratory and for generating epidemiological data, where studies are done on isolated organisms, but whether they will replace routine cultures of clinical material remains to be determined.

VIII. MATERIALS AND METHODS

A. Bacterial Strains

The *C. pylori* strain (LL) used for the preparation of probes was isolated from the antral biopsy sample of a patient undergoing gastrointestinal endoscopy. It was identified as *C. pylori* on the basis of typical microscopic and colonial morphology, strong positive reactions for oxidase, catalase, and urease, negative reactions for nitrate reduction and hippurate hydrolysis, no growth in the presence of 1% glycine and 3.5% sodium chloride, no growth at 25°C, resistance to nalidixic acid, and sensitivity to cephalothin. Apart from the UPTC strain, the other organisms used were reference cultures (see Fig. 2). The *C. pylori* strains were grown on chocolate agar plates in microaerophilic conditions (GasPak H_2 and CO_2-generating envelope, in a GasPak jar without catalyst, BBL Microbiology Systems, Cockeysville, Maryland) at 35°C for 4 days. The other *Campylobacter* species were cultured on blood agar plates under the same conditions for 2–5 days. The *W. succinogenes* strain was grown in an anaerobic atmosphere for 2 days.

B. Preparation of DNA

In order to obtain sufficient DNA, it was necessary to grow *C. pylori* LL on 50 chocolate plates. The bacteria were removed from the plates with cotton-tipped swabs and suspended in phosphate-buffered saline (PBS), pH 7.2. The suspension was centrifuged (3300 *g* for 10 min) and the pellet was suspended in 20 ml of 10 m*M* Tris-HCl (pH 8.5)–10 m*M* EDTA. The pellet was washed once and resuspended in 5 ml of this buffer. DNA was prepared essentially as described by Langenberg *et al.* (66). Briefly, lysozyme (egg white grade 1, Sigma Chemical Co., St. Louis, Missouri) was added to give a final concentration of 1 mg/ml and the solution was incubated at 37°C for 30 min. The bacteria were lysed with 1% SDS, and the lysate was incubated with 200 μg/ml of DNase-free RNase A (bovine pancreas type X11-A; Sigma) at 37°C for 1 hr. Proteinase K (Boehringer-Mannheim, Mannheim Federal Republic of Germany) was added to give a final concentration of 200 μg/ml, and the mixture was incubated at 50°C for 3 hr. The nucleic acid was extracted with an equal volume of phenol–chloroform in the presence of 1 *M* sodium perchlorate. A total of three further extractions were done with this mixture. After two extractions with chloroform, the DNA was ethanol-precipitated and resuspended in TE (10 m*M* Tris-HCl–1 m*M* EDTA, pH 7.5). The DNA concentration was determined by the diphenylamine assay, as described by Giles and Meyers (34). It was considered essentially free of contaminating proteins because the absorbance ratios 230/260 and 280/260 were 0.44 and 0.49, respectively (94).

C. Labeling of DNA

Genomic DNA was labeled by nick translation (BRL kit; Bethesda Research Laboratories, Gaithersburg, Maryland) with [α-^{32}P]dCTP [Biotechnology Research Enterprises S.A. (BRESA), Adelaide, Australia] After labeling, unincorporated nucleotides were removed by Sephadex G-50 chromatography, with 100 m*M* NaCl–20 m*M* Tris-HCl–10 m*M* EDTA buffer (pH 7.5) used for washing. The labeled DNA probes had specific activities of 4×10^7 cpm/μg. Biotin-labeled probes were prepared with Biotin-11-dUTP (BRL) and by using the BRL nick-translation kit. Unincorporated biotin was removed by ethanol precipitation and the labeled DNA was resuspended in TE. For labeling with photobin, DNA was prepared in distilled H_2O. Photobiotin acetate (BRESA) was then incorporated as recommended by the manufacturer and as described by Forster *et al.* (30). Sulfonated DNA probes were prepared by inserting an antigenic sulfone group into cytosine moieties of denatured DNA, using the Chemiprobe kit (Orgenics, Ltd., Yavne, Israel).

D. Preparation of Target DNA

1. *Blots for Sensitivity*

Dilutions of *C. pylori* LL DNA ranging from 1 μg to 1 pg were prepared in TE, heat-denatured at 100°C for 10 min, and 10-μl amounts were dotted onto NC (Bio-Rad Laboratories, Richmond, California) and nylon membranes (Pall Biodyne A, Pall Ultrafine Filtration Corp., Glen Cove, New York) by using the Bio-Dot microfiltration apparatus (Bio-Rad). Nitrocellulose membranes were presoaked in 20× SSC (1× SSC contains 0.15 *M* NaCl in 0.015 *M* sodium citrate pH 7.0) and nylon membranes in 5× SSC. After blotting, DNA on NC was washed with 20× SSC and DNA on nylon was washed with 5× SSC. The filters were air-dried at 22°C and baked *in vacuo* at 80°C for 2 hr.

Blots of whole bacteria were prepared using *C. pylori* LL, which was grown on chocolate agar. Colonies were suspended in brain heart infusion broth (Oxoid) containing 1% vitox (Oxoid), 1% glucose, 0.25% yeast extract and 10% inactivated horse serum to give the density of a McFarland standard no. 7. The total number of organisms was determined by using a Helber counting chamber, and the number of viable organisms was determined by diluting the suspension and spreading portions on chocolate agar plates, which were incubated in microaerophilic conditions. Serial dilutions of this suspension were made and 2.5-μl amounts were dotted onto nylon membranes using the Bio-Dot apparatus. After drying in air for 30 min, the membranes were treated sequentially with 0.5 *N* NaOH–1.5 *M* NaCl, proteinase K, and 3 *M* Tris-HCl (pH 8.0) as described later.

2. *Blots for Specificity*

Organisms were grown on chocolate or blood agar and incubated under appropriate atmospheric conditions. The growth from each was harvested in PBS to a density of a McFarland standard no. 7. The concentrations were checked using a Helber counting chamber and adjusted if necessary to give a uniform count of 2–4 × 10^9 bacteria per milliliter. Then, 5-μl amounts were spotted onto nylon membranes with the Bio-Dot apparatus to give a final concentration of 1–2 × 10^7 bacteria per dot. Membranes were left in air at 22°C before the bacteria were lysed and their DNA denatured. Membranes were initially floated on the surface of a hot (60°C) solution containing 0.5 *N* NaOH–1.5 *M* NaCl for 30 sec, removed, and incubated in the same solution at 22°C for 10 min. The membranes were then transferred to a solution of 3 *M* Tris-HCl (pH 8.0) and incubated at 22°C for 10 min. After drying in air for 15 min, membranes were treated with a solution of proteinase K at 200 μg/ml in 3 *M* Tris-HCl (pH 8.0) at 37°C for 30 min. The membranes were then dried in air for 15 min, treated with 3 *M* Tris-HCl (pH 8.0) for 10 min at 22°C, and baked for 1 hr *in vacuo* at 80°C.

E. Hybridization and Detection Conditions

1. ^{32}P-Labeled Probes

The strips were prehybridized for 6 hr at 65°C in hybridization solution containing 6× SSC, 5× Denhardt's solution (1× Denhardt's solution is 0.02% polyvinylpyrrolidone–0.02% Ficoll–0.02% bovine serum albumin fraction V), 20 m*M* $NaPO_4$ (pH 6.5), 0.1% SDS, 10% dextran sulfate, and 10 μg/ml denatured salmon testes DNA. The probe was denatured by heating at 100°C for 10 min and added to fresh hybridization buffer. The radioactivity counts for all probes used were 6–8 × 10^6 cpm/ml during hybridization. After 16 hr hybridization at 65°C, the membranes were washed at 50°C for 3 hr in four changes of 0.2× SSC–0.1% SDS, dried in air, and exposed to X-ray film between intensifying screens at −70°C for 20 hr. The film was developed with an X-ray film developer and examined for specific hybridization to target DNA.

2. Biotinylated Probes

Prehybridizations and hybridizations with biotin–DNA probes were done according to the manufacturer's protocol for the BluGene nonradioactive NA detection system (BRL), except that formamide was omitted from the solutions. Prehybridizations were done at 65°C and the hybridization temperature was 61°C (69). Probes were denatured at 100°C for 10 min and used at a concentration of 200 ng/ml. Hybridization was for 16 hr. Strips were washed twice with 2× SSC–0.1% SDS at 22°C for 10 min, twice more in 0.2× SSC–0.1% SDS at 22°C for 10 min, and twice more in 0.16× SSC–0.1% SDS at 65°C for 20 min. After a final wash in 2× SSC at 22°C for 5 min, hybridization reactions were detected and visualized using a streptavidin–alkaline phosphatase conjugate and a chromogenic substrate containing nitroblue tetrazolium and 5-bromo-4-chloro-3-indolyl phosphate (NBT/BCIP) supplied with the kit. The staining reaction was stopped after 4 hr, ensuring maximum color development. Hybridizations with photobiotin–DNA probes were done as described for biotin–DNA probes, including the temperature of 61°C used for hybridization. The probes were used at a final concentration of 25 ng/ml in the hybridization solution. Photobiotinylated DNA hybridized to target DNA was detected using an avidin–alkaline phosphatase conjugate and a substrate solution NBT/BCIP supplied with the BRESA kit. Color development was stopped after 4 hr.

3. *Sulfonated Probes*

Membranes were prehybridized at 65°C for 4 hr in hybridization solution containing 1 *M* NaCl, 1% SDS, 10% dextran sulfate, and 250 μg/ml denatured salmon testes DNA. The probe was denatured by heating at 100°C for 10 min and added to fresh hybridization solution (without salmon testes DNA). The concentration of the probe in the hybridization buffer was 1 μg/ml, and hybridization was for 16 hr at 65°C. Posthybridization washes consisted of two washes for 10 min each at 22°C in 2× SSC, followed by a 30-min wash at 65°C in 2× SSC–1% SDS, and a 20-min wash at 65°C in 0.1× SSC–1% SDS.

Immunological detection and visualization of the hybridized probe was done using reagents supplied in the Chemiprobe kit. Briefly, membranes were treated with a blocking solution containing 25% skimmed milk and 3.5 mg/ml of heparin. They were then incubated with a MAb to modified DNA. After washing, the membranes were incubated with an anti-species immunoglobulin–alkaline phosphatase conjugate. Following a further series of washes, the membranes were exposed to a substrate solution (NBT/BCIP) at 37°C for 1 hr.

ACKNOWLEDGMENTS

We thank S. D. Neill, Veterinary Research Laboratories, Belfast, Ireland; F. J. Bolton, Public Health Laboratory Service, Preston, England; C. S. Goodwin, Royal Perth Hospital, Perth; and B. Winter, Institute of Medical and Veterinary Science, Adelaide, for providing some of the bacterial strains. S. Evans typed this manuscript. This work was partially supported by a grant from the National Health and Medical Research Council of Australia.

REFERENCES

1. Ambinder, R. F., Charache, P., Staal, S., Wright, P., Forman, M., Hayward, S. D., and Hayward, G. S. (1986). The vector homology problem in diagnostic nucleic acid hybridization of clinical specimens. *J. Clin. Microbiol.* **24,** 16–20.
2. Archer, J. R., Romero, S., Ritchie, A. L., Hamacher, M. E., Steiner, B. M., Bryner, J. H., and Schell, R. F. (1988). Characterization of an unclassified microaerophilic bacterium associated with gastroenteritis. *J. Clin Microbiol.* **26,** 101–105.
3. Benjamin, J., Leaper, S., Owen, R. J., and Skirrow, M. B. (1983). Description of *Campylobacter laridis,* a new species comprising the nalidixic acid-resistant thermophilic *Campylobacter* (NARTC) group. *Curr. Microbiol.* **8,** 231–238.
4. Bialkowska-Hobrzanska, H. (1987). Detection of enterotoxigenic *Escherichia coli* by dot-blot hybridization with biotinylated DNA probes. *J. Clin. Microbiol.* **25,** 338–343.
5. Blaser, M. J., Barkowitz, I. D., LaForce, F. M., Cravens, J., Reller, L. B., and Wang, W.-L. L. (1979). *Campylobacter* enteritis: Clinical and epidemiological features. *Ann. Intern. Med.* **91,** 179–185.

6. Blaser, M. J., Wells, J. G., Feldman, R. A., Pollard, R. A., Allen, J. R., and The Collaborative Diarrheal Disease Study Group (1983). *Campylobacter* enteritis in the United States. A multicentre study. *Ann. Intern. Med.* **98,** 360–365.
7. Bolton, F. J., Holt, A. V., and Hutchinson, D. N. (1985). Urease-positive thermophilic campylobacters. *Lancet* **1,** 1217–1218.
8. Bolton, F. J., Hutchinson, D. N., and Parker, G. (1988). Reassessment of selective agars and filtration techniques for isolation of *Campylobacter* species from faeces. *Eur. J. Clin. Microbiol. Infect. Dis.* **7,** 155–160.
9. Borromeo, M., Lambert, J. R., and Pinkard, K. J. (1987). Evaluation of "CLO-test" to detect *Campylobacter pyloridis* in gastric mucosa. *J. Clin. Pathol.* **40,** 462–468.
10. Bryan, R. N., Ruth, J. L., Smith, R. D., and Le Bon, J. M. (1986). Diagnosis of clinical samples with synthetic oligonucleotide hybridization probes. *In* "Microbiology—1986" (L. Leive, ed.), pp. 113–116. Am. Soc. Microbiol., Washington, D.C.
11. Buck, G. E., Gourley, W. K., Lee, W. K., Subramanyam, K., Latimer, J. M., and Di Nuzzo, A. R. (1986). Relation of *Campylobacter pyloridis* to gastritis and peptic ulcer. *J. Infect. Dis.* **153,** 664–669.
12. Burnie, J. P., Lee, W., Dent, J. C., and McNulty, C. A. M. (1988). Immunoblot fingerprinting of *Campylobacter pylori*. *J. Med. Microbiol.* **27,** 153–159.
13. Butzler, J.-P., Dekeyser, P., Detroin, M., and Dehaen, F. (1973). Related vibrio in stools. *J. Pediatr.* **82,** 493–495.
14. Chevrier, D., Megraud, F., Larzul, D., and Guesdon, J.-L. (1988). A new method for identifying *Campylobacter* spp. *J. Infect. Dis.* **157,** 1097–1098.
15. Cimolai, N., Gill, M. J., Jones, A., Flores, B., Stamm, W. E., Laurie, W., Madden, B., and Shahrabadi, M. S. (1987). "*Campylobacter cinaedi*" bacteremia: Case report and laboratory findings. *J. Clin. Microbiol.* **25,** 942–943.
16. Costas, M., Owen, R. J., and Jackman, P. J. M. (1987). Classification of *Campylobacter sputorum* and allied campylobacters based on numerical analysis of electrophoretic protein patterns. *Syst. Appl. Microbiol.* **9,** 125–131.
17. Crowle, A. S. (1973). "Immunodiffusion," Chapter 2. Academic Press, New York.
18. Das, S. S., Bain, L. A., Karim, Q. N., Coelho, L. G., and Baron, J. H. (1987). Rapid diagnosis of *Campylobacter pyloridis* infection. *J. Clin. Pathol.* **40,** 701–702.
19. Donovan, R. M., Bush, C. E., Peterson, W. R., Parker, L. H., Cohen, S. H., Jordan, G. W., Vanden Brink, K. M., and Goldstein, E. (1987). Comparison of non-radioactive DNA hybridization probes to detect human immunodeficiency virus nucleic acid. *Mol. Cell. Probes* **1,** 359–366.
20. Drumm, B., O'Brien, A., Cutz, E., and Sherman, P. (1987). *Campylobacter pyloridis*—associated primary gastritis in children. *Pediatrics* **80,** 192–195.
21. Dutilh, B., Bebear, C., Taylor-Robinson, D., and Grimont, P. A. D. (1988). Detection of *Chlamydia trachomatis* by *in situ* hybridization with sulphonated total DNA. *Ann. Inst. Pasteur/Microbiol.* **139,** 115–128.
22. Eastham, E. J., Elliott, T. S. J., Berkeley, D., and Jones, D. M. (1988). *Campylobacter pylori* infection in children. *J. Infect. Dis.* **16,** 77–79.
23. Edmonds, P., Patton, C. M., Barrett, T. J., Morris, G. K., Steigerwalt, A. G., and Brenner, D. J. (1985). Biochemical and genetic characteristics of atypical *Campylobacter fetus* subsp. *fetus* strains isolated from humans in the United States. *J. Clin. Microbiol.* **21,** 936–940.
24. Edmonds, P., Patton, C. M., Griffin, P. M., Barrett, T. J., Schmidt, G. P., Baker, C. N., Lambert, M. A., and Brenner, D. J. (1987). *Campylobacter hyointestinalis* associated with human gastrointestinal disease in the United States. *J. Clin. Microbiol.* **25,** 685–691.

25. Engstrand, L., Pahlson, C., Gustavsson, S., and Schwan, A. (1986). Monoclonal antibodies for rapid identification of *Campylobacter pyloridis*. *Lancet* **2,** 1402–1403.
26. Engstrand, L., Pahlson, C., Schwan, A., and Gustavsson, S. (1988). Monoclonal antibodies against *Campylobacter pylori*. *In* "*Campylobacter pylori*" (H. Menge, M. Gregor, G. N. J. Tytgat, and B. J. Marshall, eds.), pp. 121–126. Springer-Verlag, Berlin.
27. Fahrlander, P. D., and Klausner, A. (1988). Amplifying DNA probe signals: A "Christmas tree" approach. *Bio/Technology* **6,** 1165–1168.
28. Fennell, C. L., Rompalo, A. M., Totten, P. A., Bruch, K. L., Flores, B. M., and Stamm, W. E. (1986). Isolation of *Campylobacter hyointestinalis* from a human. *J. Clin. Microbiol.* **24,** 146–148.
29. Fennell, C. L., Totten, P. A., Quinn, T. C., Patton, D. L., Holmes, K. K., and Stamm, W. E. (1984). Characterization of *Campylobacter*-like organisms isolated from homosexual men. *J. Infect. Dis.* **149,** 58–66.
30. Forster, A. C., McInnes, J. L., Skingle, D. C., and Symons, R. H. (1985). Non-radioactive hybridization probes prepared by the chemical labelling of DNA and RNA with a novel reagent, photobiotin. *Nucleic Acids Res.* **13,** 745–761.
31. Fox, J. G., Edrise, B. M., Cabot, E. B., Beaucage, C., Murphy, J. C., and Prostok, K. S. (1986). *Campylobacter*-like organisms isolated from gastric mucosa of ferrets. *Am. J. Vet. Res.* **47,** 236–239.
32. Gebhart, C. J., Edmonds, P., Ward, G. E., Kurtz, H. J., and Brenner, D. J. (1985). "*Campylobacter hyointestinalis*" sp.nov.: A new species of *Campylobacter* found in the intestines of pigs and other animals. *J. Clin. Microbiol.* **21,** 715–720.
33. Gebhart, C. J., Ward, G. E., Chang, K., and Kurtz, H. J. (1983). *Campylobacter hyointestinalis* (new species) isolated from swine with lesions of proliferative ileitis. *Am. J. Vet. Res.* **44,** 361–367.
34. Giles, K. W., and Meyers, A. (1965). An improved diphenylamine method for the estimation of deoxyribonucleic acid. *Nature (London)* **206,** 93.
35. Goodwin, C. S., Armstrong, J. A., and Marshall, B. J. (1986). *Campylobacter pyloridis*, gastritis, and peptic ulceration. *J. Clin. Pathol.* **39,** 353–365.
36. Goodwin, C. S., Blincow, E. D., Warren, J. R., Waters, T. E., Sanderson, C. R., and Easton, L. (1985). Evaluation of cultural techniques for isolating *Campylobacter pyloridis* from endoscopic biopsies of gastric mucosa. *J. Clin. Pathol.* **38,** 1127–1131.
37. Goodwin, C. S., McCulloch, R. K., Armstrong, J. A., and Wee, S. H. (1985). Unusual cellular fatty acids and distinctive ultrastructure in a new spiral bacterium (*Campylobacter pyloridis*) from the human gastric mucosa. *J. Med. Microbiol.* **19,** 257–267.
38. Goossens, H., De Boeck, M., Van Landuyt, H., and Butzler, J.-P. (1984). Isolation of *Campylobacter jejuni* from human feces. *In* "*Campylobacter* Infection in Man and Animals" (J.-P. Butzler, ed.), pp. 39–50. CRC Press, Boca Raton, Florida.
39. Gregor, M., Warrelman, M., Nieder, A., Menge, H., Hahn, H., and Riecken, E. O. (1988). Production and characterization of monoclonal antibodies against *Campylobacter pylori*. *In* "*Campylobacter pylori*" (H. Menge, M. Gregor, G. N. J. Tytgat, and B. J. Marshall, eds.), pp. 127–132. Springer-Verlag, Berlin.
40. Hartmann, D., and von Graevenitz, A. (1987). A note on name, viability and urease tests of *Campylobacter pylori*. *Eur. J. Clin. Microbiol.* **6,** 82–83.
41. Harvey, S. M., and Greenwood, J. R. (1983). Relationships among catalase-positive campylobacters determined by deoxyribonucleic acid–deoxyribonucleic acid hybridization. *Int. J. Syst. Bacteriol.* **33,** 275–284.
42. Harvey, S. M., and Greenwood, J. R. (1983). Probable *Campylobacter fetus* subsp. *fetus* gastroenteritis. *J. Clin. Microbiol.* **18,** 1278–1279.
43. Hebert, G. A., Edmonds. P., and Brenner, D. J. (1984). DNA relatedness among strains

of *Campylobacter jejuni* and *Campylobacter coli* with divergent serogroup and hippurate reactions. *J. Clin. Microbiol.* **20,** 138–140.
44. Hebert, G. A., Hollis, D. G., Weaver, R. E., Lambert, M. A., Blaser, M. J., and Moss, C. W. (1982). 30 years of campylobacters: Biochemical characteristics and a biotyping proposal for *Campylobacter jejuni. J. Clin. Microbiol.* **15,** 1065–1073.
45. Hebert, G. A., Hollis, D. G., Weaver, R. E., Steigerwalt, A. G., McKinney, R. M., and Brenner, D. J. (1983). Serogroups of *Campylobacter jejuni, Campylobacter coli,* and *Campylobacter fetus* defined by direct immunofluorescence. *J. Clin. Microbiol.* **17,** 529–538.
46. Hebert, G. A., Penner, J. L., Hennessy, J. H., and McKinny, R. M. (1983). Correlation of an expanded passive hemagglutination system for serogrouping strains of *Campylobacter jejuni* and *Campylobacter coli. J. Clin. Microbiol.* **18,** 1064–1069.
47. Hill, I. D., Sinclair-Smith, C., Lastovica, A. J., Bowie, M. D., and Emms, M. (1987). Transient protein-losing enteropathy associated with acute gastritis and *Campylobacter pylori. Arch. Dis. Child.* **62,** 1215–1219.
48. Hirschl, A., Potzi, R., Stanek, G., Wende, L., Rotter, M., Gangl, A., and Holzner, J. H. (1986). Occurrence of *Campylobacter pyloridis* in patients from Vienna with gastritis and peptic ulcers. *Infection* **14,** 275–278.
49. Hodge, D. S., Prescott, J. F., and Shewen, P. E. (1986). Direct immunofluorescence microscopy for rapid screening of *Campylobacter* enteritis. *J. Clin. Microbiol.* **24,** 863–865.
50. Hodinka, R. L., and Gilligan, P. H. (1988). Evaluation of the Campyslide agglutination test for confirmatory identification of selected *Campylobacter* species. *J. Clin. Microbiol.* **26,** 47–49.
51. Hyman, H. C., Yogev, D., and Razin, S. (1987). DNA probes for detection and identification of *Mycoplasma pneumoniae* and *Mycoplasma genitalium. J. Clin. Microbiol.* **25,** 726–728.
52. Hyypia, T. (1985). Detection of adenovirus in nasopharyngeal specimens by radioactive and nonradioactive DNA probes. *J. Clin. Microbiol.* **21,** 730–733.
53. Jones, D. M., Lessells, A. M., and Eldridge, J. (1984). *Campylobacter*-like organisms on the gastric mucosa: Culture, histological, and serological studies. *J. Clin. Pathol.* **37,** 1002–1006.
54. Karmali, M. A., and Fleming, P. C. (1979). *Campylobacter* enteritis in children. *J. Pediatr.* **94,** 527–533.
55. Karmali, M. A., and Skirrow, M. B. (1984). Taxonomy of the genus *Campylobacter. In* "*Campylobacter* Infection in Man and Animals" (J.-P. Butzler, ed.), pp. 1–20. CRC Press, Boca Raton, Florida.
56. Khan, A. M., and Wright, P. J. (1987). Detection of flavivirus RNA in infected cells using photobiotin-labelled hybridization probes. *J. Virol. Methods* **15,** 121–130.
57. King, E. O. (1957). Human infections with *Vibrio fetus* and a closely related vibrio. *J. Infect. Dis.* **101,** 119–128.
58. Klipstein, F. A., Engert, R. F., Short, H., and Schenk, E. A. (1985). Pathogenic properties of *Campylobacter jejuni:* Assay and correlation with clinical manifestations. *Infect. Immun.* **50,** 43–49.
59. Klipstein, F. A., Engert, R. F., and Short, H. B. (1986). Enzyme-linked immunosorbent assays for virulence properties of *Campylobacter jejuni* clinical isolates. *J. Clin. Microbiol.* **23,** 1039–1043.
60. Korolik, V., Coloe, P. J., and Krishnapillai, V. (1988). A specific DNA probe for the identification of *Campylobacter jejuni. J. Gen. Microbiol.* **134,** 521–529.
61. Kosunen, T. U., Bang, B. E., and Hurme, M. (1984). Analysis of *Campylobacter jejuni* antigens with monoclonal antibodies. *J. Clin. Microbiol.* **19,** 129–133.

61a. Kosunen, T. U., and Hurme, M. (1986). Monoclonal antibodies against *Campylobacter* strains. In "Monoclonal Antibodies against Bacteria" (A. J. L. Macario and E. Conway de Macario, eds.), Vol. 3, pp. 99–117. Academic Press, Inc., Orlando, Florida.
62. Krajden, S., Bohren, J., Anderson, J., Kempston, J., Fuksa, M., Matlow, A., Marcon, N., Haber, G., Kortan, P., Karmali, M., Corey, P., Petrea, C., Babida, C., and Haymen, S. (1987). Comparison of selective and nonselective media for recovery of *Campylobacter pylori* from antral biopsies. *J. Clin. Microbiol.* **525,** 1117–1118.
63. Kuritza, A. P., Getty, C. E., Shaughnessy, P., Hesse, R., and Salyers, A. A. (1986). DNA probes for identification of clinically important *Bacteroides* species. *J. Clin. Microbiol.* **23,** 343–349.
64. Lambert, M. A., Patton, C. M., Barrett, T. J., and Moss, C. W. (1987). Differentiation of *Campylobacter* and *Campylobacter*-like organisms by cellular fatty acid composition. *J. Clin. Microbiol.* **25,** 706–713.
65. Langenberg, M. L., Tytgat, G. N. J., Schipper, M. E. I., Rietra, P. J. G. M., and Zanen, H. C. (1984). *Campylobacter*-like organisms in the stomach of patients and healthy individuals. *Lancet* **1,** 1348.
66. Langenberg, W., Rauws, E. A. J., Widjojokusumo, A., Tytgat, G. N. J., and Zanen, H. C. (1986). Identification of *Campylobacter pyloridis* isolates by restriction endonuclease DNA analysis. *J. Clin. Microbiol.* **24,** 414–417.
67. Laughon, B. E., Vernon, A. A., Druckman, D. A., Fox, R., Quinn, T. C., Polk, B. F., and Bartlett, J. G. (1988). Recovery of *Campylobacter* species from homosexual men. *J. Infect. Dis.* **158,** 464–467.
68. Lawson, G. H. K., Rowland, A. C., and Wooding, P. (1975). The characterisation of *Campylobacter sputorum* subspecies *mucosalis* isolated from pigs. *Res. Vet. Sci.* **18,** 121–126.
69. Leary, J. J., Brigati, D. J., and Ward, D. C. (1983). Rapid and sensitive colorimetric method for visualizing biotin-labelled DNA probes hybridized to DNA or RNA immobilized on nitrocellulose: Bio-blots. *Proc. Natl. Acad. Sci. U.S.A.* **80,** 4045–4049.
70. Leunk, R. D., Johnson, P. T., David, B. C., Kraft, W. C., and Morgan, D. R. (1988). Cytotoxic activity in broth-culture filtrates of *Campylobacter pylori*. *J. Med. Microbiol.* **26,** 93–99.
71. Li, P., Medon, P. P., Skingle, D. C., Lanser, J. A., and Symons, R. H. (1987). Enzyme-linked synthetic oligonucleotide probes: Non-radioactive detection of enterotoxigenic *Escherichia coli* in faecal specimens. *Nucleic Acids Res.* **15,** 5275–5278.
72. Lizardi, P. M., Guerra, C. E., Lomeli, H., Tussie-Luna, I., and Kramer, F. R. (1988). Exponential amplification of recombinant-RNA hybridization probes. *Bio/Technology* **6,** 1197–1202.
73. Majewski, S. I. H., and Goodwin, C. S. (1988). Restriction endonuclease analysis of the genome of *Campylobacter pylori* with a rapid extraction method: Evidence for considerable genomic variation. *J. Infect. Dis.* **157,** 465–471.
74. Marshall, B. J., Royce, H., Annear, D. I., Goodwin, C. S., Pearman, J. W., Warren, J. R., and Armstrong, J. A. (1984). Original isolation of *Campylobacter pyloridis* from human gastric mucosa. *Microbios Lett.* **25,** 83–88.
75. Marshall, B. J., and Warren, J. R. (1984). Unidentified curved bacilli in the stomach of patients with gastritis and peptic ulceration. *Lancet* **1,** 1311–1315.
76. McClung, C. R., Patriquin, D. G., and Davis, R. E. (1983). *Campylobacter nitrofigilis* sp. nov., a nitrogen-fixing bacterium associated with roots of *Spartina alterniflora* loisel. *Int. J. Syst. Bacteriol.* **33,** 605–612.
77. McLafferty, M. A., Harcus, D. R., Weiss, A. A., Sopian, L. A., and Hewlett, E. L.

(1986). Development of a DNA probe for identification of *Bordetella pertussis*. *Am. Soc. Microbiol., Abstr. Annu. Meet.* Abstr. C168, p. 356.
78. McLaughlin, G. L., Collins, W. E., and Campbell, G. H. (1987). Comparison of genomic, plasmid, synthetic and combined DNA probes for detecting *Plasmodium falciparum* DNA. *J. Clin. Microbiol.* **25,** 791–795.
79. McLaughlin, G. L., Edline, T. D., and Ihler, G. M. (1986). Detection of *Babesia bovis* using DNA hybridization. *J. Protozool.* **33,** 125–128.
80. McNulty, C. A. M., and Dent, J. C. (1987). Rapid identification of *Campylobacter pylori* (*C. pyloridis*) by preformed enzymes. *J. Clin. Microbiol.* **25,** 1683–1686.
81. McNulty, C. A. M., and Wise, R. (1985). Rapid diagnosis of *Campylobacter*-associated gastritis. *Lancet* **1,** 1443.
82. Megraud, F., Bonnet, F., Garnier, M., and Lamouliatte, H. (1985). Characterization of "*Campylobacter pyloridis*" by culture, enzymatic profile and protein content. *J. Clin. Microbiol.* **22,** 1007–1010.
83. Megraud, F., Chevrier, D., Desplaces, N., Sedallian, A., and Guesdon, J. L. (1988). Urease-positive thermophilic campylobacter (*Campylobacter laridis* variant) isolated from an appendix and from human feces. *J. Clin. Microbiol.* **26,** 1050–1051.
84. Morris, G. K., and Patton, C. M. (1985). *Campylobacter. In* "Manual of Clinical Microbiology" (E. H. Lennette, A. Balows, W. J. Hausler, Jr., and M J. Shadomy, eds.), pp 302–308. Am. Soc. Microbiol., Washington, D.C.
85. Moss, C. W., Kai, A., Lambert, M. A., and Patton, C. (1984). Isoprenoid quinone content and cellular fatty acid composition of *Campylobacter* species. *J. Clin. Microbiol.* **19,** 772–776.
86. Nachamkin, I., and Hart, A. M. (1986). Common and specific epitopes of *Campylobacter* flagellin recognized by monoclonal antibodies. *Infect. Immun.* **53,** 438–440.
87. Neill, S. D., Campbell, J. N., O'Brien, J. J., Weatherup, S. T. C., and Ellis, W. A. (1985). Taxonomic position of *Campylobacter cryaerophila* sp. nov. *Int. J. Syst. Bacteriol.* **35,** 342–356.
88. Newell, D. G. (1986). Monoclonal antibodies directed against the flagella of *Campylobacter jejuni:* Cross-reacting and serotypic specificity and potential use in diagnosis. *J. Hyg.* **96,** 377–384.
89. Newell, D. G., Hudson, M. J., and Baskerville, A. (1988). Isolation of a gastric campylobacter-like organism from the stomach of four Rhesus monkeys, and identification as *Campylobacter pylori. J. Med. Microbiol.* **27,** 41–44.
90. Ng, L.-K., Stiles, M. E., and Taylor, D. E. (1985). Inhibition of *Campylobacter coli* and *Campylobacter jejuni* by antibiotics used in selective growth media. *J. Clin. Microbiol.* **22,** 510–514.
91. Ng, L.-K., Stiles, M. E., and Taylor, D. E. (1987). Classification of *Campylobacter* strains using DNA probes. *Mol. Cell. Probes* **1,** 233–243.
92. Ng, V. L., Hadley, W. K., Fennell, C. L., Flores, B. M., and Stamm, W. E. (1987). Successive bacteremias with "*Campylobacter cinaedi*" and "*Campylobacter fennelliae*" in a bisexual male. *J. Clin. Microbiol.* **25,** 2008–2009.
93. Oste, C. (1988). Polymerase chain reaction. *BioTechniques* **6,** 162–167.
94. Owen, R. J., and Leaper, S. (1981). Base composition, size and nucleotide sequence similarities of genome deoxyribonucleic acids from species of the genus *Campylobacter. FEMS Microbiol. Lett.* **12,** 395–400.
95. Paisley, J. W., Mirrett, S., Lauer, B. A., Roe, M., and Reller, L. B. (1982). Dark-field microscopy of human feces for presumptive diagnosis of *Campylobacter fetus* subsp. *jejuni* enteritis. *J. Clin. Microbiol.* **15,** 61–63.
96. Payne, W. J., Jr., Marshall, D. L., Shockley, R. K., and Martin, W. J. (1988). Clinical laboratory applications of monoclonal antibodies. *Clin. Microbiol. Rev.* **1,** 313–329.

97. Penner, J. L. (1988). The genus *Campylobacter:* A decade of progress. *Clin. Microbiol. Rev.* **1,** 157–172.
98. Penner, J. L., and Hennessy, J. N. (1980). Passive hemagglutination technique for serotyping *Campylobacter fetus* subsp. *jejuni* on the basis of soluble heat-stable antigens. *J. Clin. Microbiol.* **12,** 732–737.
99. Picken, R. N., Wang, Z., and Yang, H. L. (1987). Molecular cloning of a species-specific DNA probe for *Campylobacter jejuni. Mol. Cell. Probes* **1,** 245–259.
100. Quinn, T. C., Goodell, S. E., Fennell, C., Wang, S.-P., Schuffler, M. D., Holmes, K. K., and Stamm, W. E. (1984). Infections with *Campylobacter jejuni* and *Campylobacter*-like organisms in homosexual men. *Ann. Intern. Med.* **101,** 187–192.
101. Rashtchian, A., Abbott, M. A., and Shaffer, M. (1987). Cloning and characterization of genes coding for ribosomal RNA in *Campylobacter jejuni. Curr. Microbiol.* **14,** 311–317.
102. Rashtchian, A., Eldredge, J., Ottaviani, M., Abbott, M., Mock, G., Lovern, D., Klinger, J., and Parsons, G. (1987). Immunological capture of nucleic acid hybrids and application to nonradioactive DNA probe assay. *Clin. Chem.* (*Winston-Salem, N.C.*) **33,** 1526–1530.
103. Rauws, E. A. J., Langenberg, W., Houthoff, H. J., Zanen, H. C., and Tytgat, G. N. J. (1988). *Campylobacter pyloridis*-associated chronic active gastritis. A prospective study of its prevalance and the effects of antibacterial and antiulcer treatment. *Gastroenterology* **94,** 33–40.
104. Renz, M., and Kurz, C. (1984). A colorimetric method of DNA hybridization. *Nucleic Acids Res.* **12,** 3435–3444.
105. Roberts, M. C., Hillier, S. L., Schoenkrecht, F. D., and Holmes, K. K. (1984). Nitrocellulose filter blots for species identification of *Mobiluncus curtisii* and *Mobiluncus mulieris. J. Clin. Microbiol.* **20,** 826–827.
106. Roberts, M. C., McMillan, C., and Coyle, M. B. (1987). Whole chromosomal DNA probes for rapid identification of *Mycobacterium tuberculosis* and *Mycobacterium avium* complex. *J. Clin. Microbiol.* **25,** 1239–1243.
107. Roberts, M. C., Moncla, B., and Kenny, G. E. (1987). Chromosomal DNA probes for the identification of *Bacteroides* species. *J. Gen. Microbiol.* **133,** 1423–1430.
108. Rollins, D. M., and Colwell, R. R. (1986). Viable but nonculturable stage of *Campylobacter jejuni* and its role in survival in the natural aquatic environment. *Appl. Environ. Microbiol.* **52,** 531–538.
109. Romaniuk, P. J., and Trust, T. J. (1987). Identification of *Campylobacter* species by Southern hybridization of genomic DNA using an oligonucleotide probe for 16S rRNA genes. *FEMS Microbiol. Lett.* **43,** 331–335.
110. Romaniuk, P. J., Zoltowoska, B., Trust, T. J., Lane, D. J., Olsen, G. J., Pace, N. R., and Stahl, D. A. (1987). *Campylobacter pylori,* the spiral bacterium associated with human gastritis, is not a true *Campylobacter* sp. *J. Bacteriol.* **169,** 2137–2141.
111. Roop, R. M., II, Smibert, R. M., Johnson, J. L., and Krieg, N. R. (1984). Differential characteristics of catalase-positive campylobacters correlated with DNA homology groups. *Can. J. Microbiol.* **30,** 938–951.
112. Rosenthal, L. E., Smoot, D., Mobley, H. L. T., and Guisbert, W. (1988). *Campylobacter pylori* gastritis not related to peridontal disease. *Am. J. Gastroenterol.* **83,** 202.
113. Sandstedt, K., Ursing, J., and Walder, M. (1983). Thermotolerant *Campylobacter* with no or weak catalase activity isolated from dogs. *Curr. Microbiol.* **8,** 209–213.
114. Seriwatana, J., Echeverria, P., Taylor, D. N., Sakuldaipeara, T., Changchawalit, S., and Chivoratanond, O. (1987). Identification of enterotoxigenic *Escherichia coli* with synthetic alkaline phosphatase-conjugated oligonucleotide DNA probes. *J. Clin. Microbiol.* **25,** 1438–1441.

115. Skirrow, M. B. (1977). *Campylobacter* enteritis: A "new" disease. *Br. Med. J.* **2,** 9–11.
116. Skirrow, M. B. (1982). *Campylobacter* enteritis—the first five years. *J. Hyg.* **89,** 175–184.
117. Skirrow, M. B., and Benjamin, J. (1980a). '1001' Campylobacters: Cultural characteristics of intestinal campylobacters from man and animals. *J. Hyg.* **85,** 427–442.
118. Skirrow, M. B., and Benjamin, J. (1980). Differentiation of enteropathogenic campylobacter. *J. Clin. Pathol.* **33,** 1122.
119. Smibert, R. M. (1985). Genus *Campylobacter* Sebald and Veron 1963 907. *In* "Bergey's Manual of Systematic Bacteriology" (N. R. Krieg and J. G. Holt, eds.), pp. 111–118. Williams & Wilkins, Baltimore, Maryland.
120. Stahl, D. A., Flesher, B., Mansfield, H. R., and Montgomery, L. (1988). Use of phylogenetically based hybridization-probes for studies of ruminal microbial ecology. *Appl. Environ. Microbiol.* **54,** 1079–1084.
121. Steele, T. W., Sangster, N., and Lanser, J. A. (1985). DNA relatedness and biochemical features of *Campylobacter* spp. isolated in Central and South Australia. *J. Clin. Microbiol.* **22,** 71–74.
122. Stollar, B. D., and Rashtchian, A. (1987). Immunochemical approaches to gene probe assays. *Anal. Biochem.* **161,** 387–394.
123. Tanner, A. C. R., Badger, S., Lai, C.-H., Listgarten, M. A., Visconti, R. A., and Socransky, S. S. (1981). *Wolinella* gen.nov., *Wolinella succinogenes (Vibrio succinogenes* Wolin et al.) comb. nov., and description of *Bacteroides gracilis* sp. nov., *Wolinella recta* sp. nov., *Campylobacter concisus* sp. nov., and *Eikenella corrodens* from humans with peridontal disease. *Int. J. Syst. Bacteriol.* **31,** 432–445.
124. Tauxe, R. V., Patton, C. M., Edmonds, P., Barrett, T. J., Brenner, D. J., and Blake, P. A. (1985). Illness associated with *Campylobacter laridis,* a newly recognized *Campylobacter* species. *J. Clin. Microbiol.* **21,** 222–225.
125. Tchen, P., Fuchs, R. P. P., Sage, E., and Leng, M. (1984). Chemically modified nucleic acids as immunodetectable probes in hybridization experiments. *Proc. Natl. Acad. Sci. U.S.A.* **81,** 3466–3470.
126. Tee, W., Anderson, B. N., Ross, B. C., and Dwyer, B. (1987). Atypical campylobacters associated with gastroenteritis. *J. Clin. Microbiol.* **25,** 1248–1252.
127. Tenover, F. C. (1988). Diagnostic deoxyribonucleic acid probes for infectious diseases. *Clin. Microbiol. Rev.* **1,** 82–101.
128. Thompson, L. M., III, Smibert, R. M., Johnson, J. L., and Krieg, N. R. (1988). Phylogenetic study of the genus *Campylobacter. Int. J. Syst. Bacteriol.* **38,** 190–200.
129. Tompkins, L. S. (1985). Summary of DNA probes. *In* "Rapid Detection and Identification of Infectious Agents" (D. T. Kingsbury and S. Falkow, eds.), pp. 273–278. Academic Press, Orlando, Florida.
130. Tompkins, L. S., Mickelsen, P., and McClure, J. (1983). Use of a DNA probe to detect *Campylobacter jejuni* in fecal specimens. *In* "Campylobacter II: Proceedings of the Second International Workshop on *Campylobacter* Infectious Brussels, 6–9 September 1983" (A. D. Pearson, M. B. Skirrow, B. Rowe, J. R. Davies, and D. M. Jones, eds.), pp. 50–51. Public Health Laboratory Service, London.
131. Totten, P. A., Fennell, C. L., Tenover, F. C., Wezenberg, J. M., Perine, P. L., Stamm, W. E., and Holmes, K. K. (1985). *Campylobacter cinaedi* (sp.nov.) and *Campylobacter fennelliae* (sp.nov.): Two new *Campylobacter* species associated with enteric disease in homosexual men. *J. Infect. Dis.* **151,** 131–139.
132. Totten, P. A., Patton, C. M., Tenover, F. C., Barrett, T. J., Stamm, W. E., Steigerwalt, A. G., Lin, J. Y., Holmes, K. K., and Brenner, D. J. (1987). Prevalence and characterization of hippurate-negative *Campylobacter jejuni* in King County, Washington. *J. Clin. Microbiol.* **25,** 1747–1752.

133. Veron, M., and Chatelain, R. (1973). Taxonomic study of the genus *Campylobacter* Sebald and Veron and designation of the neotype strain for the type species, *Campylobacter fetus* (Smith and Taylor) Sebald and Veron. *Int. J. Syst. Bacteriol.* **23,** 122–134.
134. von Wulffen, M. (1987). Low degree of relatedness between *Campylobacter pyloridis* and enteropathogenic *Campylobacter* species as revealed by DNA–DNA blot hybridization and immunoblot studies. *FEMS Microbiol. Lett.* **42,** 129–133.
135. von Wulffen, H., Heeseman, J., Butzow, G. H., Loning, T., and Laufs, R. (1986). Detection of *Campylobacter pyloridis* in patients with antrum gastritis and peptic ulcers by culture, complement fixation test, and immunoblot. *J. Clin. Microbiol.* **24,** 716–720.
136. Wenman, W. M., Chai, J., Louie, T. J., Goudreau, C. Lior, H., Newell, D. G., Pearson, A. D., and Taylor, D. E. (1985). Antigenic analysis of *Campylobacter* flagellar protein and other proteins. *J. Clin. Microbiol.* **21,** 108–112.
137. Wetherall, B. L., McDonald, P. J., and Johnson, A. M. (1988). Detection of *Campylobacter pylori* DNA by hybridization with non-radioactive probes in comparison with a ^{32}P-labelled probe. *J. Med. Microbiol.* **27,** 257–264.
137a. Wetherall, B. L., and Johnson, A. M. (1989). Haemolytic activity of *Campylobacter pylori*. *Eur. J. Clin. Microbiol. Infect Dis.,* in press.
138. Yolken, R. H. (1988). Nucleic acids or immunoglobulins: Which are the molecular probes of the future? *Mol. Cell. Probes* **2,** 87–96.
139. Zola, H. (1985). Speaking personally: Monoclonal antibodies as diagnostic reagents. *Pathology* **17,** 53–56.

11

Detection of *Leptospira*, *Haemophilus*, and *Campylobacter* Using DNA Probes

W. J. TERPSTRA,* J. TER SCHEGGET,† AND G. J. SCHOONE*

**Department of Tropical Hygiene*
Royal Tropical Institute
Amsterdam, The Netherlands

†*Department of Medical Microbiology*
University of Amsterdam
Amsterdam, The Netherlands

I. INTRODUCTION

Culturing of specimens often preceded by staining and followed by the determination of antibiotic susceptibility of the isolate is the basis of microbiological diagnosis of most bacterial diseases. A disadvantage of this is that it may take a few days before results are available. Staining procedures, including the frequently used Gram staining, are not specific as the morphology of many bacteria is not characteristic. The lack of speed of culturing is acceptable in most cases of mild or moderate infection. The procedure is unsatisfactory when the patient is seriously ill and immediate clinical action, in particular the initiation of appropriate antibiotic or chemotherapeutic treatment, is needed. Then the search for alternative and quick diagnostic methods becomes urgent. Alternative methods are also needed when bacteria grow slowly or not at all on culture media or are easily missed or overlooked for a variety of reasons.

During infections with *Leptospira, Haemophilus,* or *Campylobacter*, clinical situations may develop in which the need for alternatives to conventional diagnostic methods arises. Various methods have been explored to improve diagnosis. The most recent of these methods are in the field of immunology, and a multitude of tests using polyclonal or monoclonal antibodies (MAb) have been employed for the specific and direct demonstration of microorganisms or their antigens in clinical materials. These techniques can vary from the specific visualization of the microorganism in smears or tissue sections to antigen capture either directly or indirectly by an inhibition test. Several of these methods have been developed for daily, routine use in the clinical microbiological laboratory. In many instances the results have been disappointing or the test has been applied as an additive to conventional methods. The search for alternative methods continues. In this respect it is only logical that with the rapid developments in the field of DNA research, methods have been explored for the specific demonstration of the nucleic acids of a microorganism thus indirectly of the microorganism itself, notably nucleic acid (NA) hybridization methods (27). With the advent of enzymatic and immunological methods as alternatives to radioactive detection systems, the prospect for the eventual application of hybridization methods in the diagnostic laboratory has become realistic. In the following sections, the development of NA hybridization techniques for *Leptospira, Haemophilus,* and *Campylobacter* will be discussed.

II. LEPTOSPIROSIS

A. Background

Leptospirosis is a disease caused by *Leptospira interrogans*. Subdivision is in serovars, which are the basic taxa (36). Analysis of leptospiral DNA has revealed the existence of different genetically interrelated groups whose classification does not completely follow the conventional classification in serogroups and serovars based on the antigenic structure as revealed by rabbit antisera (37). Presently there are ~200 recognized serovars divided into 25 serogroups. Leptospires are often adapted to a certain vertebrate host, a so-called maintenance host, which acts as a reservoir (12). Maintenance hosts may carry leptospires for a long time in their kidneys and shed them in their urine. Sometimes humans, or an animal other than the maintenance host, may accidentally become infected. The accidental host may have a subclinical infection, but more often suffers from mild to severe disease that may even result in a lethal outcome. The infection may also take a chronic course. During the disease two overlapping phases can be distinguished, a leptospiremic phase and an immune phase (36). During the leptospiremic phase the microorganisms circulate in the blood. A few days after the onset of the disease antibodies begin to appear, and the leptospires are cleared from the blood. Depending on the nature of the host–parasite relationship, leptospires may settle for a prolonged period in the kidneys where they seem to be sheltered from the damaging effects of the immune response.

The disease, presenting all signs and symptoms of Weil's syndrome, is recognized by the clinician and treated promptly. Often the symptoms of the disease are undistinguished, and laboratory support is indispensable for diagnosis. Unfortunately laboratory diagnosis is of no avail in the early stage of the disease when antibiotic treatment is most effective. The reason for this is that leptospires from the patient's blood grow slowly or not at all in culture media, and experimental animals are not always susceptible to strains that are pathogenic to humans. Direct observation of leptospires in clinical samples is possible, and leptospires can be concentrated from the blood by differential centrifugation. However, identification by dark-field microscopy or silver staining is considered unreliable, particularly when leptospires are present in low quantities. Then it may be difficult to differentiate between leptospires and artifacts. The problem with culturing from the urine, for instance from chronically infected and shedding animals, is that leptospires, apart from being fastidious, quickly die in urine. An additional problem is that urine is often badly contaminated with other microorganisms. Serology, as an indirect alternative method to culturing, may fail as it is sometimes not possible to detect antibodies in the blood of leptospira carriers (8).

The search for alternative methods to culturing focused first on the detection of leptospires in the blood in the early phase of the disease of human patients in order to aid in early diagnosis and second on the detection of leptospires in the urine of animals in order to avoid unreliable culture results.

Immunological methods have been explored because of their ability to detect leptospires with the use of antibodies reacting with specific antigens. A series of antigen-capture tests have been compared (1). They do not seem to be very sensitive, since even with radioimmunoassay the lowest detectable amount was 10^4–10^5 per milliliter. Various techniques using antibodies with a fluorescent tag (7) or an enzyme label (28) enabling subsequent development with a chromogenic substrate have been used. With these methods, considerable skill and experience are required to obtain reliable results. For all these reasons it appeared to be important to investigate the applicability of DNA–DNA hybridization to the early diagnosis of leptospirosis.

B. Results and Discussion

1. Dot-Blot Hybridization

We explored DNA–DNA hybridization on purified DNA from cultures or partially purified DNA from blood of experimentally infected animals (29). Since we aspired eventually to apply the technique on a routine basis, we compared nonradioactive biotin labeling with ^{32}P labeling. We briefly experimented (unpublished results) with *N*-acetoxy-2-acetylamino-fluorene (AAF) labeling of DNA probe followed by detection of the probe target duplex with an anti-AAF MAb and an enzymatic staining reaction (17). As the results with this method were similar to those with biotin-labeled probes (as discussed later in this chapter), we did not pursue this labeling method. In our study a DNA probe was prepared from purified total genomic DNA of strains of *L. interrogans*. We tested the DNA probe for specificity on purified target DNA from other *Leptospira* strains and from unrelated bacteria and eukaryotic human DNA (Table I; Fig. 1). With stringent washing conditions, no or hardly any cross-hybridization was

TABLE I

Smallest Quantities of Purified Target DNAs of Related and Unrelated Origin on Duplicate Filters Detected with Total Genomic DNA Probes of Two Serovars of *Leptospira interrogans*[a]

	DNA Probe			
	celledoni		copenhageni	
Target DNA	^{32}P	Biotin	^{32}P	Biotin
L. interrogans, serovar *celledoni*	4–4	5–5	20–20	50–70
L. interrogans, serovar *copenhageni*	40–100	30–70	2–4	5–10
Leptospira biflexa, serovar *patoc*	2000–6000	8000–8000	2000–5000	20,000, Negative[b]
Escherichia coli	Negative[b]	Negative[b]	Negative[b]	Negative[b]

[a] Data in picograms.
[b] Negative: no signal with 2×10^4 pg of target DNA.

observed between the DNA probes and the purified total genomic target DNAs of unrelated origins. Probably even DNA regions that are conserved between species (e.g., parts of the sequences coding for ribosomal DNA) did not form hybrids that remained stable with the washing conditions used. The degree of cross-hybridization with different serovars var-

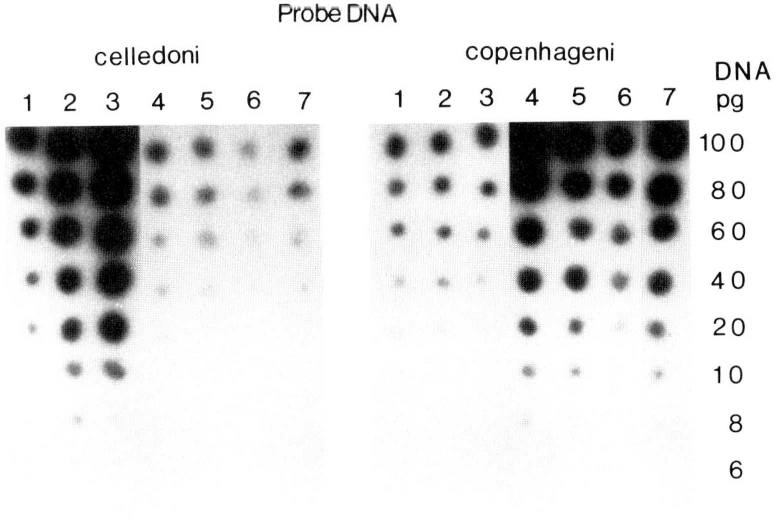

Fig. 1. Dot hybridization with ^{32}P-labeled DNA probes for serovars *celledoni* and *copenhageni* with target DNA of the following *Leptospira interrogans* serovars: (1) *ranarum,* (2) *tarassovi,* (3) *celledoni,* (4) *bataviae,* (5) *pomona,* (6) *hardjo,* (7) *copenhageni.*

ied. This probably reflects different degrees of genetic relatedness between pathogenic serovars (37). The smallest amount of homologous DNA that could be detected was 1.5 pg (which corresponds to ~750 leptospires) using the radioactive probe with target DNA fixed on nitrocellulose (NC) filters. However, when in repeated experiments cultured leptospires were mixed with blood or urine, we estimated that we were able to detect 10^3–10^4 organisms using a quick DNA purification procedure and a total genomic radioactive or biotinylated probe. The threshold detection level for biotinylated DNA probes was in general higher (Fig. 2). The radioactive detection system was apparently slightly more sensitive than the enzymatic system. Using ^{32}P- and biotin-labeled probes we could also detect leptospiral DNA in partially purified samples from the blood of experimentally infected hamsters (Fig. 3). Results with the radioactive probes were slightly better than with the biotinylated probes.

Millar *et al.* (20) were able to detect leptospires in the urine of experimentally infected pigs with a radioactive probe. They deduced from model studies in which urine samples were seeded with leptospires from a culture

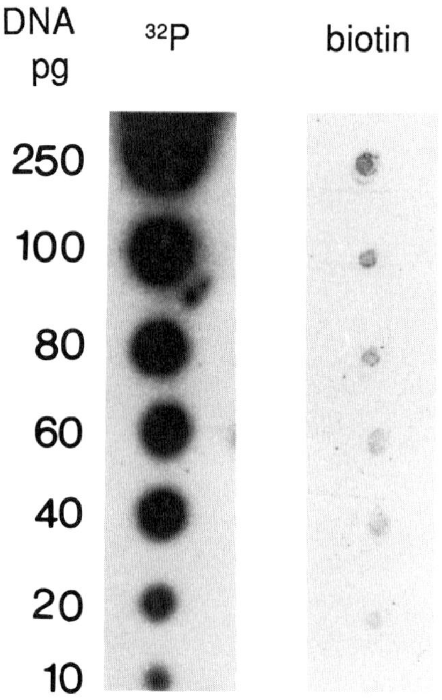

Fig. 2. Comparison of ^{32}P- and biotin-labeled DNA probe for serovar *copenhageni* on serially diluted homologous DNA dotted on a nitrocellulose membrane.

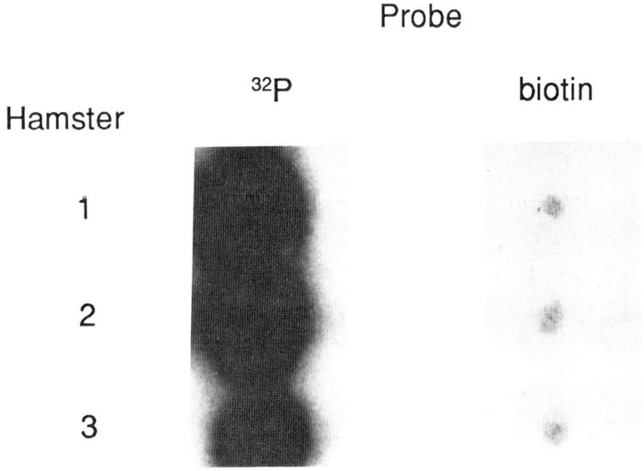

Fig. 3. Detection of leptospiral DNA in three serum samples of experimentally infected hamsters with ^{32}P- or biotin-labeled DNA probe for serovar *copenhageni.*

that the threshold detection level of their probe was 160 pg. In their experience nylon filters gave slightly better results than NC filters.

We also applied biotinylated total genomic DNA probe to partially purified urine samples from human patients with proven leptospirosis (unpublished results). It was not possible to evaluate the results with these samples, which were fixed on NC filters, due to an unacceptable degree of background staining that interfered with a reliable reading of the results.

Zuerner and Bolin (38) prepared a total genomic DNA probe of type hardjobovis and a RNA probe developed from a repetitive sequence element of this type of *L. interrogans*. Both probes were radiolabeled. The RNA probe was more specific and more sensitive than the DNA probe and was able to detect as few as 10^3 leptospires. When applied on urine samples from infected cattle, the hybridization test using the probe derived from the repetitive element performed well in comparison with culturing and antigen detection using fluorescent antibodies. The probe can probably be applied for a rapid preliminary diagnosis of an infection of cattle urine with hardjobovis.

2. In Situ *Hybridization*

To circumvent the problem of background staining on filters with urine using biotinylated DNA probe, we explored *in situ* hybridization, which is a well-known technique for eukaryotic cells, on leptospires smeared on microscope slides (30,33). The principle of this test procedure is that by gentle fixation, DNA can be kept within the cell walls and still take part in the hybridization reaction. Apparently target DNA is accessible to the DNA probe through holes in the cell walls. Though basically different from immunoperoxidase staining, *in situ* hybridization leads to the same result, a stained microorganism. Since the human eye is a very sensitive organ, we considered that the binding of a visible tag to the hybridized leptospires could lead to a high overall sensitivity of the method. Similar to dot-blot hybridization, the total genomic probe was highly specific (Table II). We found that *in situ* hybridization enabled us to detect leptospira in various clinical materials such as blood from an experimentally infected hamster (Fig. 4A), the sediment of centrifuged serum from a human patient (Fig. 4B), the urine of a naturally infected cow (Fig. 4C), a liver smear of an

TABLE II

***In situ* Hybridization with Various Bacterial Species Using DNA Probe Prepared from Total Genomic DNA**

A	B	C
Leptospira interrogans[a] serovar *copenhageni*	*H. influenzae*[a] types a–f and five noncapsulated strains	*C. jejuni*[a]
L. *interrogans*[a] serovar *hardjo*	*Haemophilus parainfluenzae*[a]	*Campylobacter* fetus subsp. *fetus*
Borrelia burgdorferi	*Haemophilus haemolyticus*[a]	C. *coli*[a]
Campylobacter jejuni	*Haemophilus parahaemolyticus*[a]	*V. cholerae*
Vibrio cholerae	*Haemophilus aprhophilus*	*Y. enterocolitica*
Yersinia enterocolitica	*Haemophilus haemoglobinophilus*	*E. coli*
Escherichia coli	*Haemophilus ducreyi*	*Shigella flexneri*
Klebsiella pneumoniae	*Bordetella pertussis*	*Shigella boydii*
Pseudomonas aeruginosa	*Bordetella parapertussis*	*Salmonella typhi*
Haemophilus influenzae	*E. coli*	*Salmonella* sp. group B
Staphylococcus aureus	*Proteus mirabilis*	*P. mirabilis*
Streptococcus pneumoniae	*Kl. pneumoniae*	*Kl. pneumoniae*
Neisseria gonorrhoeae	*Ps. aeruginosa*	*Ps. aeruginosa*
	Campylobacter coli	*H. influenzae*
	Staphylococcus aureus	*H. parainfluenzae*
	Streptococcus pneumoniae	*Streptococcus pneumoniae*
	Neisseria meningitidis	*N. gonorrhoeae*
	Branhamella catarrhalis	*L. interrogans*
	Y. enterocolitica	*Leptospira biflexa*
	L. interrogans	

[a] Positive reaction between target DNA and DNA probe from *Leptospira interrogans, Haemophilus influenzae,* or *Campylobacter jejuni* in columns A, B, or C, respectively.

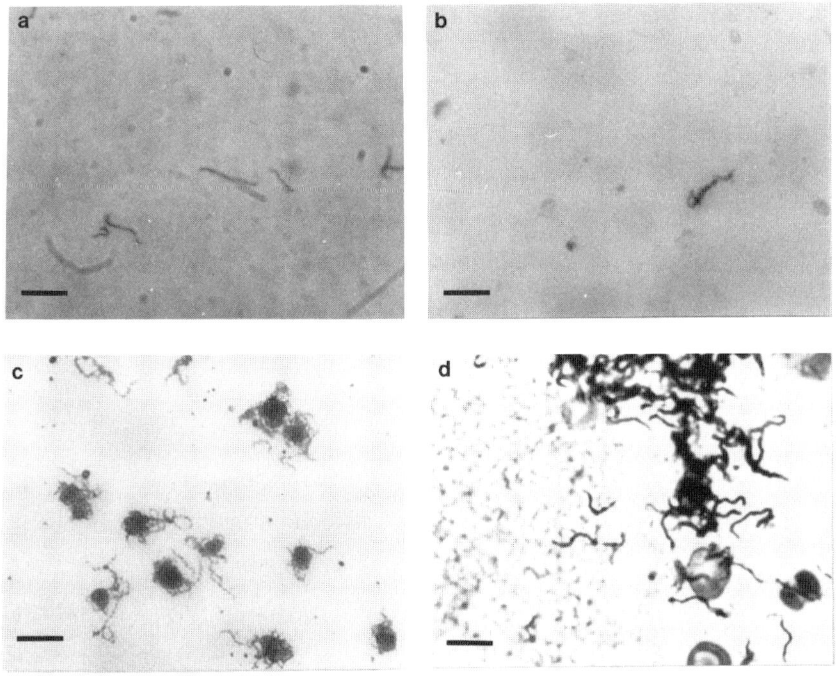

Fig. 4. Visualization by *in situ* hybridization of leptospires in blood of (A) an experimentally infected hamster, (B) a human serum sediment, (C) cattle urine, and (D) human kidney tissue. Staining of kidney tissue section was silver-enhanced. Bars = 10 μm.

experimentally infected hamster, and a post mortem kidney section of a human patient (Fig. 4D).

We investigated the value of *in situ* hybridization on a small series of blood (thick smears) and urine samples (smears of sediments) of human patients with infections proven by culture or serology (W.J. Terpstra and C. O. R. Everard, unpublished results). From 20 urine samples we found 7 positive by *in situ* hybridization and 1 by culture. From 18 blood samples none was positive by hybridization and 5 by culture. These preliminary results need substantiation with more observation, but one can assume that many leptospires were shed in the urine where they were detected by hybridization, although not by culture because leptospires die quickly in urine. In contrast, even a few leptospires in the blood were able to grow in culture, but they were undetectable by hybridization because of their low concentration. Contrary to humans in whom (Fig. 4B) probably even in the acute stage of the disease leptospires in the blood are scarce, experimentally infected hamsters undergo an overwhelming infection in which the blood is flooded with leptospires (Fig. 4A).

In general, leptospires can be easily detected and identified either by *in situ* hybridization or immunoperoxidase staining. Both methods possess specificity and allow subsequent visualization of the characteristic morphology of the leptospires. However, this only applies to the situation in which leptospires are present in large quantities. When only a few leptospires are present, it is difficult to avoid confusion with artifacts. Similar problems are faced as with dark-field microscopy or silver staining. The signal of diaminobenzidine tetrahydrochloride (DAB), which was used as a coloring substrate, could be considerably amplified with the use of a DAB silver-enhancement method (DAB Enhancement Kit, Amersham, United Kingdom, RPN 1174). Particularly for the staining of leptospires in paraffin-embedded tissue sections, the enhancement method was successful. However, the drawback of this method is that background staining may occur as a result of an interaction between tissue particles and heavy metal leading to nonspecific silver precipitation.

For taxonomic or epidemiological purposes specific probes may be needed. With cloned DNA fragments it is possible to differentiate between serovars and strains of *L. interrogans* (9,18). We were able to specifically stain type hardjobovis of serovar *hardjo* with such a probe (Fig. 5). The probe did not lead to staining of the serologically closely related strain hardjoprajitno. This observation suggests that specific probes prepared from recombinant clones can be used for the detection and identification of strains.

III. *HAEMOPHILUS*

A. Background

A wide variety of pathogenic and commensal species of *Haemophilus* live in organs of humans and animals (14). For humans, *Haemophilus influenzae* is by far the most important of these commensal species. It causes acute and chronic infectious conditions, particularly in childhood (11). *Haemophilus influenzae* is widely prevalent and may, in small numbers, occur in the throat of apparently healthy humans as part of the commensal flora. The observation that *H. influenzae* type b causes disease chiefly and most severely in the younger age groups suggests that after the weaning of maternal antibodies in the first few months after birth, the child slowly develops specific immunological resistence. During this process the

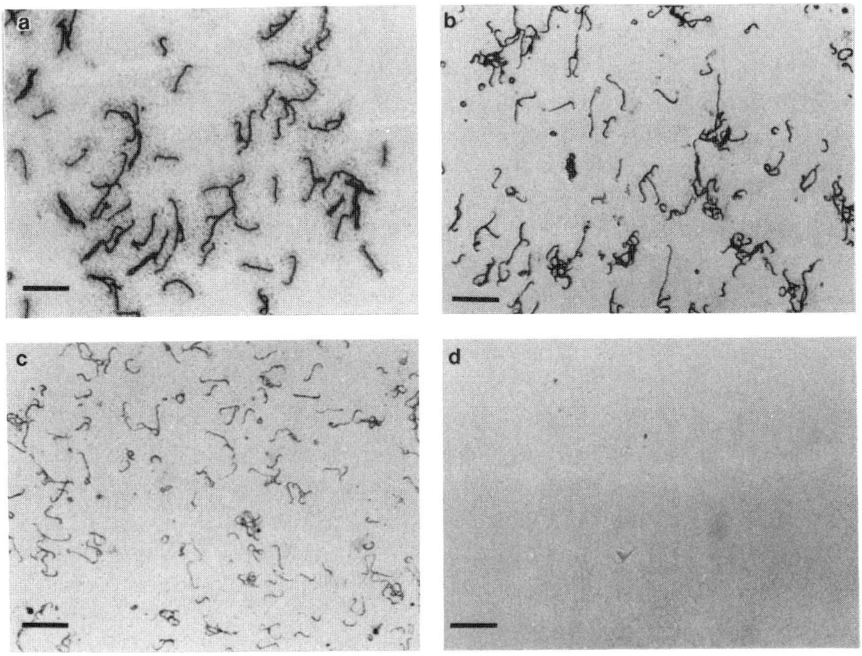

Fig. 5. *In situ* hybridization on hardjobovis (A and C) and hardjoprajitno (B and D) from a culture using biotinylated probes of total genomic DNA (A and B) and recombinant cloned DNA of hardjobovis (C and D). In all preparations staining was silver-enhanced. Bars = 25 μm.

status of the microorganisms, so far as systemic infections are concerned, gradually changes from pathogenic to commensal. *Haemophilus influenzae* is important chiefly as a cause of respiratory tract infection. A rather rare but very dramatic manifestation of *Haemophilus* infection is meningitis. This serious condition demands undelayed clinical intervention including the administration of effective antibiotics. *Haemophilus* is readily cultivable on appropriate culture media, but in the case of meningitis quick results are necessary and culturing takes too long. Examination of a Gram-stained sediment of spinal fluid may yield, or, as well, fail to yield, gram-negative rods likely to be *Haemophilus*. Obviously there is a need for a quick alternative method to specifically detect *Haemophilus* in cerebrospinal fluid (CSF). Numerous immunological tests were developed for the detection of antigen in CSF, and some of these are used routinely as quick additional tests to the slower culture method (35). A quite different situation exists with cystic fibrosis (CF), in which *H. influenzae* plays a role as a major pathogen. One of the prominent features of this hereditary disease is the chronic infection of the respiratory tract, mainly by *Pseudomonas* spp., *Staphylococcus aureus*, and *H. influenzae* (13). Cystic fibrosis patients with chronic respiratory infection usually produce sputum copiously. From this sputum *Pseudomonas* spp. often grow profusely on culture media, with typical mucoid colonies that tend to obscure other microorganisms by overgrowth. It is particularly difficult to retrieve *H. influenzae* from these cultures because the growth of *Pseudomonas* spp. is not selectively suppressed, while neither the macroscopic colony morphology of *H. influenzae* nor its microscopic appearance are particularly helpful for differentiation from *Pseudomonas*. Since there is a metabolic dysfunction underlying the respiratory infection, those CF patients severely affected need continuous monitoring, both clinically for lung function and microbiologically for prevalent microorganisms and their antibiotic susceptibility patterns. Rapidity is not the main issue in respiratory tract infection in CF, but the problems with culturing justify the search for alternative methods for the detection of *Haemophilus* in these patients. Evidently, in severe respiratory tract infection a quick alternative test to culturing would be useful.

B. Results and Discussion

1. Dot-Blot Hybridization

Much research has recently been done on the use of DNA technology in *Haemophilus* in the fields of gene expression (5,16) and classification (2). Recently these techniques were also explored for detection.

Malouin *et al.* (19) developed a highly specific radioactive probe from a 5-kb fragment of *H. influenzae* DNA that reacted with *H. influenzae* and the related *H. aegyptius*. A slight degree of cross-reactivity with *H. parainfluenzae* could be largely eliminated by changing hybridization conditions. Clinical samples containing bacteria were directly spotted on NC filters and lysed with NaOH. They were able to demonstrate the presence of *H. influenzae* in samples of spinal fluid in a period of 4 hr when the radioactive signal was measured with a scintillation counter. After a processing time of 8 hr, they found the hybridization test highly specific and sensitive for the detection of *H. influenzae* in sputum samples and, in fact, found that the test was more often positive than with culturing. A culture may fail to detect *H. influenzae* for several reasons, as will be discussed in the next section on *in situ* hybridization. In addition, they were able quickly to identify *H. influenzae* in blood cultures in which the microorganisms had multiplied. The threshold level of detection for their probe was 10^5 microorganisms per 10 μl. They considered radioactive probes superior to biotinylated probes.

2. In Situ *Hybridization*

We have explored *in situ* hybridization as a tool to detect specifically *Haemophilus* in sputum (Fig. 6). For this purpose we prepared a total genomic DNA probe extracted from *H. influenzae* type b in the same way as described for *Leptospira* (31,33) When a large number of different bacterial species were tested for specificity, cross-hybridization was observed with *H. parainfluenzae, H. haemolyticus,* and *H. parahaemolyticus,* all throat-inhabiting bacteria (Table II). We considered this useful for diagnostic purposes because of the potential pathogenicity of these species (15). However, for taxonomic or epidemiological studies it may be necessary to prepare specific probes. Alternatively, more stringent washing conditions can be used. When examining a hybridized sputum smear one has to allow for the fact that *Haemophilus* spp. may occur in low numbers in the throats of healthy persons and may inadvertently contaminate sputum samples even when these are washed. Therefore, we only considered those smears that were clearly positive according to a set of criteria empirically formulated by us. When *in situ* hybridization was applied to large numbers of sputum samples from patients with bronchitis we found that the test was more often positive than the culture. We could argue that

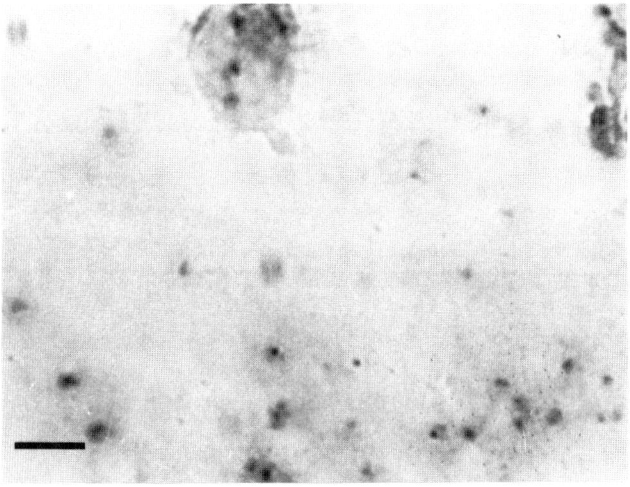

Fig. 6. *In situ* DNA hybridization using biotinylated total *Haemophilus influenzae* DNA on a sputum smear of a patient with cystic fibrosis. Bar = 10 μm.

in fact the hybridization test was probably more sensitive than culturing mainly because it avoids the problem of overgrowth and obscuring of *Haemophilus* by other bacteria and, in addition, because hybridization can detect *Haemophilus* that is damaged by antibiotic treatment prior to sampling. The advantage of hybridization, in comparison with culturing of sputum from patients with conditions other than CF, was surprising to us, yet the ability of the test to detect *Haemophilus* was most striking in the sputum of CF patients with *Pseudomonas* infections (Table III). Van Alphen and co-workers (32) prepared a MAb that reacted specifically with the outer-membrane protein P6 of *H. influenzae*. This MAb was more specific than our total genomic probe. In a collaborative study, we compared immunoperoxidase staining using this MAb with *in situ* hybridization. We examined a large number of sputum samples with both techniques. *In situ* hybridization and immunoperoxidase staining gave almost equal results and both methods were superior to culturing in detecting a larger number of infections with *Haemophilus*. Immunoperoxidase staining is slightly easier and quicker (3 hr) than *in situ* hybridization (4 hr). Both techniques are applicable to clinical situations, notably the monitoring of *Haemophilus* infections in CF patients. This direct applicability of the tests is by virtue of the relatively easy detection and identification of *Haemophilus*, which are usually present in sufficiently large numbers in the sputum smears to enable quick detection. This is in contrast to the situation with *Leptospira* and *Campylobacter*, which will be discussed in the next section, in which detection of even a single bacterium may have

TABLE III

Results of Culture of *Haemophilus influenzae* and *Pseudomonas* spp. in Comparison with *in Situ* Hybridization Using a *Haemophilus* Probe on Sputum Samples of Nine Patients with Cystic Fibrosis

Patient	Number of samples	Culture positive for		Positive hybridization
		H. influenzae	*Pseudomonas* spp.	
P	4	—	4	2
E	6	—	5	6
G	6	3	6	5
E	3	—	3	3
G	2	—	2	1
L	2	2	—	2
H	2	1	1	2
S	1	—	—	—
O	1	1	—	—
	27	7	21	21

clinical significance. Actually, when present in small quantities, *Haemophilus* is difficult to distinguish from artifacts. The problem with *Leptospira* and *Campylobacter* is significant because *Haemophilus* lacks a characteristic morphology.

IV. *CAMPYLOBACTER*

A. Background

The genus *Campylobacter* is widely prevalent in animals and humans (21). Many species cause a variety of different infections in humans. The most important human pathogen is *Campylobacter jejuni*. The most frequent clinical manifestation of infection with this species is diarrhea. *Campylobacter jejuni* is widely distributed among animals. Although it is a frequent cause of diarrhea, symptomless human carriers are rare. *Campylobacter jejuni* is a fastidious microorganism requiring critical growth conditions. Since its discovery, standardized procedures, including selective culture conditions and selective culture media, have been introduced that allow its ready isolation in the microbiological diagnostic laboratory. Isolation, identification, and determination of the pattern of antibiotic sensitivity take a few days. The observation of typical "gull wing"-shaped bacteria in a stained stool smear of a suspected patient allows only the presumptive diagnosis of campylobacteriosis. *Campylobacter jejuni* is just one of numerous viral, bacterial, and parasitic causes of enteric infection. In diarrhea, the diagnostic laboratory usually applies a battery of different tests to cover as many pathogens as possible. If the clinical condition of the patient with intestinal infection is serious enough to warrant antibiotic treatment, the issue of the nature of the causative agent becomes urgent, since different bacterial species have different antibiotic susceptibility patterns. Knowing the causative agent may at least give a presumptive indication to the selection of a certain antibiotic. This problem explains and justifies the search for alternative methods for the quick identification of *Campylobacter* in feces.

B. Results and Discussion

1. Dot-Blot Hybridization

As with *Haemophilus*, much research has been done on DNA technology in campylobacters, particularly in the field of classification (23). Claus *et al.* (4) prepared radioactive DNA probes from *C. jejuni*, which cross-hybridized only with *Campylobacter coli*. They report that they were able to detect *Campylobacter* in stool directly spotted on NC to a minimal number of 10^3 bacteria. They do not provide data on culturing. One may argue that culturing as a method of *in vivo* DNA amplification and subsequent rapid identification of multiplied bacteria is a step in direct identification in clinical samples. Chevrier *et al.* (3) made AAF-labeled probes from total genomic DNA of several *Campylobacter* species. Cross-hybridizations were observed within the genus. The most interesting feature of their study for our subject was that they were able to identify *Campylobacter* spp. in mixed cultures that were grown overnight. Echeverria *et al.* (6) explored gene probes for the detection of various pathogenic enteric bacteria, among these *Campylobacter* spp. They hybridized directly on fecal material spotted on NC filters, using a probe prepared of chromosomal fragments of *C. jejuni*. In a few hundred diarrheal stool specimens, they found that the hybridization test was positive in 44% of the culture-positive samples, while an additional 5% was positive on culture-negative samples. Apparently the hybridization method falls far behind the culturing method. Tompkins and Krajden (34) were able to detect 10^3–10^4 organisms with a radioactive probe of total genomic DNA of *C. jejuni* in a simulated-infection experiment seeding *Campylobacter* in feces. They experimented with extraction procedures aiming at the decrease of nonspecific background signal, which they consider to be caused mainly by protein. They suggest eventual clinical application. An interesting approach to the detection of hybridization was by Stollar and Rashtchian (26), who captured rRNA–DNA hybrids using a goat anti-poly(A)–poly(dT) antibody bound to the solid phase of polystyrene plates. The probes were radiolabeled DNA sequences complementary to ribosomal RNA of *C. jejuni* and synthetic oligonucleotides labeled with biotin. Bound biotin-labeled probes were detected with an enzymatic system and a chromogenic substrate. This method is expected to be quick and to suffer minimally from background signals and false positives, but still needs to be applied on clinical specimens.

2. In Situ *Hybridization*

In a manner similar to that described for *Leptospira* and *Haemophilus*, we prepared a biotinylated probe from total genomic DNA extracted from *C. jejuni* (33). During introductory studies the specificity of this probe appeared to be restricted to *C. jejuni* and to *C. coli*, which causes in humans diseases similar to those of *C. jejuni* (Table II). We considered this probe diagnostically useful. For taxonomic and epidemiological studies more specific probes are needed. With the technique of *in situ* hybridization we were able to detect *Campylobacter* in stool smears (Fig. 7). The specifically stained bacteria sometimes showed the characteristic "gull

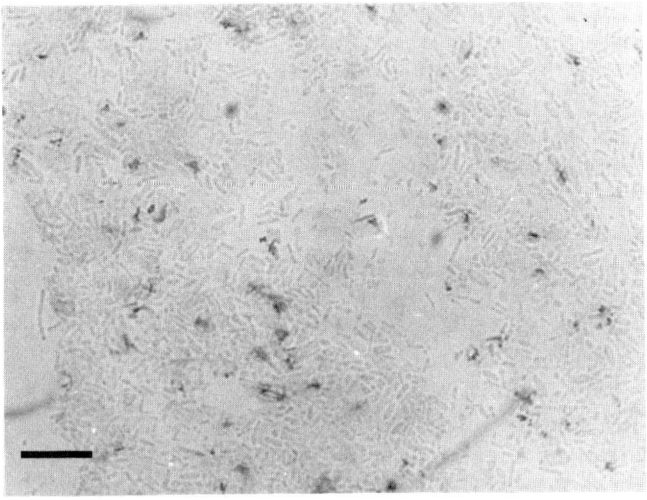

Fig. 7. *In situ* DNA hybridization using biotinylated total *Campylobacter jejuni* DNA on a fecal smear from which *C. jejuni* was isolated. Bar = 10 μm.

wing" shape. Notwithstanding the addition of the morphological characteristic to the specificity of the hybridization reaction, we found it difficult to differentiate campylobacters from artifacts when the bacteria were present in small numbers in the feces smears. Here we faced the same problem as with *Leptospira* and *Haemophilus*. In addition, the examination of feces smears after *in situ* hybridization is more difficult than of blood or sputum smears, since feces contain many small particles that are naturally colored or particles that are easily stained nonspecifically by the chromogenic substrate used during the hybridization procedure. Perhaps the problem can be overcome with the application of different enzymes or different chromogens than the ones that we used, but we did not pursue this line of research.

V. CONCLUSIONS

1. It is important to define for what purpose one wants to use NA hybridization. For taxonomy or epidemiology one would prefer specific, sometimes even highly specific probes. It was shown that molecularly cloned fragments of leptospiral DNA could be applied in *in situ* hybridization for simultaneous detection and identification of a strain. This principle probably applies to other bacterial species as well. For epidemiology,

rapidity is not highly important. The applicability of the hybridization test on dead bacteria in preserved specimens can be an advantage in comparison with culturing in case there are transport problems. For diagnostic investigations, one would prefer broadly reactive probes that cover a group of bacterial strains or perhaps even species that, as a group, are clinically important. For clinical reasons the emphasis may be on a short hybridization procedure giving quick results. Reliability rather than rapidity is the issue in case the hybridization test aims to detect fastidious, slow-growing, or uncultivable organisms.

2. *Leptospira, Haemophilus*, and *Campylobacter* can be detected in clinical samples with hybridization tests. So far the best results were obtained in *Haemophilus* infections probably because bacteria usually are present in large quantities in sputum. In CF patients hybridization on *Haemophilus* can be used as an additive, perhaps even an alternative, to selective culturing methods. The limited results obtained so far suggest that hybridization tests are insufficiently sensitive to detect small numbers of *Leptospira*, for instance in blood of human patients.

3. In the case of *Leptospira* and *Haemophilus* the scope of the issue as described in the preceding sections was limited to a single microorganism and a highly particular clinical condition. In the case of diarrhea the scope is wide. If one aims to use hybridization for the detection of enteric pathogens, panels of probes have to be developed to cover the whole range of pathogens, and *Campylobacter* is just one of many causes. Either simultaneous or serial application of a panel of probes will mean a considerable complication in terms of labeling, labor, and costs.

4. Nucleic acid hybridization is still largely experimental. Many methods are still in the phase of model studies, and observations on clinical materials are often anecdotal. More observations on clinical materials must be done to allow appropriate evaluation.

5. The observation on the detection of *Haemophilus* in sputum with *in situ* hybridization and immunoperoxidase staining in comparison with culturing suggests that hybridization tests can be as sensitive as immunological ones. This is not surprising, since both techniques detect colored bacteria.

6. Culturing is time-consuming, but not elaborate. Hybridization, at least in a rapid version, is quick but elaborate. In order to be eventually used in diagnostic laboratories, the test procedures must be further simplified. It is imperative that the laborious, time-consuming, and potentially dangerous radioactive detection systems be replaced by simpler, quicker, and less dangerous enzymatic or immunological methods.

7. Enzymatic detection methods are less sensitive than radioactive methods, especially when unpurified DNAs are used. Their sensitivity must be increased and the amount of background signal decreased.

8. *In situ* hybridization appears to be suitable for application to smears or tissue sections in which morphological integrity is important.

VI. PROSPECTS FOR THE FUTURE

Methods in nucleic acids research will eventually find wide application in diagnostical microbiology together with MAb (14a). At present the need to search for alternatives to conventional methods is sustained by two important problems: a serious clinical condition requiring a quick diagnosis, and the occurrence of a pathogenic microorganism that cannot or can only with difficulty be detected by conventional methods. Ideally, the medical microbiologist would place a ''magic stick' into the clinical sample to obtain an instant diagnosis. One can imagine an ultimate and ideal situation in which hybridization solves all the key issues in diagnostic microbiology in a quick and reliable way, that is, giving information, preferably quantitative, on the presence of a pathogenic microorganism in clinical material, on its identity, or, in case of mixed infections, on their identities and susceptibilities to drugs. For the last issue, the presence of resistance genes should be known. At the present state of hybridization technology, each separate part of the issue is in the process of being slowly unraveled. One cannot yet foresee how an integrated system, using a multitude of probes providing complete information, will be shaped. Present systems are much too cumbersome, and much more development has to be done. Once these problems have been solved and hybridization procedures have become quick and simple, the interest will grow to solve less urgent diagnostic problems as well.

Culturing is in many respects the ''golden standard'' in medical microbiology, but has a few flaws. The common claim that culturing may even detect a single microorganism is theoretically true. Doubtlessly culturing will remain one of the most sensitive detection methods for many microorganisms, but culturing may fail for a variety of reasons. We have shown for example, that culturing sometimes failed to detect *Haemophilus* because of overgrowth by other microorganisms, while the hybridization test was clearly positive. However, when microorganisms are present in small quantities, the hybridization test may fail because the number of microorganisms is below the detection level. It is possible to increase the sensitivity of the detection systems by lowering background staining and increasing the specific signal with a variety of sophisticated tags and gauging apparatus. It is doubtful if even the best of these systems will be able to detect a few microorganisms. In the future much is to be expected from a fundamentally new approach: the *in vitro* specific primer-directed amplification of DNA with a thermostable DNA polymerase, or the polymerase chain reaction (PCR) (25). Using PCR, human immunodeficiency virus type 1 (HIV-1) DNA was detected in DNA isolated from peripheral blood mononuclear cells, even when the virus culture was negative (22). Van Eys *et al.* (10) using a set of oligonucleotide primers developed from *L. interrogans* serovar *hardjo* type hardjobovis obtained with the PCR a positive signal even in urine samples seeded with less than 10 leptospires. In addition, positive results were obtained with these primers in the PCR applied on urine samples of naturally infected cattle. These examples indicate that this new technique opens new possibilities in microbial diagnosis.

VII. SUMMARY

DNA probes for known microorganisms can be used in NA hybridization tests to detect and identify corresponding target DNA of the same or related microorganisms. Probes can be prepared from total genomic DNA or from cloned DNA fragments. The formation of DNA probe–target DNA duplexes can be visualized by radioactive, enzymatic, or immunological detection systems. In dot-blot hybridization, the test is performed on purified target DNA, partially purified DNA, or directly on patient specimens applied on a solid surface, usually NC or nylon. In *in situ* hybridization, the test is performed directly on intact microorganisms in smears or tissue sections fixed to microscope slides.

With dot-blot and *in situ* hybridization leptospires can be detected in various clinical samples. Cloned DNA probes allow not only detection but also more specific identification of target DNA than total genomic DNA probes. Preliminary observations suggest that hybridization can be used to detect leptospires in human urine samples, but the test appears to be insufficiently sensitive to detect leptospires in human blood. Dot-blot hybridization can be used to detect and identify *H. influenzae* in various clinical samples. Both dot-blot and *in situ* hybridization seem to be suitable for the detection of *Haemophilus* in sputum samples, in particular in sputum from CF patients. Various hybridization methods (dot-blot and *in situ* hybridization) have been described to detect *Campylobacter* in clinical samples. A single series of observations on fecal samples suggests that these methods are not very sensitive.

In general, additional observations on various clinical samples must be available to allow a reliable evaluation of the several hybridization methods used. Although NA hybridization is a promising new tool in microbiology, its sensitivity must be further increased for large-scale use as an addition, or even as an alternative, to conventional diagnostic methods in the routine diagnostic laboratory. *In vitro* amplification of the target sequences by the newly developed PCR will probably be helpful in improving the sensitivity. In addition, radioactive detection systems need to be replaced and the test procedures simplified and expedited.

VIII. MATERIALS AND METHODS

A. Extraction of Genomic DNA from Bacterial Cells

1. Collect the cells by centrifugation (10 min, 3000 *g* for *Haemophilus* and *Campylobacter*, or 30 min, 10,000 *g* for *Leptospira*).
2. Resuspend the cells in phosphate buffered-saline (PBS).
3. Add sodium dodecyl sulfate (SDS) to a final concentration of 0.5% and shake gently until the solution is viscous and clear.
4. Add pronase (Sigma type I) to a final concentration of 100 μg/ml and incubate for 2 hr at 60°C.
5. Cool to 37°C and add 20 μg/ml proteinase K (Sigma type XI). Incubate for 2 hr at 37°C.
6. Add one volume of buffered phenol and shake gently for 30 min. Centrifuge 15 min at 3000 rpm in a tabletop centrifuge (Beckman JT6) and transfer the water phase to a clean tube.
7. Repeat step 6. Reextract phenol with a small volume of H_2O.
8. Extract twice with an equal volume of chloroform–isoamyl alcohol (24 : 1). Centrifuge and transfer water phase to a clean tube.
9. Add sodium acetate, pH 5.2, to a final volume of 0.3 *M*.
10. Add 2.5 volumes of ice-cold ethanol. Let stand for 10 min.
11. Spool the DNA precipitate with the bent end of a Pasteur pipette.
12. Wash the DNA precipitate in 70% ethanol and transfer to a tube containing 10 m*M* Tris-HCl–1 m*M* EDTA, pH 7.5 (TE).
13. Dissolve DNA completely. Determine DNA concentration spectrophotometrically (OD_{260} of 1 is equal to ~50 μg/ml DNA).
14. This DNA preparation also contains RNA. If necessary, digest the sample (for 30 min at 37°C) with DNase-free RNase, final concentration 100 μg/ml.
15. Extract once with phenol and once with chloroform–isoamyl alcohol (24 : 1) and precipitate DNA with ethanol. Resuspend DNA in TE.

B. Rapid Procedure for DNA Extraction of Bacterial Cells from Blood or Urine for Use as Target DNA in a Dot-Blot Assay

1. Centrifuge 1 ml serum or urine in a microcentrifuge (10,000 *g*, 10 min).

2. Add to the pellet 250 μl 1% SDS in TE.
3. Vortex briefly and add NaCl to a final volume of 0.1 *M*.
4. Extract the lysate with a phenol–chloroform–isoamyl alcohol (25 : 24 : 1) mixture. Transfer the water phase to a clean tube.
5. Add 600 μl ice-cold ethanol and precipitate DNA for 30 min at −20°C.
6. Centrifuge DNA 15 min in a microcentrifuge (10,000 *g*).
7. Suspend the pellet in 5 μl 0.5 *M* NaOH.
8. Neutralize denatured DNA by the addition of 5 μl 0.5 *M* HCl in 0.5 *M* Tris-HCl, pH 7.5. This DNA preparation is suitable for spotting on a NC membrane.

C. Standard Nick-Translation Procedure

A typical reaction contains 0.1–1.0 μg DNA in a volume of 50 μl. However, the reaction can be scaled down to volumes as small as 5 μl [see Rigby *et al.* (24)].

Reagents

10× buffer:	500 m*M* Tris-HCl (pH 7.2) 100 m*M* $MgSO_4$ 1 m*M* dithiothreitol (DTT) 500 μg/ml Bovine serum albumin (BSA)
DNase I:	Prepare a stock solution of 1 mg/ml in 0.15 *M* NaCl–50% glycerol. Store at −20°C.
dNTPs:	Stock solutions of deoxynucleotide triphosphates with concentrations of 1 m*M*.

Procedure

Keep all ingredients on ice.

Mix:	5 μl	10× Buffer
	x μl (as wanted)	DNA 1 μg
	1 μl	0.1 m*M* dATP
	1 μl	0.1 m*M* dGTP
	10 μl	[α-^{32}P]dCTP, 100 μCi
	10 μl	[α-^{32}P]dTTP, 100 μCi
	1 μl	0.1 μg/ml DNase I
	1 μl	5 U/μl DNA polymerase I

Add water to 50 μl.

Incubate 90 min at 16°C.

Stop the reaction by adding 2 μl of 0.5 *M* EDTA.

Separate the labeled DNA from unincorporated dNTPs by chromatography on Sephadex G-50.

Specific activities of $>10^8$ cpm/μg DNA can be achieved.

D. Nick-Translation Procedure for Labeling with Bio-11-dUTP

Reagents are the same as described for labeling with [^{32}P]dNTPs.

Procedure

Keep all ingredients on ice.

Mix:	5 μl	10× Buffer
	5 μl	0.3 m*M* dATP
	5 μl	0.3 m*M* dGTP
	5 μl	0.3 m*M* dCTP
	1 μl	0.3 m*M* dTTP
	2.5 μl	0.4 m*M* Bio-11-dUTP
	2.0 μl	0.1 μg/ml DNase
	1.0 μl	5 U/μl DNA polymerase I
	x μl (as wanted)	DNA (1 μg)

Add water to 50 μl.

Incubate 18 hr at 16°C.

Stop the reaction by adding 2 μl of 0.5 *M* EDTA. Labeled DNA can be purified by ethanol precipitation.

Note: Trace amount of [^{32}P]dATP can be added to the reaction mixture to calculate the incorporation of Bio-11-dUTP.

Note: At present we usually label DNA with the Random Primed DNA labeling procedure. A complete kit is supplied by Boehringer-Mannheim. We perform the labeling as recommended by the manufacturer. Higher specific activities can be achieved with lower amounts of labeled dNTPs.

E. Dot-blot Assay with Dot-Blot Apparatus

1. Float the NC membrane (Schleicher and Schuell, BA 85) 20 min in distilled H_2O then immerse the membrane in 10× SSC for another 20 min.
2. Denature sample DNA by heating for 3 min at 100°C and quench quickly in ice water. Add an equal volume of 20× SSC.
3. Prepare the minifold. Put supporting filter on the dot-blot device (Schleicher and Schuell, SRC 9) and wet it with 10× SSC. Layer the membrane on top of the supporting filter and cover with the top of the dot-blot device.
4. Transfer the denatured DNA samples to the wells and apply slight vacuum until the wells are dry.
5. Fill the wells with a same volume of 10× SSC and apply vacuum.
6. Disassemble the dot-blot device and bake the dry filters 2 hr at 80°C.

F. Spot Assay on a Dry Filter

Use this method if a dot-blot device is not available or when very small DNA spots are desirable.

1. Prepare filter as in step 1 (Section VIII, E).
2. Dry the filter between sheets of Whatman filter paper.
3. Spot 1-μl amounts of denatured DNA on filter.
4. Rinse the filter in 4× SSC and dry and bake for 2 hr at 80°C.

Note: For some nonradioactive detection systems it is advisable to miniaturize filters to allow hybridization with a high concentration (>200 μg/ml) of DNA probe.

G. Hybridization Procedure for Filters

1. Seal the filters in a plastic bag.
2. Add prehybridization solution and incubate for 2 hr at 68°C. Prehybridization solution contains 6× SSC, Ficoll 0.1%, polyvinylpyrrolidone 0.1%, BSA 0.01%, low molecular weight denatured 100 μg salmon sperm DNA/ml, and 0.5% SDS.
3. Denature DNA probe and quench on ice.
4. Add DNA probe to hybridization fluid, which contains 6× SSC, 100 μg salmon sperm DNA/ml, 1 m*M* EDTA, and 20% dextran sulfate.
5. Discard prehybridization solution and add hybridization fluid containing DNA probe to the filter in the plastic bag and seal.
6. Hybridize for 18 hr at 68°C.
7. Wash the filters 30 min in 2× SSC containing 0.1% SDS at 68°C.
8. Wash the filters 30 min in 0.1× SSC containing 0.1% SDS at 68°C.
9. Dry the filters and proceed according to the method necessary for detection of the labeled DNA.

Note: The stringency of the hybridization procedure can be varied by increasing or lowering the temperature and salt concentration of the washing solutions. Hybridization time can be shortened to 1 hr depending on the complexity of the DNA probe.

H. *In Situ* DNA Hybridization Protocol for Smears of Clinical Materials on Microscope Slides

Preparation of slides

1. Allow the smear to dry on the slide.
2. Fix the slides with methanol for 10 min.

Pretreatment of the sample preparation

3. Drop 100–200 μl of a freshly prepared solution of 0.5 mg pronase per milliliter of PBS on each slide. Incubate for 10 min.
4. Rinse the slides with PBS and incubate for 10 min in 1% H_2O_2 in PBS.
5. Wash for 2 min in PBS and dry the slides.

Hybridization

6. Drop 5 μl hybridization solution (see below) containing DNA probe (see below) on the sample preparation of the slide.
7. Cover the slide with a piece of plastic sealing tape to prevent evaporation of hybridization fluid.
8. Incubate slides in waterbath for 10 min at 90°C.
9. Incubate slides for ≥2 hr in a wet incubation chamber at 37°C.

Posthybridization washing

10. Wash the slides for 15 min in 0.1× SSC–50% formamide at room temperature.
11. Wash the slides twice for 5 min in PBS.

Detection with streptavidin peroxidase method

12. Incubate slides for 30 min at 37°C with a 1 : 200 diluted streptavidin biotinylated–horseradish peroxidase complex (Amersham) in PBS containing 1% (w/v) BSA.
13. Wash the slides three times for 3 min in PBS.
14. Incubate slides for 6 min with 200 μl of a freshly prepared solution of 0.05% DAB in PBS containing 0.01% H_2O_2.
15. Wash the slides twice with double-distilled H_2O and mount the slides with Aquamount.

The staining signal of DAB can be considerably amplified by the DAB enhancement method. All the reagents are available in a kit that is supplied by Amersham. Follow the protocol of the manufacturer. Use the method only if necessary, for example, when cloned fragments are used as DNA probe or for detection of bacteria in paraffin-embedded tissue sections.

Instead of detection of hybridization with peroxidase, one can also use the more sensitive alkaline phosphatase. However, the substrate needed for alkaline phosphatase gives a more diffuse precipitate than DAB, which makes it sometimes unsuitable for *in situ* hybridization of bacteria.

Hybridization fluid

50% Formamide

100 μg low molecular weight denatured salmon sperm DNA per milliliter

2× SSC (sodium saline citrate, 0.15 *M* NaCl–15 m*M* sodium citrate, pH 7.0)

20% Dextran sulfate
0.01 *M* Tris-HCl, pH 7.0

DNA Probe

At least 20% of the available deoxythymidine residues in the DNA were substituted by biotin-labeled deoxyuridine. Concentration of probe DNA varies between 100 and 2000 μg DNA per milliliter of hybridization fluid.

REFERENCES

1. Adler, B., Chappel, R. J., and Faine, S. (1982). The sensitivities of different immunoassays for detecting leptospiral antigen. *Zentralbl. Bakteriol. Mikrobiol. Hyg., Abt. 1, Orig. A* **252,** 405–413.
2. Casin, I., Grimont, F., and Grimont, P. A. D. (1986). Deoxyribonucleic acid relatedness between *Haemophilus aegyptius* and *Haemophilus influenzae. Ann. Inst. Pasteur/Microbiol.* **137B,** 155–163.
3. Chevrier, D., Megraud, F., Larzul, D., and Guesdon, J. L. (1988). A new method for identifying *Campylobacter* spp. *J. Infect. Dis.* **157,** 1097–1098.
4. Claus, P., Moseley, S. L., and Falkow, S. (1983). The use of gene-specific DNA probes for the identification of enteric pathogens. *Prog. Food Nutr. Sci.* **7,** 139–142.
5. Deich, R. A., Metcalf, B. J., Finn, C. W., Farley, J. E., and Green, B.A. (1988). Cloning of genes encoding a 15,000-dalton peptidoglycan-associated outer membrane lipoprotein and an antigenically related 15,000-dalton protein from *Haemophilus influenzae. J. Bacteriol.* **170,** 489–498.
6. Echeverria, P., Seriwatana, J., Sethabutr, O., and Taylor, D. N. (1985). DNA hybridization in the diagnosis of bacterial diarrhea. *Clin. Lab. Med.* **5,** 447–462.
7. Ellis, W. A., O'Brien, J. J., Neill, S. D., Ferguson, H. W., and Hanna, J. (1982). Bovine leptospirosis: Microbiological and serological findings in aborted fetuses. *Vet. Rec.* **110,** 147–150.
8. Ellis, W. A., O'Brien, J. J., Neill, S. D., and Hanna, J. (1982). Bovine leptospirosis: Serological findings in aborting cows. *Vet. Rec.* **110,** 178–180.
9. Eys, G. J. J. M. Van, Zaal, J., Schoone, G. J., and Terpstra W. J. (1988). DNA hybridization with hardjobovis-specific recombinant probes as a method for type discrimination of *Leptospira interrogans* serovar *hardjo. J. Gen. Microbiol.* **134,** 567–574.
10. Eys Van, G. J. J. M., Gravekamp, C., Gerritsen, M. J., Quint, W., Cornelissen, M. T. E., Ter Schegget, J., and Terpstra, W. J. (1989). Detection of leptospira in urines by polymerase chain reaction. *J. Clin. Microbiol.* In press.
11. Granoff, D. M., and Munson, R. S. (1986). Prospect for prevention of *Haemophilus influenzae* type b disease by immunization. *J. Infect. Dis.* **153,** 448–461.
12. Hathaway, S. C. (1981). Leptospirosis in New Zealand: An ecological view. *N. Z. Vet. J.* **29,** 109–112.
13. Høiby, N., Friis, B., Jensen, K., Koch, C., Møller, N. E., Stovring, S., and Szaff, M. (1982). Antimicrobial chemotherapy in cystic fibrosis patients. *Acta Paediatr. Scand., Suppl.* **301,** 75–100.
14. Kilian, M. (1985). *Haemophilus. In* "Manual of Clinical Microbiology" (E. H. Lennette, A. Balows, W. J. Hausler, Jr., and H. J. Shadomy, eds.) pp. 387–393. Am. Soc. Microbiol., Washington, D.C.

14a. Kosunen, T. W. and Hurme, M. (1986). Monoclonal antibodies against *Campylobacter* strains *In* "Monoclonal Antibodies against Bacteria" (A. J. L. Macario and E. Conway de Macario, eds.), Vol. 3, pp. 99–117. Academic Press, Orlando, Florida.

15. Kilian, M., Heine-Jensen, J., and Bülow, P. (1972). *Haemophilus* in the upper respiratory tract of children. A bacteriological. serological and clinical investigation. *Acta Pathol. Microbiol. Scand., Sect. B* **80,** 571–578.
16. Kroll, J. S., and Moxon, E. R. (1988). Capsulation and gene copy number at the cap locus of *Haemophilus influenzae* type b. *J. Bacteriol.* **170,** 859–864.
17. Landegent, J. E., Jansen in de Wal, N., Baan, R. A., Hoeymakers, J. H. J., and Van der Ploeg, M. (1984). C_2-acetylaminofluorene-modified probes for the indirect hybridocytochemical detection of specific nucleic-acid sequences. *Exp. Cell Res.* **153,** 61–72.
18. Le Febvre, R. B. (1987). DNA probe for the detection of *Leptospira interrogans* serovar *hardjo* genotype hardjobovis. *J. Clin. Microbiol.* **25,** 2236–2238.
19. Malouin, F., Bryan, L. E., Shewciw, P., Douglas, J., Li, D., Van den Elzen, H., and Lapointe, J. R. (1988). DNA probe technology for rapid detection of *Haemophilus influenzae* in clinical specimens. *J. Clin. Microbiol.* **26,** 2132–2138.
20. Millar, B. D., Chappel, R. J., and Adler, B. (1987). Detection of leptospires in biological fluids using DNA hybridization. *Vet. Microbiol.* **15,** 71–78.
21. Morris, G. K., and Patton, C. M. (1985). Campylobacter. *In* "Manual of Clinical Microbiology" (E. H. Lennette, A. Balows, W. J. Hausler, Jr., and H. J. Shadomy, eds.), pp. 302–308. Am. Soc. Microbiol., Washington, D.C.
22. Ou, C. Y., Kwok, S., Mitchell, S. W., Mack, D. H., Sninsky, J. J., Krebs, J. W., Feorino, P., Warfield, D., and Schochetman, G. (1987). DNA amplification for direct detection of HIV-1 in DNA of peripheral blood mononuclear cells. *Science* **239,** 295–297.
23. Owen, R. J. (1983). Nucleic acids in the classification of campylobacters. *Eur. J. Clin. Microbiol.* **2,** 367–377.
24. Rigby, P. W. J., Dieckmann, M., Rhodes, C., and Berg, P. (1977). Labeling deoxyribonucleic acid to high specific activity in vitro by nick translation with DNA polymerase I. *J. Mol. Biol.* **113,** 237–251.
25. Saiki, R. K., Gelfand, D. H., Stoffel, S., Scharf, S. J., Higuchi, R., Horn, G. T., Mullis, K. B., and Erlich, H. A. (1988). Primer-directed enzymatic amplification of DNA with a thermostable DNA polymerase. *Science* **239,** 487–491.
26. Stollar, B. D., and Rashtchian, A. (1986). Immunochemical approaches to gene probe assays. *Anal. Biochem.* **161,** 387–394.
27. Tenover, F. C. (1988). Diagnostic deoxyribonucleic acid probes for infectious diseases. *Clin. Microbiol. Rev.* **1,** 82–101.
28. Terpstra, W. J., Jabboury-Postema, J., and Korver, H. (1983). Immunoperoxidase staining of leptospires in blood and urine. *Zentralbl. Bakteriol. Mikrobiol. Hyg., Abt. 1, Orig. A* **254,** 534–539.
29. Terpstra, W. J., Schoone, G. J., and Ter Schegget, J. (1986). Detection of leptospiral DNA by nucleic acid hybridization with ^{32}P- and biotin-labelled probes. *J. Med. Microbiol.* **22,** 23–28.
30. Terpstra, W. J., Schoone, G. J., Ligthart, G. S., and Ter Schegget, J. (1987). Detection of *Leptospira interrogans* in clinical specimens by *in situ* hybridization using biotin-labelled DNA probes. *J. Gen. Microbiol.* **133,** 911–914.
31. Terpstra, W. J., Schoone, G. J., Ter Schegget, J., Van Nierop, J. C., and Griffioen, R. W. (1987). *In situ* hybridization for the detection of *Haemophilus* in sputum of patients with cystic fibrosis. *Scand. J. Infect. Dis.* **19,** 641–646.
32. Terpstra, W. J., Groeneveld, K., Eijk, P. P., Geelen, L. J., Schoone, G. J., Ter Schegget, J., Van Nierop, J. C., Griffioen, R. W., and Van Alphen, L. (1988). Comparison of two

nonculture techniques for detection of *Hemophilus influenzae* in sputum. *Chest* **94,** 126S–129S.

33. Terpstra, W. J., Schoone, G. J., Ligthart, G. S., and Ter Schegget, J. (1988). Detection of *Leptospira, Haemophilus* and *Campylobacter* in clinical specimens by in situ hybridization using biotin-labelled DNA probes. *Isr. J. Vet. Med.* **44,** 19–24.
34. Tompkins, L. S., and Krajden, M. (1986). Approaches to the detection of enteric pathogens, including *Campylobacter,* using nucleic acid hybridization. *Diagn. Microbiol. Infect. Dis.* **4,** 71S–78S.
35. Turk, D. C. (1984). Towards better understanding of *Haemophilus influenzae* infections. *Abst. Hyg. Communicable Dis.* **59,** R1–R15.
36. Turner, L. H. (1967). Leptospirosis I. *Trans. R. Soc. Trop. Med. Hyg.* **61,** 842–855.
37. Yasuda, P. H., Steigerwalt, A. G., Sulzer, K. R., Kaufman, A. F., Rogers, F., and Brenner, D. J. (1987). Deoxyribonucleic acid relatedness between serogroups and serovars in the family Leptospiraceae with proposals for seven new *Leptospira* species. *Int. J. Syst. Bacteriol.* **37,** 407–415.
38. Zuerner, R. L., and Bolin, C. A. (1988). Repetitive sequence element cloned from *Leptospira interrogans* serovar hardjo type Hardjo-Bovis provides a sensitive diagnostic probe for bovine leptospirosis. *J. Clin. Microbiol.* **26,** 2495–2500.

12

Nucleic Acid Probes for the Identification of *Salmonella*

FRAN A. RUBIN
Department of Bacterial Immunology
Walter Reed Army Institute of Research
Washington, D.C.

I. INTRODUCTION

Salmonellae are gram-negative bacilli that belong to the family Enterobacteriaceae. Members of this genus infect many animal species, resulting in clinical infections or asymptomatic colonization. Thus, animals serve as reservoirs for the transmission of *Salmonella* either through contaminated animal products or fecally contaminated water. Salmonellae cause a wide range of human disease from mild gastroenteritis to life-threatening typhoid fever. Gastroenteritis is a common problem in all areas of the world; it is associated with outbreaks of food poisoning, which have increased in developed countries as a result of mass production and distribution of food products. In addition, enteric fever continues to be a global health problem, endemic in many developing countries where contaminated water represents a major source of infection. There has been an increasing upward trend in the number of nontyphoid *Salmonella* isolates from humans in the United States with an average of 30,000–40,000 cases reported each year while only 500–600 isolates of *S. typhi* are reported yearly (11). In 1986 at the International Workshop on Typhoid Fever, it was estimated that 6.98 million cases of typhoid fever occur per year in south and east Asia, 749,000 in west Asia, 4.36 million in Africa, 15,000 in Egypt, and 406,000 in Latin America and the South Pacific Islands (18). These figures probably underestimate the incidence of typhoid fever, since many cases go undetected because of limited detection capability in developing countries.

Control of the diseases caused by *Salmonella* depends on public health and socioeconomic factors. Detection and identification of *Salmonella* in clinical samples, food, water, and other environmental samples is necessary in order to define endemic patterns, identify trends in disease transmission, and monitor control efforts. Rapid identification is important for initiating treatment to the patient and to prevent spread of the disease (11). In addition, time is a factor in industry where food, feed, and raw products are stored while being screened for presence of *Salmonella*.

The best method for identifying *Salmonella* would be rapid, sensitive, specific, and simple. Culture of the etiological organism is the "gold standard" for the definitive diagnosis of *Salmonella* infections. This method requires 2–5 days and includes isolation of bacteria, and biochemical and serological characterization. Although many immunological tests have been developed for detecting antigens or antibodies, none have gained widespread acceptance other than the Widal test. Nucleic acid (NA) probes offer a new approach to the detection and identification of *Salmonella*. Advances in molecular biology provide us with the tools to prepare NA probes from unique sequences in *Salmonella*. What kinds of probes have been developed? What kinds of detection systems are available? How do probes compare with immunological methods using antisera and monocolonal antibodies (MAb)? How complex is the methodology? Is special equipment necessary? Can the test be standardized so that results of different laboratories can be compared? Is the method adaptable to different kinds of samples? These questions will be evaluated.

II. BACKGROUND

A. Taxonomy

The classification of *Salmonella* has been controversial. Traditionally, strains were named after the form of disease it caused or the place where it was first isolated. There are >2400 strains in this genus based on serological typing of the somatic (O), flagellar (H), and capsular (Vi) antigens. Edwards and Ewing (20) proposed that there are only three species of *Salmonella: Salmonella cholerae-suis, Salmonella typhi,* and *Salmonella enteritidis,* with the other antigenic types being serotypes of *S. enteritidis*. However, the Kauffmann–White scheme (20) gives species status to each antigenic type. Serotypes with common O antigens are grouped together. These systems are based on biochemical and serological differences of the bacterial strains, but differ in the way strains are named. A third system divides salmonellae into subgenera by biochemical tests and is independent of serological differences. Subgroup I contains salmonellae that have been isolated from humans and warm-blooded animals; 99% of the salmonella cultures derived from human disease are in this subgroup. Salmonellae found in cold-blooded animals are in subgroups II–VI. Arizona strains are placed in subgroup III. Previous studies demonstrated that members of the genus *Salmonella* are >70% related by DNA hybridization studies (17,95) and suggest one species of *Salmonella* with six subspecies (60). Thus, the biochemical and DNA hybridization analyses are in agreement and the classification of salmonellae into six subgroups is the currently accepted taxonomic system. Table I illustrates the most recent classification system and includes the serotypes discussed in this review.

B. Epidemiology

Several clinical syndromes are seen with salmonellae: gastroenteritis, bacteremia (with or without localized infections), enteric fever, and a carrier state. *Salmonella* gastroenteritis is usually a mild, self-limited illness occurring after ingestion of contaminated food or water. The major source of nontyphoid human salmonellae is poultry and domestic livestock. Major outbreaks have occurred from contaminated dried and whole milk, raw milk, poultry, and poultry products such as eggs (11). Dogs and cold-blooded animals can carry *Salmonella* and have been implicated as sources of human infection. Acute *Salmonella* gastroenteritis is characterized by vomiting and diarrhea with a slight rise in temperature; a small proportion of the patients may become bacteremic. Salmonellae are the most common cause of food-borne outbreaks in the United States (12); the

TABLE I

***Salmonella* Classification System**[a,b]

Subgroup	Name of strain	Serogroup	Somatic (O) antigens	Flagellar (H) antigens	
				Phase 1	Phase 2
I	*S. paratyphi* A	A	1,2,12	a	[1,5]
	S. paratyphi B	B	1,4,[5],12	b	1,2
	S. typhimurium	B	1,4,[5],12	i	1,2
	S. heidelberg	B	1,4,[5],12	r	1,2
	S. paratyphi C	C_1	6,7,Vi	c	1,5
	S. choleraesuis	C_1	6,7	c	1,5
	S. newport	C_2	6,8	e,h	1,2
	S. muenchen	C_2	6,8	d	1,2
	S. typhi	D	9,12,Vi	d	—
	S. enteritidis	D	1,9,12	g,m	[1,7]
	S. dublin	D	1,9,12,[Vi]	g,p	—
II	*S. salamae*		1,9,12	l,w	e,n,x
IIIa	*Arizona arizonae*		51	z_4,z_{23}	—
IIIb	*Arizona diarizonae*		6,7	l,v	z_{53}
IV	*S. houtenae*		45	g,z_{51}	—
V	*S. bongori*		66	z_{41}	—
VI	*S. indica*		1,6,14,25	a	e,n,x

[a] Antigenic factors from Kauffman–White scheme as described in Edwards and Ewing (20) and Le Minor *et al.* (60).

[b] Numbers underlined represent phage-determined antigenic factors. Numbers in brackets represent antigens present only in some strains.

most frequently reported serotypes are *S. typhimurium, S. enteritidis, S. heidelberg,* and *S. newport* (11). Patients with *Salmonella* gastroenteritis may shed the organisms in their stool for several weeks and thus can be responsible for person-to-person spread of salmonellosis. Complications of self-limited gastroenteritis relate directly to dehydration and immunological status, and occur in the very young, the elderly, and the immunocompromised, for example, patients with malignancies and AIDS. The diagnosis is confirmed by identification of a *Salmonella* strain from stool culture.

Salmonella bacteremia can occur with and without gastrointestinal disease. Once bacteria have entered the bloodstream, localized infections can occur such as arthritis, cholecystitis, osteomyelitis, and meningitis. *Salmonella cholerasuis* has been the isolate most frequently associated with this clinical pattern. The diagnosis is made by isolating a strain of *Salmonella* from blood culture.

Enteric fevers result typically from infection with *S. typhi, S. paratyphi* A, or *S. paratyphi* B. These organisms are invasive, causing bacteremia and systemic disease. Although paratyphoid fever is generally milder than typhoid fever, the only method to distinguish typhoid from paratyphoid infection is isolation and identification of the etiological organism. Contaminated food or water is the usual source of infection. The pathogenesis of enteric fever in humans is not fully understood. The proposed sequence of events (42) has been based on information derived from studies on human volunteers, chimpanzees experimentally infected with *S. typhi*, and mice infected with *S. typhimurium*. After ingestion, bacteria invade the intestinal epithelium, enter the lymphatic and circulatory systems, and infect the reticuloendothelial organs. Salmonellae then multiply within mononuclear phagocytic cells of the lymphoid follicles, lymph nodes, liver, and spleen. At the end of the incubation period, 7–14 days, bacteria reenter the bloodstream causing a prolonged bacteremia and the onset of symptoms: fever, headache, malaise, and lethargy. Infection of other organs occurs at this time, with bacteria found most commonly in the liver, spleen, bone marrow, gallbladder, and Peyer's patches in the terminal ileum. Bacteria in the biliary system reinfect the intestinal tract and are excreted in the stool. Gastrointestinal symptoms include constipation or diarrhea. Rose-colored spots may appear. Bacteria are also excreted in the urine in a small proportion of cases. Complications include hemorrhage, intestinal perforation, and abscess formation. Delirium and other neurological signs indicate severe disease with a high mortality. Although mortality is generally ≤1% in nontreated patients, most patients are ill for 8–20 days and lose time from school or work (S. Hoffman, unpublished data; M. Finch, unpublished data). Bacteria can be recovered from blood (early in the disease), bone marrow aspirates, intestinal secretions (string capsule), urine, and stool (later in the disease). Culture and nonculture methods have been used with varying success for the diagnosis of enteric fever (Sections II,C and III).

Salmonella typhi is uniquely adapted to humans, and thus human carriers are the only known reservoir for these organisms. Approximately 3% of untreated patients with typhoid fever become chronic intestinal carriers of *S. typhi* who excrete the organism for ≥1 year. Asymptomatic transient carriage of <1 year occurs more frequently. The chronic carrier is characteristically an older woman with gallbladder disease. The typhoid bacilli reside in gallbladder stones and in scars in the biliary tree and are excreted in large numbers. Carriers contaminate food and water, spreading the infection to other individuals. The detection of carriers has been difficult since intermittent shedding can occur, resulting in negative stool cultures during nonshedding periods. Serological tests have been useful (see Section II,C,2,b). Identification and treatment of carriers is thought to be essential for the control of enteric fever in endemic areas. In addition, methods for identifying carriers are important in developed countries where food handlers from endemic areas may be carriers and cause sporadic outbreaks by directly contaminating food served to others (64).

C. Laboratory Detection of *Salmonella*

1. Culture and Immunodiagnostic Methods

Conventional bacterial culture methods are the "gold standard" for determining the presence of *Salmonella* in clinical specimens, foods, and feeds. These procedures generally require 2–5 days or longer. After bacterial colonies are isolated, biochemical and serological testing is necessary to identify the species of *Salmonella*. Since *Salmonella* strains frequently possess multiple O and H antigens, some in different antigenic phases, complete serotype identification is complex. Phage typing (38) and plasmid profile analysis (45) are sometimes employed to characterize further strains of the same serotype for epidemiological purposes. This process is very time-consuming, and improved methods need to be developed. Although salmonellae can usually be cultured from stools of patients with gastroenteritis, isolation of bacteria from patients with bacteremia and enteric fevers is not always successful. In clinical samples as well as food and feed, bacteria may be present in low numbers which would require enrichment or concentration of bacteria for detection. Cultural and immunodiagnostic approaches have been summarized in a previous monograph (62), and therefore only material pertinent to a comparison with NA probes will be discussed (Sections II,C,2, and V). Detection of salmonellae in food and feed will be discussed in Chapter 13 (Wernars and Notermans) of this monograph.

2. Typhoid Fever

a. Detection of *Salmonella typhi* in Typhoid Patients. Improvements are needed in the diagnosis of typhoid fever. Currently, bacterial culture of clinical specimens provides the definitive diagnosis. Not only is culture methodology time-consuming, but additional factors involved in the detection and identification of *S. typhi* in clinical samples require consideration. Although culture of a bone marrow aspirate is currently the most sensitive method for the diagnosis of typhoid fever (34,37,43), it is not practical to obtain these samples in many endemic areas. Another sensitive method, not always well tolerated by the patient, involves the culture of *S. typhi* from intestinal secretions obtained with a duodenal string capsule (5,7,44). Blood culture is generally the standard method for the diagnosis of typhoid fever in areas where bacteriology facilities are available, but a wide range of sensitivity has been reported (5,23,34,37,43). Some difficulties in culture methods relate to the low concentration of typhoid bacilli in the clinical specimen. The number of *S. typhi* in the blood of typhoid patients is generally <15 organisms per milliliter (68a,88,88a,104); therefore, blood culture may require long periods of incubation for isolation of typhoid bacilli (23). Culture of *S. typhi* from stool requires enrichment by employing media that selectively inhibit growth of fecal flora. The time that the specimen was collected, in relation to the length of illness, is another factor in isolating *S. typhi*. Positive blood cultures are common during the first 2 weeks of illness, but less common later; it becomes more difficult to isolate the organism from blood as the disease progresses. The opposite situation is found when recovering *S. typhi* from stool, with less positive cultures at the beginning of the illness. In addition, prior exposure of the patient to antibiotics may be a factor contributing to the lower sensitivity of blood culture as compared to bone marrow in isolating *S. typhi* from patients in certain geographic locations where antibiotics are readily available.

A number of assays that detect *S. typhi* antigens and antibodies have been described for the diagnosis of typhoid fever. These tests are summarized in a recent review (62). At this time, no nonculture method can be used to replace culture of *S. typhi* from clinical specimens.

b. Detection of Typhoid Carriers. As mentioned previously, chronic carriers may shed *S. typhi* intermittently. Therefore multiple stool cultures over a period of several months may be necessary to culture typhoid bacilli to identify a carrier. Since this is impractical, other methods have been and continue to be evaluated, including improved serological assays and NA probes (discussed here and in Sections III and V).

Vi serology has been used in the detection of chronic carriers since 1938, when the relationship between high serum titers of Vi antibody and the chronic carrier state was demonstrated (24). These methods have been controversial, with disagreements based on the interpretation of results. Titers of Vi antibody vary in different geographic locations, depending on the prevalence of typhoid fever. In addition, antibody titers are determined to be significant for a particular assay, thus it is difficult to compare results from laboratories using different assay systems.

The first tests used a *S. typhi* strain, expressing high levels of Vi, as the antigen in direct agglutination reactions (24), but were not specific nor sensitive when tested in an endemic area (8). Although improved sensitivity could be obtained by adsorbing crude or partially purifed Vi to erythrocytes in a passive hemagglutination assay (PHA), problems with specificity remained until a highly purified Vi antigen became available (57,73,74). In a comparison of PHA, using a crude extract of Vi versus purified Vi antigen, for detection of current typhoid carriers in a nonendemic area, the sensitivity was about the same (71–74%), but the specificity was increased from 72% to 100% with purified antigen, by reducing the number of false positives from 18% to 0% (73). The PHA with highly purified Vi antigen had a sensitivity of 75% and a specificity 92% when used to detect typhoid carriers in an endemic area (56).

Another serological method for identifying typhoid carriers involves detection of Vi antibodies using an enzyme-linked immunoassay (ELISA) (6), which was used to identify a typhoid carrier as the cause of an outbreak of typhoid fever on a southwestern United States Indian reservation (22). Recently, an asymptomatic carrier from an endemic area was identified as the source of a restaurant-associated outbreak of typhoid fever by measuring serum Vi antibodies using a radioimmunoassay (64). In addition, methods using fluorescent antibodies (13,16), counterimmunoelectrophoresis (14,39), and crossed immunoelectrophoresis (15) have been described for analyzing sera of chronic typhoid carriers. However, these techniques require expensive equipment, which restricts their use for detecting carriers in developing countries where typhoid fever is endemic.

D. Virulence Factors

1. Surface Antigens

Although attachment of *Salmonella* to receptor sites in host cells has not yet been delineated, it is likely that the surface O antigens have a role in adherence and virulence (25,61). The Vi capsular antigen has been recognized as a virulence factor (25,33,46,48,65). Assays measuring Vi antibodies have been developed that have been useful for the identification of typhoid carriers (see Section II,C,2,b). Immunological assays to detect Vi antigen in clinical specimens have been reported (see Section II,C). A DNA sequence encoding structural genes of the Vi antigen has been used as a DNA probe for detection and identification of *S. typhi* (86–88; see Section III).

2. Invasiveness

Like the invasive *Shigella, S. typhi* must invade the human gastrointestinal tract to cause disease. Perhaps outer-membrane proteins, similar to those described in *Shigella* (70), are necessary for invasion. Although the invasive mechanism(s) of *S. typhi* is not known, methods in molecular biology are being utilized to analyze this aspect of pathogenesis (21). This work may lead to the development of NA probes based on the invasion system of *S. typhi*.

3. *Plasmids and Other Factors*

Plasmids have been implicated in the virulence of nontyphoid salmonellae including *S. typhimurium* (52), *S. enteritidis* (72), and *S. dublin* (98). A common virulence region on plasmids from different *Salmonella* serotypes has been identified (106). Studies with *S. dublin* and *S. typhimurium* indicate that the virulence plasmid is not required for invasion, but is necessary for establishing a progressive infection in the reticuloendothelial system of infected hosts (40,67). Other factors influencing virulence that are currently being investigated include survival of salmonellae in macrophages (26) and serum resistance (36,41). These studies may result in the development of additional NA probes.

E. Nucleic Acid Probes

Genetic probes are defined NA sequences that are used to detect complementary sequences within a sample. Recombinant DNA technology provides the tools for cloning desired regions of plasmid or chromosomal NA, for the detection of certain pathogenic traits or pathogenic organisms. Some advantages of NA probes include (1) a high degree of specificity with a unique NA sequence, (2) the ability to detect a pathogen in a mixed bacterial population, and (3) the independence of detection from gene expression; that is, the presence of an intact antigen is not required, the NA sequence encoding the antigen is sufficient. In addition, NA probes should be able to detect nonviable organisms or viable bacteria in phagocytic cells. In the past decade, probes have been used successfully by molecular biologists, taxonomists, and epidemiologists. Recently, probes have been used to detect bacteria in clinical samples (19,55,71,85,88,101) as well as in food (80). Advances in DNA hybridization detection technology have resulted in several commercially available NA probes, with improvements and new probe systems not far behind. Nucleic acid probes have the potential for detecting and identifying clinically significant organisms in simple assays with a high degree of specificity and sensitivity, with detection systems that are more rapid than classical microbiological methods. This chapter will review results obtained with currently available *Salmonella* probes as well as those being developed.

III. RESULTS AND DISCUSSION

A. Nucleic Acid Probes for Detection and Identification of *Salmonella typhi*

1. Diagnosis of Typhoid Fever Using Vi DNA Probe

Vi capsular antigen is relatively unique, encoded by strains of *S. typhi*, *S. paratyphi* C, and a few atypical but genetically related *Citrobacter* and *Salmonella dublin*. Two distinct chromosomal loci, *viaA* and *viaB*, are necessary for Vi antigen expression (51,93,94); the *viaB* region appears to encode the structural genes for this antigen (50). Selected fragments of a *Citrobacter freundii viaB* locus, contained in a recombinant cosmid, were tested as potential probes for identification of *S. typhi* (86). Colony blot hybridization assays, with >180 stains representing a variety of enteric bacteria and including *Salmonella* groups A–Z, 51–55, 66, and 67, demonstrated that an 8.6-kb *Eco*RI fragment was highly specific for laboratory-maintained strains encoding Vi antigen. Next, this 8.6-kb Vi DNA probe was used to study a variety of microbiologically characterized, gram-negative bacteria freshly isolated from febrile patients in endemic areas of Lima, Peru and Jakarta, Indonesia (87). Initial studies in Peru indicated the importance of adequate filter preparation and compared the use of nitrocellulose (NC) and Whatman no. 541 filters for colony hybridization. The sensitivity of the Vi probe for the detection of bacteriologically identified *S. typhi* increased from 81% (295/365) to 96% (112/117) with modifications in the protocol that include filter preparation. In addition, the first set of colony hybridizations were done on NC filters while the second set included NC and Whatman no. 541 filters, demonstrating that Whatman filters are as sensitive as NC. In addition, Whatman no. 541 filters are inexpensive, do not require baking for fixation of DNA, have a high wet strength that makes handling easier, and can be reprobed easily. For all these reasons, only Whatman no. 541 filters were used to evaluate the Vi probe with Indonesian isolates; the results are summarized in Table II. The Vi probe was shown to hybridize with >99.7% (608/610) of bacteriologically identified *S. typhi* isolates. The specificity of the Vi probe was analyzed with respect to nontyphoid salmonellae as well as other gram-negative bacteria. The Vi probe did not react with 226 nontyphoid salmonellae from Indonesia or 133 nontyphoid isolates from Peru. Although no hybridization was detected with 90 non-*Salmonella* isolates from Indonesia, the Vi probe did react with 4 of 377 gram-negative isolates from Peru. Two of these strains were identified in Peru as *C. freundii* and the others as *Proteus mirabilis* and *Enterobacter*. Unfortunately, these strains were not available for further analysis. Of 469 gram-negative, non-*Salmonella* isola-

TABLE II

Vi DNA Probe Analysis of Bacterial Isolates from Indonesia[a]

Isolates tested	Number of hybridization-positive colonies per total number of colonies tested
Salmonella typhi	608/610
Nontyphoid salmonellae	0/226
Other gram-negative bacteria[b]	0/90

[a] Colony blots using Whatman no. 541 filters hybridized with radiolabeled Vi DNA probe, as described previously (86).

[b] Including *Escherichia coli, Enterobacter* spp., *Klebsiella* spp., and *Shigella* spp.

tes evaluated in Peru and Indonesia, 50 colonies were identified bacteriologically as *C. freundii* (44 from stool and 6 from blood); only two of these *C. freundii* stool isolates reacted with the Vi DNA probe Problems with Vi+ *Citrobacter* in blood specimens are not anticipated. This can be supported by recent studies in which *C. freundii* was not cultured from any of 144 patients in Peru and Indonesia with symptoms of typhoid fever who had conventional blood cultures within the last year (F.A. Rubin, unpublished results).

Although the Vi probe was highly specific and sensitive for the detection of *S. typhi* among bacterial isolates, demonstrating its usefulness for epidemiological studies, it could not be used to probe blood from patients directly because of sensitivity limitations. The probe cannot detect <500 bacterial cells (86), but typhoid patients generally have <15 bacterial per milliliter of blood (104). We developed a method that concentrates bacteria in a blood sample and subsequently amplifies the total amount of bacterial DNA. Our next study was designed to determine if the Vi probe could be used with this method to detect bacteria in the blood of patients with typhoid fever (88). In this study, clinical specimens including blood, bone marrow aspirate, and stool were collected from febrile patients suspected of having typhoid fever in Indonesia. Blood was processed for the probe assay as follows. After lysis–centrifugation (Dupont Isolator tube) of 10 ml of blood, a 1-ml concentrate was obtained that consisted of lysed blood cells and any bacteria that may have been in the blood sample. A 250-μl sample of this concentrate was spotted onto a nylon filter placed on top of a nutrient agar plate. After overnight incubation at 37°C, filters were processed for hybridization. These filters were then hybridized with the Vi DNA probe. In addition to processing blood for the probe study, standard blood cultures using 8 ml blood as well as cultures of bone marrow aspirates and rectal swabs were done on each patient. The Vi probe detected *S. typhi* DNA from the blood of 14 of 33 culture-confirmed typhoid fever patients, using the equivalent of 2.5 ml of blood. The blood culture was positive for *S. typhi* in 17 of the same 33 patients, using 8 ml of blood. The probe hybridized to specimens from 4/47 patients from whom *S. typhi* was not isolated; filters from 3 of these 4 patients had only a few reactive spots, which may represent low levels of bacteremia that were picked up by the probe. This study demonstrated that a DNA probe could be used to detect *S. typhi* that are present in blood when bacteria in blood are concentrated and the bacteria DNA is amplified by overnight incubation of bacteria on nylon filters (88). Although the sensitivity of the Vi probe was less than the sensitivity of standard blood culture, only the equivalent of 2.5 ml of blood was probed as compared to the culture of 8 ml of blood. Currently, a study is being conducted in Lima, Peru to determine the sensitivity of probing the entire 1-ml concentrate obtained after lysis–centrifugation of 10 ml of blood, as compared to culturing 10 ml of blood using standard bacteriological methods. Preliminary results in this study indicate that DNA probing is as effective as blood culture in detection of *S. typhi* from blood of typhoid patients (88a).

2. *Determination of Typhoid Chronic Carriers Using Gene Probes*

A chronic carrier of typhoid fever is defined as an individual from whom *S. typhi* has been isolated for at least 1 year after an episode of bacteriologically confirmed typhoid fever. It is generally assumed that the necessity for stool cultures at multiple intervals to isolate *S. typhi* from chronic carriers is explained by intermittent shedding. However, an issue to be considered is the success of isolating *S. typhi* in relation to the amount of *S. typhi* in stool samples. The numbers of bacilli in the stool from chronic carriers varies widely. Although selective and differential media are used to inhibit growth of the fecal flora, this may not result in enrichment of growth for low numbers of typhoid bacilli.

Nucleic acid probes offer a new approach to the identification of typhoid carriers. Numerous studies have demonstrated the use of gene probes for the detection and identification of pathogenic *Escherichia coli* in stool in the presence of normal intestinal flora [reviewed in Rubin and Kopecko (85)]. Although stool blots have been hybridized with NA probes for identification of specific pathogens in stool (55,71), this technique may not be appropriate for detection of *S. typhi*. In a study with a small group of culture-positive stool samples from a restaurant-associated typhoid fever outbreak, a positive hybridization reaction with Vi probe was only observed with the stool blot from one of six culture-confirmed specimens

(F.A. Rubin, unpublished results). When dilutions of these stool specimens were spread onto selective and differential media, varying numbers of *S. typhi* were isolated. It appeared that *S. typhi* present in these samples were overgrown by the normal fecal flora and/or affected by inhibitors in stool. Therefore, a modification in sample preparation is necessary for NA probes to be used for detection of *S. typhi* in the stool of chronic carriers. Small numbers of verotoxin-producing bacteria in stool samples were detected using a NA probe (92); colony blots for hybridization were prepared by lifting bacterial growth from the dilution of stool, which when spread on an agar plate resulted in the growth of isolated colonies. This colony-lift technique is currently being used to compare probe technology (for detecting *S. typhi* in stool) with a serological assay (PHA for Vi antibody) in a group of chronic typhoid carriers in Lima, Peru (88b). In a preliminary study with a small number of carriers, some of the colony blots hybridized with the Vi probe resulted in hybridization-positive spots (5–40), but some colony blots had only a single equivocal spot. Therefore, to confirm the presence of *S. typhi,* colony blots will be hybridized initially with the Vi DNA probe, the Vi probe will be removed, and then the blots will be hybridized with a NA probe directed at the d flagellar antigen, as described later.

Salmonellae are motile. More than 60 types of flagella have been identified, based on antigenic determinants present on these filaments (20); serotypes of salmonellae alternate between the production of two flagellar antigenic forms, phase 1 and phase 2. Although there are separate structural genes for each phase, phase 2 antigens are shared by various salmonellae and thus are nonspecific as compared to phase 1 antigens. The specificity of flagellar antigens is determined by flagellin, the monomeric protein of the flagellar filament. The NA sequences that encode phase 1 flagellins a, c, d, and i, have been cloned (53;105; G. Frankel and S. Newton, unpublished results). A comparison of the flagellins indicated that the amino acid sequences are conserved at both ends of the molecule, with variable regions between them (105). Subclones from the variable region are a potential source of gene probes for use in clinical specimens. Recently, an oligonucleotide probe was developed from a NA sequence cloned from *S. muenchen* chromosomal DNA encoding phase 1, d flagellar antigen (G. Frankel and S. Newton, unpublished results). This flagellar probe is designated Hd and is being tested in conjuction with the Vi probe. Although the Hd probe will hybridize with other strains of salmonellae that are capable of synthesizing d flagellar antigen, *S. typhi* is the only stain of *Salmonella* that is capable of expressing both Vi antigen and d flagellar antigen. Thus, colony blots prepared from chronic carriers in Lima, Peru will be hybridized sequentially with the Vi probe and the Hd probe to confirm the identification of *S. typhi*. Orientation of filters will be maintained for autoradiography. We expect that the presence of *S. typhi* will be detected by positive signals at identical locations on filters, following independent reactions with the Vi probe and the Hd probe (88b).

B. Nucleic Acid Probes for Nontyphoid Salmonellae

1. Genus-Specific Probes

Salmonella-specific probes are currently being used for quality control in industry (e.g., detection of *Salmonella* in foods). Although detection of food-borne pathogens will be discussed in another chapter in this monograph, a brief description of this successful gene probe system is included in this chapter for comparison to other probes for salmonellae.

Fragments prepared from cloned *S. typhimurium* DNA that are specific for identifying all salmonellae have been incorporated into the commercially available Gene-Trak *Salmonella* Assay (Framingham, Massachusetts). Initial testing involved hybridization reactions against a variety of *Salmonella* isolates, representing >180 serotypes and >340 individual strains; the probes were found to detect all salmonellae tested (27,28). In addition, when the probes were screened against a panel of nonsalmonellae representing related (enteric) and unrelated bacteria, no hybridization was detected outside the genus *Salmonella* (27). The sensitivity of the assay has been established at 5×10^6 *Salmonella* per milliliter in the final enrichment broth. Although the Gene-Trak procedure involves cultural enrichment steps, this system allows identification of *Salmonella*-free samples within 48 hr instead of the minimum of 4 days required for the standard method, BAM/AOAC [Bacteriological Analytical Manual of the FDA (31)/Association of Official Analytical Chemists (4)]. Results of a field trial (30) as well as an AOAC collaborative study (29) demonstrated that the Gene-Trak assay was as effective as the BAM/AOAC reference method in detection of *Salmonella* in the food products tested and significantly better in some cases (i.e., raw turkey). A second-generation *Salmonella* test employing ribosomal RNA (rRNA) target hybridization and a colorimetric detection system has recently been introduced.

The SNAP *Salmonella* culture identification diagnostic test is a salmonellae-specific assay being developed by Molecular Biosystems Inc. (MBI), San Diego, California. In this nonisotopic probe system, the target(s) is chromosomal DNA. In a study analyzing 515 stool culture isolates, 230 isolates that were bacteriologically identified as salmonellae gave a positive reaction in the SNAP test. All but 1 of the 285 nonsalmonellae isolates were SNAP-negative. The remaining isolate was biochemically identified as *Edwardsiella tarda,* which may have been an atypical strain, since the ATCC type strain for *E. tarda* tested by this probe was consistently negative. In addition, other studies demonstrate that no cross-reactivity was observed when colonies from 21 other genera were tested, including enteric bacteria (81). In a comparison of the SNAP *Salmonella* test to a commercially available latex agglutination test evaluating 378 culture specimens, the specificity, sensitivity, and overall agreement were 92%, 96%, and 94%. Of the discrepant results, 12 specimens were SNAP-positive and latex-negative; 11 were determined to be *Salmonella*-positive by the culture reference method and 1 was determined to be non-*Salmonella*. All 9 of the SNAP-negative, latex-positive specimens were determined to be non-*Salmonella* by the reference culture method (L. Risen, unpublished results).

2. Species-Specific Probes

Species-specific NA probes are not currently available for detection of nontyphoid salmonellae. However, species-specific patterns were observed when cloned random chromosomal sequences of *S. enteritidis* were used to probe restriction endonuclease digests of whole-cell DNA from strains of *S. enteritidis, S. dublin,* and *S. typhimurium* (100), indicating regions for further analysis and for probe development. In addition, the probes used in this study detected many unique chromosomal fingerprint patterns among strains of the same serotype, some that appeared to have identical restriction endonuclease digestion patterns, demonstrating the power of gene probes to detect rearrangements in chromosomal fragments.

IV. CONCLUSIONS

Nucleic acid probes for identifying salmonellae are highly specific. Extensive testing of the genus-specific probes in the Gene-Trak *Salmonella* assay and the SNAP^R *Salmonella* test has demonstrated hybridization to salmonellae and no cross-reactivity with unrelated and related (enteric) bacteria. The Vi probe has been shown to be highly specific for clinical use. However, the Vi probe hybridizes to all laboratory strains encoding Vi antigen, including *S. typhi, S. paratyphi* C, atypical Vi$^+$ *S. dublin,* and the *C. freundii* strain initially used for cloning. The Vi antigen is only found in these related strains, and is unique compared to other bacterial antigens. It is unusual to culture atypical Vi$^+$ *S. dublin* or *S. paratyphi* C strains from the blood of patients with symptoms of typhoid fever, making such bacteria an unlikely cause for false positives.

However, Vi$^+$ *C. freundii* may be present in fecal flora, and hybridization of these bacteria with the Vi probe in stool samples processed with new methods [i.e., extractor columns and/or polymerization chain reaction (PCR); discussed below and in Section VI], could interfere with the detection of *S. typhi*. As discussed before, sequential hybridization of bacterial DNA from stool samples with two probes, the Vi probe and the Hd probe, is likely to be quite specific for the identification of *S. typhi*. Nucleic acid probes and sequences that have been used for hybridization with salmonellae are summarized on Table III.

The use of NA probes for identifying salmonellae in clinical samples is currently limited by the number of bacteria in the sample and the sensitivity of the detection system. Thus, sample preparation becomes a critical issue; for example, samples need to be processed in a way that results in sufficient salmonellae DNA being available for hybridization with a gene probe. New approaches to sample preparation and more sensitive detection systems are discussed in Section VI. For epidemiological studies, methods that result in growth of salmonellae may be all that is needed. For example, large quantities of potentially contaminated water could be filtered, the filter placed on an nutrient agar plate at 37°C overnight, and any resulting colonies probed for the presence of salmonellae. In addition, as mentioned earlier, gene probes can be used to hybridize colony blots prepared by lifting isolated colonies on agar plates that resulted from growth of diluted stool samples (0.1 ml of 10^{-4} or 10^{-5} dilution of stool spread onto the plate).

Time is another factor that needs to be considered when assessing the role of NA probes in identifying *Salmonella* in clinical specimens. Although the Gene-Trak *Salmonella* assay provides the food industry with a more rapid method than what was currently available, the time must be reduced even further if a gene probe assay is to be useful in clinical medicine for initiation of treatment. The Vi probe study, using lysis–centrifugation for concentration of the typhoid bacilli from blood of typhoid patients and subsequent amplification of bacterial DNA by overnight growth, demonstrated that a NA probe could be used to identify *S. typhi* directly in the blood of typhoid patients in Indonesia. A rapid assay with a gene probe could be developed by replacing the culture step of this protocol with another method for amplifying bacterial DNA. The polymerization chain reaction (PCR) is a new method for amplifying DNA (89), which has been used to detect human immunodeficiency virus in the blood of AIDS patients (58,76). PCR has also been used with other clinical specimens (7a,75) and hopefully will provide the technology for replacing bacterial culture.

TABLE III

Defined Nucleic Acid Probes and Other DNA Sequences Used for Hybridization with Salmonellae

Specificity	Source of DNA sequence[a]	Use of probe (target)	Reference
Vi⁺ Strains including *S. typhi*	*C. freundii*, encoding Vi antigen (C)	Clinical samples	86–88b
Hd flagellar strains including *S. typhi*	*S. myenchen*, encoding d flagellar antigen (C)	Clinical samples	G. Frankel and S. Newton (unpubl results)
S. typhi	*S. typhi*, encoding epithelial cell invasion (C)	Research	21
S. enteritidis,[b] *S. dublin*,[b] and *S. typhimurium*[b]	*S. enteritidis*, random fragments (C)	Fingerprint analysis	100
Salmonella serotypes[c]	*S. typhimurium*, associated with virulence (P)	Research	106
Salmonella spp.	*S. typhimurium*, fragments (C)	Food	27–30
Salmonella spp.	Information proprietary.	Clinical samples	81

[a] C, Chromosome; P, Plasmid.

[b] Strains probed in these limited studies; the specificity of these sequences has not been defined.

[c] Eleven of 34 *Salmonella* serotypes tested.

V. GENE PROBES VERSUS ANTISERA AND MONOCLONAL ANTIBODIES

How do gene probes compare with immunological detection systems (i.e., antisera and MAb) for identification of salmonellae? Nucleic acid probes offer increased specificity for detection of salmonellae over immunological methods. The specificity of gene probes is based on the unique sequence that has been cloned directly from the nucleic acid of salmonellae. Genus-specific NA probes react only with salmonellae and do not cross-react with other Enterobacteriaceae. Since MAb are derived from single clones of cells [from hybridomas (54) or myelomas (77)], they have an increased specificity over polyclonal antisera. However, only a limited number of MAb are currently available for salmonellae and none is genus-specific as are NA probes. Myeloma 467 (M467) is a MAb that recognizes a flagellar antigenic determinant common to many salmonellae (82). When M467 was tested in a colorimetric enzyme immunoassay (EIA), positive reactions were detected in 93% of 800 different salmonellae serotypes as well as a few serotypes of *Yersinia enterocolitica* and one strain of *C. freundii* (1). Salmonellae serotypes not detected by M467 include *S. typhi, S. paratyphi* A, *S. paratyphi* B, and *S. paratyphi* C. Thus, M467 has been combined with 6H4, a complementary MAb that recognizes a nonflagellar salmonellae antigen (68), in a commercially available EIA (*Salmonella* Bio-EnzaBead Screening Kit, OrganonTeknika Corp., Durham, North Carolina). When this kit was used for detection of salmonellae in naturally contaminated foods and animal feed, results showed an 82% agreement between the EIA and cultural methods (99).

Vi MAb, produced by immunization to purified *Citrobacter* Vi, detected Vi-positive bacteria with the same specificity as the Vi gene probe (102). Thus, the Vi MAb represent an improved serological reagent that could be used in place of commercially available Vi-agglutinating serum.

Although gene probes and MAb are more specific for detecting salmonellae than polyclonal antisera, conventional antisera will continue to be used as an important diagnostic reagent because of the limited availability of MAb and gene probes. Specific antisera, prepared by hyperimmunizing rabbits, needs to be extensively adsorbed with appropriate serotypic strains to remove agglutinins for the many common cross-reacting antigens in salmonellae species and related bacteria. The use of polyclonal antiserum, in several different assay systems developed for specific detection of *S. typhi* antigens, has resulted in limited application of these methods because of problems with cross-reacting antibodies (2,49,84,96).

Affinity chromatography can be used to increase the specificity of polyclonal antiserum. An affinity-purified polyclonal antibody that recognizes common structural salmonellae antigens has been used in an enzyme-linked immunosorbent assay (ELISA; BacTrace *Salmonella* antibody (CSA-1), Kirkegaard and Perry Laboratories, Inc., Gaithersburg, Maryland). In testing specimens that were not cultured enriched nor concentrated from patients with cultures that were positive for *Salmonella*, the sensitivity of this ELISA for the detection of *Salmonella* antigen in serum and urine was 48% and 81%, respectively (3). There are potential problems relating to the affinity of an antibody for its antigenic epitope, which will vary among the *Salmonella* strains; this issue represents a factor that may limit the use of immunoassays in general. Under certain conditions BacTrace *Salmonella* antibodies have been shown to cross-react with low affinity to some strains of *E. coli* and related Enterobacteriaceae. Since the quantitative interpretation of this colorimetric assay is subjective, one needs to be careful in analyzing the results as the aforementioned factors will contribute to color development. At present, information from this ELISA could be useful in conjunction with culture methods.

Despite having increased specificity, the successful use of gene probes and MAb for identifying salmonellae in clinical specimens will ultimately depend on the sensitivity of the assay system and time required to complete the assay. Factors to be considered include the amount of bacterial target (antigens/nucleic acid) present in the clinical specimen, sample preparation (processing the clinical specimen), and the limits of detection of the assay system. An advantage of gene probes is that the integrity of salmonellae antigens at the time the clinical sample is collected and during processing of the sample is not a factor, since the target in the clinical specimen is bacterial DNA.

The amount of target in the clinical specimen is a concern with both gene probes and MAb and will have a direct effect on the length of time needed to complete the assay, and possibly sample preparation. As described in Section III, in order to use the Vi DNA probe to detect *S. typhi* in the blood of typhoid patients, sample preparation included lysis–centrifugation and overnight culture because the amount of bacterial DNA in the clinical sample (average bacteremia 11 organisms per milliliter) was below the limits of detection of the hybridization assay (500 bacterial cells). As discussed in Sections IV and VI, incorporation of new technology (PCR and oligonucleotide probes) into the gene probe system is expected to reduce the time required for detection of *S. typhi* in clinical specimens from 2–3 days to within 8 hr of specimen collection.

As with gene probes, the use of MAb for detection of salmonellae is also dependent on the amount of target (antigen) in the sample in relation to the limit of sensitivity of the assay system. Monoclonal antibodies produced against the O-9 antigen of *Salmonella* bind only to salmonellae in serogroup D and thus have diagnostic potential. However, limited sensitivity was demonstrated for detection of *S. typhi* when latex agglutination reactions (latex particles sensitized with *Salmonella* O-9 MAb) were performed on blood culture broths (63). Although 100% detection was achieved in broth cultures after 5 days of incubation, most (73%) were detected after 3 days, but only 18% were positive after the first day. These results could be explained by the limit of sensitivity of the latex agglutination (sensitized with *Salmonella* O-9 MAb) test, which had been determined at 10^8 organisms per milliliter (62). Similar antigen detection studies, using polyvalent sera in a coagglutination (COAG) test for detection of *S. typhi* in blood cultures, have been reported 22% positive in the first day with blood cultures (83) but 60% positive in the first day with blood clot culture (69); both COAG tests resulted in 95% positive detection after 7 and 5 days of culture, respectively. However, these COAG tests included antisera against Vi and antisera against salmonellae somatic antigens. Perhaps better results could be obtained if the *Salmonella* O-9 MAb was combined with the Vi MAb in a serological test to detect *S. typhi* in clinical specimens or blood cultures, using a detection system with more sensitivity.

Although the amount of antigen, associated or unassociated with the bacterial cell, in clinical specimens is a important factor, few data are presently available. If antigens are associated with the bacterial cell, increased sensitivity in assays with MAb may be obtained by concentrating bacteria in the sample. Assays with MAb have the potential to generate quantitative data on the amount of antigen present in clinical specimens. If assays with MAb can directly detect antigens at levels present in clinical specimens, there would be a great advantage in using MAb as diagnostic reagents. Monoclonal antibodies STP13 and STP14, produced against Barber protein (Bp, a 34-kd protein antigen prepared from *S. typhi*), could detect Bp antigen at $\sim$0.6 μg/ml in a modified double-antibody sandwich ELISA (91). This sensitivity was determined with various concentrations of Bp; evaluation of this assay with clinical samples will determine the diagnostic potential of the Bp MAb.

Monoclonal antibodies have been produced against several protein antigens of *S. typhi* (62,63,79,91); these MAb should be useful in studies of antigens (chemical structure and immunological properties), of immunological response of host to these antigens, and for diagnosis (serotyping and detection of antigens in clinical specimens). Gene probes have been used the detection of salmonellae in the food industry and several are being developed for diagnostic and epidemiological applications. Additional gene probes and MAb will be available in the future. It is expected that this new technology will make significant contributions to the field of public health by increasing our knowledge of the pathogenesis and immunology of salmonellae infections.

VI. PROSPECTS FOR THE FUTURE

Nucleic acid probes offer a new approach for the identification and analysis of salmonellae. In the research laboratory, gene probes are used for bacterial classification as well as for genetic studies of samonellae that include proteins, antigens, virulence, invasion, and pathogenesis. The limiting factor is the creativity of investigators for finding unique sequences that can be applied to their specific investigations. Methods are available and continue to be developed for using NA probes to detect and identify salmonellae in food, feeds, and clinical and epidemological samples. At the present time there are a limited number of gene probes; however, one can expect a large increase in the future. What kinds of probes and problems will be encountered?

Genus-specific probes have been used for the detection of salmonellae in the food industry (30). However, in clinical medicine genus-specific probes would have limited use without species-specific probes for further identification. When a sample is being prepared for hybridization, a positive reaction with a genus-specific probe would need to be followed up by tests with species-specific probes to determine patient management. In the case of typhoid fever, a gene probe is already available (85–87). Other species-specific probes are being developed. One expected source of new probes will be virulence studies. For example, Elsinghorst (21) has cloned a DNA sequence from *S. typhi* that is responsible for invasion into epithelial cells, a portion of which could serve as a gene probe similar to the invasion-associated probes that have been constructed for detection of *Shigella* spp. and enteroinvasive *E. coli* (103).

In epidemiological studies, gene probes have been used for large-scale studies that would have been impractical with standard bacteriological methods [reviewed in Rubin and Kopecko (85)]. Genus-specific probes could be used in screening patients, food, and environmental samples in outbreaks of salmonellae to identify sources of infection. Although chromosomal probe fingerprint analysis has been demonstrated as a valuable epidemiological tool (100), it is time-consuming and currently limited to a few probes.

Since the number of salmonellae in a sample (e.g., blood), is a limiting factor in the use of gene probes for identifying salmonellae, new methods for sample preparation will be required. As described earlier, obtaining sufficient bacterial DNA from the clinical sample to react with a gene probe is imperative for a rapid assay system. Although we expect that amplification of bacterial DNA in clinical samples will be possible with PCR, some unique processing of samples may be necessary. For example, extractor columns (MBI, San Diego, California) are now available for obtaining nucleic acid from stool samples. Bacterial DNA produced from extractor eluates of stool samples and PCR-amplified has been successfully hybridized with gene probes (75). In addition, bacterial DNA obtained from rectal swabs has been amplified using PCR and hybridized with gene probes (G. Frankel, unpublished results). Thus, PCR or similar techniques for amplifying target DNA copy number should now allow one to detect small numbers of pathogenic bacteria in a sample within relatively short periods of time (e.g., 6–8 hr).

Although DNA amplification methods look very promising, additional methods for the preparation of clinical samples need further exploration. For example, new techniques for concentrating salmonellae in blood should be developed, since components of blood such as heme may interfere with NA amplification. In addition to lysis–centrifugation, we have tested another method to concentrate *S. typhi* from the blood of typhoid patients (68a). This method employed continuous-gradient centrifugation of blood with a colloidal silica (Sepracell, Sepratech Corp., Oklahoma City, Oklahoma), which resulted in layers of different blood cell types, which were cultured. Growth of bacterial colonies was associated with culture of the mononuclear cell layer; colonies of *S. typhi* were identified using a 10-min COAG assay. This protocol resulted in identification of typhoid bacilli within 18 hr after specimen acquisition. It is clear that newly developed methods for concentrating bacteria from blood could be used in a variety of different protocols, for antigen detection as well as for gene probes.

The practical application of gene probes to clinical medicine, in addition to specificity of the assay, depends on how rapid a result can be obtained. As discussed earlier, new methods for sample preparation are being explored to reduce the time to obtain target DNA. The next consideration is to decrease the amount of time for completion of hybridization of target DNA from the sample with the gene probe. Since the time necessary for hybridization is directly proportional to the concentration of reactants (66), this time factor can be effectively reduced by increasing the amount of target bacterial DNA from the clinical sample with methods such as PCR. The use of oligonucleotide probes will also reduce the time required for hybridization, as discussed later.

Oligonucleotide probes are short gene probes, 10–50 bp in length, that are synthesized in the laboratory. Once the specificity and sensitivity of a gene probe has been demonstrated, the NA sequence of the probe is determined and used for synthesizing oligonucleotide probes. Oligonucleotide probes have several advantages: (1) uniformity, in that they are synthesized using a specific sequence, (2) stability over time, and (3) relative ease of preparation as compared to the preparation and purification of probes from recombinant plasmids. Because of their small size, oligonucleotide probes hybridize to target DNA at very rapid rates, with reaction times <30 min as compared to 4–16 hr required for longer probes to complete hybridization (9).

Many new detection systems for DNA–DNA hybrids are becoming available, which need to be evaluated. Radiolabeled probes have been used extensively, since autoradiography offers a very sensitive method for detecting radiolabeled probes bound to target DNA. All the studies with the Vi probe were done with isotopically labeled probe. However, there are problems with the use of radiolabeled probes, especially in developing countries. Radiolabeled probes are expensive, have a short half-life, are hazardous to personnel, and create hazardous waste. A number of non-isotopic labeling and detection methods have been described (reviewed in 59 and 97), which are sensitive and promising; it is expected that they will soon replace isotopic methods.

The need for rapid, sensitive, specific, and relatively simple assays for detecting salmonellae is constantly increasing. Although contaminated animals, food, and water remain sources of infection, there has been an increase in salmonellae bacteremia that is associated with AIDS patients (10,78). In the area of typhoid fever, detection of *S. typhi* in clinical specimens is essential for diagnosis and patient management. Chronic typhoid carriers are a public health concern, especially in endemic areas. An important component of any program to control typhoid fever is the identification of typhoid carriers so that they can be treated and reduce the transmission of *S. typhi*. Although cholecystectomy is sometimes advocated, this is unacceptable for most asymptomatic carriers and may fail in >20% of those patients treated surgically (32). Management of chronic carriers usually involves antimicrobial therapy. A new group of quinolones, including ciprofloxacin and norfloxacin, have been shown to be effective and well tolerated in the treatment of chronic carriers (35,47,90). This class of antimicrobial agents is the most promising to date. Thus, it is even more important than before to identify and treat carriers so as to eradicate the carrier state.

VII. SUMMARY

Gene probes offer exquisite specificity for the identification of salmonellae. Genus-specific probes are available for detecting the presence of salmonellae in food, in an assay system that is more rapid than conventional culture methods. Although a limited number of gene probes are currently available for detection and identification of salmonellae in clinical specimens, improved protocols for sample preparation and detection need to be developed before this technology can be used to replace conventional culture methods. The Vi DNA probe is specific and sensitive for detection and identification of *S. typhi* in bacterial isolates, demonstrating its usefulness for epidemiological studies. The Vi probe has been used to detect *S. typhi* in the blood of typhoid fever patients, and increased sensitivity should be possible by incorporating new technologies into the assay system. In addition, the Vi probe shows great potential for detecting carriers. Nonisotopic labeling and detection methods are now available, which will be essential for hybridization reaction systems to be used in developing countries. A number of other gene probes are being developed, which should be helpful for the identification of other salmonellae. Thus, in the future gene probes to detect salmonellae should be useful in epidemiological studies, for example to investigate disease outbreaks, determine reservoirs of infection, and screen food handlers. It is only a matter of time before NA probe systems will be available to provide rapid, sensitive, specific, and simple assays for identification of specific pathogenic salmonellae.

ACKNOWLEDGMENTS

I would like to thank Dennis Kopecko, Steve Hoffman, and Mark Finch for editorial assistance and helpful discussions, Larry Risen for use of information about the SNAP *Salmonella* test cited in the text, Mark Mazola for use of information about the Gene-Trak *Salmonella* assay cited in the text, Susan Wetherell for use of information cited about the BacTrace *Salmonella* antibody cited in the text, Salete Newton and Gadi Frankel for unpublished data, and Alma Murlin for information and discussions about classification.

REFERENCES

1. Aleixo, J. A. G., and Swaminathan, B. (1988). A fluorescent enzyme immunoassay for *Salmonella* detection. *J. Immunoassay* **9,** 83–95.
2. Appassakij, H., Bunchuin, N., Sarasombath, S., Rungpitarangsi, B., Manatsathit, S., Komolpit, P., and Sukosol, T. (1987). Enzyme-linked immunosorbent assay for detection of *Salmonella typhi* protein antigen. *J. Clin. Microbiol.* **25,** 273–277.
3. Araj, G. F., and Chugh, T. D. (1987). Detection of *Salmonella* spp. in clinical specimens by capture enzyme-linked immunosorbent assay. *J. Clin. Microbiol.* **25,** 2150–2153.
4. Association of Official Analytical Chemists (1984). "Official Methods of Analysis for the Association of Official Analytical Chemists". AOAC, Arlington, Virginia.
5. Avendano, A., Herrera, P., Horwitz, I., Duarte, E., Prenzel, I., Lanata, C., and

Levine, M. L. (1986). Duodenal string cultures: Practicality and sensitivity for diagnosing enteric fever in children. *J. Infect. Dis.* **153,** 359–361.
6. Barrett, T. J., Blake, P. A., Brown, S. L., Hoffman, K., Liort, J. M., and Feeley, J. D. (1983). Enzyme-linked immunosorbent assay for detection of human antibodies to *Salmonella typhi* Vi antigen. *J. Clin. Microbiol.* **17,** 625–627.
7. Benavente, L., Gotuzzo, E., Guerra, J., Grados, O., Guerra, H., and Bravo, N. (1984). Diagnosis of typhoid fever using a string capsule device. *Trans. R. Soc. Trop. Med. Hyg.* **78,** 404–406.
7a. Bobo, L., Coutlee, F., Quinn, T., Yolken, R., and Viscidi, R. (1989). Detection of *Chlamydia trachomatis* DNA from cervical specimens by PCR and nonisotopic hybridization. *Program Abstr. Intersci. Conf. on Antimicrob. Agents and Chemother., 29th,* Abstr. no. 130.
8. Bokkenheuser, V., Smit, P., and Richardson, N. (1964). A challenge to the validity of the Vi test for the detection of chronic typhoid carriers. *Am. J. Public Health* **54,** 1507–1513.
9. Bryan, R. N., Ruth, J. L., Smith, R. D., and Le Bon, J. M. (1986). Diagnosis of clinical samples with synthetic oligonucleotide hybridization probes. *In* "Microbiology—1986" (L. Leive, ed.), pp. 113–116. Am. Soc. Microbiol., Washington, D.C.
10. Celum, C. L., Chaisson, R. E., Rutherford, G. W., Barnhart, J. L., and Echenberg, D. F. (1987). Incidence of salmonellosis in patients with AIDS. *J. Infect. Dis.* **156,** 998–1002.
11. Centers for Disease Control (1988). *Salmonella* isolates from humans in the United States, 1984–1986. *CDC Surveill. Summ.* **37,** 25–31.
12. Centers for Disease Control (1986). Foodborne disease outbreaks, Annual Summary, 1982. *CDC Surveill. Summ.* **35,** 7–16.
13. Chau, P. Y., and Chan, A. C. H. (1976). Modified Vi tests in the screening of typhoid carriers. *J. Hyg* **77,** 97–104.
14. Chau, P. Y., and Tsang, R. S. W. (1982). Vi serology in screening of typhoid carriers: Improved specificity by detection of Vi antibodies by counterimmunoelectrophoresis. *J. Hyg.* **89,** 261–267.
15. Chau, P. Y., Wan, K. C., and Tsang, R. S. W. (1984). Crossed immunoelectrophoretic analysis of anti-*Salmonella typhi* antibodies in sera of typhoid patients and carriers: Demonstration of the presence of typhoid-specific antibodies to a non-O, non-H, non-Vi antigen. *Infect. Immun.* **43,** 1110–1113.
16. Chitkata, Y. K., and Urquhart, A. E. (1978). Fluorescent Vi antibody test in the screening of typhoid carriers. *Am. J. Clin. Pathol.* **72,** 87–89.
17. Crosa, J. H., Brenner, D. J., Ewing, U. H., and Falkow, S. (1973). Molecular relationships among salmonellae. *J. Bacteriol.* **115,** 307–315.
18. Edelman, R., and Levine, M. M. (1986). Summary of an international workshop on typhoid fever. *Rev. Infect. Dis.* **8,** 329–349.
19. Edelstein, P. H., Bryan, R. N., Enns, R. K., Kohne, D. E., and Kacian, D. L. (1987). Retrospective study of Gen-Probe rapid diagnostic system for detection of *Legionellae* in frozen clinical respiratory tract samples. *J. Clin. Microbiol.* **23,** 1022–1026.
20. Edwards, P. R., and Ewing, W. H. (1972). "Identification of Enterobacteriaceae," 3rd ed. Burgess, Minneapolis, Minnesota.
21. Elsinghorst, E. A., Baron, L. S., and Kopecko, D. J. (1989). Penetration of human intestinal epithelial cells by *Salmonella:* Molecular cloning and expression of *Salmonella typhi* invasion determinants in *Escherichia coli. Proc. Natl. Acad. Sci. U.S.A.* **86,** 5173–5177.
22. Engleberg, N. C., Barrett, T. J., Fisher, H., Porter, B., Hurtado, E., and Hughes, J. M. (1983). Identification of a carrier by using Vi enzyme-linked immunosorbent assay

serology in an outbreak of typhoid fever on an Indian reservation. *J. Clin. Microbiol.* **18,** 1320–1322.

23. Escamilla, J., Santiago, L. T., Uylangco, C. V., and Cross, J. H. (1983). Evaluation of sodium polyanethanol sulfonate as a blood culture additive for recovery of *Salmonella typhi* and *Salmonella paratyphi* A. *J. Clin. Microbiol.* **18,** 380–383.
24. Felix, A. (1938). Detection of chronic typhoid carriers by agglutination tests. *Lancet* **2,** 738–741.
25. Felix, A., and Pitt, R. M. (1951). The pathogenic and immunogenic activities of *Salmonella typhi* in relation with its antigenic constituents. *J. Hyg.* **49,** 92–110.
26. Fields, P. I., Swanson, R. V., Haidaris, C. G., and Heffron, F. (1986). Mutants of *Salmonella typhimurium* that cannot survive within the macrophage are avirulent. *Proc. Natl. Acad. Sci. U.S.A.* **83,** 5189–5193.
27. Fitts, R. (1985). Development of a DNA–DNA hybridization test for the presence of *Salmonella* in foods. *Food Technol.* **39,** 95–102.
28. Fitts, R., Diamond, M., Hamilton, C., and Nerik, M. (1983). DNA–DNA hybridization assay for detection of *Salmonella* spp. in foods. *Appl. Environ. Microbiol.* **46,** 1146–1151.
29. Flowers, R. S., Klatt, M. J., Mozola, M. A., Curiale, M. S., Gabis, D. A., and Silliker, J. H. (1987). A DNA hybridization assay for the detection of *Salmonella* in foods: Collaborative study. *J. Assoc. Off. Anal. Chem.* **70,** 521–529.
30. Flowers, R. S., Mozola, M. A., Curiale, M. S., Gabis, D. A., and Silliker, J. H. (1987). Comparative study of a DNA hybridization method and the conventional method for detection of *Salmonella* in foods. *J. Food Sci.* **52,** 781–785.
31. Food and Drug Administration (1984). "Bacteriological Analytical Manual." FDA, Association of Official Analytical Chemists, Arlington, Virginia.
32. Freitag, J. L. (1964). Treatment of chronic typhoid carriers by cholecystectomy. *Public Health Rep.* **79,** 567–570.
33. Gaines, S., Tully, G. J., and Tiggert, W. D. (1961). Enhancement of the mouse virulence of a non-Vi variant of *Salmonella typhosa* by Vi antigen. *J. Immunol.* **86,** 543–551.
34. Gilman, R. H., Terminel, M., Hernandez-Mendoza, P., and Hornick, R. B. (1975). Relative efficiency of blood, urine, rectal swab, bone marrow and rose-spot cultures for recovery of *Salmonella typhi* in typhoid fever. *Lancet* **1,** 1211–1213.
35. Gotuzzo, E., Guerra, J. G., Benavente, L., Palomino, J. C., Carrillo, C., Lopera, J., Delgado, F., Nalin, D. R., and Sabbaj, J. (1988). Use of norfloxacin to treat chronic typhoid carriers. *J. Infect. Dis.* **157,** 1221–1225.
36. Grossman, N., Schmetz, M. A., Foulds, J., Klima, E., Jimenez-Lucho, V. E., Leive, L., and Joiner, K. A. (1987). Lipopolysaccharide size and distribution determine serum resistance in *Salmonella montevideo*. *J. Bacteriol.* **169,** 856–863.
37. Guerra-Caceres, J. G., Gotuzzo-Herencia, E., Crosby-Dagnino, E., Miro-Quesada, J., and Carillo-Parodi, C. (1979). Diagnostic value of bone marrow culture in typhoid fever. *Trans. R. Soc. Trop. Med. Hyg.* **73,** 680–683.
38. Guinee, P. A. M., and van Leeuwen, W. J. (1978). Phage typing of *Salmonella*. *In* "Methods in Microbiology" (T. Bergan and J.R. Norris, eds.), Vol. 11, pp. 157–191. Academic Press, New York.
39. Gupta, A. K., and Rao, K. M. (1979). Simultaneous detection of *Salmonella typhi* antigen and antibody in serum by counterimmunoelectrophoresis for an early and rapid diagnosis of typhoid fever. *J. Immunol. Methods* **30,** 349–353.
40. Hackett, J., Kotlarski, I. K., Mathan, V., Francki, K., and Rowley, D. (1986). The colonization of Peyer's patches by a strain of *S. typhimurium* cured of the cryptic plasmid. *J. Infect. Dis.* **153,** 1119–1125.

41. Hackett, J., Wyk, P., Reeves, P., and Mathan, V. (1987). Mediation of serum resistance in *Salmonella typhimurium* by an 11-kilodalton polypeptide encoded by the cryptic plasmid. *J. Infect. Dis.* **155,** 540–549.
42. Hoffman, S. L. (1984). Typhoid fever. *In* "Hunter's Tropical Medicine" (G. T. Strickland, ed.), pp. 282–297. Saunders, Philadelphia, Pennsylvania.
43. Hoffman, S. L., Edman, D. C., Punjabi, N. H., Lesmana, M., Cholid, A., Sundah, S., and Harahap, J. (1986). Bone marrow aspirate culture superior to streptokinase clot culture and 8 ml 1 : 10 blood-to-broth blood culture for diagnosis of typhoid fever. *Am. J. Trop. Med. Hyg.* **35,** 836–839.
44. Hoffman, S. L., Punjabi, N. H., Rockhill, R. C., Sustomo, A., Rivai, A. R., and Pulungsih, S. P. (1984). Duodenal string-capsule culture compared with bone-marrow, blood and rectal swab cultures for diagnosing typhoid and paratyphoid fever. *J. Infect. Dis.* **149,** 157–161.
45. Holmberg, S. D., and Wachsmuth, K. (1989). Plasmid and chromosomal DNA analyses in the epidemiology of bacterial diseases. *In* "Nucleic Acid and Monoclonal Antibody Probes" (B. Swaminathan and G. Prakash, eds.), pp. 105–129. Dekker, New York.
46. Hornick, R. B., Greisman, S. E., Woodward, T. E., DuPont, H. L., Dawkins, A. T., and Synder, M. J. (1970). Typhoid fever: Pathogenesis and immunologic control. *N. Engl. J. Med.* **283,** 686–691, 739–746.
47. Hudson, S. J., Ingham, H. R., and Snow, M. H. (1985). Treatment of *Salmonella typhi* carrier state with ciprofloxacin. *Lancet* **1,** 1047.
48. Jimenez-Lucho, V. E., and Foulds, J. (1988). Vi antigen prevents alternative-pathway mediated opsono-phagocytosis of *Salmonella typhi. Am. Soc. Microbiol., 86th Annu. Meet.,* Abst. B80.
49. John, T. J., Sivadasan, K., and Kurien, B. (1984). Evaluation of passive bacterial agglutination for the diagnosis of typhoid fever. *J. Clin. Microbiol.* **20,** 751–753.
50. Johnson, E. M., Krauskopf, B., and Baron, L. S. (1965). Genetic mapping of VI and somatic antigenic determinants in *Salmonella J. Bacteriol.* **90,** 302–308.
51. Johnson, E. M., Krauskopf, B., and Baron, L. S. (1966). Genetic analysis of the viaA-his chromosomal region in *Salmonella J. Bacteriol.* **92,** 1457–1463.
52. Jones, G. W., Rabert, D. K., Svinarich, D. M., and Whitfield, H. J. (1982). Association of adhesive, invasive and virulent phenotypes of *Salmonella typhimurium* with autonomous 60-megadalton plasmids. *Infect. Immun.* **38,** 376–386.
53. Joys, T. M. (1985). The covalent structure of the phase-1 flagellar filament protein of *Salmonella typhimurium* and its comparison with other flagellins. *J. Biol. Chem.* **260,** 15758–15761.
54. Kohler, G., and Milstein, C. (1975). Continuous cultures of fused cells secreting antibodies of predefined specificity. *Nature (London)* **256,** 495–497.
55. Lanata, C. F., Kaper, J. B., Baldini, M. M., Black, R. E., and Levine, M. M. (1985). Sensitivity and specificity of DNA probes with the stool blot technique for detection of *E. coli* enterotoxins. *J. Infect. Dis.* **152,** 1087–1090.
56. Lanata, C. F., Ristori, C., Jimenez, L., Garcia, J., Levine, M. M., Black, R. E., Salcedo, M., and Sotomayor, V. (1983). Vi serology detection of chronic *Salmonella typhi* carriers in an endemic area. *Lancet* **2,** 441–443.
57. Landy, M., and Lamb, E. (1953). Estimation of Vi antibody employing erythrocytes treated with purified antigen. *Proc. Soc. Exp. Biol. Med.* **82,** 593–598.
58. Laure, F., Courgnaud, V., Rouzioux, C., Blanche, S., Veber, F., Burgard, M., Jacomet, C., Griscelli, C., and Brechot, C. (1988). Detection of HIV1 DNA in infants and children by means of the polymerase chain reaction. *Lancet* **2,** 538–541.
59. Leary, J. J., and Ruth, J. L. (1989). Nonradioactive labeling of nucleic acid probes. *In*

"Nucleic Acid and Monoclonal Antibody Probes" (B. Swaminathan and G. Prakash, eds.), pp. 33–57. Dekker, New York.

60. Le Minor, L., Vernon, M., and Popoff, M. (1982). Proposition pour une nomenclature des *Salmonella*. *Ann. Microbiol. (Paris)* **133B,** 245–254.
61. Lieve, L., and Jimenez-Lucho, V. E. (1986). Lipopolysaccharide O-antigen structure controls alternative pathway activation of complement: Effects on phagocytosis and virulence of salmonellae. *In* "Microbiology—1986" (L. Leive, ed.), pp. 14–17. Am. Soc. Microbiol., Washington, D.C.
62. Lim, P. L. (1986). Diagnostic uses of monoclonal antibodies to Salmonella. *In* "Monoclonal Antibodies against Bacteria" (A. J. L. Macario and E. Conway de Macario, eds.), Vol. 3, pp. 29–75. Academic Press, Orlando, Florida.
63. Lim, P. L., and Fok, Y. (1987). Detection of group D salmonellae in blood culture broth and of soluble antigen by tube agglutination using an O-9 monoclonal antibody latex conjugate. *J. Clin. Microbiol.* **25,** 1165–1168.
64. Lin, C. F., Becke, J. M., Groves, C., Lim, B. P., Israel, E., Becker, E. F., Helfirch, R. M., Swetter, D. S., Cramton, T., and Robbins, J. (1988). Restaurant-associated outbreak of typhoid fever in Maryland: Identification of carrier facilitated by measurement of serum Vi antibodies. *J. Clin Microbiol.* **26,** 1194–1197.
65. Looney, J. R., and Steigbigel, R. T. (1986). Role of the Vi antigen of *Salmonella typhi* in resistance to host defense in vitro. *J. Lab. Clin. Med.* **108,** 506–516.
66. Maniatis, T., Fritsch, E. F., and Sambrook, J. (1982). "Molecular Cloning: A Laboratory Manual." Cold Spring Harbor Lab., Cold Spring Harbor, New York.
67. Manning, E. J., Baird, G. D., and Jones, P. W. (1986). The role of plasmid genes in the pathogenicity of *Salmonella dublin*. *J. Med. Microbiol.* **21,** 239–243.
68. Mattingly, J. A. (1984). An enzyme immunoassay for the detection of all *Salmonella* using a combination of a myeloma protein and a hybridoma antibody. *J. Immunol. Methods* **73,** 147–156.

68a. McWhirter, P. D., Rubin, F. A., Punjabi, N. H., Lane, E., Kumala, S., Sudarmono, P., Punlungsih, S. P., Lesmana, M. and Hoffman, S. H. (1989). Rapid diagnosis of typhoid fever: Identification of *Salmonella typhi* within 18 hours of specimen acquisition by culture of the mononuclear cell/platelet fraction of blood. (1989). Submitted for publication.

69. Mikhail, I. A., Sanborn, W. R., and Sippel, J. E. (1983). Rapid, economical diagnosis of enteric fever by a blood clot culture coagglutination procedure. *J. Clin. Microbiol.* **17,** 564–565.
70. Mills, J. A., Buysse, J. M., and Oaks, E. (1988). *Shigella flexneri* invasion plasmid antigens B and C: Epitope location and characterization with monoclonal antibodies. *Infect. Immun.* **56,** 2933–2941.
71. Moseley, S. L., Echeverria, P., Seriwatana, J., Tirapat, C., Chaicumpa, W., Sakuldaipeara, T., and Falkow, S. (1982). Identification of enterotoxigenic *Escherichia coli* by colony hybridization using three enterotoxin gene probes. *J. Infect. Dis.* **145,** 863–869.
72. Nakamura, M., Sato, S., Ohya, T., Suzuki, S., and Ikeda, S. (1985). Possible relationship of a 36-megadalton *Salmonella enteritidis* plasmid to virulence in mice. *Infect. Immun.* **47,** 831–833.
73. Nolan, C. M., Feeley, J. C., White, P. C., Jr., Hambie, E. A., Brown, S. L., and Wong, K. H. (1980). Evaluation of a new assay for Vi antibody in chronic carriers of *Salmonella typhi*. *J. Clin. Microbiol.* **12,** 52–56.
74. Nolan, C. M., White, P. C., Feeley, J. C., Hambie, E. A., Brown, S. L., and Wong, K. H. (1981). Vi serology in the detection of typhoid carriers. *Lancet* **1,** 583–586.
75. Olive, D. M. (1989). Detection of enterotoxigenic *Escherichia coli* following polymerase

chain reaction amplification with a thermostable polymerase. *J. Clin. Microbiol.* **27,** 261–265.
76. Ou, C. Y., Kwock, S., Mitchell, S. W., Mack, D. H., Sninsky, J. J., Krebs, J. W., Feorino, P., Warfield, D., and Schochetman, G. (1987). DNA amplification for direct detection of human immunodeficiency virus-1 (HIV-1) in DNA of peripheral mononuclear cells. *Science* **239,** 295–297.
77. Potter, M. (1977). Antigen-binding myeloma proteins of mice. *Adv. Immunol.* **25,** 141–211.
78. Profeta, S., Forrester, C., Eng, R. H. K., Liu, R., Johnson, E., Palinkas, R., and Smith, S. M. (1985). *Salmonella* infections in patients with acquired immunodeficiency syndrome. *Arch. Intern. Med.* **145,** 670–672.
79. Qadri, A., Gupta, S. K., and Gursaran, T. P. (1988). Monoclonal antibodies delineate multiple epitopes on the O antigens of *Salmonella typhi* lipopolysaccharide. *J. Clin. Microbiol.* **26,** 2292–2296.
80. Rashtchian, A., and Curiale, M. S. (1989). DNA probe assays for detection of *Campylobacter* and *Salmonella. In* "Nucleic Acid and Monoclonal Antibody Probes" (B. Swaminathan and G. Prakash, eds.), pp. 221–239. Dekker, New York.
81. Risen, L., Marich, J. E., and Roszak, D. (1988). Detection of *Salmonella* from clinical samples using a colorimetric nucleic acid probe. *Program Abstr., Intersci. Conf. Antimicrob. Agents Chemother., 28th,* Abstr. no. 1052.
82. Robison, B. J., Pretzman, C. I., and Mattingly, J. A. (1983). Enzyme immunoassay in which a myeloma protein is used for detection of salmonellae in foods. *Appl. Environ. Microbiol.* **45,** 1816–1821.
83. Rockhill, R. C., Lesmana, M., Moechtar, M. A., and Sutomo, A. (1980). Detection of *Salmonella* C_1, D and Vi antigens by coagglutination in blood cultures from patients with *Salmonella* infections. *Southeast Asian J. Trop. Med. Public Health* **11,** 441–445.
84. Rockhill, R. C., Rumans, L. W., Lesmana, M., and Dennis, D. T. (1980). Detection of *Salmonella typhi* D, Vi, and d antigens, by slide coagglutination in urine from patients with typhoid fever. *J. Clin. Microbiol.* **11,** 213–216.
85. Rubin, F. A., and Kopecko, D. J. (1989). Nucleic acid probes for detection of clinically significant bacteria. *In* "Nucleic Acid and Monoclonal Antibody Probes." (B. Swaminathan and G. Prakash, eds.), pp. 185–219. Dekker, New York.
86. Rubin, F. A., Kopecko, D. J., Noon, K. F., and Baron, L. S. (1985). Development of a DNA probe to detect *Salmonella typhi. J. Clin. Microbiol.* **22,** 600–605.
87. Rubin, F. A., Kopecko, D. J., Sack, R. B., Sudarmono, P., Yi, A., Maurta, D., Meza, R., Moechtar, M. A., Edman, D. C., and Hoffman, S. L. (1988). Evaluation of a DNA probe for identifying *Salmonella typhi* in Peruvian and Indonesian bacterial isolates. *J. Infect. Dis.* **157,** 1051–1053.
88. Rubin, F. A., McWhirter, P. D., Punjabi, N. H., Lane, E., Sudarmono, P., Pulungsih, S. P., Lesmana, M., Kumala, S., Kopecko, D. J., and Hoffman, S. L. (1989). Use of a DNA probe to detect *Salmonella typhi* in the blood of patients with typhoid fever. *J. Clin. Microbiol.* **27,** 1112–1114.
88a. Rubin, F. A., Gotuzzo, E., Carillo, C., Kopecko, D. J., and Finch, M. J. (1989). Detection of *Salmonella typhi* in the blood of typhoid patients using a lysis-centrifugation method and the Vi DNA probe. *Program Abstr. Intersci. Conf. on Antimicrob. Agents and Chemother., 29th,* Abstr. no. 860.
88b. Rubin, F. A., Finch, M. J., Carillo, C., Kopecko, Frankel, G., and Gotuzzo, E. (1989). Combination of Vi and Hd probes to detect *Salmonella typhi* in stool to identify chronic carriers. *Program Abstr. Intersci. Conf. on Antimicrob Agents and Chemother., 29th,* Abstr. no. 861.
89. Saiki, R. K., Scharf, S., Faloona, F., Mullis, K. B., Horn, G. T., Erlich, H. A., and

Arnheim, N. (1985). Enzymatic amplification of beta-globin genomic sequences and restriction site analysis for diagnosis of sickle cell anemia. *Science* **230,** 1350–1354.

90. Sammalkorpi, K., Lahdevirta, J., Mäkelä, T., and Rostila, T. (1987). Treatment of chronic salmonella carriers with ciprofloxacin. *Lancet* **1,** 164–165.
91. Sarasombath, S., Lertmemongkolchai, G., and Banchuin, N. (1988). Characterization of monoclonal antibodies to protein antigen on *Salmonella typhi*. *J. Clin. Microbiol.* **26,** 508–512.
92. Smith, H. R., Rowe, B., Gross, R. J., Fry, N. K., and Scotland, S. M. (1987). Haemorrhagic colitis and vero-cytotoxin-producing *Escherichia coli* in England and Wales. *Lancet* **1,** 1062–1064.
93. Snellings, N. J., Johnson, E. M., and Baron, L. S. (1977). Genetic basis of Vi antigen expression in *Salmonella paratyphi* C. *J. Bacteriol.* **131,** 57–62.
94. Snellings, N. J., Johnson, E. M., Kopecko, D. J., Collins, H. H., and Baron, L. S. (1981). Genetic regulation of variable Vi antigen expression in a strain of *Citrobacter freundii*. *J. Bacteriol.* **145,** 1010–1017.
95. Stoleru, L., LeMinor, L. and Lhéritier, A. M. (1976). Polynucleotide sequence divergence among strains of *Salmonella* subgenus IV and closely related organisms. *Ann. Microbiol. (Paris)* **127,** 477–486.
96. Taylor, D. N., Harris, J. R., Barrett, T. J., Hargrett, N. T., Prentzel, I., Valdivieso, C., Palomino, C., Levine, M. M., and Blake, P. A. (1983). Detection of urinary Vi antigen as a diagnostic test for typhoid fever. *J. Clin. Microbiol.* **18,** 872–876.
97. Tenover, F. C. (1988). Diagnostic deoxyribonucleic acid probes for infectious diseases. *Clin. Microbiol. Rev.* **1,** 82–101.
98. Terakado, N., Sehizaki, T., Hashimoto, K., and Naitoh, S. (1983). Correlation between the presence of a fifty-megadalton plasmid in *Salmonella dublin* and virulence for mice. *Infect. Immun.* **41,** 443–444.
99. Todd, L. S., Roberts, D., Bartholomew, B. A., and Gilbert, R. J. (1987). Assessment of an immunoassay for the detection of salmonellas in foods and animal feeding stuffs. *Epidemiol. Infect.* **98,** 301–310.
100. Tompkins, L. S., Troup, N., Labigne-Roussel, A., and Cohen, M. L. (1986). Cloned, random chromosomal sequences as probes to identify *Salmonella* species. *J. Infect. Dis.* **154,** 156–162.
101. Totten, P. A., Holmes, K. K., Handsfield, H. H., Knapp, J. S., Perine, P. L., and Falkow, S. (1983). DNA hybridization technique for the detection of *Neisseria gonorrhoeae* in men with urethritis. *J. Infect. Dis.* **148,** 462–471.
102. Tsang, R. S. W., and Chau, P. Y. (1987). Production of Vi monoclonal antibodies and their application as diagnostic agents. *J. Clin. Microbiol.* **25,** 531–535.
103. Venkatesan, M., Buysse, J. M., Vandendries, E., and Kopecko, D. J. (1988). Development and testing of invasion-associated DNA probes for detection of *Shigella* spp. and enteroinvasive *Escherichia coli*. *J. Clin. Microbiol.* **26,** 261–266.
104. Watson, K. C. (1959). Isolation of *Salmonella typhi* from the blood stream. *J. Lab. Clin. Med.* **47,** 329–332.
105. Wei, L., and Joys, T. M. (1985). Covalent structure of three phase-1 flagellar filament proteins of *Salmonella*. *J. Mol. Biol.* **186,** 791–803.
106. Williamson, C. M., Baird, G. D., and Manning, E. J. (1988). A common virulence region on plasmids from eleven serotypes of *Salmonella*. *J. Gen. Microbiol.* **134,** 975–982.

13

Gene Probes for Detection of Food-Borne Pathogens

K. WERNARS AND S. NOTERMANS
Laboratory of Water and Food Microbiology
National Institute of Public Health and Environmental Protection
Bilthoven, The Netherlands

I. INTRODUCTION

In food microbiology some microorganisms are important because they are used for fermentation, but they can also cause food spoilage or even disease. In judging the quality of food and food products, the microbial condition is of great importance, and various laboratory methods have been developed to detect, enumerate, and classify the microorganisms present. Classical techniques are based mainly on culturing and isolation of microorganisms. Most of these techniques, however, are laborious and time-consuming. Continuous efforts are made to develop new methods that are both rapid and reliable.

Improvements introduced in the last decade include more sophisticated plating systems such as computer-assisted plate counts (21) and spiral plating (130), ATP assays (90), electrical impedance methods (45), fluorescent microscopy techniques (119), and various immunoassays. Thanks to the rapid progress in the field of molecular biology in the past few years (98,99,160), recombinant DNA techniques now offer possibilities for practical application as analytical tools in food microbiology. Nucleic acid (NA) hybridization, especially, is an exciting new approach for the detection of microorganisms in food, and the number of research papers demonstrating the potential of this method is increasing rapidly. The presence of microorganisms is demonstrated by hybridization of their genetic material to specific gene probes (e.g., probes for detecting genes involved in virulence or toxin production or genes that are characteristic for a microorganism, in particular, or a group of microorganisms). With the help of the hybridization technique, food-borne microorganisms may be identified in a simple way without the use of test animals, and it offers an opportunity to avoid other, more laborious methods. This article focuses on the use and potential of gene probes for the detection and identification of bacterial food-borne pathogens; *Escherichia coli, Staphylococcus aureus, Clostridium perfringens, Clostridium botulinum, Salmonella* spp., and *Listeria monocytogenes* are chosen for illustration. Each of these microorganisms possesses the ability to cause disease on consumption of food in which they are present or have produced their toxins. The symptoms of food poisoning they cause can range from vomiting (*S. aureus*), diarrhea of temporary nature (*C. perfringens*), persistent diarrhea (*E. coli* and *Salmonella* spp.), meningitis and abortion (*L. monocytogenes*), to lethal intoxication (*C. botulinum*). For some of these food-borne pathogens, for example *S. aureus, C. perfringens,* and *C. botulinum*, the basis of their pathogenic behavior is well understood. The others mentioned previously possess virulence factors that have been elucidated only partially until now. Among these factors are the production of toxins, the ability to colonize the intestine, invasiveness, and hemolytic activity. The combination of knowledge concerning virulence factors and the use of gene probes offers unique possibilities for concomitant detection and identification of pathogenic microorganisms. We believe that a discussion on the application of the NA hybridization technique to the detection and identification of the aforementioned pathogens will reveal possibilities for the use of this method in food microbiology. Section II of this Chapter presents background information on the role of these pathogens in food poisoning, pathogenicity, and conventional detection methods. Section III contains an overview on the use of gene probes for the detection of food-borne pathogens, with results of recent work carried out in our laboratory.

II. BACKGROUND

A. *Escherichia coli*

Escherichia coli is a gram-negative, facultative, anaerobic microorganism and is part of the normal flora of human and animal intestines. The presence of *E. coli* in food is usually caused by fecal contamination, and their concentration in a food product is a measure of the success of "good manufacturing practices" during processing. Although the majority of strains within this species are nonpathogenic, some *E. coli* strains can cause diarrhea on ingestion of contaminated food and water. These diarrheogenic *E. coli* have been divided into four main categories: enterotoxinogenic (ETEC), enteropathogenic (EPEC), enteroinvasive (EIEC), enterohemorrhagic (EHEC), and a fifth less well-defined enteroadherent (EAEC) category. They display distinct clinical symptoms on infection, and differ in their pathogenesis and serogroups. The epidemiology and pathogenesis of these diarrheogenic *E. coli* categories have been reviewed (94). Production of toxin(s) is a common virulence factor for diarrheogenic *E. coli*. Exterotoxinogenic *E. coli* strains, which are a major cause of diarrhea, especially in developing countries, produce heat-labile or heat-stabile enterotoxins (LT or ST, respectively), and frequently both (124). A number of toxins other than LT and ST are produced by EPEC, EIEC, and EHEC strains, such as the *Shiga* toxin. Essential for the pathogenic character of *E. coli* strains is their ability to colonize the intestinal tract, and a number of different factors have been found to be involved in the adhesion of *E. coli* cells.

In EIEC a virulence factor has been recognized that enables the microorganism to invade the epithelial cells of the intestine, thus emulating the pathogenic behavior of *Shigella*. For both organisms invasiveness depends on the possession of large plasmids that encode various membrane proteins, and they are closely related in EIEC and *Shigella*. Because diarrheogenic *E. coli* has been a subject of many studies over the past three decades, the basis of the pathogenic behavior of these microorganisms has been clarified to some extent. This knowledge has proved of great value in the development of specific DNA probes for their detection.

1. Detection and Enumeration of Diarrheogenic Escherichia Coli *in Food*

At present no conventional procedure is available for the direct enumeration of diarrheogenic *E. coli* in food. This is due to the lack of biochemical tests to distinguish pathogenic strains from others. Moreover, the food microbiologist may need to identify very small numbers of microorganisms among a large competing flora to judge a health risk. This is quite different from the detection of diarrheogenic *E. coli* in clinical samples. Stool samples from an individual suffering from an enteric infection may contain millions of cells per gram of the pathogenic strain. Enumeration of diarrheogenic *E. coli* in food employs cultivation of total *E. coli* and testing for (potential) pathogenicity. For the enumeration and isolation of *E. coli* from food, an international standard procedure has been described (ISO/DIS 6391). It starts with the inoculation of cellulose acetate membranes, using a suspension of the test sample. These membranes are then overlaid on a resuscitation medium and incubated for 4 hr at 37°C. Subsequently the membranes are transferred to tryptone–bile agar and incubated at 44°C for 18 hr. Testing for the production of indole for each colony is used as a confirmation for *E. coli*. Serotyping of the isolated strains according to the O : H system (83) can be used as a first indication of pathogenicity since the strains within each category of diarrheogenic *E. coli* tend to fall within distinct O : H serogroups (94).

Various techniques have been developed for pathogenicity testing, both *in vitro* and *in vivo*. Methods for the detection of LT and ST enterotoxins have been reviewed by Linggood (96), and involve animal tests (ligated gut, suckling mouse, rabbit skin, and rat perfusion assays), tissue culture techniques (Chinese hamster ovary cell and mouse adrenal cell assays), and immunoassays [passive hemagglutination assay (PHA), radioimmunoassay (RIA), and enzyme-linked immunosorbent assay (ELISA)]. Over the past few years the ELISA has become the most convenient test for LT and ST, especially because of the availability of monoclonal antibodies (MAb) [for an overview, see Svennerholm *et al.* (151)]. Similar tests have been developed for the identification of EPEC [reviewed by Law (91)] and EIEC (117). Identification methods for EAEC and EHEC are less well developed.

B. *Staphylococcus aureus*

Staphylococcus aureus is a major cause of bacterial food poisoning all over the world (14). Food poisoning follows the ingestion of enterotoxins produced in foods by some strains of *S. aureus* (15). A number of enterotoxins (SE) have been differentiated by immunological methods, and they are designated alphabetically into type SEA through SEE (102). Three subtypes of SEC, numbered 1, 2, and 3, have been described (7,18,121). The immunological differences between these subtypes are small, and complete neutralization occurs with heterologous antibodies (121). *Staphylococcus aureus* enterotoxins are a group of relatively low molecular weight proteins (~25,000–30,000) that elicit emesis and sometimes diarrhea. The toxins are marked by their heat stability (28). The illness typically results from the ingestion of contaminated foods in which the bacterium has been allowed to proliferate and to produce toxin. Symptoms are usually observed within 2–6 hr of eating of SE-containing food. The site of emetic action of SE lies within the gastrointestinal tract (40,149) and generates impulses that react with the subcortical vomiting center via vagal and sympathic afferents (149). Detection of SE in epidemiologically implicated food has a high evidential value to confirm an outbreak of *S. aureus* food poisoning.

1. Enterotoxin Production by **Staphylococcus aureus**

Only some *S. aureus* strains produce SE. In Table I, production of SE by *S. aureus* isolated from different sources is presented. From these results it is clear that ~10% of the strains isolated from food are enterotoxigenic, whereas ~50% of human isolates are enterotoxigenic. In contrast, *S. aureus* isolated from food poisoning outbreaks are usually all enterotoxigenic. Growth of *S. aureus* and production of SE depend on several factors such as temperature, pH, and water activity (a_W). No growth occurs at pH <4.5 or at an a_W <0.88 (109,155). Growth is possible in a temperature range of 8°–46°C (2). It was demonstrated by Notermans and Heuvelman (109) that SEA is produced under all conditions that allow growth. Ewald and Notermans (41) found identical results for the production of SED. On the other hand, production of SEB and SEC is highly influenced by a_W (109). No SEB or SEC is produced at a_W <0.92 and <0.94, respectively.

2. Detection and Enumeration of Enteropathogenic **Staphylococcus aureus** *Strains in Food*

At this time, there is no technique available for the direct enumeration of enterotoxigenic *S. aureus* in food. The techniques applied involve isolation of *S. aureus* by using selective agar plating techniques, followed by culturing of isolated strains and testing of the culture fluid for the presence of SE.

A selective enumeration procedure for *S. aureus* from food is described by ISO/DIS 6888. Using this standard technique, dilutions of food samples are inoculated on the agar medium of Baird-Parker (8). After incubation at

TABLE I

Enterotoxin Production by *Staphylococous aureus* Strains Isolated from Various Substrates

Source of strains	Number of strains tested	Number of strains with enterotoxin production	Type of enterotoxin					Reference
			SEA	SEB	SEC	SED	SEE	
Beef	361	56	29	7	16	16	4	116
Milk	50	5	0	0	0	3	0	161
Poultry	104	5	0	1	0	5	0	113
Human clinical infections	59	22	10	9	3	4	0	105
Human carriers	48	31	10	4	5	5	0	105
Food poisoning incidents	120	113	88	2	18	48	3	59

37°C for 48 hr, typical colonies (black, shining, and convex, and surrounded by a clear zone that may be partially opaque) are confirmed by testing coagulase activity. Beckers *et al.* (12) introduced an improved selective agar medium that combines enumeration and confirmation by the addition of rabbit plasma fibrinogen to Baird Parker agar base. Isolated strains are then tested for SE production for which various assays have been described in the past (93,137). Introduction of the ELISA and the latex agglutination techniques has simplified the detection of SE. For this purpose, strains of *S. aureus* are inoculated in Erlenmeyer flasks containing brain-heart infusion (BHI) broth. After incubation with shaking at 37°C for 48 hr, the culture fluid is tested for the presence of SE.

3. Detection of Enterotoxin

Different enzyme immunoassays (ELISA) have been described in the last decade. The so-called sandwich ELISA is very suitable for the detection of SE (44,106). With this technique antibodies are coated onto a solid matrix (wells of a microtiter tray). After incubation with the test sample, the amount of SE adsorbed is measured using enzyme-labeled antibody. The enzyme activity is proportional to the concentration of the enterotoxin. For labeling, different enzymes can be used via different conjugation procedures. With ELISA, the minimal detectable quantity of SE is 0.1–1.0 ng/ml. An early latex agglutination assay for detecting SE was described by Salomon and Tew (127). With this test latex particles are coated with antibody. A drop of coated latex particles is then mixed with a drop of the test sample. If SE is present in the sample agglutination occurs. The latex agglutination assay is easy to perform and ~1 ng of SE/per milliliter can be detected (118).

C. *Clostridium perfringens*

Clostridium perfringens is an anaerobically growing and spore-forming microorganism. It is found in soil, air, water, and the intestinal tracts of humans and animals. Several groups are distinguished (A–D), based on the production of various toxins. *Clostridium perfringens* is a major cause of food poisoning in humans (154). All symptoms associated with the poisoning are due to a heat-labile enterotoxin (CPE) (144), which is produced mainly by *C. perfringens* type A strains (73). The poisoning is caused by the ingestion of food that contains large numbers of vegetative cells of enterotoxigenic strains (82). On sporulation in the intestine, these cells release CPE (53). Abdominal pain and diarrhea are the typical symptoms, which occur 12–24 hr after eating the contaminated food. Confirmation of an outbreak requires the detection of high levels of *C.*

perfringens possessing a common serotype, in the suspected food and in the feces of patients. In addition, fecal CPE may be detected if appropriate samples are available (10,31,110). Besides ileal loop activity, CPE exerts several other biological activities, namely erythemal activity and lethality (69,144). The toxin has been purified and identified as a heat-labile protein with MW 35,352 and a pI of 4.6 (60,61).

Although rare reports mention production of CPE in nonsporulating cultures of *C. perfringens* (53,61), the CPE responsible for illness is generally thought to be synthesized only during sporulation (34,67,82), and has been described as a sporulation-specific gene product (34) and structural component of the spore coat (53,54).

1. Enumeration of **Clostridium perfringens** *and Testing for CPE Production*

For enumeration of *C. perfringens* in food, a general method has been described (ISO 7937). With this method 1 ml of a diluted food sample is poured into petri dishes and mixed with tryptose sulfite cycloserine agar of Harmon *et al.* (65,66). After setting it is overlaid with the same agar medium. Then the dishes are incubated anaerobically at 37°C. After 20 hr the black colonies are counted and are confirmed by testing some of them for the absence of motility, nitrate reduction, and for the ability to metabolize lactose and gelatin.

To test for the presence or absence of *C. perfringens* in food, the sample is incubated in a liquid medium such as sulfite polymyxin iron broth (100). After anaerobic incubation at 37°C for 20 hr, culture fluid with gas formation and a black precipitate are streaked on blood agar plates. After incubation of these plates anaerobically at 37°C for 20 hr colonies surrounded by hemolysis are confirmed as described previously.

Identification of strains that produce CPE is difficult because *C. perfringens* sporulates poorly in ordinary culture media (33,67,120). For CPE production various sporulation media have been developed in attempts to obtain maximum toxin production. However, as yet no medium has been found that is suitable for all strains (9). *Clostridium perfringens* enterotoxin produced in culture fluid can be detected by ELISA (10,110), latex agglutination (16), or cell assays (75). Routine testing for CPE production is not practical as many false negative results are obtained due to poor sporulation.

D. *Clostridium botulinum*

Clostridium botulinum is an anaerobically growing and spore-producing microorganism. The organisms produce botulinum toxin and are classified alphabetically into type A through type G, depending on the immunological type of botulinum toxin produced. The botulinum toxin causes a neuroparalytic effect, affecting both humans and animals, and is the most dangerous form of bacterial food poisoning with a high mortality rate. Food-borne botulism is caused by ingestion of the toxin preformed in food. Usually two or more persons are intoxicated at a time as a result of the consumption of a common toxin source. Types A, B, and E toxins, especially, have been involved in human outbreaks. The foods most frequently implicated include raw, smoked, or fermented fish products, home-bottled vegetables, meat, and fish, and homemade sausages. There are various symptoms of botulism, but they usually include disturbance of vision and difficulty in speaking and swallowing. The mucous membranes of the mouth, tongue, and pharynx are usually extremely dry. These symptoms are followed by progressive weakness and respiratory failure. Onset of symptoms begins 18–36 hr after consumption of the contaminated food. On absorption into the circulatory system, the toxin attaches to the presynaptic terminals, causing cholinergic blockade by inhibiting the release of the transmitter substance, acetylcholine (138,139).

Laboratory diagnosis of food-borne botulism is dependent on the detection and identification of the toxin in the blood serum of the patient and in the food implicated (23,30,55). Detection of high numbers of toxin-producing microorganisms in the implicated food is also useful. In food microbiology detection of *C. botulinum* organisms is important both for prevention and epidemiological purposes. It is especially of interest to know in which raw materials *C. botulinum* is present. However, at this time, detecting *C. botulinum* microorganisms and distinguishing them from other clostridia are based solely on the detection of the toxin in enrichment cultures.

1. Detection of **Clostridium botulinum**

Detection of *C. botulinum* is based on enrichment of the microorganism and the subsequent detection of the botulinum toxin in the enrichment fluid. For this a sample, usually 25 g, is homogenized for 1–2 min in a homogenizer by adding 1- to 5-fold of physiological saline solution. Parts of the homogenate are then transferred into two test tubes containing the enrichment fluid. As enrichment fluid, fortified cooked meat medium (108), fortified egg meat medium (134), TPGY (68), and liver broth (63) are used. In order to remove oxygen the media used must either be freshly prepared or boiled before use. One test tube is heated at 70°C for 20 min to kill all vegetative bacterial cells. The heated and the nonheated tubes are then incubated anaerobically at 30°C for 5 days. After incubation the culture fluid is tested for the presence of botulinum toxin. If botulinum toxin is present, then the original sample contained *C. botulinum* organisms.

2. *Assays for Botulinum Toxin*

The quantity of toxin produced in the enrichment fluid depends on several factors, such as the type of sample, the presence of competitive microorganisms, and incubation temperatures (108). Generally, only small quantities of toxin are produced. In addition, production of toxic components by microorganisms other than *C. botulinum* has to be taken into account. Therefore only ultrasensitive assays are of interest. These assays include the bioassay in mice and the highly sensitive immunoassays, such as the ELISA. The most sensitive and widely used assay for botulinum toxin is the intraperitoneal (ip) injection of material into mice (126). The test however, is unsuitable for the examination of samples containing other toxic substances that may cause interference or nonspecific death in mice. To stabilize botulinum toxin, samples are diluted in 0.05 *M* phosphate buffer, pH 6.0, containing 0.2% gelatin. After centrifugation of the homogenized samples, the botulinum toxin can be potentiated considerably by the addition of trypsin, which causes limited proteolysis of the toxin (26,150). When test samples are injected ip into mice of 18–22 g, the symptoms usually develop within 4 hr and include characteristic vibration of the abdominal wall, followed by a wasp-shaped abdomen reaction, and labored breathing with or without paralysis of the limbs. Heating of the sample (80°C for 5 min) or neutralization by specific antibodies results in negative mouse bioassays. Identification of the toxin type by neutralization reactions requires the use of many test animals as seven immunologically different types of botulinum toxin exist. Immunoassays for botulinum toxins have been summarized by Notermans and Kozaki (112). The sensitivity of all *in vitro* immunological methods is less than that of the mouse ip injection method, although some investigators have claimed techniques with a comparable sensitivity. Using MAb in the so-called amplified ELISA, Appleton *et al.* (4) reported a sensitivity approaching the level of the mouse bioassay. A general disadvantage of immunoassays is that only the antigenicity is determined, and this may differ from the actual toxicity (112). Another drawback is that immunoassays, such as ELISA, are potentially sensitive for nonspecific reactions (114).

E. *Salmonella* spp.

Salmonella organisms are the cause of acute gastroenteritis (salmonellosis). The disease is food-borne and is common all over the world. In Europe, about one case of salmonellosis per 100 inhabitants yearly is reported (11). Besides salmonellosis the organisms can cause enteric fever, which includes typhoid and paratyphoid. Enteric fever causes more serious illness and is endemic in several geographic regions.

Meat and meat products are a common source of *Salmonella* organisms. Poultry and pork are frequently contaminated with *Salmonella* organisms. Control of salmonellosis depends largely on public health measures and surveillance programs to eradicate *Salmonella* from livestock (39,115).

The cause of diarrhea by nontyphoid salmonellae has been investigated for >20 years. It appears that *Salmonella*-induced diarrhea is a complex phenomenon that may involve several mechanisms (147). The demonstration that enterotoxins play an important role in diarrheal diseases caused by *Vibrio cholerae* and *E. coli* reawakened interest in the possible role of enterotoxins in salmonellosis. Enterotoxic activity has been reported by several groups of workers who have used various tissue culture assays and animal tests (ileal loop tests). The properties of the enterotoxic entity are said to resemble those of cholera toxin (128), *E. coli* heat-stable toxin (77,87), *Shiga* toxin (88), and even a hybrid of cholera–*Shiga* toxin (85). Until now, however, the role of *Salmonella* enterotoxin (or whatever) is far from clear. In most cases enterotoxic activity has been found only in bacterial extracts and not in culture supernates. Although the pathogenic activity of *Salmonella* is not clear, the infective dose for *Salmonella* is known to be low and has been well documented for a wide variety of food products (5,24,50,51).

A detailed background of *Salmonella* bacteria, including the antigens (lipopolysaccharides, Vi antigens, H antigens, lipoproteins), laboratory detection of *Salmonella* and its infections (culture, immunodiagnosis), and antibodies to *Salmonella*, has been presented previously (93a).

1. Detection and Enumeration of **Salmonella** *in Food*

In order to prevent salmonellosis, screening of foods and feeds for the absence of *Salmonella* has become routine. For decades, the detection of *Salmonella* in foods has been based on the application of culture methods originally designed to detect this pathogen in clinical specimens. However, the detection of *Salmonella* in foods presents a greater challenge than in clinical samples because the level of this pathogen in naturally contaminated foods is generally much lower than that in clinical samples. It is not unusual to find *Salmonella* in food at a level as low as one organism in 25 g of sample. The detection of low numbers of organisms is made even more difficult by the debilitated state of the *Salmonella* cells that have survived processing. A general method for the detection of *Salmonella* is described by ISO 6579. With this method a test sample is preenriched in a nonselective liquid medium (buffered pepton–water). This preenrichment allows detection of low numbers of *Salmonella* and of even injured *Salmonella*. After incubation at 37°C for 24 hr, 1 ml of the preenrichment broth is inoculated into a selective medium (tetrathionate medium or selenite cystine medium). This selective enrichment is necessary since *Salmonella* is often accompanied by considerably larger numbers of other *Enterobacteriaceae*. After incubation of the selective medium at 43°C for 24 hr and for 48 hr, the medium is streaked on brilliant green–phenol red agar plates. After incubation at 37°C for 24 hr the plates are checked for the presence of presumptive *Salmonella*. Colonies of presumptive *Salmonella* are confirmed by means of appropriate biochemical tests (fermentation of glucose, fermentation of lactose, formation of hydrogen sulfide, splitting of urea, and decarboxylation of lysine).

Recent studies have shown that better results are obtained if the selective medium proposed by ISO 6579 is replaced by Rappaport–Vassiliadis broth (156,157).

The primary disadvantage of using the aforementioned culture methods is the amount of time required to determine whether a sample contains *Salmonella*. A sample cannot be reported as negative for *Salmonella* until the fourth day after the initiation of analysis. Therefore, in the course of time several rapid methods such as the fluorescent-antibody technique (52), other immunoassays (1,48,101,143), and impedance microbiology (6) have been introduced to speed up the detection of *Salmonella* in foods. However, even with these methods a preenrichment followed by an enrichment in a selective medium is still necessary for obtaining reliable results. De Smedt and Bolderdijk (141) developed a rapid *Salmonella* detection method in foods based on the motility of the organism in semisolid selective enrichment medium. After preenrichment, 0.1 ml of the preenrichment culture is inoculated on the surface of the semisolid medium (modified Rappaport–Vassiliadis medium). After incubation at 42°C motile *Salmonella* will migrate and produce a growth circle. Serological confirmation tests may be done directly from the migrated culture. The method, however, does not work for nonmotile salmonellae.

F. *Listeria monocytogenes*

Listeria monocytogenes is a gram-positive rod, catalase-positive, and motile. The organism has long been recognized as a veterinary pathogen (79). In humans bacterial meningitis is the most common form of listeriosis, whereas perinatal infections may result in abortion, stillbirth, and infant death. Conditions predisposing nonpregnant adults to *Listeria* infections include diabetes, alcoholism, cancer, and treatment with corticosteroids or other immunosuppressive drugs (19,58,79). Until 1967 listeriosis was considered a rare disease in healthy adult humans, but recent food-borne outbreaks involving *L. monocytogenes* have highlighted its importance in public health (62,76,131). Foods such as milk, coleslaw, and cheese have been involved in *Listeriosis* outbreaks (47,76,131).

The genus *Listeria* consists of a heterogeneous group of microorganisms including *L. monocytogenes, L. innocua, L. welshimeri, L. seeligeri, L. ivanovii, L. grayi,* and *L. murrayi*. These species are differentiated on the basis of hemolytic activity, nitrate reduction, and the ability to metabolize distinct sugars. A classification scheme of biotypes of *Listeria* strains, according to Rocourt *et al.* (123), is presented in Table II. For epidemiological purposes identification according to Seeliger and Höhne (133) has been developed. The different serotypes are presented in Table III. Also in this table the different biotypes occurring in the different serotypes as found by us (107) are presented. Phage typing has also been suggested to be useful in elucidating epidemic outbreaks of listeriosis (122).

Only bacteria belonging to the species *L. monocytogenes* are pathogenic for humans. The mechanisms of pathogenesis of *L. monocytogenes* is not completely understood. Among the factors that contribute to the virulence of *L. monocytogenes* are hemolysins (57,74,80,81), invasive factors (56), factors that facilitate intracellular survival (20) and causing delayed-type hypersensitivity (DTH) (3,97), and camp factors (164). Pathogenicity of strains can be tested by injection of the test strains in mice using the mouse bioassay as described by Kaufmann (84). We found that strains of all serotypes of *L. monocytogenes* showed pathogenic activity, whereas all other biotypes (*L. seeligeri, L. innocua,* and *L. welshimeri*) were negative.

Listeria monocytogenes is found in the soil, in decaying and dead vege-

TABLE II

Classification Scheme of Biotypes of *Listeria* Strains Unable to Reduce Nitrate[a]

Biotype	Hemolysis	D-Xylose	L-Rhamnose
L. monocytogenes	+	−	+
L. seeligeri	+	+	−
L. innocua	−	−	+/−
L. welshimeri	−	+	+/−
L. ivanovii	+++	+	−

[a] From Rocourt *et al.* (123).

TABLE III

Serotypes of *Listeria* Strains and the Occurring Biotypes

Serotypes	Biotypes
1/2a	*L. monocytogenes*
1/2b	*L. monocytogenes*
	L. seeligeri
	L. welshimeri
1/2c	*L. monocytogenes*
3a	*L. monocytogenes*
3b	*L. monocytogenes*
	L. welshimeri
	L. seeligeri
3c	*L. monocytogenes*
	L. innocua
4a	*L. monocytogenes*
4b	*L. monocytogenes*
4c	*L. monocytogenes*
	L. seeligeri
4d	*L. monocytogenes*
4e	*L. monocytogenes*
4ab	*L. monocytogenes*
5	*L. ivanovii*
6a	*L. innocua*
	L. welshimeri
6b	*L. innocua*
	L. welshimeri
7	*L. monocytogenes*

tation, and in the intestinal tracts of humans, domestic and wild birds, and other animals (13,42,79). As a consequence the organism is common in all raw food materials such as raw milk and vegetables (13,43). A special feature of this organism is its ability to grow rapidly at refrigerator temperatures (29). As of this writing, however, no simple, reliable, sensitive, and rapid isolation or identification procedures are available for *L. monocytogenes*. The procedures currently used include (selective) enrichment in liquid media, subsequent plating of the enrichment fluid on selective agar media, and testing of suspected colonies by a number of biochemical tests followed by biotyping and serotyping (13,32,64,92). Biochemical tests and biotyping are essential since the media used are not selective for *L. monocytogenes* only.

III. RESULTS AND DISCUSSION

A. *Escherichia coli* and *Shigella* spp.

Most reports on the detection of *E. coli* by hybridization probes are in the field of clinical microbiology and deal with ETEC strains. Almost a decade ago Moseley *et al.* (103) demonstrated that a ^{32}P-labeled fragment from the *E. coli* LT gene could be used as a probe to identify ETEC colonies grown overnight on a nitrocellulose membrane and lysed subsequently by alkaline treatment. Since then many studies have been published describing the application of LT and ST gene probes for the detection of ETEC using both cloned gene fragments and synthetic oligonucleotides (35–37,70–72,89,95,104,125,135). To obtain adequate hybridization signals, most authors use an enrichment procedure. However, since the plasmids carrying the ST and LT genes may be lost during the enrichment culturing (70,153), this may result in false negative results. From this point of view a method for the direct detection of ETEC in samples is preferable. Lanata *et al.* (89) have shown that the detection of ETEC in stool samples without enrichment is possible if a minimum of 10^4 target cells is present per gram of sample. Hill *et al.* (71) were able to detect 10^3 ETEC colony-forming units per gram of food sample without enrichment using a colony hybridization procedure.

Since the practical use of DNA probes as an analytical tool strongly depends on its sensitivity, many of the initial experiments were carried out with ^{32}P-radiolabeled probes. Although the use of ^{32}P-labeled ETEC probes allow low detection limits, they are not suitable for routine diagnosis because of radiation hazards and short half-lives. Efforts have been made to develop other sensitive labeling and detection methods without the use of radioactive compounds. Seriwatana *et al.* (135) used synthetic LT and ST oligonucleotide probes covalently linked to an alkaline phosphatase marker molecule and compared their sensitivity and specificity to that of the same probes labeled with ^{32}P. They found that for the detection of ETEC in stool samples the overall sensitivity of alkaline phosphatase-labeled probes is somewhat lower than that found with ^{32}P-labeled probes. Similar results were obtained by Echeverria *et al.* (38), who used biotinylated oligonucleotide probes for the detection of ETEC in stools. Nevertheless the use of these nonradioactive DNA probes allows laboratories with minimal equipment to carry out ETEC identifications by hybridization without the need to overcome the problems imposed by the use of radiochemicals.

Probes have also been developed for another class of pathogenic *E. coli* strains that is, EIEC, and *Shigella* spp. Boileau *et al.* (17) have isolated

DNA fragments from a *Shigella flexneri* plasmid carrying genetic information involved in the invasiveness of this organism. A 17-kb DNA fragment was found to hybridize with EIEC strains and *Shigella* spp. The same probe was tested for its applicability for the detection of EIEC and *Shigella* spp in stools (136,152). With both ^{32}P-labeled and biotinylated probe DNA, Sethabutr *et al.* (136) were able to identify all of the 52 EIEC-positive samples and 10 of 12 *Shigella*-positive samples. Results obtained by Taylor *et al.* (152) were similar. They found that although EIEC and *Shigella* colonies could be reliably detected with this probe, it was less sensitive for use directly on stool blots identifying 76% of the EIEC and 45% of the *Shigella*-positive samples.

Another probe consisting of a 2.5-kb DNA fragment from an EIEC invasion plasmid has been isolated by Small and Falkow (140). It was tested for its diagnostic value by Wood *et al.* (162) and compared with the 17-kb probes (17). Although all invasive *Shigella* and EIEC strains tested were identified correctly, both probes showed a positive hybridization with 3 of 32 noninvasive *E. coli* and *Shigella* samples. From this it can be concluded that the probes are not completely invasive-specific, thus limiting their diagnostic value.

These problems may be overcome using a set of probes based on the invasive plasmid antigen genes from *S. flexneri*. Venkatesan *et al.* (159) have demonstrated that these probes distinguish between EIEC and invasive *Shigella* spp. and their noninvasive counterparts. Moreover, these probes did not react with 300 other nondysenteric, pathogenic gram-negative bacteria.

It should be possible to use gene probes for the detection of diarrheogenic *E. coli* in food as has been done for clinical samples. Since *E. coli* is a common inhabitant of the intestinal tract of humans and animals, its presence in food usually indicates fecal contamination, and compared with clinical samples, like stools, their concentration is much lower. Moreover, only a small proportion of these contaminating *E. coli* cells will have diarrheogenic properties. For these reasons the use of gene probes for routine detection of diarrheogenic *E. coli* strains in food samples is less obvious than for clinical samples.

B. *Staphylococcus aureus*

In our laboratory efforts have been made to develop DNA probes that are suitable for the detection and identification of food-borne enteropathogenic *S. aureus* (111). A set of three oligonucleotides was synthesized, based on different parts of the nucleotide sequence of the gene coding for the enterotoxin type B (SEB) (78). In the three corresponding parts of the toxin molecule, SEB and SEC1 have an identical amino acid sequence (132). To test the specificity of these probes, they were ^{32}P-labeled and used in a colony hybridization assay. All 22 *S. aureus* strains that were known to produce SEB, from immunological characterization, were positive with all three probes, whereas 111 enterotoxin-negative *S. aureus* strains did not hybridize. Among *S. aureus* strains producing enterotoxins other than SEB, one probe detected all 69 SEC strains, a second probe 21 of 69 SEC strains, and a third none of the SEC strains. The first of these three probes, therefore, seems suitable for the identification of both SEB- and SEC-producing strains. We recently found that the SEC-producing strains hybridizing with the second probe all belong to subtype SEC1, whereas SEC2 and SEC3 do not react. Thus with this set of three probes, SEB, SEC1, and SEC2 + SEC3-producing strains can be distinguished. The finding in this study that the presence of these genes in all strains tested leads to the actual production of SEB or SEC adds to the value of the probe for the detection of enterotoxinogenic *S. aureus*.

C. *Clostridium perfringens*

In our laboratory we have developed a DNA probe for the detection and identification of food-borne enterotoxinogenic *C. perfringens*. We aimed at the gene for CPE, which is the causative agent for food poisoning by this microorganism. As no nucleotide sequence data were available we cloned the entire CPE gene and determined its sequence (158). Based on this information, four oligonucleotide probes were synthesized covering various parts of the 320-amino acid open reading frame of the CPE gene. These oligonucleotides and a cloned 0.8-kb DNA fragment of the CPE gene were ^{32}P-labeled and used as probes in a *C. perfringens* colony hybridization assay. All strains tested were *C. perfringens* isolated from feces, food, and the environment associated with outbreaks of human food-borne infections. Colonies were lysed on nylon membranes by an alkaline steaming procedure described by Notermans *et al.* (111). The hybridization results obtained are presented in Table IV. The probes showed identical specificity: 130 of 226 strains were positive with all five, whereas the others reacted with none of them. Although all strains isolated were associated with outbreaks of food poisoning, only 57% contained the CPE gene. Since *C. perfringens* is believed to produce only one enterotoxin (146), the conclusion is obvious that all hybridization-negative strains that were isolated are not capable of producing enterotoxin. *Clostridium perfringens* is a ubiquitous microorganism and is frequently present in the intestinal tract of nonsymptomatic individuals (148,163). Food causing poisoning may well be contaminated with several *C. perfringens* strains from various

TABLE IV

Hybridization Results of *Clostridium perfringens* Strains with Various Probes[a]

Origin of strains	Number of strains	Number of strains hybridizing with				
		Probe 1	Probe 2	Probe 3	Probe 4	Probe 5
Feces	174	99	99	99	99	99
Food	48	31	31	31	31	31
Environment	4	0	0	0	0	0
Total	226	130	130	130	130	130

[a] Strains were isolated from stool, food, and environmental samples after human food-borne infections. Probes 1–4 are synthetic oligonucleotides with a sequence identical to regions of the CPE gene (158). Probe 1, Amino acid codon (aac) 1–14; probe 2, aac 101–112; probe 3, aac 213–225; probe 4, aac 294–307. Probe 5 is a 0.8-kb cloned fragment consisting of 0.7 kb of CPE-coding region and a 0.1-kb region upstream of the CPE gene.

sources. Further studies will be undertaken to evaluate the use of these CPE probes for investigating the distribution of enterotoxinogenic *C. perfringens* strains in the environment.

D. *Clostridium botulinum*

Within the group of clostridia, *C. botulinum* is solely defined by its ability to produce botulinum toxin. Therefore detection of *C. botulinum* by hybridization can only be specific using a gene probe encoding the neurotoxin. To determine the nucleotide sequence of this gene a strategy was used similar to that described for the CPE gene (158). A gene bank was constructed in *E. coli* containing small DNA fragments (<1 kb) from the toxin B-producing *C. botulinum* strain Okra. This bank was screened with a synthetic oligonucleotide probe mixture (41-mers), its sequence being based on reverse translation of the amino acid sequence at the N-terminal end of the light chain of the *C. botulinum* type B toxin molecule (25,129,142). Several positive clones were identified and sequenced. These clones were found to contain the N-terminal start of the toxin B gene (Table V). Based on these sequence data obtained, an oligonucleotide was synthesized and used as probe in a colony hybridization experiment. The results of this preliminary study are shown in Table VI. The probe used can distinguish toxin B-producing *C. botulinum* from others producing types A, E, and F. A toxin B-producing strain that had lost the ability to produce the toxin during propagation in our laboratory did not hybridize. We now are in the process of sequencing other parts of the toxin B gene and cloning parts of various other types of *C. botulinum* toxin genes. The

TABLE V

Partial Nucleotide Sequence and Deduced Amino Acid Sequence of the Start of the *Clostridium botulinum* Toxin B L-Chain (Strain Okra)[a]

ATG	CCA	GTT	ACA	ATA	AAT	AAT	TTT	AAT	TAT	AAT	GAT	CCT	ATT	GAT	AAT	AAT	AAT
MET	Pro	Val	Thr	Ile	Asn	Asn	Phe	Asn	Tyr	Asn	Asp	Pro	Ile	Asp	Asn	Asn	Asn
1									10								
ATT	ATT	ATG	ATG	GAG	CCT	CCA	TTT	GCG	AGA	GGT	ACG	GGG	AGA	TAT	TAT	AAA	GCT
Ile	Ile	Met	Met	Glu	Pro	Pro	Phe	Ala	Arg	Gly	Thr	Gly	Arg	Tyr	Tyr	Lys	Ala
	20										30						
TTT	AAA. . . .																
Phe	Lys . . .																

[a] The underlined part indicates the nucleotide sequence of a synthetic oligonucleotide used as a probe for the identification of toxin B-producing *C. botulinum* strains.

TABLE VI

Hybridization Results of an Oligonucleotide Probe Encoding Part of the Botulinum Toxin B Gene and Various *Clostridium botulinum* Strains[a]

Number of strains	Type of toxin produced	Number of strains hybridization with toxin B probe
5	A	0
14	B	14
8	E	0
1	F	0
1	B, atox	0
1	E, atox	0

[a] The nucleotide sequence of the probe is indicated in Table III. Strain B, atox and E, atox are strains that were able to produce toxin B and E, respectively, but lost this ability during laboratory propagation.

aims of our future work are the development of probes and practical methods that will eliminate the need for the mouse bioassay for detection and identification of *C. botulinum*.

E. *Salmonella* spp.

Development of DNA probes for the specific detection of pathogenic *Salmonella* is hindered by a lack of knowledge of *Salmonella* virulence factors. Characterization of genes that are involved in *Salmonella* infection is an important step to be taken for the future use of specific *Salmonella* probes. Gene probes that have been reported to date are either cloned, random by selected DNA fragments, or genes encoding factors that are not directly involved in virulence.

Various reports have indicated the potential of the use of gene probes for the detection of *Salmonella* in food. In 1983 Fitts *et al.* (46) developed a DNA hybridization test for the presence of *Salmonella* spp. in food. They used ^{32}P-labeled, cloned random DNA fragments from *S. typhimurium* as probes and were able to demonstrate the presence of *Salmonella* spp. in enrichment cultures of various food products. Rubin *et al.* (124) have described a gene probe specific for *S. typhi*. This probe consists of a cloned fragment from the *viaB* gene, which encodes for a capsular antigen known to be present only in *S. typhi, S. paratyphi,* and some *Citrobacter freundii* strains. No data are available on the use of this probe in food samples.

In a collaborative study (49) it was shown that the use of a *Salmonella* spp-specific DNA probe is a practical tool for detection of this microorganism in foods. The data indicated that the hybridization method was at least as productive as standard culturing methods.

F. *Listeria monocytogenes*

Despite many improvements of culture methods, recovery and characterization of *L. monocytogenes* from food and environmental samples remains laborious and time-consuming. To overcome these problems various attempts have been made to apply the hybridization technique to the detection of *Listeria*. Klinger *et al.* (86) have described the use of an assay based on the detection of unique *Listeria* 16 S ribosomal RNA (rRNA) sequences with a ^{32}P-labeled synthetic oligonucleotide. In a colony hybridization assay all of the 139 *Listeria* strains of all biotypes gave a positive result, whereas 73 non-*Listeria* strains did not hybridize. The probe detected *Listeria* spp. in artificially contaminated food and environmental samples after enrichment culturing.

The fact that rRNA is used as a target has a positive effect on the sensitivity due to the presence of many target molecules in each cell. On the other hand, the specificity of this probe is such that no distinction can be made between the pathogenic species (*L. monocytogenes* and *L. ivanovii*) and nonpathogenic species of *Listeria* (*L. innocua, L. seeligeri, L. welshimeri, L. grayi,* and *L. murrayi*). The presence of these nonpathogenic species in food samples provides information concerning preprocessing and postprocessing hygiene.

Other investigators (27) have tested a gene probe coding for the *L. monocytogenes* β-hemolysin for the detection of *L. monocytogenes* colonies. Hybridization was carried out with filter paper sheets on which bacterial colonies had been lysed by alkaline denaturation and irradiation in a microwave oven. The 0.5-kb probe reacted with 10 of 18 *L. monocytogenes* strains isolated from various outbreaks of listeriosis. Small numbers of other *Listeria* spp. were also tested and found not to hybridize with the probe. Although this probe only reacts with *L. monocytogenes* strains, a drawback in its use for detection of virulent *Listeria* might be the observation that 8 of 18 *L. monocytogenes* did not hybridize.

In our laboratory detection of virulent *Listeria* strains was investigated using a gene probe encoding the *L. monocytogenes* DTH factor. This protein factor is capable of eliciting a DTH reaction in immune mice, and preliminary hybridization experiments indicated that the gene is present in the pathovars *L. monocytogenes* and *L. ivanovii* and absent in all other bacteria of this genus (22).

We used a 1.1-kb DTH gene fragment for colony hybridization with 284 *Listeria* strains, including all known serovars and biovars (107). The gene was found to be present in all 117 tested *L. monocytogenes* of serotype 1/2a, 1/2b, 1/2c, 3a, 3b, 3c, 4ab, 4c, 4d, 4e, 7, and in *L. ivanovii*. Of 78 *L. monocytogenes* strains of serotype 4b, 77 contained the gene, whereas it was absent in all of 10 *L. monocytogenes* strains of serotype 4a.

The gene was absent in all other *Listeria* spp. tested (*L. seeligeri, L. grayi, L. murrayi, L. welshimeri,* and *L. innocua*). These results indicated that the DTH gene probe could be a valuable tool for specific detection of *L. monocytogenes*. However, the strains tested in this experiment were chosen to represent all known *Listeria* biotypes and serotypes, and they do not necessarily reflect the normal distribution of *Listeria* spp. in field samples. To investigate this distribution and to test the utility of the probe in this setting, we isolated *Listeria* strains from naturally contaminated food samples by cultural methods and determined their biotype and serotype. Hybridization results are presented in Table VII. Only 47% of the isolated strains were *L. monocytogenes* and 48% typed as *L. innocua*. Among all *L. monocytogenes* isolates 78% were of serotype 1/2a. Using the DTH probe for hybridization, all *L. monocytogenes* were identified. The probe did not react with the nonpathogenic *Listeria* spp. (*L. seeligeri, L. welshimeri,* and *L. innocua*). These results demonstrate that the DTH probe is very promising for the specific detection of *Listeria* pathovars in food and environmental samples.

TABLE VII

Hybridization Results Obtained with a Delayed-Type Hypersensitivity Probe of *Listeria* Strains Isolated from Naturally Contaminated Food Samples

		Number of strains	
Serotype	Biotype	Isolated	Positive with DTH probe
1/2a	*L. monocytogenes*	122	122
1/2b	*L. monocytogenes*	14	14
	L. seeligeri	2	0
1/2c	*L. monocytogenes*	5	5
3a	*L. monocytogenes*	1	1
4b	*L. monocytogenes*	13	13
4c	*L. seeligeri*	7	0
4ab	*L. monocytogenes*	1	1
6a	*L. innocua*	108	0
6b	*L. welshimeri*	8	0
	L. innocua	53	0

IV. CONCLUSIONS

The use of gene probes for the detection and identification of pathogenic microorganisms in food is at this time just beginning. However, the number of applications is increasing rapidly. Those probes encoding gene products involved in pathogenic behavior, especially, are developing into very useful tools that allow direct screening for virulence without laborious *in vivo* assays. As shown, cloned DNA sequences and synthetic oligonucleotides are useful as gene probes in hybridization assays, and the specific detection and identification of pathogens are possible.

Since food-borne microorganisms usually occur in relatively low concentrations in food samples, direct detection with gene probes is not yet possible. Therefore, enrichment culturing or growth of colonies is a necessary part of detection procedures using gene probes for food-borne pathogens.

V. HYBRIDIZATION ASSAY VERSUS CONVENTIONAL ASSAYS

The genetic material of each living organism contains a blueprint of that organism and all its properties. When pathogenic behavior is one of these properties, the genetic material contains genes that encode the factors involved in virulence. The use of gene probes that encode these factors allows detection and identification of pathogens directly through the presence of this genetic information. This is a clear advantage compared to the use of other assays. When, for example, a pathogenic microorganism is identified by means of its toxin production during culturing, immunological or animal testing procedures will fail if no toxin is produced as a result of experimental conditions. A DNA probe, however, will give a positive result since hybridization does not depend on expression of the gene, thus increasing the reliability of the assay. The application of DNA probes encoding virulence factors will also allow a strong decrease in the use of laboratory test animals. At present as many as 15 mice need to be sacrificed for the detection and characterization of *C. botulinum* in a single sample. Development of a set of DNA probes encoding the different immunological types of *C. botulinum* neurotoxin has the clear potential of making the mouse bioassay redundant for the detection of *C. botulinum*.

Especially for the routine testing of large numbers of samples, the use of DNA probes may reduce the labor time per sample compared to conventional methods. Using a colony hybridization procedure the DNA or RNA of at least 100 colonies can be fixed on a membrane the size of a cigarette box, and many of these can be processed in a single hybridization experiment. However, as the time required to carry out the hybridization assay is not determined by the number of samples tested, the gain in time decreases if fewer samples are assayed.

The use of gene probes offers an interesting new tool in the field of food microbiology. However, a decision to employ them for the detection and identification of food-borne microorganisms should always be preceded by a careful consideration of the advantages and disadvantages of other conventional techniques such as MAb (114a,129a,151,153a), available to the food microbiologist.

VI. PROSPECTS FOR THE FUTURE

Conventional methods for the detection and identification of food-borne pathogenic microorganisms are laborious and time-consuming (*Salmonella, L. monocytogenes*) or require a well-equipped laboratory (*C. botulinum,* enterotoxinogenic *S. aureus* and *C. perfringens,* diarrheogenic *E. coli*). In order to circumvent these problems and yet be able to judge or predict the safety of food products, "indicator organisms" are determined. The presence of high numbers of coliform organisms, for instance, indicates a fecal contamination of the raw materials, whereas the presence of gram-negative microorganisms demonstrates insufficient pasteurization. High numbers of staphylococci may be a result of poor hygienic conditions during food processing. Detection of "indicator organisms" is very simple. A suspension of the food product is plated on a defined agar medium and incubated for 18–22 hr under standard conditions allowing rapid determination of a colony count. At present, manufacturers of food products are continuously introducing rules of "good manufacturing practice" and "codes of production" to ensure the safety of their products. As a general rule, only raw ingredients are processed that are free of pathogens. In addition, processing conditions are well defined, known to kill all microorganisms, and avoid recontamination. Precise setting of intrinsic factors such as pH, water activity, and redox potential and extrinsic factors such as gas-packaging and storage temperature will prevent outgrowth of pathogens that might eventually be present in the final product. A general introduction of these rules and codes may drastically decrease the need for routine detection procedures of pathogens in the near future. Testing for microbial specifications will then be limited to raw materials and end-products.

The need for rapid and reliable procedures for the detection and identification of food-borne microbial pathogens is important in the field of epidemiology. The use of gene probes has proved to have the potential to fulfill this need, especially those probes that encode well-understood virulence factors. The pathogenesis of a number of microorganisms is understood and the virulence factors are known (*C. botulinum, C. perfringens,* and *S. aureus*). Others, however, display a pathogenic behavior based on unknown mechanisms (*Salmonella*) or a complex of virulence factors (*E. coli, L. monocytogenes*). Development of new gene probes for the detection of these pathogens will profit from future research in this field.

A serious drawback in the practical application of DNA probes is the use of radiolabels such as ^{32}P. Their use is limited to skilled personnel in specially equipped laboratories. In the last few years, however, an increasing number of nonradioactive alternatives has become available (98), which will certainly stimulate a more common use of DNA probes. Another drawback in the use of gene probes in food microbiology is the inability to detect very low numbers of pathogens among large numbers of contaminating microorganisms. Using conventional hybridization methods, at least 10^4–10^5 target microorganisms are required to obtain a positive hybridization result, thus limiting the use of DNA probes for the direct detection of pathogens in food samples. However, the recently developed polymerase chain reaction (PCR) technique (160) may drastically improve the sensitivity of hybridization assays. It is based on the *in vitro* amplification of target DNA sequences. With the PCR method, it has been possible to detect a target microorganism at a concentration of one cell per gram of soil sample against a background of 10^9 nontarget organisms (145). The use of the PCR method by food microbiologists may, however, pose new problems due to its extreme sensitivity. Extensive studies are needed to test the effect of killed target organisms in the sample on the results of the hybridization assay.

The use of gene probes in food microbiology will improve and simplify epidemiological studies and will enable the screening of large numbers of samples for the presence of pathogenic microorganisms. Hybridization tests are not likely to replace all conventional detection methods but certainly will add an excellent new tool to those already available to the food microbiologist.

VII. SUMMARY

Conventional techniques for the detection and identification of food-borne pathogenic microorganisms are often laborious, time-consuming, and unreliable. The use of specific gene probes is a new analytical tool that has the potential to overcome many of these problems. It allows detection of pathogens by screening of their genetic material for the presence of genes that encode the virulence factors. For some food-borne pathogens (*C. botulinum,* enterotoxinogenic *C. perfringens,* and enterotoxinogenic *S. aureus*) these factors are known, whereas for others (diarrheogenic *E. coli, L. monocytogenes,* and *Salmonella*) the virulence factors are less well understood. At present gene probes are mainly used as an analytical tool in clinical microbiology, although the number of applications in the field of food microbiology is rapidly increasing. Compared with conventional methods, the use of virulence-specific gene probes allows more reliable and rapid testing of large numbers of samples for food-borne pathogens without the need for *in vivo* assays. These properties make the use of gene probes suitable for epidemiological studies on food-borne pathogens.

VIII. MATERIALS AND METHODS

An essential step in the hybridization procedures described is the preparation of the membranes carrying the immobilized bacterial nucleic acid. The method described here has been used by us for a wide variety of gram-positive and gram-negative microorganisms with excellent results.

A. Preparation of Membranes for Colony Hybridization

Inoculate a petri dish containing a suitable agar-solidified growth medium.

Incubate until colonies have appeared 1–2 mm in diameter.

Cover the colonies with a sheet of dry Gene Screen Plus membrane (DuPont Co.) and allow the membrane to moisten completely.

Peel off the membrane now carrying the colonies.

Prepare a dish containing three layers of filter paper (Whatmann 3 MM) soaked in a 0.5 *M* NaOH solution and place the membrane on it.

Incubate the dish and its contents in a closed boiling-water bath just above the water level for 10 min.

Then immerse the membrane in a solution containing 1.0 *M* Tris-HCl (pH 7.5) for 1 min.

Transfer the membrane to a solution containing 5× SSC (1× SSC contains 0.15 *M* NaCl and 0.015 *M* sodium citrate).

While immersed, thoroughly rub off all bacterial debris using a Kleenex tissue.

Allow the membrane to dry completely at room temperature and use for hybridization.

REFERENCES

1. Andrews, W. H. (1985). A review of culture methods and their relation to rapid methods for the detection of *Salmonella* in foods. *Food Technol* **39,** 77–82.
2. Angelotti, R., Foster, M. J., and Lewisk, N. (1961). Time–temperature effects on *Salmonellae* and *Staphylococci* in foods. *Appl. Microbiol.* **9,** 308–315.
3. Antonissen, A. C. J. M., Lemmens, P. J. M. R., van den Bosch, J. F., and van Boven, C. P. A. (1986). Purification of a delayed hypersensitivity-inducing protein from *Listeria monocytogenes*. *FEMS Microbiol. Lett.* **34,** 91–95.
4. Appleton, N., Wilton-Smith, P., Hambleton, P., Modi, N., Gatley, S., and Melling J. (1986). *In vitro* assays for botulinum toxin and antitoxins. *Dev. Biol. Stand.* **64,** 141–145.
5. Armstrong, R. W., Fodor, T., Curlin, G. T., Cohen, A. B., Morris, G. K., Martin, W. T., and Feldman, J. (1970). Epidemic *Salmonella* gastroenterititis due to contaminated imitation ice cream. *Am. J. Epidemiol.* **91,** 300–307.
6. Arnott, M. L., Gutteridge, C. S., Pugh, S. J., and Griffiths, J. L. (1988). Detection of *Salmonella* in confectionery products by conductance. *J. Appl. Bacteriol.* **64,** 409–420.
7. Avena, R. M., and Bergdoll, M. S. (1967). Purification and some physiochemical properties of enterotoxin B, *Staphylococcus aureus* strain 361. *Biochemistry* **6,** 1474–1480.
8. Baird-Parker, A. C. (1962). An improved diagnostic and selective medium for isolating coagulase-positive *Staphylococci*. *J. Appl. Bacteriol.* **25,** 12–19.
9. Bartholomew, B. A., and Stringer, M. F. (1984). *Clostridium perfringens* enterotoxin: A brief review. *Biochem. Soc. Trans.* **12,** 195–197.
10. Bartholomew, B. A., Stringer, M. F., Watson, G. N., and Gilbert, R. J. (1985). Development and application of an enzyme-linked immunosorbent assay for *Clostridium perfringens* type A enterotoxin. *J. Clin. Pathol.* **38,** 222–228.
11. Beckers, H. J. (1986). Foodborne diseases in the Netherlands. *J. Food Prot.* **49,** 924–931.
12. Beckers, H. J., van Leusden, F. M., Bindschedler, O., and Guarraz, D. (1983). Evaluation of a pour-plate system with rabbit plasma–bovine fibrinogen agar for the enumeration of *Staphylococcus aureus* in food. *Can. J. Microbiol.* **30,** 470–474.
13. Beckers, H. J., Soentoro, P. S. S., and Delfgou-van Asch, E. H. M. (1987). The occurrence of *Listeria monocytogenes* in soft cheeses and raw milk and its resistance to heat. *Int. J. Food Microbiol.* **4,** 249–256.
14. Bergdoll, M. S. (1983). Enterotoxins *In "Staphylococci* and Staphylococcal Infections" (C. S. F. Easmon, and C. Adlam, eds.), pp. 559–598. Academic Press, New York.
15. Bergdoll, M. S. (1979). Staphylococcal intoxications. *In* "Foodborne Infections and Intoxications" (H. Rieman, and F. L. Bryan, eds.), pp. 443–494. Academic Press, New York.
16. Berry, P. R., Rodhouse, J. C., Hugler, S., Bartholomew, B. A., and Gilbert, R. J. (1988). Evaluation of ELISA, RPLA and Vero all assays for detecting *Clostridium perfringens* enterotoxin in faecal specimens. *J. Clin. Pathol.* **41,** 458–461.
17. Boileau, C. R., d'Hauteville, H. M., and Sansonetti, P. J. (1984). DNA hybridization technique to detect *Shigella* species and enteroinvasive *Escherichia coli*. *J. Clin. Microbiol.* **20,** 959–961.
18. Borja, C. R., and Bergdoll, M. S. (1967). Purification and partial characterization of enterotoxins produced by *Staphylococcus aureus* strain 137. *Biochemistry* **6,** 1467–1473.
19. Bortolussi, R., Schlech, W. F., and Albritton, W. L. (1985). *Listeria*. *In* "Manual of

Clinical Microbiology" (E. H. Lennette, A. Balows, W. J. Haasler, Jr. and H. J. Shadomy, eds.), Vol. 4, pp. 205–208. Microbiol. Washington, D.C.

20. Bortolussi, R., Vandenbroucke-Grauls, C. M. J. E., van Asbeck, B. S., and Verhoef, J. J. (1987). Relationship of bacterial growth phase to killing of *Listeria monocytogenes* by oxidative agents generated by neutrophils and enzyme systems. *Infect. Immun.* **55,** 3197–3203.
21. Brodsky, M. H., Cieben, B. W., and Schieman, D. A. (1979). A critical evaluation of automatic bacterial colony counters. *J. Food Prot.* **42,** 138–143.
22. Chakraborty, T., Leimeister-Wächter, M., Goebel, W., Wernars, K., and Notermans, S. (1988). Letter. *Lancet* (to be published).
23. Craig, J. M., Iida, H., and Inoue, K. (1970). A recent case of botulism in Hokkaido, Japan. *Jpn. J. Med. Sci. Biol.* **23,** 193–198.
24. Craven, P. C., Mackel, D. C., Baine, W. B., Barker, W. H., Gangarossa, E. J., Goldfield, M. Rosenfeld, M., Altman, R., Lachapelle, G., Davies, J. W., and Swanson, R. C. (1975). International outbreak of *Salmonella eastborne* infection traced to contaminated chocolate. *Lancet* **1,** 788–792.
25. Dasgupta, B. R., and Datta, A. (1978). Botulinum Neurotoxin type B (strain 657); partial sequence and similarity with tetanus toxin. *Fed. Proc., Fed. Am. Soc. Exp. Biol.* **46,** 2289.
26. Dasgupta, B. R., and Sugiyama, H. (1972). A common subunit structure in *Clostridium botulinum* types A, B and E toxins. *Biochem. Biophys. Res. Commun.* **48,** 108–112.
27. Datta, A. R., Wentz, B. A., and Hill, W. E. (1987). Detection of hemolytic *Listeria monocytogenes* by using DNA colony hybridization. *Appl. Environ. Microbiol.* **53,** 2256–2259.
28. Denney, C. P., Tan, P. L., and Bokrer, C. W. (1966). Heat inactivation of staphylococcal enterotoxin A. *J. Food Sci.* **31,** 762–767.
29. Donnely, C. W., and Briggs, E. H. (1986). Psychrotrophic growth and thermal inactivation of *Listeria monocytogenes* as a function of milk composition. *J. Food Prot.* **49,** 994–998.
30. Dowell, V. R., McCrosky L. M., Hatheway, C. L., Lombard, G. L., Hughes, J. M., and Merson M. H. (1977). Coproexamination for botulinal toxin and *Clostridium botulinum*. *J. Am. Med. Assoc.* **238,** 1829–1832.
31. Dowell, V. R., Jr., Torres-Anjel, M. J., Riemann, H. P., Merson, M., Whaley, D., and Darland, G. (1975). A new criterion for implicating *Clostridium perfringens* as the cause of food poisoning. *Rev. Latinoam. Microbiol.* **17,** 137–142.
32. Doyle, M. P., and Schoeni, J. L. (1986). Selective-enrichment procedure for isolation of *Listeria monocytogenes* from fecal and biologic specimens. *Appl. Environ. Microbiol.* **51,** 1127–1129.
33. Duncan, C. L., and Strong, D. H. (1968). Improved medium for sporulation of *Clostridium perfringens*. *Appl. Microbiol.* **16,** 82–89.
34. Duncan, C. L., Strong, D. H., and Sebald, M. (1972). Sporulation and enterotoxin production by mutants of *Clostridium perfringens*. *J. Bacteriol.* **110,** 378–391.
35. Echeverria, P., Seriwatana, J., Chityothin, O., Chaicumpa, W., and Tirapat, C. (1982). Detection of enterotoxinogenic *Escherichia coli* in water by filter hybridization with three enterotoxin gene probes. *J. Clin. Microbiol.* **16,** 1086–1090.
36. Echeverria, P., Seriwatana, J., Leksomboom, U., Tirapat, C., Chaicumpa, W., and Rowe, B. (1984). Identification by DNA hybridization of enterotoxinogenic *Escherichia coli* in homes of children with diarrhea. *Lancet* **1,** 63–66.
37. Echeverria, P., Seriwatana, J., Sethabutr, O., and Taylor, D. N. (1985). DNA hybridization in the diagnosis of bacterial diarrhea. *Clin. Lab. Med.* **5,** 447–462.

38. Echeverria, P., Taylor, D. N., Seriwatana, J., Chatkaeomorakot, A., Khungvalert, V., Sakuldaipeara, T., and Smith, R. D. (1986). A comparative study of enterotoxin gene probes and tests for toxin production to detect enterotoxinogenic *Escherichia coli. J. Infect. Dis.* **153,** 255–260.
39. Edel, W., van Schothorst, M., and Kampelmacher, E. H. (1977). Salmonella and salmonellosis: The present situation. *Proc. Int. Symp. Salmonella Prospects Control, 1977,* pp. 1–26.
40. Elwell, M. R., Liu, C. T., Spertzel, R. O., and Beizel, W. R. (1975). Mechanism of oral staphylococcal enterotoxin B-induced emesis in monkey. *Proc. Soc. Exp. Biol. Med.* **148,** 424–427.
41. Ewald, S., and Notermans, S. (1988). Effect of water activity on growth and enterotoxin D production of *Staphylococcus aureus. Int. J. Food Microbiol.* **6,** 25–30.
42. Farber, J. M., Johnston, M. A., Purvis, V., and Loit, A. (1987). Surveillance of soft and semi-soft cheeses for the presence of *Listeria* spp. *Int. J. Food Microbiol.* **5,** 157–163.
43. Fernandez Garayzabal, J. F., Dominguez Rodriquez, L., Vasquez Boland, J. A., Blanco Cancelo, J. L., and Suarez Fernandez, S. (1986). *Listeria monocytogenes* dans le lait pasteurisé. *Can. J. Microbiol.* **32,** 149–150.
44. Fey, H., Pfister, H., and Rüegg, O. (1984). Comparative evaluation of different enzyme-linked immunosorbent assay systems for the detection of staphylococcal enterotoxins A, B, C and D. *J. Clin. Microbiol.* **19,** 34–38.
45. Firstenberg-Eden, R. (1986). Electrical impedance for determining microbial quality of foods. *In* "Foodborne Microorganisms and their Toxins: Developing Methodology" (M. D. Pierson, and N. J. Stern, eds.), pp. 129–144. Dekker, New York.
46. Fitts, R., Diamond, M., Hamilton, C., and Neri, M. (1983). DNA–DNA hybridization assay for detection of *Salmonella* spp. in foods. *Appl. Environ. Microbiol.* **46,** 1146–1151.
47. Fleming, D. W., Cochi, S. L., MacDonald, L. K., Brondum, J., Hayes, P. S., Plikaytis, B. D., Holmes, M. B., Audurier, A., Broome, C. V., and Reingold, R. L. (1985). Pasteurized milk as a vehicle of infection in an outbreak of listeriosis. *N. Engl. J. Med.* **312,** 404–407.
48. Flowers, R. S., Klatt, M. J., and Keelan, S. (1988). Visual immunoassay for detection of *Salmonella* in foods: Collaborative study. *J. Assoc. Off. Anal. Chem.* **71,** 973–980.
49. Flowers, R. S., Klatt, M. J., Mozola, M. A., Curiale, M. S., Gabis, D. A., and Silliker, J. H. (1985). DNA hybridization assay for detection of *Salmonella* in foods: Collaborative study. *J. Assoc. Off. Anal. Chem.* **70,** 521–529.
50. Fontaine, R. E., Arnon, S. Martin, W. T., Vernon, T. M., Gangarosa, E. J. Farmer, J. J., Moran, A. B., Silliker, J. H., and Decker, D. L. (1978). Raw hamburger: An interstate common source of human salmonellosis. *Am. J. Epidemiol.* **107,** 36–45.
51. Fontaine, R. E., Cohen, M. L., Martin, W. T., and Vernon, T. M. (1980). Epidemic salmonellosis from cheddar cheese: Surveillance and prevention. *Am. J. Epidemiol.* **111,** 247–253.
52. Food and Drug Administration (1978). "Bacteriological Analytical Manual," 5th ed. Association of Official Analytical Chemists, Washington, D.C.
53. Frieben, W. R., and Duncan, C. L. (1975). Heterogeneity of enterotoxin-like protein extracted from spores of *Clostridium perfringens* type A. *Eur. J. Biochem.* **55,** 455–463.
54. Frieben. W. R., and Duncan, C. L. (1973). Homology between enterotoxin protein and spore structural protein in *Clostridium perfringens* Type A. *Eur. J. Biochem.* **39,** 393–401.
55. Fukuda, T., Kitao, T., Tamikawa, H., and Sakaguchi, G. (1970). An outbreak of type B botulism occurring in Miyazaki prefecture. *Jpn. J. Med. Sci. Biol.* **23,** 243–248.

56. Gaillard, J. L., Berche, P., Mounier, J., Richard, S., and Sansonetti, P. (1987). Penetration of *Listeria monocytogenes* into the host: A crucial step of the infection process. *Ann. Inst. Pasteur/Microbiol.* **138,** 259–264.
57. Gaillard, J. L., Berche, P., and Sansonetti, P. (1987). Transposon mutagenesis as a tool to study the role of hemolysin in the virulence of *Listeria monocytogenes. Infect. Immun.* **52,** 50–55.
58. Gantz, N. M., Meyerowitz, R. L., Medeiros, A. A., Garrera, G. F., Wilson, R. E., and O'Brien, T. F. (1975). Listeriosis in immunosuppressed patients. *Am. J. Med.* **58,** 637–643.
59. Gilbert, R. J. (1974). Staphylococcal food poisoning and botulism. *Postgrad. Med. J.* **50,** 603–611.
60. Goldner, S. B., Solberg, M., Jones, S., and Post, L. S. (1986). Enterotoxin synthesis by nonsporulating cultures of *Clostridium perfringens. Appl. Environ. Microbiol.* **52,** 407–412.
61. Granum, P. E., and Whitaker, J. P. (1980). Improved method for purification of enterotoxin from *Clostridium perfringens* type A. *Appl. Environ. Microbiol.* **39,** 1120–1122.
62. Gravani, R. B. (1987). Bacterial foodborne diseases. *Dairy Food Sanit.* **7,** 137–141.
63. Haagsma, J. (1973). Ph.D dissertation, University of Utrecht, Utrecht.
64. Hao, D. Y. Y., Beuchat, L. R., and Bracket, R. E. (1987). Comparison of media and methods for detecting and enumerating *Listeria monocytogenes* in refrigerated cabbage. *Appl. Environ. Microbiol.* **53,** 955–957.
65. Harmon, S. M., Kautter, D. A., and Peeler, J. T. (1971). Comparison of media for the enumeration of *Clostridium perfringens. Appl. Microbiol.* **21,** 922–927.
66. Harmon, S. M., Kautter, D. A., and Peeler, J. T. (1971). Improved medium for enumeration of *Clostridium perfringens. Appl. Microbiol.* **22,** 688–692.
67. Harmon, S. M., and Kautter, D. A. (1986). Improved media for sporulation and enterotoxin production by *Clostridium perfringens. J. Food Prot.* **49,** 706–711.
68. Hauschild, A. H. W., and Hilsheimer, R. (1983). Prevalence of *Clostridium botulinum* in commercial liver sausage. *J. Food Prot.* **46,** 242–244.
69. Hauschild, A. H. W., Hilsheimer, R., and Rogers, C. G. (1971). Rapid detection of *Clostridium perfringens* enterotoxin by a modified ligated intestinal loop technique in rabbits. *Can. J. Microbiol.* **17,** 1475–1476.
70. Hill, W. E. (!981). DNA hybridization method for detecting enterotoxinogenic *Escherichia coli* in human isolates and its possible application to food samples. *J. Food Saf.* **3,** 233–247.
71. Hill, W. E., Madden, J. M., McCardell, B. A., Shah, D. B., Jagow, J. A., Pakne, W. L., and Boutin, B. K. (1983). Foodborne enterotoxinogenic *Escherichia coli:* Detection and enumeration by DNA colony hybridization. *Appl. Environ. Microbiol.* **45,** 1324–1330.
72. Hill, W. E., Wentz, B. A., Pakne, W. L., Jagow, J. A., and Zon, G. (1986). DNA colony hybridization method using synthetic oligonucleotides to detect enterotoxinogenic *Escherichia coli:* Collaborative study. *J. Assoc. Off. Anal. Chem.* **69,** 531–536.
73. Hobbs, B. C. (1965). *Clostridium welchii* as a food poisoning organism. *J. Appl. Bacteriol.* **28,** 74–82.
74. Hof, H. (1984). Virulence of different strains of *Listeria monocytogenes* serovar 1/2a. *Med. Microbiol. Immunol.* **173,** 207–219.
75. Horiguchi, Y., Uemura, T., Kozaki, S., and Sakaguchi, G. (1985). The relationship between cytotoxic effect and binding to mammalian cultured cells of *Clostridium perfringens* enterotoxin. *FEMS Microbiol. Lett.* **28,** 131–135.
76. James, S. M., Fannin, S. L., Agee, B. A., Hall, B., Parker, E., Vogt, J. Run, G., Williams, J., Lieb, L., Salminen, C., Pendergast, T., Werner, S. B., and Chin, J. (1985).

Listeriosis outbreak with Mexican-style cheese. *Morbid. Mortal. Wkly. Rep.* **34,** 357–359.
77. Jiwa, S. F. N., and Mansson, I. (1983). Hemagglutinating and hydrophobic surface properties of salmonellae producing enterotoxin neutralized by cholera antitoxin. *Vet. Microbiol.* **8,** 443–458.
78. Jones, C. L., and Khan, S. A. (1986). Nucleotide sequence of the enterotoxin B gene from *Staphylococcus aureus. J. Bacteriol.* **166,** 29–33.
79. Kampelmacher, E. H., and van Noorle-Jansen, L. M. (1980). Listeriosis in humans and animal in the Netherlands (1958–1987). *Zentralbl. Bakteriol., Mikrobiol. Hyg. Abt. 1, Orig. A* **246,** 211–227.
80. Kathariou, S., Köhler, S., Kuhn, S., and Goebel, W. (1987). Identification of the virulence components of *Listeria monocytogenes* by transposon (Tn *916*) mutagenesis. *Ann. Inst. Pasteur/Microbiol.* **138,** 256–258.
81. Kathariou, S., Metz, P., Hof, H., and Goebel, W. (1987). Tn *916* induces mutations in the hemolysin determinant affecting virulence of *Listeria monocytogenes. J. Bacteriol.* **169,** 1291–1297.
82. Katsaras, K., and Hildebrandt, G. (1979). Ursachen bakterieller Lebensmittelvergiftungen: Enterotoxin von *Clostridium perfringens* type A. *Fleischwirtschaft* **59,** 955–958.
83. Kauffman, F. (1947). The serology of the coli group. *J. Immunol.* **57,** 71–100.
84. Kaufmann, S. H. E. (1984). Acquired resistance to facultative intracellular bacteria: Relationship between persistence, cross-reactivity at the T-cell level, and capacity to stimulate cellular immunity of different *Listeria* strains. *Infect. Immun.* **45,** 234–241.
85. Kétyi, I., Pásca, S., Emödy, L., Verténui, A., Kocsis, B., and Kuch, B. (1979). *Shigella dysentriae* 1-like cytotoxic enterotoxins produced by *Salmonella* strains. *Acta Microbiol. Acad. Sci. Hung.* **26,** 217–223.
86. Klinger, J. D., Johnson, A., Croan, D., Flynn, P., Whipple, K., Kimball, M., Lawric, J., and Curiale, M. (1988). Comparative studies of nucleic acid hybridization assay for *Listeria* in foods. *J. Assoc. Off. Anal. Chem.* **71,** 669–673.
87. Koupal, L. R., and Deibel, R. H. (1975). Assay, characterization and localization of an enterotoxin produced by *Salmonella. Infect. Immun.* **11,** 14–22.
88. Kov, F. C. W., Peterson, J. W., Houston, C. W., and Molina, N. C. (1984). Pathogenesis of experimental salmonellosis: Inhibition of protein synthesis by cytotoxin. *Infect. Immun.* **43,** 93–100.
89. Lanata, C. F., Kaper, J. B., Baldini, M. M., Blackard, R. E., and Levine, M. M. (1985). Sensitivity and specificity of DNA probes with the stool blot technique for detection of *Escherichia coli* enterotoxins. *J. Infect. Dis.* **152,** 1087–1090.
90. La Rocco, K. A., Littel, K. J., and Pierson, M. D. (1986). The bioluminescent ATP assay for determining the microbial quality of foods. *In* "Foodborne Microorganisms and Their Toxins: Developing Methodology" (M. D. Pierson, and N. Stern, eds.), pp. 145–174. Dekker, New York.
91. Law, D. (1988). Virulence factors of enteropathogenic *Escherichia coli. J. Med. Microbiol.* **26,** 1–10.
92. Lee, W. H., and McClain, D. (1986). Improved *Listeria monocytogenes* selective agar. *Appl. Environ. Microbiol.* **52,** 1215–1217.
93. Lentz, W., Theeler, R., Pickenhahn, P., and Brandis, H. (1983). Detection of enterotoxin in cultures of *Staphylococcus aureus* by the enzyme-linked immunosorbent assay (ELISA) and the microslide immunodiffusion. *Zentralbl. Bakteriol. Mikrobiol. Hyg. Abt. 1, Orig. A* **253,** 466–475.
93a. Leong Lim, P. (1986). Diagnostic uses of monoclonal antibodies to *Salmonella. In*

"Monoclonal Antibodies against Bacteria" (A. J. L. Macario, and E. Conway de Macario, eds.), Vol. 3, pp. 29–75. Academic Press, Orlando, Florida.

94. Levine, M. M. (1987). *Escherichia coli* that cause diarrhea: Enterotoxinogenic, enteropathogenic, enteroinvasive, enterohemorrhagic and enteroadherent. *J. Infect. Dis* **155,** 377–389.

95. Li, P., Medon, P. P., Skingle, D. C., Lanser, J. A., and Symons, R. H. (1987). Enzyme-linked synthetic oligonucleotide probes: Non-radioactive detection of enterotoxinogenic *Escherichia coli* in faecal specimens. *Nucleic Acids Res.* **15,** 5275–5287.

96. Linggood, M. A. (1982). *Escherichia coli:* Detection of enterotoxins. *In* "Isolation and Identification Methods for Food Poisoning Organisms" (J. E. L. Corry, D. Roberts, and F. A. Skinner, eds.), pp. 227–238. Academic Press, London.

97. Mackaness, G. B. (1967). The relationship of delayed hypersensitivity to acquired cellular resistance. *Br. Med. Bull.* **23,** 52–54.

98. Matthews, J. A., and Kricka, L. J. (1988). Analytical strategies for the use of DNA probes. *Anal. Biochem.* **169,** 1–25.

99. Maxam, A. M., and Gilbert, W. (1980). Sequencing end-labeled DNA with base-specific chemical cleavages. *In* "Methods in Enzymology" (L. Grossman and K. Moldave, eds.), Vol. 65, pp. 499–560. Academic Press, New York.

100. Mead, G. C. (1985). Selective and differential media for *Clostridium perfringens*. *Int. J. Food Microbiol.* **2,** 89–98.

101. Minnich, S. A., Hartman, P. A., and Heimsch, R. C. (1982). Enzyme immunoassay for detection of *Salmonella* in foods. *Appl. Environ. Microbiol.* **43,** 877–833.

102. Minor, T. E., and Marth, E. M. (1976). "*Staphylococci* and their Significance in Foods." Elsevier/North-Holland Publ., Amsterdam.

103. Moseley, S. L., Huq, I., Alim A. R. M. A., So, M., Sampaour-Motalebi, M., and Falkow, S. (1980). Detection of enterotoxinogenic *Escherichia coli* by DNA colony hybridization. *J. Infect. Dis.* **142,** 892–898.

104. Murray, B. E., Mathewson, J. J., Dupont, H. L., and Hill, W. E. (1987). Utility of oligodeoxyribonucleotide probes for detecting enterotoxinogenic *Escherichia coli. J. Infect. Dis.* **155,** 809–811.

105. Nooy, M. de, Leeuwen W. J. van, and Notermans, S. (1987). Enterotoxin production by strains of *Staphylococcus aureus* isolated from clinical and non-clinical specimens with special reference to enterotoxin F and toxic shock syndrome. *J. Hyg.* **89,** 499–505.

106. Notermans, S., Boot, R., Tips, P. D., and de Nooy, M. P. (1983). Extraction of staphylococcal enterotoxins (SE) from minced meat and subsequent detection of SE with ELISA. *J. Food Prot.* **46,** 238–241.

107. Notermans, S., Chakraborty, T., Leimeister-Wächter, M., Dufrenne, J., Heuvelman, C. J., Maas, H., Jansen, W., Wernars, K., and Guinée, P. A. M. (1989). A specific gene probe for detection of bio- and serotyped *Listeria* strains. *Appl. Environ. Microbiol.* **55,** 902–906.

108. Notermans, S., Dufrenne, J., and van Schothorst, M. (1979). Recovery of *Clostridium botulinum* from mud samples incubated at different temperatures. *Eur. J. Appl. Microbiol. Biotechnol.* **6,** 403–407.

109. Notermans, S., and Heuvelman, K. J. (1983). Combined effect of water activity, pH and sub-optimal temperature on growth and enterotoxin production of *Staphylococcus aureus* strains. *J. Food Sci.* **48,** 1832–1835.

110. Notermans, S., Heuvelman, K. J., Beckers, H. J., and Uemura, T. (1984). Evaluation of the ELISA as tool in diagnosing *Clostridium perfringens* enterotoxins. *Zentralbl. Bakteriol. Mikrobiol Hyg., Abt. 1, Orig. B* **179,** 225–234.

111. Notermans, S., Heuvelman, K. J., and Wernars, K. (1988). Synthetic enterotoxin B

DNA probes for detection of enterotoxinogenic *Staphylococcus aureus*. *Appl. Environ. Microbiol.* **54,** 531–533.

112. Notermans, S., and Kozaki, S. (1987). *In vitro* techniques for detecting botulinal toxins. *In* "Avian Botulism: An International Perspective" (M. W. Eklund, and V. R. Dowel, eds.), pp. 323–336. Thomas, Springfield, Illinois.
113. Notermans, S., Leeuwen J. A. van, and Rost, J. A. (1983). *Staphylococcus aureus* indigenous to poultry processing plants; persistence, enterotoxigenicity and biochemical characteristics. *In* "Quality of Poultry Meat" (C. Lahellec, F. H. Ricard, and P. Colin, eds.), pp. 255–261.
114. Notermans, S., Timmermans, D., and Nagel, J. (1982). Interaction of staphylococcal protein A in ELISA for detecting staphylococcal antigens. *J. Immunol. Methods* **55,** 35–41.
114a. Oguma, K., Syuto, B., Kubo, S., and Iida, H. (1985). Analysis of antigenic structure of *Clostridium botulinum* Type C_1 and D toxins by monoclonal antibodies *In* "Monoclonal Antibodies against Bacteria" (A. J. L. Macario and E. Conway de Macario, eds.), Vol. 2, pp. 159–184. Academic Press, Orlando, Florida.
115. Oosterom, J., and Notermans, S. (1983). Further research into the possibility of *Salmonella*-free fattening and slaughter of pigs. *J. Hyg.* **91,** 59–69.
116. Othmann, Y. M., and May, E. (1981). The occurrence and significance of enterotoxigenic *Staphylococci* in raw beef. *In* "*Staphylococci* and Staphylococcal Infections" (J. Jeljaszewicz, ed.), Zentralbl. Bakteriol., Suppl. 10. Fischer, Stuttgart.
117. Pál, T., Pásca, A. S., Emödy, L., Vörös, S., and Sélley, E. (1985). Modified enzyme-linked immunosorbent assay for detecting enteroinvasive *Escherichia coli* and virulent *Shigella* strains. *J. Clin. Microbiol.* **21,** 415–418.
118. Park, C. E., and Szabo, R. (1986). Evaluation of the reversed passive latex agglutination (RPLA) test kits for detection of staphylococcal enterotoxins A, B, C and D in foods. *Can. J. Microbiol.* **32,** 723–727.
119. Pettipher, G. L., Mansell, R., McKinnon, C. H., and Cousins, C. M. (1980). Rapid membrane filtration–epifluorescent microscopy technique for direct enumeration of bacteria in raw milk. *Appl. Environ. Microbiol.* **39,** 423–429.
120. Phillips, K. D. (1986). A sporulation medium for *Clostridium perfringens*. *Lett. Appl. Microbiol.* **3,** 77–79.
121. Reiser, R. F., Robbins, R. N., Noleto, A. L., Khoe, G. P., and Bergdoll, M. S. (1984). Identification, purification and some physiochemical properties of staphylococcal enterotoxin C3. *Infect. Immun.* **45,** 625–630.
122. Rocourt, J., Catimel, B., and Schrettenbrunner, A. (1985). Isolement de bacteriophage de *Listeria seeligeri* et *L. welshimeri*. Lysotypie de *L. monocytogenes, L. ivanovii, L. innocua, L. seeligeri* et *L. welshimeri*. *Zentralbl. Bakteriol., Mikrobiol. Hyg., Abt. 1, Orig. A* **259,** 341–350.
123. Rocourt, J., Schrettenbrunner, A., and Seeliger, H. P. R. (1983). Differentiation biochemique des groupes génomiques de *Listeria monocytogenes* sensu lato. *Ann. Inst. Pasteur/Microbiol.* **134A,** 65–71.
124. Rubin, F. A., Kopecko, D. J., Noon, K. F., and Baron, L. S. (1985). Development of a DNA probe to detect *Salmonella* typhi. *J. Clin. Microbiol.* **22,** 600–605.
125. Sack, B. B. (1975). Human diarrheal disease caused by enterotoxinogenic *Escherichia coli*. *Annu. Rev. Microbiol.* **29,** 333–353.
126. Sakaguchi, G. (1983). *Clostridium botulinum* toxins. *Pharmacol. Ther.* **19,** 165–194.
127. Salmon, L. L., and Tew, R. W. (1968). Assay of staphylococcal enterotoxin B by latex agglutination. *Proc. Soc. Exp. Biol. Med.* **129,** 539–542.
128. Sandufur, P. D., and Peterson, J. W. (1977). Neutralization of *Salmonella* toxin-induced

elongation of Chinese hamster ovary cells by cholera antitoxin. *Infect. Immun.* **15,** 988–992.

129. Sathyamoorthy, V., and Dasgupta, B. R. (1985). Separation, purification, partial characterization and comparison of the heavy and light chains of botulinum neurotoxin types A, B and E. *J. Biol. Chem.* **260,** 10461–10466.

129a. Sato, H. (1986). Monoclonal antibodies against *Clostridium perfringens* Toxin (Perfringolysino) *In* "Monoclonal Antibodies against Bacteria" (A. S. L. Macario and E. Conway de Macario, eds.), Vol. 3, pp. 203–228. Academic Press, Orlando, Florida.

130. Schalkowsky, S. (1986). Plating systems. *In* "Foodborne Microorganisms and their Toxins: Developing Methodology" (M. D. Pierson, and N. J. Stern, eds.), pp. 107–128. Dekker, New York.

131. Schlech, W. F., Lavigne, P. M., and Bortolussi, R. A., Allen, A. C., Haldane, E. V., Wort, A. J., Hightower, A. W., Johnson, S. E., King, S. H., Nicholls, E. S., and Broome, C. V. (1983). Epidemic listeriosis—evidence for transmission by food. *N. Engl. J. Med.* **308,** 203–206.

132. Schmidt, J. J., and Spero, L. (1983). The complete amino acid sequence of staphylococcal enterotoxin C_1. *J. Biol. Chem.* **258,** 6300–6306.

133. Seeliger, H. P. R., and Höhne, K. (1979). Serotyping of *Listeria monocytogenes* and related species. *In* "Methods in Microbiology" (T. Bergan, and J. R. Norris, eds.), Vol. 13, pp. 31–49. Academic Press, New York.

134. Segner, W. P., Schmidt, C. F., and Boltz, J. K. (1971). Enrichment, isolation and cultural characteristics of marine strains of *Clostridium botulinum* type C. *Appl. Microbiol.* **22,** 1017–1022.

135. Seriwatana, J., Echeverria, P., Taylor, D. N., Sakuldaipeara, T., Changchawalt, S., and Chivoratanond, O. (1987). Identification of enterotoxinogenic *Escherichia coli* with synthetic alkaline phosphatase-conjugated oligonucleotide DNA probes. *J. Clin. Microbiol.* **25,** 1438–1441.

136. Sethabutr, O., Echeverria, P., Hanchalay, S., Tailor, D. N., and Leksomboon, U. (1985). A non-radioactive DNA probe to identify *Shigella* and enteroinvasive *Escherichia coli* in stools of children with diarrhea. *Lancet* **2,** 1095–1097.

137. Simkovicova, M., and Gilbert, R. J. (1971). Serological detection of enterotoxin from food poisoning strains of *Staphylococcus aureus*. *J. Med. Microbiol.* **4,** 19–30.

138. Simpson, L. L. (1974). Studies on the binding of botulinum toxin type A to the rat phrenic nerve–hemidiaphragm preparation. *Neuropharmacology* **13,** 683–691.

139. Simpson, L. L. (1973). The interaction between divalent cations and botulinum toxin type A in the paralysis of the rat phrenic nerve–hemidiaphragm preparation. *Neuropharmacology* **12,** 165–176.

140. Small, P. L. C., and Falkow, S. (1986). Development of a DNA probe for the virulence plasmid of *Shigella* spp. and enteroinvasive *Escherichia coli*. *In* "Microbiology—1986" (L. Leive, ed.), pp. 121–124. Am. Soc. Microbiol., Washington, D.C.

141. Smedt J. M. de, and Bolderdijk, R. F. (1987). Dynamics of *Salmonella* isolation with modified semi-solid Rappaport–Vassiliadis medium. *J. Food Prot.* **50,** 658–661.

142. Smidt, J. J., Sathyamoorthy, V., and Dasgupta, B. R. (1985). Partial amino acid sequences of botulinum neurotoxins types B and E. *Arch. Biochem. Biophys.* **238,** 544–548.

143. Sperber, W. H., and Deibel, R. H. (1969). Accelerated procedure for *Salmonella* detection in dried foods and feeds involving only broth cultures and serological reactions. *Appl. Microbiol.* **17,** 533–539.

144. Stark, R. I., and Duncan, C. L. (1971). Biological characteristics of *Clostridium perfringens* type A enterotoxin. *Infect. Immun.* **4,** 89–96.

145. Steffan, R. J., and Atlas, R. M. (1988). DNA amplification to enhance detection of genetically engeneered bacteria in environmental samples. *Appl. Environ. Microbiol.* **54,** 2185–2191.
146. Stelma, G. N., Crawford, R. C. Spaulding, P. L., and Twedt, R. M. (1985). Evidence that *Clostridium perfringens* produces only one enterotoxin. *J. Food. Prot.* **48,** 232–233.
147. Stephan, J., Wallis, T. S., Starkey, W. G., Candy, D. C. A., Osbourne, M. P., and Haddon, S. (1985). Salmonellosis: In retrospect and prospect. *Ciba Found. Symp.* **112,** 175–192.
148. Stringer, M. F., Watson, G. N., and Gilbert, R. J. (1985). Faecal carriage of *Clostridium perfringens*. *J. Hyg.* **95,** 277–288.
149. Sugiyama, H., and Hayama, T. (1965). Abdominal viscera as site of emetic action for staphylococcal enterotoxin in the monkey. *J. Infect. Dis.* **115,** 330–336.
150. Sugiyama, H., von Mayeruauser, B., Gogat, G., and Hesmsch, R. C. (1967). Immunological reactivity of trypsinized *Clostridium botulinum* type E toxin. *Proc. Soc. Exp. Biol. Med.* **126,** 690–694.
151. Svennerholm, A. M., Wikström, M., Lindholm, L., and Holmgren, J. (1986). Monoclonal antibodies and immunodetection methods for *Vibrio cholerae* and *Escherichia coli* enterotoxins. *In* "Monoclonal Antibodies against Bacteria" (A. J. L. Macario, and E. Conway de Macario, eds.), Vol. 3, pp. 77–97. Academic Press, Ortando, Florida.
152. Taylor, D. N., Echeverria, P., Sethabutr, O., Pitarangsi, C., Leksomboon, U., Blacklow, N. R., Rowe, B., Gross, R., and Cross, J. (1988). Clinical and microbiologic features of *Shigella* and enteroinvasive *Escherichia coli* infections detected by DNA hybridization. *J. Clin. Microbiol.* **26,** 1362–1366.
153. Thompson, M. R., Jordan, R. L., Luttrell, M. A., Brandwein, H., Kaper, J. B., Levine, M. M., and Gianella, R. (1986). Blinded two-laboratory comparative analysis of *Escherichia coli* heat-stable enterotoxin production by using monoclonal antibody enzyme-linked immunosorbent assay and gene probes. *J. Clin. Microbiol.* **24,** 753–758.
153a. Thompson, N. E., Bergdoll, M. S., Meyer, R. F., Bennet, R. W., Miller, L., and Macmillan J. D. (1985). Monoclonal antibodies to the enterotoxins and the toxic shock syndrome toxin produced by *Staphylococcus aureus*. *In* "Monoclonal Antibodies against Bacteria" (A. J. L. Macario and E. Conway de Macario, eds.), Vol. 2, pp. 23–59. Academic Press, Orlando, Florida.
154. Todd, E. C. D. (1978). Foodborne diseases in six countries—a comparison. *J. Food Prot.* **41,** 559–565.
155. Troller, J. A. (1972). Effect of water activity on enterotoxin A production and growth of *Staphylococcus aureus*. *Appl. Microbiol.* **24,** 440–443.
156. Vassiliadis, P. (1983). The Rappaport–Vassiliadis (RV) enrichment medium for the isolation of *Salmonellae:* An overview. *J. Appl. Bacteriol.* **54,** 69–76.
157. Vassilidiadis, P., Tricholpoulos, D., Papadakis, J., Kalapothaki, V., Zavitsamos, X., and Serie, C. (1981). Salmonella isolation with Rappaport's enrichment medium of different compositions. *Zentralbl. Bakteriol., Mikrobiol. Hyg. Abt. 1, Orig. B* **173,** 388–389.
158. Van Damme-Jongsten, M., Wernars, K., and Notermans, S. (1988). Cloning and sequencing of the *Clostridium perfringens* enterotoxin gene. *Antonie van Leeuwenhoek* (to be published).
159. Venkatesan, M., Buysse, J. M., Vandendries, E., and Kopecko, D. J. (1988). Development and testing of invasion-associated DNA probes for detection of *Shigella* spp. and enteroinvasive *Escherichia coli*. *J. Clin. Microbiol.* **26,** 261–266.
160. Vries M. Verlaan, de, Bogaard, M. E., van den Elst, M., van Boom, J. H., van der Eb,

A. J., and Bos, J. L. (1986). A dot-blot screening procedure for mutated *ras* oncogenes using synthetic oligonucleotidese. *Gene* **50,** 313–320.
161. Wieneke, A. A. (1974). Enterotoxin production by strains of *Staphylococcus aureus* isolated from foods and human beings. *J. Hyg.* **73,** 255–262.
162. Wood, P. K., Morris, J. G., Jr., Small, P. L. C., Sethabutr, O., Regina, M., Toledo, F., Trabulsi, L., and Kaper, J. B. (1986). Comparison of DNA probes and the Sereny test for identification of invasive *Shigella* and *Escherichia coli* strains. *J. Clin. Microbiol.* **24,** 498–500.
163. Yamagishi, T., Serikawa, T., Morita, R., Nakamura, S., and Nishida, S. (1976). Persistent high numbers of *Clostridium perfringens* in the intestines of Japanese aged adults. *Jpn. J. Microbiol.* **20,** 397–403.
164. Young, D. B., Kent, L., and Young, R. A. (1987). Screening of a recombinant mycobacterial library with polyclonal antiserum and molecular weight analysis of expressed antigens. *Infect. Immun.* **55,** 1421–1425.

14

The Use of Gene and Antibody Probes in Identification and Enumeration of Rumen Bacterial Species

J. D. BROOKER, R. A. LOCKINGTON, G. T. ATTWOOD, AND S. MILLER
Department of Animal Sciences
Waite Agricultural Research Institute
Glen Osmond, South Australia, Australia

I. INTRODUCTION

Microbial fermentation of plant material is an essential feature of ruminant nutrition; cellulose and other insoluble plant products are degraded with the products being metabolized by the microbial biomass. Breakdown of the microbial flora provides essential nutrients for the animal (26). Ruminant productivity is therefore dependent on effective fermentation by microbial populations. However, under many naturally occurring grazing conditions, optimum rates of microbial fermentation and maximum yields of metabolizable energy are not always achieved (53); potentially this activity can be improved by genetic engineering of individual bacterial species (21,52). The major barrier to this objective, however, is that unlike commonly used bacterial species such as *Escherichia coli*, very little is known about the genetics and population dynamics of rumen bacterial populations. The ability to monitor accurately, genetically engineered organisms is critical in determining their potential impact on the rumen environment. Understanding rumen microbiology and ecology is therefore of considerable importance.

There are at least 30 predominant species of anaerobic bacteria in the rumen, and at least another 30 minor species making up a very complex microbial ecosystem in which bacterial populations interact and compete for their survival (26). The behavior of any particular bacterial species within this ecosystem is therefore the subject of considerable interest, especially when it concerns selective enrichment of individual species over others as a consequence of dietary change. The need for techniques to measure components of this natural microbial ecosystem has also been focused on by interest in the possible release of genetically engineered microorganisms into the rumen for improving production efficiency in domestic ruminants (21).

Despite the importance of microbial competition and survival in rumen function, there have been few detailed studies of the fluctuation in numbers of individual bacterial species in the rumen. This is because of inherent difficulties in analyzing changes in microbial populations using imprecise morphological classification techniques and the fact that phenotypic descriptions frequently do not satisfactorily differentiate between related but different isolates. This is particularly true for the pleiomorphic *Bacteroides* species. Moreover, culturing anaerobic rumen bacteria for the purpose of determining the representation of a specific species is difficult methodologically, and suffers from the criticism that selection for any one species may change when a sample is transferred from the rumen to a culture dish where media and conditions will be clearly different from those in the rumen. No selective or differential media exist for many rumen bacterial species, and it would not be unusual for ≤1% of a particular species to survive and proliferate under artificial culture conditions. Even then, this proportion would most likely represent a subpopulation of cells that have adapted to, or are tolerant of the culture conditions. Thus detailed population descriptions in the rumen have been almost impossible, leaving a large gap in our knowledge of bacterial interactions in this complex and dynamic ecosystem. Similar problems have been encountered in analyzing colonic anaerobes, particularly *Bacteroides* species in human fecal material (33,49).

This chapter will consider previously used and new methods for identification and enumeration of bacteria in the rumen. In particular we will consider the impact of recombinant DNA technology on microbial analysis and compare data obtained using gene probes with serological measurements of bacterial populations, especially those using monoclonal antibodies (MAb).

II. BACKGROUND

A. Bacterial Identification

There are many important groups of microorganisms for which no simple method of species identification has previously been possible. This includes most of the anaerobic species of rumen bacteria as well as some more common species such as those in the genus *Bacteroides*, the most predominant member of the microbial flora of the human colon (39). When colonic *Bacteroides* strains were first studied, they appeared to be morphologically similar so were all classified as subspecies of *B. fragilis* (39). Even fermentation tests, which have been the major means of species identification, could not easily distinguish between different isolates. Classification of rumen bacteria is also based on morphological studies and fermentation product profiles (6,43). However, as with the colonic *Bacteroides*, there are many examples of bacteria previously assigned to one species on the basis of their fermentation products and morphology that have now been shown to be members of entirely different genera. The rumen bacterial species *B. succinogenes* has recently been demonstrated to be unrelated to the true *Bacteroides* spp. and has no close association with any other organism so far characterized (46). This is an extreme example, but nevertheless, with the limitations inherent in extrapolating from data on bacterial morphology under culture conditions and from the analyses of fermentation products using various substrates, bacterial classification based solely on these principles can be in error. Other methods of bacterial classification include cataloguing of lipid composition, especially volatile and total cell fatty acids, carotenoids, cytochromes, and isoprenoid quinones (17). Regardless of the basis for these classification schemes, however, phenotypic characterizations can often obscure microbial diversity or miss important relationships. This is particularly true for complex ecosystems such as the rumen, and it is very likely that even now there may be distinct rumen microbial species that have been historically lumped together in the one group despite their performing very different functions *in vivo* and reacting differently as populations under various dietary regimes (14). For more accurate bacterial identification and classification, phenotypic studies of bacterial interactions with the environment must be complemented with genotypic determinations. Recognition of microbial diversity both within pure cultures through mutation and selection, and in the environment should therefore contribute to a better understanding of how microbial ecosystems work (1).

Fortunately, through advances in molecular biology and immunology, microbial genotyping is now possible. This is largely as a result of the development of species-specific and even strain-specific gene and antibody probes. A variety of different procedures have been developed, some of which are limited in their applications because of their extreme sensitivity and others that are of a more general nature. Where gene techniques are to be used for population studies and bacterial enumeration, care must be taken to ensure that the genotype probe is not so specific that even closely related strains are excluded and not so general that members of different species react with a positive signal to the probe.

1. Gene Techniques

One of the earliest techniques developed for the genetic identification of microbial species was that of base composition analysis (22). Broad species characterizations were made on the basis of the guanosine + cytosine (G+C) content of total bacterial DNA. For rumen bacteria these values can range from 25 mol% for *Clostridium cellobioparum* to 64.5 mol% for *Bifidobacterium pseudolongum* (43). Most rumen bacteria, however, had average values around 35–45 mol%. Therefore, although G+C values were of some use in establishing that a particular bacterial isolate was not a member of any given species, the values were of less use in establishing which species that isolate did belong to. In practice the technique was used more for confirmation of identity than for genetic characterization.

As an alternative to this technique, several workers have demonstrated the highly conserved character of the genes coding for ribosomal RNA (rRNA) compared with those forming the bulk of the bacterial genome. The ribosomal RNAs are universally distributed, exhibit constancy of function, and appear to change in sequence very slowly, far more so than most proteins (14). Moreover, they are easily isolated. Therefore, rRNA sequence comparisons between apparently related organisms should reflect their evolutionary relationships. Within the rRNA sequences there are highly conserved and highly variable regions. The sequence of nucleotides within the variable regions are characteristic of particular bacterial species. Oligonucleotide cataloguing and comparative sequencing of the rRNAs, principally the 16 S-like rRNAs, has been an important development for establishing microbial phylogeny between closely or even distantly related organisms (14,44,51). Disadvantages of the original method of RNA sequencing, particularly where the organism was fastidious or slow-growing have been overcome with newer methods. The procedure is greatly simplified by the fact that the 16 and 23 S rRNAs have both highly conserved regions and variable regions; differences in nucleotide sequences, particularly across conserved–variable region boundaries can be used to determine the relationships of microbial species (44). For example, rRNA sequencing, together with rRNA cistron homology studies have shown that the *Pseudomonas* species represent a phylogenetically heterogeneous group of bacteria covering nearly the whole range of one main phylum of the eubacteria (45,54). In contrast, there are many organisms not called pseudomonads that are nevertheless closely related to the *Pseudomonas* group of species.

Another method of phylogenetic analysis, and one that has been developed for situations in which the relationship between two apparently related strains is in question, involves the technique of restriction mapping of the bacterial chromosome. By digesting chromosomal DNA from a standard bacterial strain with a range of restriction enzymes and then analyzing the gene fragments for homology to a strain-specific gene probe, the pattern of hybridization can be used as an indication of strain identity. This is often referred to as DNA "fingerprinting," and the pattern serves as a benchmark for analyzing other potentially related strains. This procedure is probably easier to perform than DNA sequencing but both techniques are complex and time-consuming. For purposes such as population studies, it is usually sufficient to identify particular species within mixtures rather than establish phylogenetic relationships.

2. *Antibody Techniques*

An alternate technique to the use of gene probes, and one that has found many applications in the medical sciences, is that of immunoassays. Antibody detection systems, particularly fluorescent-antibody systems have been developed (4) primarily for the detection of bacterial and viral pathogens and genera- and species-specific antibodies directed against unique surface components of individual bacterial species have been used both to identify and to enumerate those species (20,25,48). This use of antibodies is largely the result of the development of selectively enriched polyclonal antibody preparations and highly specific MAb that recognize single antigenic epitopes (32). Antibodies have been used in both direct and indirect assay techniques. The direct procedure involves the use of antibody that has been raised against some surface determinant of the target microorganism and has been conjugated with a specific label. The indirect procedure involves a sandwich technique in which the first antibody, specific for the microorganism, is reacted in a second stage of the assay with an antiimmunoglobulin antibody that has been conjugated with a fluorescent label (fluorescein isothiocyanate, FITC) or an enzyme such as alkaline phosphatase or horseradish peroxidase (20). The latter procedure is the more commonly used procedure because of its heightened sensitivity and flexibility.

For the major cell surface groupings of bacteria (i.e., gram-positive and gram-negative), each group has many common antigenic determinants associated with the bacterial wall, particularly the presence of proteins and lipopolysaccharides (5). Unenriched polyclonal antisera raised against any particular bacterial species within these groups is likely to cross-react with many other members of the same group. Nevertheless, some enriched species-specific polyclonal antisera preparations have been used in bacterial enumeration experiments (48). In particular, many bacterial strains of medical significance have been identified and quantified serologically (15), methanogenic flora of waste digesters have been quantified using polyclonal antisera (36), and relationships between various isolates of some rumen bacterial species have been studied using serological techniques (20,48). These include *B. ruminicola, Selenomonas ruminantium, Butyrivibrio, Bacteroides* and *Ruminococcus* spp. In practice, however, polyclonal antisera are only of limited use for population studies without selective absorption of cross-reactive antibodies. This problem has been overcome by taking advantage of the fact that within individual species, there are subtle differences in the structures of some of the polysaccharide linkages and some cell surface proteins. These differences can be highlighted by the production of MAb that recognize these specific determinants (32). Monoclonal antibodies with defined specificities can be developed against a range of different bacterial species, and used in enzyme-linked immunosorbent assays (ELISA) or immunofluorescence assays to quantify particular bacterial species in unselected microbial samples. Increasingly, MAb are finding their way into microbial enumeration procedures particularly in the medical sciences (11). These probes should also prove invaluable in ecological studies in the rumen.

A MAb raised against a membrane protein from *Pseudomonas aeruginosa* has been described (42) that recognizes both the *Pseudomonas fluorescens* group and *Azotobacter* species. Monoclonal antibodies have also been used to characterize and enumerate various strains of methanogens (9), *Butyrivibrio* (20), and *Bacteroides* spp. from the rumen (unpublished data), and many bacterial species of medical importance (Table I). For bacterial population studies, however, especially in complex ecosystems such as the rumen, the use of MAb is still in its infancy.

B. Bacterial Quantification

Quantification of microbial species in mixed populations is an extension of phylogenetic studies and has been achieved using defined genotypic probes. In these assays, total DNA is usually extracted from the bacteria and analyzed *in vitro* for the presence of specific gene sequences by DNA–DNA hybridization

The rRNA comparative sequencing technique has recently been extended to examining changes in particular bacterial populations using species-specific synthetic oligodeoxynucleotide gene sequences complementary to unique regions of 16 S or 23 S rRNA. For example, species- and group-specific 16 S rRNA-targeted oligonucleotide hybridization probes were developed to enumerate various strains of *B. succinogenes* and *Lachnospira multiparus*-like organisms from the bovine rumen before,

TABLE I

Some Examples of Monoclonal Antibodies Used in Microbial Identification or Enumeration[a]

Bacterial species	Reference
Listeria monocytogenes	8
Butyrivibrio sp.	20
Methanogenic bacteria	9
Chlamydia trachomatis	3
Toxoplasma gondii	19
Streptococcus sp.	11

[a] *Editor's note:* A comprehensive list can be found in the treatise "Monoclonal Antibodies against Bacteria" (A. J. L. Macario and E. Conway de Macario, eds., Vols. 1–3. Academic Press 1985–1986, and in *Biotechnology Advances,* **6,** 135–150 (1988).

during, and after perturbation of that ecosystem by the addition of the ionophore antibiotic monensin (50). Randomly cloned chromosomal fragments have also been used to differentiate *Bacteroides thetaiotaomicron* in human fecal samples (49) and to enumerate *Bacteroides ruminantium* in mixed culture and in the rumen (2), an enterotoxin gene has been used to identify *Vibrio cholerae* (30), and rRNA genes have been used to identify mycoplasmas (16)(Table II). In each of these examples, no prior culture or enrichment of the bacterial population was required before DNA extraction, and representations down to 1% of the total population could be readily detected. In contrast, other techniques of microbial analysis using gene probes have involved the bacteria being plated out and cultured before *in situ* lysis on the plate (18). However, the major drawback in both procedures is the difficulty in obtaining direct quantification of the data. In most cases,the assays have involved hybridization of bacterial DNA with a radioactively labeled gene probe followed by some form of densitometric analysis of silver grains on an exposed X-ray film. This is time-consuming and, above all, inaccurate compared with a direct quantitative assay. The use of biotin-tagged deoxynucleotides in probe-labeling reactions rather than radioactive deoxynucleotides may be the answer to this problem, although it has been reported (33) that for *Bacteroides* sp. this was not

TABLE II

Some Enteric and Ruminal Bacterial Species That Have Been Studied Using Gene Probes

Bacterial species	Gene probe for:	Reference
Bacteroides thetaiotaomicron	Chromosomal DNA	49
Bacteroides uniformis	Chromosomal DNA	34
Bacteroides distasonis	Chromosomal DNA	34
Bacteroides ovatus	Chromosomal DNA	34
Bacteroides vulgatus	Chromosomal DNA	34
Bacteroides fragilis	Chromosomal DNA	28
Bacteroides ruminicola	Chromosomal DNA	2
Vibrio cholerae	Enterotoxin gene	30
Pseudomonas fluorescens	rRNA Genes	13
Bacteroides succinogenes	rRNA Oligonucleotides	50
Listeria multiparus	rRNA Oligonucleotides	50
Escherichia coli	rRNA Genes	40
	Enterotoxin gene	41
Mycoplasma sp.	rRNA Genes	16
Yersinia enterocolitica	Enterotoxin gene	23

successful because these extracts contained material that bound the streptoavidin–peroxidase detection reagent.

In addition to quantitation problems, accurate sampling of rumen populations poses an even greater problem, especially where a bacterial population may be wholly or partly bound to particulate plant material rather than being free in the rumen liquor (7,26). Although extensive washing can release a proportion of these cells, nevertheless, a significant number are protected within plant fragments and remain behind. Quantitation of particle-bound rumen microorganisms therefore poses special problems not encountered with free organisms. Difficulties in quantitation are further compounded by the contamination of bacterial DNA with plant DNA derived from minute particles of plant material in the bacterial samples. The smallest of these particles are not easily removed without also removing larger bacteria and protozoans from the sample. Therefore, quantification of particular species should be verified by other techniques.

III. RESULTS AND DISCUSSION

The need for quantitative measurements of bacterial populations in the rumen has been emphasized recently by the interest in developing genetically engineered ruminal strains for modifying production in domestic ruminants (21,52). Both gene and serological probes have been used.

A. Gene Probes for Bacterial Quantification

The development of techniques utilizing DNA–DNA hybridization with labeled gene probes has resulted in many recent advances in microbial genetics, particularly in the biomedical sciences. Among the techniques based on this approach are Southern blot analysis (37), dot-blot hybridization (2), and *in situ* hybridization (18). All of these are based on the same underlying principle: the complementarity of related single-stranded molecules of RNA or DNA. A number of techniques have been developed for the labeling of gene probes, the most commonly used of which are nick translation (37) and randomly primed oligolabeling (12). In both cases, the procedure involves DNA polymerase-mediated copying of DNA template strands in the presence of deoxynucleotides, one of which is radiolabeled. In nick translation the startpoints are generated by random nicks in the template DNA, while oligolabeling involves the use of random oligonucleotides to hybridize with denatured single-stranded template DNA thus providing a polymerase startpoint. More recently, the development of asymmetric RNA probes through selective transcription of the DNA has resulted in a dramatic increase in hybridization efficiency compared with labeled DNA probes (38). This is largely achieved through the elimination of self-hybridization in the probe and the concentration of label only in the region of the probe and not in the vector. Probe–target interactions are usually visualized by autoradiography (see Fig. 1).

A number of studies have utilized gene probes for identifying and quantifying microorganisms in various environments (Table II). Gene probes that are specific for particular bacterial genera can be generated by random selection of chromosomal fragments cloned into an appropriate vector or by the use of a known gene or fragment of that gene. Species or genus specificity must be determined for each potential gene probe.

Recombinant plasmids containing chromosomal DNA inserts from the rumen bacterial species *Bacteroides ruminicola* subsp. *brevis* B14 were screened in dot-blot assays for hybridization to DNA isolated from various *Bacteroides* strains as well as other bacterial species including *Selenomonas* and *E. coli* (2). Although in these studies, the gene library was not large nor fully representative of the genome of *B. ruminicola*, some clones were isolated that demonstrated specificity for *Bacteroides* sp. in dot-blot hybridization experiments. Several clones showed hybridization to two different rumen-derived strains of *B. ruminicola* but did not hybridize with the human enteric species *Bacteroides fragilis*. One clone hybridized only with DNA from rumen strain B14 of *B. ruminicola* and not with any other strain. Nonrecombinant plasmid DNA showed no homology with DNA isolated from any of the ruminal anaerobes. The cloned DNA was verified as being derived from *B. ruminicola* by Southern blot analysis of genomic DNA fractionated by gel electrophoresis.

In these experiments, recombinant plasmids containing *B. ruminicola*-specific inserts were tested for their ability to hybridize to total DNA isolated from a mixed bacterial population freshly isolated from the rumen of a sheep maintained on a high-protein diet. In dot-blot assays containing up to 10 μg of total ruminal DNA per dot, there was little detectable hybridization to either of the recombinant plasmids. Only purified control DNA from *B. ruminicola* subsp. *brevis* B14 reacted in this assay. The minimum amount of DNA detectable on control dots was 50 ng, equivalent to 2×10^7 bacteria. Quantitation was by liquid scintillation counting of radioactive spots cut from the hybridization membrane (Fig. 1). In contrast to these results, a *Selenomonas ruminantium*-specific DNA probe (isolated from a genomic library of this species) showed positive hybridization to total ruminal DNA, and by reference to standards, indicated a population of selenomonads of at least 10^8 per milliliter of rumen sample (results not shown). This value could be an underestimation of total rumen selenomonads by as much as 10-fold because the authors did not establish

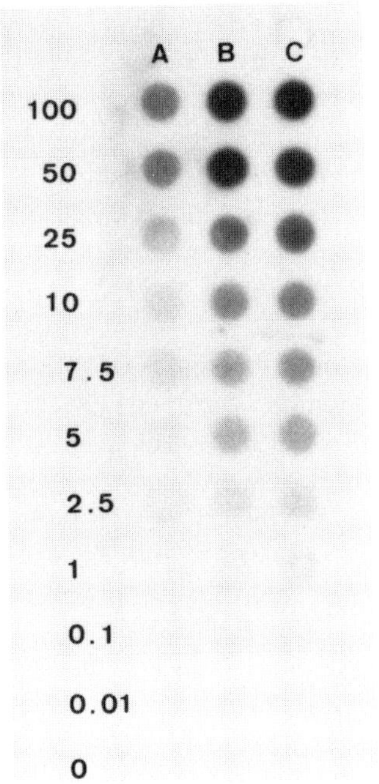

Fig. 1. Sensitivity of *Bacteroides ruminicola* subsp. *brevis* B14 detection in a mixture of total ruminal microbial DNA. Strain B14 cells were mixed in various proportions with total ruminal bacteria, and 1 μg (Lane A), 5 μg (Lane B), and 10 μg samples (Lane C) of DNA from each mixture were spotted onto nylon membranes. The membranes were hybridized to a radiolabeled pB-18 DNA probe specific for *B. ruminicola* B14. The numbers on the left represent the proportions of strain B14 cells in the mixtures.

whether the gene probe recognized all members of this species or only a subpopulation. Furthermore, new gene probes isolated for *S. ruminantium* strains isolated freshly from the rumen show that not all isolates react with all gene probes. There is almost certainly a degree of genetic diversity between isolates of the same apparent species. Nevertheless, the results did support the concept of rumen population studies using cloned gene probes.

Genomic DNA probes have also been used to enumerate human *Bacteroides* species in pure culture, mixed culture, and in samples of feces (33,34,49). These authors used a similar dot-blot technique to that of Attwood *et al.*, (2) and, within the limits of the filtering capacity of nitrocellulose membranes, were able to obtain values of *Bacteroides* sp. expressed per gram dry weight of fecal material that agreed well with estimates obtained previously by conventional methods (39).

In a similar way, rRNA gene probes have been used by several groups to enumerate bacterial populations. The use of a gene probe for the 23 S rRNA of *Ps. aeruginosa* was described for *in situ* and dot-blot hybridizations to identify members of the related group *Ps. fluorescens* (13). This probe comprised a 360-bp fragment from the 23 S rRNA gene cloned into an *E. coli* vector and allowed clear differentiation of *Ps. fluorescens* from other pseudomonads and closely related or unrelated bacterial strains. These experiments were performed on DNA extracts from pure cultures. Gene probes for other bacterial species have also been obtained for 23 S rRNA. A number of subclones from a 7500-bp fragment of 23 S rRNA were isolated and tested for specificity for the *Micrococcus luteus–Micrococcus lylae* group, the *Arthrobacter–Micrococcus* group, and eubacteria (47). In each case specific gene probes were isolated.

In other studies, probes for 16 S rRNA sequences were used to enumerate populations of *B. succinogenes* and *Lachnospira multiparus* in the bovine rumen following perturbation of that ecosystem with the ionophore antibiotic monensin (50). Quantification was by densitometric analysis of autoradiographs after exposure to X-ray film. In this work, the probes were synthetic oligodeoxynucleotide sequences. For *B. succinogenes*, the signature probe was complementary to the 16 S rRNA of all but one strain so far characterized by comparative sequencing; two other probes identified either of two natural groups within the larger *Bacteroides* assemblage. The *L. multiparus* probe was synthesized to be complementary to a 22-base sequence of the 16 S rRNA for this species in a region that is known to be the most variable between bacterial species.

B. Bacteriophages as Gene Probes

An alternative to the use of chromosomal gene probes is to quantify specific bacterial species on the basis of sensitivity to infection with a species-specific bacteriophage. Bacteriophage sensitivity was used to identify *Salmonella* spp. (24), and the same technique has been used to identify *Pseudomonas syringae* infections on tomato plants (10). There have been several reports of bacteriophage present in isolates from the rumen (31), and the isolation and genetic characterization of a temperate bacteriophage from the bacterial species *S. ruminantium* has been reported (35). Infectivity studies have established that this bacteriophage only infects the one species, although it has not been established whether all strains within the species are sensitive. Current work in progress will clarify this situation and may lead to an easy method for quantification of this species in crude rumen samples. Since the bacteriophage is temperate, most infected cells will contain the phage DNA integrated into the genome.

Therefore the bacteriophage DNA can be used as a specific gene probe in conventional hybridization assays. Limitations to this approach arise from the difficulty in determining accurately the numbers of infected bacteria because, at any particular point in time, there will always be a proportion of cells that are about to proceed into the lytic cycle. These cells will therefore contain multiple copies of the bacteriophage whereas other bacteria in the lysogenic phase will usually contain single copies of the bacteriophage gene integrated into the bacterial chromosome. In addition, the lack of information on other species-specific bacteriophages in the rumen has, so far, restricted this technique to the *Selenomonas* sp.

C. Sensitivity of Gene Probes for Bacterial Quantification

The experiments just outlined have made use of randomly cloned fragments of genomic DNA, bacteriophages that have been demonstrated to be species-specific, or probes complementary to regions of 23 S or 16 S rRNA. The major difference between these is the copy number of the target sequence and therefore the sensitivity of the probe. The randomly cloned probes were derived from gene sequences that appear in the genome infrequently, perhaps only once. In contrast, the DNA sequences corresponding to rRNA are repeated, often tandemly, and occur in the genome of bacterial cells at least seven times (5). This difference in target copy number is reflected in the differing sensitivities in the experiments just discussed.

A randomly cloned fragment of *B. ruminicola* DNA, when used as a probe, could detect a minimum of 50 ng of target DNA (2). Based on the figure of 4–5 μg of DNA in 10^9 bacteria, the sensitivity of the probe would be of the order of 10^7 cells. Since the capacity of the membranes is a maximum of 10 μg of DNA per spot, and this would be derived from 2×10^9 bacteria, the overall sensitivity of the technique is $\sim$0.5% of a mixed population. The detection of organisms representing $<$0.5% of the rumen population is possible, but to achieve the necessary 50 ng of species-specific DNA in the sample, $>$100 μg of DNA would need to be bound to the nitrocellulose membrane. Hybridization results using this concentration of DNA are unlikely to be as reproducible as lower concentrations because of interference effects and uneven DNA distribution over the area of a dot. These values also do not take into account the presence of plant nucleic acid in the rumen samples. Even with careful washing, it is difficult to remove small particles of plant material and these will contribute to the DNA recovered in rumen extractions. Therefore, the practical sensitivity of random gene probes for rumen bacterial ecology studies will be less than that calculated for pure culture studies.

Nevertheless, gene probes have been used for population studies in mixed culture or the rumen. *Bacteroides ruminicola* subsp. *brevis* B14 cells could be quantified after adding them to a mixed culture of rumen bacteria *in vitro* or after inoculation into the rumen (2). For the *in vitro* experiment, 25-ml cultures of ruminal bacteria were inoculated with various amounts of strain B14 cells to yield initial proportions of 10% and 50% of the total bacterial cell number. Samples were taken at intervals over a 48-hr period, and total bacterial DNA was extracted from these samples and analyzed for B14 levels by hybridization to the B14-specific gene probe. For the *in vivo* experiments, 3×10^{12} or 1×10^{13} B14 cells were introduced, in separate experiments, through a fistula into the rumen of a sheep, yielding an initial inoculum of 3 or 10% B14 cells, respectively. Samples of ruminal bacteria were taken after 30 min and then at intervals over the next 2 days. Total bacterial DNA from these samples was analyzed by hybridization to the B14-specific gene probe. Hybridization signals were compared with those obtained for standard DNA mixtures on the same hybridization membrane.

In the mixed culture, *in vitro*, total bacterial cell numbers decreased by 50% over 48 hr but strain B14 cell numbers decreased by the same proportion in 9 hr (Fig. 2a). This represents a loss of 3×10^{8} cells per hour, a rate at least 3-fold greater than the average for the total population. By contrast, *in vivo* the decrease in B14 cell numbers was precipitous such that by 30 min, 25% of the inoculum had disappeared. After 3 hr there was no detectable strain B14 DNA in any of the samples (Fig. 2b). These data suggest a very rapid fall in strain B14 numbers *in vivo*, far greater than that observed in mixed culture *in vitro*. The results of these experiments, though perhaps discouraging for the future of introducing genetically engineered bacterial strains into the rumen, clearly show the value of specific quantification from mixed bacterial populations. This type of assay will be of even greater importance when genetically engineered rumen strains become available.

In similar experiments on human enteric *Bacteroides* populations (33,49), an assay procedure was reported similar to that previously described. In these studies, the effect of increasing bacterial cell numbers rather than DNA concentration was also investigated. The results suggested that the saturation point of nitrocellulose membrane for bacterial lysates rather than purified DNA may also pose a problem, particularly for the detection of species that make up a very low relative percentage ($<0.1\%$) of total bacteria in a complex mixture of organisms.

Increased sensitivity in these assays can be achieved by using single-stranded RNA rather than DNA probes. In the latter, both vector and gene sequence are usually labeled. However, an alternative is to clone the gene

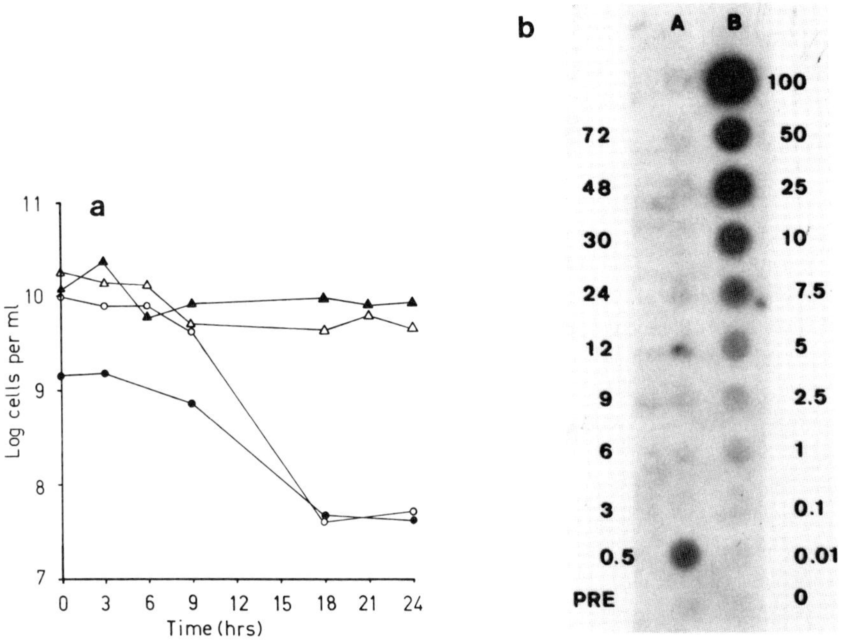

Fig. 2. Survival of *Bacteroides ruminicola* subsp. *brevis* B14 in mixed culture *in vitro* and when introduced into the rumen. (a) *In vitro* culture. Strain B14 cells were inoculated into mixed cultures of ruminal bacteria at levels of 10% and 50%. Cultures were incubated and samples were taken at regular intervals for DNA analysis using a radiolabeled RNA hybridization probe derived from clone B-18. Parallel samples were used for cell number determinations. Data are expressed as equivalent cell numbers based on DNA standard curves. ▲, Total cell number 10% inoculum; △, total cell number 50% inoculum; ●, strain B14 cell number 10% inoculum; ○, strain B14 cell number 50% inoculum. (b) *In vivo* culture. Strain B14 cells were introduced into the rumen at initial inocula of 3% and 10%. Samples were taken at intervals over a 48-hr period and analyzed as just described for B14 DNA content. Only results for the 10% inoculum are shown. (Lane A), 10 μg of sample DNA taken at the times (in hours) shown on the left; (Lane B), DNA standards of strain B14-ruminal bacterial mixtures in the proportion (percentages) shown on the right. PRE, Preinoculation.

probe sequence into a transcription vector (38) and, using RNA polymerase, produce a single-stranded RNA probe that contains only the gene probe sequence. In the work described by Attwood *et al.* (2), RNA probes were used instead of DNA probes. By selective transcription, the incorporated label was restricted to the probe sequence with very little runover into the vector. Hence the specific radioactivity of the probe and assay sensitivity was greater than with a DNA–vector probe.

In contrast to results using random chromosomal probes, it would be expected that the use of probes for rRNA genes would result in a sensitivity increase of up to 7-fold due to the number of copies of these genes in bacterial cells. In their phylogenetic studies of *Pseudomonas* sp. sensitivity was not a problem, and Festl *et al.* (13) did not report on the sensitivity of their probe; the level of *Pseudomonas* DNA added to the filters was 2–3 μg, much greater than the lower limit of sensitivity (50 ng) reported for the genomic probes and less than the capacity for the hybridization membranes. Although in this work the dot-blot technique was not used to study mixed populations, the technique of *in situ* hybridization (28) was applied to mixed plate cultures of *Pseudomonas* sp. Results of this experiment showed that, at least on plates, individual bacterial colonies could be readily detected. Although this technique is not universally applicable to all bacterial populations, particularly the fastidious ones that do not grow well in culture, it can be used where cell numbers of target species are initially too low for the dot-blot hybridization technique but can be amplified by plating. In contrast, 16 S rRNA sequences were used to quantify bacteria in rumen samples (50). Rather than relating the results to absolute bacterial numbers in the rumen, organism abundance was expressed as a fractional contribution to the total ribosome population. This was done to remove any uncertainties due to variable recoveries of nucleic acids. Since cellular ribosome abundance is correlated with growth rate, a relative measure may better reflect the contribution by each organism to total metabolic activity. The lower limit for detecting a unique 16 S rRNA sequence in the 2 μg of DNA spotted on the membrane was ~0.005% of total 16 S rRNA. The detection limit was probably higher than this for nucleic acid recovered from ruminal contents (0.01%) because of the contribution from plant nucleic acid to these samples. Quantification in these experiments was related to bacterial numbers by traditional culture and identification experiments. Of 600 colonies picked from cellobiose–agar media only 1 (0.17%) was identified as *B. succinogenes*. This value is probably a conservative estimate because it depends on culture conditions; the ribosomal probes are therefore at least 4-fold more sensitive than random genomic probes for quantifying bacterial numbers in mixed populations.

D. Gene Probes for Bacterial Phylogeny Studies

In addition to quantification studies on rumen bacteria, there is also a pressing need to establish phylogenetic relationships between apparently related bacterial species. Early methods involved determinations and comparisons of average base compositions between bacterial strains (43), and values for rumen bacteria ranged from 25 to 64.5 mol%. For many rumen species, however, the values tended to cluster around the 35–45 mol%, and strain–strain variations were often greater than species–species variations. Results of these comparisons were therefore indicative of a relationship but not definitive enough for reliable phylogenetic determinations. In practice, percentage G+C analyses were used to establish that an isolate was not related to another defined strain rather than to identify it. Nevertheless, percentage G+C values are useful in confirming identification that has been made by some other technique. The development of rRNA sequencing using defined synthetic oligonucleotide primers has replaced base composition analyses in phylogeny studies and is now a relatively direct procedure that can reliably determine both close and very distant relationships. A systematic survey of the rumen microbial ecosystem by comparative 16 S rDNA sequencing of dominant ruminal bacterial has begun (50). These studies will be able simultaneously to evaluate the validity of existing classifications and identify signature regions within these rRNAs that can serve as hybridization targets for oligonucleotide probes. Previous 16 S rRNA cataloguing of the neotype strain S85 of *B. succinogenes* showed it to be unrelated to the true *Bacteroides* species associated with cecal material. The 16 S rRNA sequence differences separating the ruminal and cecal strains (~90% similarity) are greater than those separating *Proteus vulgaris* and *E. coli* (93% similarity). This genetic diversity between bacterial strains may underlie the possible differences in ecological functions.

Where the relationship between apparently identical ruminal strains may need to be established, the rRNA sequencing technique is probably too complicated and time-consuming. Another approach that is being developed in our laboratory for *Selenomonas* sp. is to establish a map of restriction enzyme cutting sites around particular genes of one strain as a benchmark and then to compare this map with other apparently related strains. This technique has been referred to as "fingerprinting." Using random chromosomal gene fragments as probes, restriction digests (minimum of three) of DNA from test bacterial isolates are fractionated on an agarose gel, transferred to nitrocellulose or nylon membranes and hybridized with the radioactively labeled probes. An identical restriction–hybridization pattern for each digest and with several different probes would be good evidence that the strains were identical. Any differences would imply a strain difference. Although this technique is not applicable to accurate phylogenetic studies, for population studies where it is important to establish how many different strains of a particular species are present, this procedure is useful. The analogous technique in eukaryotic genetics is the use of restriction fragment length polymorphisms (RFLP).

E. Antibodies for Rumen Bacterial Identification and Quantification

Bacterial serology has played a major role in medical microbiology for many years, particularly where there has been a need for rapid identification of pathogenic species in tissue or blood samples, or where food contamination may be present. The potential for serological analysis of rumen microorganisms has, however, only recently been realized (20,48). Serological techniques using polyclonal antisera have shown that different morphological types of rumen bacteria observed *in vivo* were related to pure cultures of known microorganisms cultured *in vitro* (25). Serological evidence also suggested the existence of two different species within the genus *Ruminococcus* (27), and fluorescent antiserum was used to describe diurnal variations in ruminal numbers of cellulolytic cocci. Different serotypes have also been described for the genus *Butyrivibrio* (20), but within this large and heterogeneous group, it was not possible to define taxa on the basis of antigenic relatedness. More recently, antisera and MAb have been used to establish antigenic relationships and a rapid identification procedure for the methanogenic archaebacteria, some of which compose part of the normal rumen microflora (29). Serological probes, developed in hens against ruminal bacterial *Bacteroides* and selenomonads have also been described (48) and, in ELISA, these have been shown to distinguish between related and unrelated strains. However, for more detailed ecosystem studies it is likely that only MAb will have the specificity to distinguish between related but different bacterial strains. The production of MAb against strains of the rumen bacterial genus *Butyrivibrio* has been reported (20). In this work a number of MAb were screened and, of these, five were classified into three types on the basis of their competition for binding to cells of one particular strain of *Butyrivibrio,* indicating that there are at least three different antigenic determinants on the cell surface. Cross-reactivity studies with these showed that with one exception, the MAb probes were specific for this genera and could be used to complement biochemical identification procedures in taxonomic studies and possibly to permit rapid identification and enumeration of selected species in natural populations or mixed laboratory cultures.

In similar types of experiments, we have prepared and screened a number of MAb directed against surface determinants on rumen isolates of *Selenomonas* and *Bacteroides* (J. D. Brooker, unpublished data). Cross-reactivity studies have shown that several of the monoclonals are species-specific and in immunofluorescence assays, can be used to identify these species in pure and mixed culture. Quantitative ELISA (Fig. 3) in which total cell numbers in each well of a microtiter plate were kept constant but

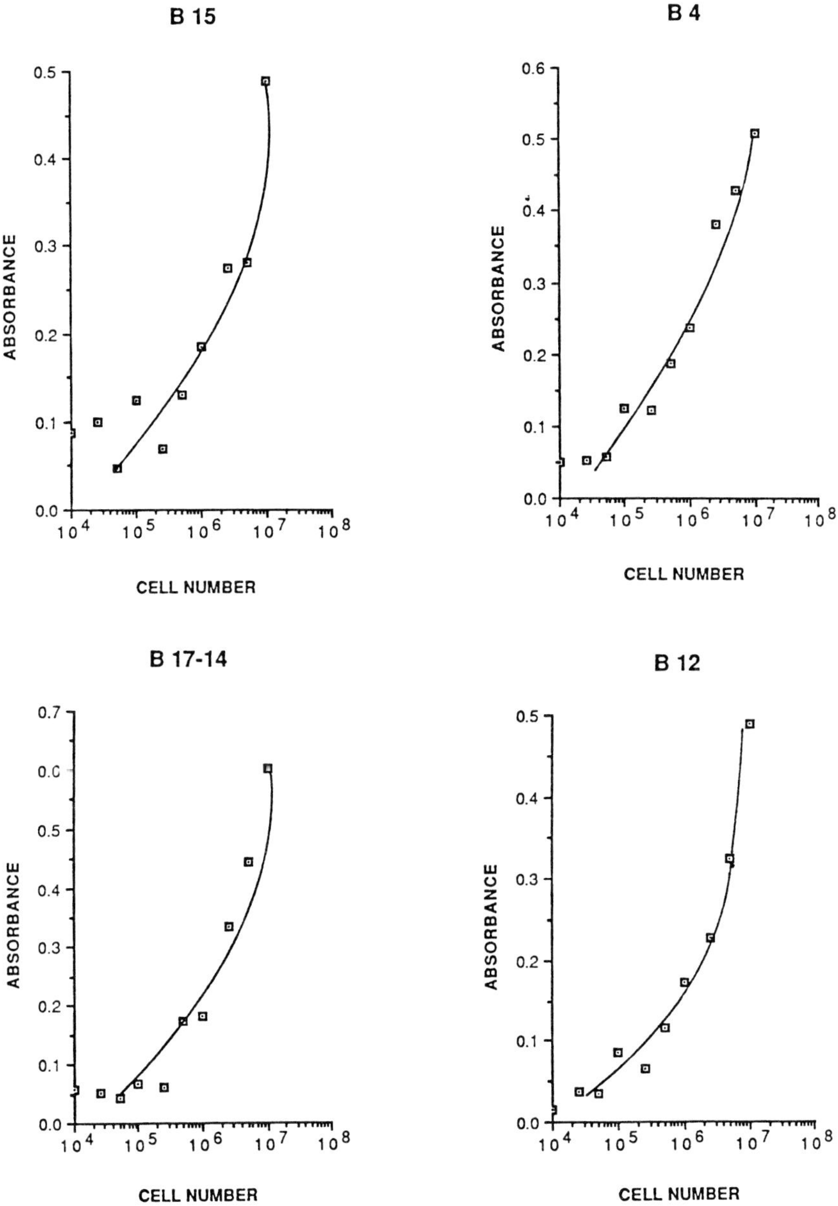

Fig. 3. Competition ELISA for *Bacteroides ruminicola* subsp. *ruminicola*. ELISA plates, pretreated with poly-L-lysine, were set up with varying proportions of *B. ruminicola* subsp. *ruminicola* cells in a total of 10^8 cells per well. The nonspecific competitor was *Escherichia coli*. Cells were fixed with glutaraldehyde and assayed by ELISA using *B. ruminicola* subsp. *ruminicola*-specific monoclonal antibodies (MAb) and anti-mouse IgG–alkaline phosphatase conjugate as the second antibody. The results are expressed as absorbance at 405 nm. B15, B4, B17-14, and B12 , are different *B. ruminicola*-specific MAb.

the proportion of *B. ruminicola* varied from 100 to 0.01% showed that the sensitivity of the antibody assay is much greater than with gene probes, the lower limit being 10^4 cells detectable as a positive signal compared with 10^7 cells in gene probe assays. These results were reproducible using four different MAb and represent a practical sensitivity of 0.1% of bacteria in a mixture. This is 5- to 10-fold more sensitive than that obtained using gene probes. Moreover, since this procedure does not require the isolation of DNA, the problem of plant contamination in rumen samples is largely avoided. Serological methods of microbial identification in crude samples of rumen fluid may therefore be the method of choice for reasons of sensitivity, directness of quantitation, and the analytical procedure's avoidance of radioactive labeling or photographic development.

IV. CONCLUSIONS

The differentiation of bacterial species in ecological or clinical studies or the establishment of phylogenetic relationships using classical morphological and metabolic characteristics is difficult and time-consuming. For fastidious organisms, studies are made more difficult because the organisms either do not grow well in culture or require special conditions. This is particularly true for rumen bacteria that not only are anaerobic but also require various media supplements for survival. They can also be difficult to lyse for *in situ* hybridization experiments. These problems make classical approaches to ecological studies of little use for rumen studies. Limitations to ecological and phylogenetic studies have now largely been overcome by the introduction of molecular and serological techniques. Through the development of DNA–DNA hybridization procedures and recombinant DNA methodology, the isolation of species-specific gene sequences is readily achieved. Gene sequences can be either unique to the particular species or be more highly repeated genes such as the rRNA genes that contain regions of specificity within them. In the latter case, the preparation of specific gene probes relies on the chemical synthesis of defined deoxyribonucleotide sequences that can be used as primers for DNA sequencing reactions in phylogenetic studies or for DNA–DNA hybridization probes in population estimates.

A newly developed technique that also utilizes synthetic oligodeoxynucleotides is that of the polymerase chain reaction (PCR) (41a). In this technique, primers ~20 bases in length anneal to opposite strands of genomic DNA flanking a particular target sequence. The primers are positioned so that the DNA polymerase-catalyzed template-directed extension of one primer can serve as a template strand for the other. In this configuration, repeated cycles of denaturation, primer annealing, and primer extension result in the exponential accumulation of the DNA target sequence with a length defined by the distance between the 5′ ends of the PCR primers. For example, 20 cycles of PCR ampification can produce a 220,000-fold increase in the amount of a 110-bp target sequence. In the present of isotopically labeled deoxynucleotides, this technique can be used to detect even rare components of rumen bacterial DNA in a complex mixture, provided that the DNA sequence flanking the target gene is known or can be obtained.

In DNA–DNA homology experiments, the most commonly used technique is that of dot-blot hybridization. This is easily done and the results are quite reproducible. The major disadvantage of this technique, however, is the need for prior preparation of DNA from the bacterial sample and the inherent inaccuracies that accompany these preparations, particularly with a rumen sample where there may be varying amounts of plant material contaminating the sample but where removal of this material may result in significant losses of certain subpopulations of microorganisms. For some bacteria, this difficulty can be overcome by using an *in situ* hybridization assay instead of the dot blot. This assay certainly avoids the problems of DNA preparation and contamination with plant material. However, it is limited by the ability to grow the particular organisms in culture, not an easy task for some fastidious rumen microorganisms, and is a more complex assay than the dot blot. Clearly, the decision to use one technique or the other must be made with full cognizance of these problems.

Nevertheless, some useful data can be obtained using gene probes to assess bacterial populations in the rumen. In the data described earlier, DNA–DNA homology studies showed that some strains of rumen bacteria that have been cultured extensively in the laboratory for many years have lost the capacity to compete effectively and survive, either in a mixed culture of rumen bacteria *in vitro* or when inoculated into the rumen. The reason for the rapid decline in bacterial numbers in these experiments was not established, although there was a possibility that some form of bacteriocinlike activity may have been present. Whether this was derived from the animal or the feed was not established, although later experiments (J. D. Brooker, unpublished data) have suggested that the *B. ruminicola* B14 bactericidal activity may have been a component of alfalfa animal feed. This observation has not been followed up further.

Studies of bacterial phylogeny do not suffer from the same difficulties because quantitation is not required. In these studies, absolute specificity is more important and the technique of comparative gene sequencing is probably the most powerful and direct method available. Because the variable regions of genes for rRNA are species-specific, gene sequence data across these regions and into more conserved regions can clearly differentiate between bacterial species. This technique is a very powerful tool for phylogenetic studies and, for complex ecosystems such as the rumen, will continue to play a vital role in establishing the relationship between apparently similar bacterial isolates.

In our present studies of the rumen selenomonads, we have shown that randomly selected *Selenomonas* chromosomal gene probes do not necessarily hybridize with DNA from all isolates. This suggests that there may be a number of similar yet different strains of *Selenomonas* in the rumen. In order to establish how these strains contribute to the overall function of the rumen ecosystem, it is important to establish how their numbers fluctuate with changes in their environment. Identification of the different strains can now be made, relatively simply, by establishing benchmark DNA fingerprints of the regions bordering particular genes. Once strains have been identified as being different, gene probes that are specific for individual strains could be isolated by differential screening of cloned gene libraries. Therefore a combination of gene cloning, restriction mapping, and DNA–DNA hybridization procedures can make a set of very powerful techniques for ecological studies in the rumen.

V. GENE PROBES VERSUS ANTISERA AND MONOCLONAL ANTIBODIES: PROSPECTS FOR THE FUTURE

In contrast to procedures using gene probes, serological analysis of bacterial species is a well-established technique. However, specificity has usually been a problem because of the nature of polyclonal antisera, and special efforts must be made to remove any cross-reactive antibodies from the preparation before use. This difficulty has now been overcome, largely with the development of MAb that can be selected for their reactivity with only defined antigenic determinants on the target organism. The use of MAb in the serological analysis of microbial populations has therefore provided a new impetus for ecological studies, particularly in complex systems such as the rumen. In addition, our studies using MAb species-specific probes have shown that they can form the basis of an assay that may be more sensitive than assays using gene probes. Sensitivity assays suggest that population levels as low as 0.1% of the total rumen biomass can be readily detected using antibodies whereas a limit of 0.5–1.0% was detectable using gene probes. (This limit may be reduced with the development of PCR primers for rumen bacterial species.) Relatively minor bacterial species in the rumen ecosystem may therefore be accessible to analysis provided they can be isolated, cultured, and used as immunogens. Moreover, through the use of immunofluorescence techniques, individual bacteria of a target species can be visually identified under the microscope, even when present in relatively small numbers in a mixed bacterial sample. This technique may be valuable for rapid screening of rumen samples when determining the relative proportions of any particular species over others. However, there is a limit to the use of serological analysis of complex ecosystems. Despite their sensitivity and specificity, it is unlikely that antibodies, even monoclonals, can be used in the same way as DNA probes to distinguish between closely related strains. Although the antibodies may be species-specific, relatively small genetic differences between isolates of the same species are unlikely to be manifest as changes on the outside of the bacterial cell. DNA probes probably provide the only reliable method of detecting these differences. However, for a rapid and sensitive assay of total bacterial numbers in a particular species, immunological assays may be the method of choice.

In conclusion, therefore, the application of genetic and serological techniques to the field of rumen ecology has opened the way for a new phase of studies on the interaction and competition between bacterial species in the rumen. With the current interest in genetically manipulating particular rumen bacterial strains and reintroducing these into the rumen for the purpose of improving production performance in domestic ruminants, techniques for enumerating bacterial populations *in vivo* have become very important. This is especially so following our observation of the rapid destruction of a laboratory strain reintroduced into the rumen. The future interaction between wild-type and recombinant bacterial species in the rumen will therefore be of considerable interest.

VI. SUMMARY

For many important groups of microorganisms, especially those in the rumen, there have previously been no simple methods for bacterial classification and enumeration. With the introduction of recombinant DNA techniques and MAb procedures, this limitation has now been removed. Gene probes that hybridize to specific bacterial species have been prepared and used both in phylogenetic analyses of bacterial genotypes and for direct enumeration of bacterial species within the complex rumen ecosystem. Monoclonal antibodies that recognize specific antigenic determinants on the surface of particular bacterial species have also been developed and used to establish relationships between bacterial strains. These have been used in the direct quantitation of bacterial species. The combination of gene probes and MAb in assays of bacterial populations paves the way for a new phase of research into bacterial interactions in the rumen.

VII. MATERIALS AND METHODS

Detailed descriptions for the preparation of gene and antibody probes can be found in the following references: preparation of gene probes (2,12,33,37,38,49); DNA–DNA hybridization (2,18,41a,50); preparation of antibodies (9,20); and use of bacteriophages (35).

ACKNOWLEDGMENTS

We thank D. Jose for excellent technical help in the preparation of gene probes and B. Stokes for the preparation of monoclonal antibodies. The work was supported in part by the Rural Credits Development Fund of the Reserve Bank of Australia and the Australian Wool Corporation.

REFERENCES

1. Atlas, R. M. (1983). Use of microbial diversity measurements to assess environmental stress. *In* "Current Perspectives in Microbial Ecology" (M. J. Klug and C. A. Reddy, eds.), pp. 540–545. Am. Soc. Microbiol., Washington, D.C.
2. Attwood, G. T., Lockington, R. A., Xue, G.-P., and Brooker, J. D. (1988). Use of a unique gene sequence as a probe to enumerate a strain of *Bacteroides ruminicola* introduced into the rumen. *Appl. Environ. Microbiol.* **54,** 534–539.
3. Baehr, W., Zhang, Y.-X., Joseph, T., Su, H., Nano, F. E., Everett, K. D., and Caldwell, H. D. (1988). Mapping antigenic domains expressed by *Chlamydia trachomatis* major outer membrane protein genes. *Proc. Natl. Acad. Sci. U.S.A.* **85,** 4000–4004.
4. Bohlool, B. B., and Schmidt, E. L. (1980). The immunofluorescence approach in microbial ecology. *Adv. Microb. Ecol.,* **4,** 203–236.
5. Brock, T. D., and Madigan, M. T. (1988) "Biology of Microorganisms," 5th ed. Prentice-Hall International Editions, Englewood, New Jersey.
6. Bryant, M. P. (1959). Bacterial species of the rumen. *Bacteriol. Rev.* **23,** 125–153.
7. Bryant, M. P. (1977). Microbiology of the rumen. *In* "Dukes Physiology of Domestic Animals" (M. J. Swenson, ed.), pp. 287–304. Cornell Univ. Press, Ithaca, New York.
8. Butman, B. T., Plank, M. C., Durham, R. J., and Mattingly, J. A. (1988). Monoclonal antibodies which identify a genus-specific *Listeria* antigen. *Appl. Environ. Microbiol.* **54,** 1564–1569.
9. Conway de Macario, E., Macario, A. J. L., and Kandler, O. (1982). Monoclonal antibodies for immunochemical analysis of methanogenic bacteria. *J. Immunol.* **129,** 1670–1674.
10. Cuppels, D. A. (1984). The use of pathovar-indicative bacteriophages for rapidly detecting *Pseudomonas syringae* pv. *tomato* in tomato leaf and fruit lesions. *Phytopathology* **74,** 891–894.

11. Doyle, G., and Everhart, D. (1985). Characterisation of monoclonal antibodies raised to *Streptococcus mutans*. *Curr. Microbiol.* **12,** 197–202.
12. Feinberg, A. P., and Vogelstein, B. (1984). A technique for radiolabelling DNA restriction endonuclease fragments to high specific activity. *Anal. Biochem.* **137,** 226–267.
13. Festl, H., Ludwig, W., and Schleifer, K. H. (1986). DNA hybridisation probe for the *Pseudomonas fluorescens* group. *Appl. Environ. Microbiol.* **52,** 1190–1194.
14. Fox, G. E., Stackebradt, E., Hespell, R. B., Gibson, J., Maniloff, J., Dyer, T. A., Wolfe, R. S., Gupta, R., Bonen, L., Lewis, B. J., Stahl, D. A., Luehrson, K. R., Chen, K. N., and Woese, C. R. (1980). The phylogeny of prokaryotes. *Science* **209,** 457–463.
15. Garvey, J. S., Cremer, N. F., and Susdorf, D. H. (1977). "Methods in Immunology," 3rd ed. Benjamin, Reading, Massachussets.
16. Gobel, U. B., and Stanbridge, E. J. (1984). Cloned *Mycoplasma* ribosomal RNA genes for the detection of mycoplasma contamination in tissue cultures. *Science* **226,** 1211–1213.
17. Gottschalk, G., ed. (1985). "Methods in Microbiology," Vol. 18. Academic Press, Orlando, Florida.
18. Grunstein, M., and Hogness, D. S. (1975). Colony hybridisation: A method for the isolation of cloned DNAs that contain a specific gene. *Proc. Natl. Acad. Sci. U.S.A.* **72,** 3961–3965.
19. Handman, E., and Remington, J. (1980). Serological and immunochemical characterisation of monoclonal antibodies to *Toxoplasma gondii*. *Immunology* **40,** 579–588.
20. Hazlewood, G. P., Theodorou, M. K., Hutchings, A., Jordan, D. J., and Galfre, G. (1986). Preparation and characterisation of monoclonal antibodies to a *Butyrivibrio* sp. and their potential use in the identification of rumen *Butyrivibrio*, using an ELISA. *J. Gen. Microbiol.* **132,** 43–52.
21. Hespell, R. B. (1985). Potential for manipulating rumen bacteria using recombinant DNA technology. *Rev. Rural Sci.* **6,** 95–100.
22. Hill, L. R. (1966). An index to DNA base composition of bacterial species. *J. Gen Microbiol.* **44,** 419–437.
23. Hill, W. E., Payne, W. L., and Aulisio, C. C. G. (1983). Detection and enumeration of virulent *Yersinia enterocolitica* in food by DNA colony hybridisation. *Appl. Environ. Microbiol.* **46,** 636–641.
24. Hirsch, D., and Martin, L. D. (1983). Rapid detection of *Salmonella* spp. by using Felix-01 bacteriophage and high performance liquid chromatography. *Appl. Environ. Microbiol.* **46,** 260–264.
25. Hobson, P. N., and Mann, S. O. (1977). Some studies on the identification of rumen bacteria with fluorescent antibodies. *J. Gen. Microbiol.* **16,** 463–471.
26. Hungate, R. E. (1966). "The Rumen and Its Microbes." Academic Press, New York.
27. Jarvis, B. D. W., Williams, V. J., and Annison, E. F. (1967). Enumeration of cellulolytic cocci in the sheep rumen by using a fluorescent antibody technique. *J. Gen. Microbiol.* **48,** 161–169.
28. Johnson, J. L. (1978). Taxonomy of the *Bacteroides*. I. DNA homologies among *Bacteroides fragilis* and other saccharolytic *Bacteroides* species. *Int. J. Syst. Bacteriol.* **28,** 245–256.
29. Jovell, R. J., Conway de Macario, E., Alito, A. E., Wolin M. J., and Macario, A. J. L. (1981). Analysis of methane synthesising bacteria by means of antisera and monoclonal antibodies. *Fed. Proc. Fed. Am. Soc. Exp. Biol.* **40,** 1124.
30. Kaper, J. B., Bradford, H. B., Roberts, N. C., and Falkow, S. (1982). Molecular epidemiology of *Vibrio cholerae* in the U.S. gulf coast. *J. Clin. Microbiol.* **16,** 129–134.
31. Klieve, A. V., and Bauchop, T. (1988). Morphological diversity of ruminal bacteriophages from sheep and cattle. *Appl. Environ. Microbiol.* **54,** 1637–1641.

32. Kohler, G., and Milstein, C. (1975). Continuous culture of fused cells secreting antibody of predefined specificity. *Nature (London)* **256,** 495–497.
33. Kuritza, A. P., Getty, C. E., Shaughnessy, P., Hesse, R., and Salyers, A. A. (1986). DNA probes for identification of clinically important *Bacteroides* species. *J. Clin. Microbiol.* **23,** 343–349.
34. Kuritza, A. P., Shaughnessy, P., and Salyers, A. A. (1986). Enumeration of polysaccharide-degrading *Bacteroides* species in human faeces by using species-specific DNA probes. *Appl. Environ. Microbiol.* **51,** 385–390.
35. Lockington, R. A., Attwood, G. T., and Brooker, J. D. (1988). Isolation and characterisation of a temperate bacteriophage from the ruminal anaerobe, *Selenomonas ruminantium*. *Appl. Environ. Microbiol.* **54,** 1575–1580.
36. Macario, A. J. L., and Conway de Macario, E. (1988). Quantitative immunologic analysis of the methanogenic flora of digestors reveals a considerable diversity. *Appl. Environ. Microbiol.* **54,** 79–86.
37. Maniatis, T., Fritsch, E. F., and Sambrook, J. (1982). "Molecular Cloning: A Laboratory Manual." Cold Spring Harbor Lab., Cold Spring Harbor, New York.
38. Melton, D. A., Kreig, P. A., Rebagliati, M. R., Maniatis, T., Zinn, K., and Green, M. R. (1984). Efficient *in vitro* synthesis of biologically active RNA and RNA hybridisation probes containing a bacteriophage SP6 promoter. *Nucleic Acids Res.* **12,** 7035–7056.
39. Moore, W. E. C., and Holdeman, L. V. (1974). Human fecal flora: The normal flora of 20 Japanese-Hawaiians. *Appl. Microbiol.* **27,** 961–979.
40. Mordarski, M. (1985). Detection of ribosomal nucleic acid homologies. *In* "Chemical Methods in Bacterial Systematics" (M. Goodfellow and D. E. Minnikin, eds.), Soc. Appl. Bacteriol. Tech. Ser., pp. 41–65. Academic Press, London.
41. Moseley, S. L., Echeverria, P., Seriwatana, J., Tirapat, C., Chaicumpa, W., Sakuldaipaera, T., and Falkow, S. (1982). Identification of enterotoxic *E. coli* by colony hybridisation using 3 enterotoxin gene probes. *J. Infect. Dis.* **145,** 863–869.
41a. Mullis, K. B., and Faloona, F. A. (1987). The polymerase chain reaction. *In* "Methods in Enzymology" (R. Wu, ed.), Vol. 155, pp. 335–350. Academic Press, San Diego, California.
42. Mutharia, L. M., and Hancock, E. W. (1985). Monoclonal antibody for an outer membrane lipoprotein of the *Pseudomonas fluorescens* group of the family Pseudomonadaceae. Int. J. Syst. Bacteriol. **35,** 530–532.
43. Ogimoto, K., and Imai, S. (1984). "Atlas of Rumen Microbiology." Jpn. Sci. Soc. Press, Tokyo.
44. Pace, N. R., Stahl, D. A., Lane, D. J., and Olsen, G. J. (1986). The use of rRNA sequences to characterise natural microbial populations. *Adv. Microb. Ecol.* **9,** 1–55.
45. Palleroni, N. J., Kunisawa, R., Contopoulou, R., and Doudoroff, M. (1973). Nucleic acid homologies in the genus *Pseudomonas*. *Int. J. Syst. Bacteriol.* **23,** 333–339.
46. Paster, B. J., Ludwig, W., Weisberg, W. G., Stackebrandt, E., Hespell, R. B., Hahn, C. M., Reichenbach, H., Stetter, K. O., and Woese, C. R. (1985). A phylogenetic grouping of the *Bacteroides, Cytophagas* and certain flavobacteria. *Syst. Appl. Microbiol.* **6,** 34–42.
47. Regensburger, A., Ludwig, W., and Schleifer, K-H. (1988). DNA probes with different specificities from a cloned 23S rRNA gene of *Micrococcus luteus*. *J. Gen. Microbiol.* **134,** 1197–1204.
48. Ricke, S. C., Schaeffer, D. M., Cook, M. E., and Kang, K. H. (1988). Differentiation of ruminal bacterial species by ELISA using egg yolk antibodies from immunised chicken hens. *Appl. Environ. Microbiol.* **54,** 596–599.

49. Salyers, A. A., Lynn, S. P., and Gardner, J. F. (1983). Use of randomly cloned DNA fragments for identification of *Bacteroides thetaiotaomicron*. *J. Bacteriol.* **154,** 287–293.
50. Stahl, D. A., Flesher, B., Mansfield, H. R., and Montgomery, L. (1988). Use of phylogenetically based hybridisation probes for studies of ruminal microbial ecology. *Appl. Environ. Microbiol.* **54,** 1079–1084.
51. Stahl, D. A., Lane, D. J., Olsen, G. J., and Pace, N. R. (1985). Characterisation of a Yellowstone hot spring microbial community by 5S ribosomal RNA sequences. *Appl. Environ. Microbiol.* **49,** 1379–1384.
52. Teather, R. M. (1985). Application of gene manipulation to rumen microflora. *Can. J. Anim. Sci.* **65,** 563–574.
53. Warner, A. C. (1962). Some factors influencing the rumen microbial population. *J. Gen. Microbiol.* **28,** 129–133.
54. Woese, C. R., Balnz, P., and Hahn, C. M. (1984). What isn't a Pseudomonad: The importance of nomenclature in bacterial classification. *Syst. Appl. Microbiol.* **5,** 179–195.

15

Gene Probe Detection of Human and Cell Culture Mycoplasmas

RAM DULAR
Ontario Ministry of Health
Regional Public Health Laboratory
Ottawa, Ontario, Canada

I. INTRODUCTION

The term mycoplasmas is used rather loosely, but widely, to denote any species included in the class Mollicutes. They are characterized by a lack of a cell wall and are the most prevalent parasites of human, animal, plant, insect, and *in vitro* tissue cell cultures. Many cause a variety of diseases. According to their host association, they are generally referred to as human, animal, plant, insect, or cell culture mycoplasmas. Since the field

GENE PROBES
FOR BACTERIA

of mycoplasmology is very vast, the coverage of this chapter will be limited to mycoplasmas of interest to clinical microbiologists involved with the laboratory diagnosis of human mycoplasmal infections. Mycoplasmas in this category are those associated with human diseases, for example, *Mycoplasma pneumoniae* (respiratory mycoplasma) and *Ureaplasma urealyticum, M. hominis* and *M. genitalium* (urogenital mycoplasmas), and those of both human and animal origin that affect the laboratory diagnosis of viral infections by infecting cell cultures used as the viral propagation media (cell-infecting or cell-contaminating mycoplasmas).

Present techniques for laboratory diagnosis of human mycoplasmal infections depend primarily on the isolation and identification of the organisms and/or the demonstration of antibody development by serological means. Although culture is the method of choice for the detection of mycoplasmas, diagnostic resources for the isolation and identification of these organisms are not widely available. This is partly due to the complexities of media, culture, and identification procedures. Furthermore, in the case of some mycoplasmas (e.g., *M. pneumoniae*), primary isolation from clinical specimens may take one to several weeks. Culture results thus become available too late to have much impact on the management of the vast majority of cases of *M. pneumoniae* infections. Detection of cell-infecting mycoplasmas by culture isolation is more complicated than human mycoplasmas in the sense that too many species of varying nutritional requirements are involved with cell culture infections and many strains cannot be cultured by conventional methods (20). Indirect methods are relatively insensitive. Serological tests available for human mycoplasma infections suffer from a significant degree of variability and lack of sensitivity and specificity. Furthermore, serological tests become meaningful only on demonstration of a rise in antibody titer, requiring the testing of acute- and convalescent-phase sera, again causing a delay in delivery of laboratory results to the physician. Cross-reactivities of polyclonal antibodies among members of the class Mollicutes (15,58,59,71,108,122,150) have been the major limiting factor in the development of reliable immunodiagnostic techniques. Monoclonal antibodies (MAb) against several cell-infecting mycoplasmas have been described and used for the identification of these organisms (10,98), but a pool of these antibodies has not been tried for the routine screening of cell cultures for mycoplasmal infections.

In the past, major emphasis has been on developing new media formulations, culture procedures, and specific immunodiagnostic techniques for the rapid detection of mycoplasmas from clinical specimens as well as from cell cultures. Progress made in these areas, however, has been very limited. Recent advances in the development of deoxyribonucleic acid

(DNA) and MAb probes have created great hope among mycoplasmologists for their possible application as rapid, specific, and sensitive diagnostic tools in mycoplasmology. This chapter briefly reviews the general characteristics of human and cell-infecting mycoplasmas and their pathogenesis, as well as present techniques of laboratory diagnosis. The use of genetic probes in diagnostic mycoplasmology will also be detailed. Also, a brief summary of recent advances on the use of antibody probes, especially monoclonals, as rapid diagnostic tools will be presented for comparison purposes.

II. BACKGROUND

A. The Organisms

Mycoplasmas were first discovered as the etiological agent of contagious bovine pleuropneumonia in 1898 (89). Since then various other mycoplasmas have been isolated and shown to cause disease in animals, but these organisms attracted little interest among microbiologists until 1937, when a mycoplasma was isolated from a Bartholin's gland abscess of a human subject (22). Subsequently, numerous reports appeared on the frequent isolation of these organisms from humans, especially from the urogenital tract and the oropharynx, and occasionally from other sites as well (33). The classification and nomenclature of mycoplasmas were not established until the mid-1950s, and thus all isolates were referred to as pleuropneumonialike organisms (PPLO). Presently, >90 species of mycoplasmas from human, animal, plant, insect, and cell cultures have been identified and characterized. All mycoplasmas are placed in a single class, Mollicutes (*mollis,* soft; *cutis,* skin). The present classification of the Mollicutes into orders, families, and genera and some of their major distinguishing characteristics are summarized in Table I. Detailed description of the various taxa and species is given elsewhere (60,99,142).

Mycoplasmas are the smallest, most primitive, and simplest of self-replicating prokaryotes. The cells are highly pleomorphic, varying in shape from coccoid to helical or filamentous, usually 300–800 nm in diameter, resembling larger viruses in size, but multiply in cell-free media and form colonies on solid media. Colonies on solid media are minute, usually much smaller than 1 mm in diameter. Under optimal conditions, almost all species form colonies that have a characteristic "fried-egg" appearance. They lack a cell wall, and ultrastructural studies show that they are constructed of only three organelles: a plasma membrane to separate the cytoplasm from the environment; ribosomes to assemble cell proteins; and

TABLE I

Taxonomy and Major Characteristics of Members of the Class Mollicutes[a,b]

Classification	Number of species	Genome size (Da)	Mol% G + C of DNA	Sterol requirement	Location of NADH oxidase	Distinctive properties	Habitat
Division: Tenericutes							
Class: Mollicutes							
Order I: Mycoplasmatales		5×10^8		+	Cytoplasm		
Family I: Mycoplasmataceae		5×10^8					
Genus I: *Mycoplasma*	~76		23–41			Ferment glucose and/or hydrolyze arginine	Human and animals
Genus II: *Ureaplasma*	2		27–30			Hydrolyze urea	Human and animals
Family II: Spiroplasmataceae		1×10^9				Helical filaments	Arthropods and plants
Genus I: *Spiroplasma*	5		25–31				
Order II: Acholeplasmatales		1×10^9		–	Membrane		
Family I: Acholeplasmataceae							
Genus I: *Acholeplasma*	~10		27–36				Human, animals, plants, insects
Mollecutes of uncertain taxonomic position							
Genus: *Anaeroplasma*	2	1×10^9	29–34	Some + Some –	Unknown	Anaerobic; some digest bacteria	Bovine and ovine rumen

[a] Details can be found in Tully and Taylor-Robinson (142), Razin (99), and Krieg and Holt (60).
[b] G + C, Guanine + cytosine; NADH, reduced nicotinamide adenine dinucleotide.

a double-stranded DNA to provide the information for protein synthesis. In some mycoplasmas unique polar organelles shaped as tapered tips or blobs built around a central striated rod have been observed. These tips appear to play a role in the attachment of mycoplasmas to host cells and to inert surfaces. The mycoplasma genome is typically prokaryotic, consisting of a circular double-stranded DNA molecule that replicates semiconservatively (120). However, the size of the genome is much smaller than that of eubacteria, ranging from 5×10^8 D in *Mycoplasma* and *Ureaplasma* species to $\sim 1 \times 10^9$ D in *Acholeplasma* and *Spiroplasma* species. The small size of the genome and its extremely low guanine + cytosine (G + C) content (23–43 mol% G + C) (Table I) restrict the amount of genetic information available to these organisms. Consequently the number of protein species synthesized by these organisms is significantly smaller than in eubacteria. This apparently results in limited biosynthetic capabilities of these organisms and their dependence on the host for the exogenous supply of essential nutrients for growth and multiplication. As in other prokaryotes, the ribosomes of mycoplasmas are made up of 50 S and 30 S subunits, consisting of ~50 protein species, and 23 S, 16 S, and 5 S ribosomal RNA (rRNA) (99,120). Information on the genetics and molecular biology of mycoplasmas is still fragmentary. It is partially due to lack of genetic tools, such as selectable markers, and methods to transfer DNA, such as conjugation, transformation, and transduction. However, the recent report on the transposition of streptococcal transposon T_n916 in *Acholeplasma laidlawii* and *M. pulmonis* (27) should serve as a powerful genetic tool for the study of these organisms.

So far, mycoplasmas of 12 different species have been isolated from human subjects, mainly from mucous membrane surfaces (6,11,12, 19,26,134,140) (Table II). They can be divided into respiratory mycoplasmas and urogenital mycoplasmas depending on the sites from which the organisms can be isolated most frequently. More than 20 mycoplasmas of both human and animal origin are shown to be associated with cell infection. They are listed in Table III. Some mycoplasmas not listed in the table but which have been infrequently isolated from cell culture are *U. urealyticum* (112), *M. pirum* (21), and *M. pneumoniae* (58,94).

B. Pathogenesis

*1. Respiratory Mycoplasmas (*Mycoplasma pneumoniae*)*

Mycoplasma pneumoniae is a well-recognized human respiratory tract pathogen associated with a wide spectrum of both pulmonary and extrapulmonary syndromes that may be benign, self-limited, moderately troublesome, or sometimes life-threatening (Table IV) (6,11,12,17,19,

TABLE II

Human Mycoplasmas, Their Occurrence, and Association with Disease[a]

Mycoplasma	Frequency of isolation from the: Respiratory tract	Urogenital tract	Cause of disease
Acholeplasma laidlawii	Rare	—[b]	No
Mycoplasma buccale	Rare	—	No
M. faucium	Rare	—	No
M. fermentans	Rare[c]	Rare	No
M. genitalium	Rare[d]	Rare[d]	Yes
M. hominis	Rare	Common	Yes
M. lipophilum	Rare	—	No
M. orale	Common	—	No
M. pneumoniae	Common	Very rare	Yes
M. primatum	—	Rare	No
M. salivarium	Common	Rare	No
Ureaplasma urealyticum	Rare	Common	Yes

[a] See references in text.

[b] No report of isolation.

[c] During an outbreak of *M. pneumoniae* infection in the Ottawa area, <20% of respiratory specimens yielded *M. fermentans* (R. Dular, unpublished observation).

[d] Because of a lack of appropriate cultural procedures, frequency of occurrence is questionable.

26,33,62,95,140). In some cases extrapulmonary manifestations may actually overshadow or occur in the absence of symptomatic respiratory tract involvement. Most clinical infections are related to nothing more serious than upper respiratory tract infections, tracheitis, or bronchitis (86). Clinically apparent pneumonia develops in only 3–10% of infected persons. Nevertheless, it probably accounts for ≤20% of all pneumonia in the general population and for ≤50% in closed populations (e.g., family, military recruits, and schoolchildren) (86).

Mycoplasma pneumoniae infections occur throughout the year in the general population, but epidemics are most likely to occur in colder months and in institutions (17). *Mycoplasma pneumoniae* infects people of all ages, however, infection rates are much higher in school-aged children and young adults (32). Epidemics show a cyclic pattern, occurring every 4–6 years. However, this pattern seems to be changing in some European countries. The 4-year cycle observed in Denmark and Sweden has been

TABLE III

Mycoplasmas Associated with Cell Infection[a]

Source and species	Natural habitat
Human and primate	
Mycoplasma orale	Oropharynx
M. hominis	Oropharynx, urogenital tract
M. fermentans	Oropharynx, urogenital tract
M. salivarium	Oropharynx
M. buccale	Oropharynx
Bovine	
Acholeplasma sp. (unspeciated)	Presumably bovine
A. laidlawii[b]	Oropharynx, urogenital tract
A. axanthum[c]	Nasal passage, oropharyngeal tracts
Mycoplasma strain HRC70-159	Cell cultures only
M. arginini[d]	Oropharynx, urogenital tract
M. bovis	Oropharynx, joints, udder
M. bovoculi	Conjunctiva
Swine	
M. hyorhinis	Nasal passage
A. oculi[e]	Conjunctiva
Murine	
M. arthritidis	Rat tissues
M. pulmonis	Mouse and rat oropharynx
Avian (chickens and turkeys)	
M. gallisepticum	Oropharynx
M. gallinarium	Oropharynx
Canine	
M. canis	Oropharynx, genital tract
Mycoplasma sp. serogroup HRC689	Urogenital tract

[a] Based on data from Barile (3).
[b] Also isolated from avian, caprine, canine, equine, feline murine, ovine, porcine, primates, human (rare), and plant tissues.
[c] Also from swine and equine tissues.
[d] Also isolated from sheep, goats, chamois, swine, and wild cats.
[e] Also from caprine and equine tissues.

obliterated, and rates are now similar from year to year (70). Although the epidemiological pattern in the United States for the last decade is unknown, in neighboring Canada the old pattern still exists (32).

The mechanism of pathogenesis of *M. pneumoniae* is poorly understood. Adherence of mycoplasmas to cells of the respiratory tract is the initial step in successful colonization and pathogenesis. The adherence is mediated by a protein that is located on the tip structure of *M. pneumoniae* cells (31) and is believed to be an essential virulence factor (11). Infection

TABLE IV

Clinical Spectrum of *Mycoplasma pneumoniae* Infection[a]

Pulmonary	Extrapulmonary
Tracheobronchitis	Hematological disorders (autoimmune hemolytic anemia, thrombocytopenia, intravascular coagulation)
Pharyngitis	Gastrointestinal disorders (anorexia, nausea, vomiting, transient diarrhea, anicteric hepatitis, pancreatitis)
Otitis media	Musculoskeletal symptoms (myalgia, arthralgia, polyarthritis)
Pneumonia	Dermatological symptoms (erythema multiforme, Stevens–Johnson syndrome, and other rashes)
Multiagent pneumonia	Cardiovascular abnormalities (myocarditis, pericarditis, pericardial effusion, conduction defects)
Wheezing in infants	Neurological symptoms (meningitis, meningoencephalitis, transverse myelitis, peripheral and cranial neuropathy, cerebellar ataxia)
Rhinitis	Miscellaneous disorders (acute glomerulonephritis, general lymphadenopathy, splenomegaly, interstitial nephritis)
Myringitis	

[a] Details on frequency of occurrence of various syndromes and specific references can be found in Cassel and Cole (12) and Biberfeld (6).

is associated with ciliostasis, altered cellular metabolism, and cytonecrosis. The exact mechanism of cell injury is unknown, but it has been suggested that superoxides, hydrogen peroxide, and nucleic acid (NA) depletion may lead to damage of cellular components (11,12). Inherent to the pathogenicity of *M. pneumoniae* is their ability to modulate host immunity. These organisms are able nonspecifically to activate B-cell production, suppress the initiation of immune response, and stimulate the production of autoantibodies against lymphocytes, lung, heart, liver, kidney, smooth muscle, and brain. Production of immune complexes and autoantibodies may well be responsible for many of the extrapulmonary manifestations associated with *M. pneumoniae* infections, including rashes, arthritis, hemolytic anemia, thrombocytopenia, myocarditis, hepatic necrosis, and neural lesions (6,11,12).

2. *Urogenital Mycoplasmas (***Ureaplasma urealyticum, Mycoplasma hominis,** *and* **Mycoplasma genitalium***)*

Ureaplasma urealyticum, M. hominis, and *M. genitalium* are the genital mycoplasmas of clinical importance for humans. The association of genital mycoplasmas with human disease is complicated by the fact that they may be found as part of the normal genitourinary tract flora. About one-third of healthy infants are colonized at birth with *U. urealyticum,* presumably acquired during passage through the birth canal. *M. hominis* is much less common. The rates of isolation of both mycoplasmas decreases during the first year of life, and are extremely low before puberty. Following puberty, the frequency of colonization by both mycoplasmas depends on a variety of factors including sexual experience, race, socioeconomic status, contraception, menstruation, and menopausal changes, of which, perhaps, the most important is sexual experience (75,134). Over 50% of sexually active men and women are colonized with them.

Ureaplasma urealyticum has been implicated as an etiological agent in various urogenital and extragenital tract infections since its discovery in 1954 (Table V) (114). Formerly described as T-strain mycoplasmas or T-mycoplasmas, they were later placed in a new genus and species, *U. urealyticum,* because of their unique ability of metabolizing urea (116). *Ureaplasma urealyticum* is a proven etiological agent of nongonococcal urethritis (130) and may cause at least 10% of the existing cases (12,133). With the exception of nongonococcal urethritis, however, a causal relationship between ureaplasmal colonization and other diseases has not been convincingly demonstrated. There is, however, an increasing body of evidence supporting a causative role of *U. urealyticum* in chorioamnionitis, perinatal morbidity and mortality (28,97), spontaneous and recurrent abortion (87), neonatal meningitis (147), and septic arthritis (131).

Studies on the role of *M. hominis* as an etiological agent of human disease began soon after its discovery as the first human mycoplasma isolated in pure culture in 1937 (22). The organism is a common isolate from the genital tract of asymptomatic men and women, and has been occasionally isolated from extragenital sources. It is now known as a potential pathogen in a variety of situations (Table V). Several epidemio-

TABLE V

Diseases Associated with Urogenital Mycoplasmas[a]

Organism	Diseases
Ureaplasma urealyticum	Pneumonia, respiratory distress, urethritis, prostatitis, septic arthritis, spontaneous abortion and stillbirth, infertility, chorioamnionitis, low birthweight, urinary calculi, and neonatal meningitis
Mycoplasma hominis	Pyelonephritis, abscess of Bartholin's gland, pelvic inflammatory disease, arthritis, postabortal fever, postpartum fever, postsurgical complications, and neonatal pneumonia and meningitis
Mycoplasma genitalium	Urethritis, salpingitis

[a] Specific references and other details related to causal relationship between urogenital mycoplasmas and various diseases can be found in several reviews (12, 75, 128, 129, 132–134).

logical, microbiological, serological, and animal model studies have shown that it is potentially pathogenic and may be involved in, for example, acute pelvic inflammatory diseases, acute pyelonephritis, and postpartum fever (12,75,129,133,134). There is strong evidence that *M. hominis* is one cause, among others, of acute pelvic inflammatory disease (74,84). Various other extragenital infections due to *M. hominis* have been reported, but the frequency with which they occur is not known. The involvement of *M. hominis* in septicemia (117), intracerebral abscess (91), neonatal meningitis (82,147), wound sepsis (113), peritonitis (92), and septic arthritis (131) has been described.

Mycoplasma genitalium is a new mycoplasma isolated from urethral specimens of human patients with nongonococcal urethritis (137,143). It shows a remarkable resemblance to the well-established human pathogen *M. pneumoniae* in cell shape, possession of a tip structure, complex nutritional requirements, and antigenic structure (129,136). Seroepidemiological data (85) and experimental animal infections (129,136,144) suggest that *M. genitalium* may be responsible for some cases of nongonococcal urethritis and salpingitis. Although technical problems have impeded large-scale survey for the prevalence of this organism in the past, current studies utilizing DNA probes show that this organism is a common inhabitant of the urethra in men with and without urethritis (44). A recent report on the isolation of this organism from throat specimens identifies the human oral cavity as an additional locale for the organism (5). What role *M. genitalium* may play in human respiratory disease remains to be determined.

Pathogenic mechanisms of urogenital mycoplasmas are not well established. Like many other mycoplasmas, they adhere to host cell surface and replicate there. *Mycoplasma pneumoniae*-type adhesion proteins have not been demonstrated in *U. urealyticum* or *M. hominis*. *Ureaplasma urealyticum,* through its urease activity, produces ammonium ions, which induce cytopathology in a variety of established cell lines, and cytonecrosis and ciliostasis in bovine fallopian tube organ cultures (12). *Mycoplasma hominis* produces no cytopathology or ciliostasis, but causes marked ciliary swelling (12). *Mycoplasma genitalium,* as *M. pneumoniae,* possesses a predominant surface nap extending distally from the tip organelle (141). This structure appears to mediate the adherence of this organism to Vero monkey kidney cell monolayers (14). A protein designated as MgPa (140 kDa), located on the surface of the terminal structure of *M. genitalium,* is believed to be the counterpart of the P1 protein of *M. pneumoniae* associated with cell adherence (46).

3. *Cell-Infecting Mycoplasmas*

Mycoplasmal infection of cell cultures was first discovered in 1956 (107), and since then intensive studies have been done on the effects mycoplasmas have on their cell culture hosts. The importance of mycoplasmal cell infection has been recognized for many years (41). It constitutes a serious threat to every laboratory utilizing cell cultures because of its drastic effects on the physiology and structural integrity of the infected host cells (Table VI). This has resulted in numerous erroneous reports throughout the scientific literature on findings of new enzyme systems, cellular metabolites, antigenic structures, cytopathic agents, and instability of hybridomas. As shown in Table VI, a very disparate group of metabolic changes involving DNA, RNA, and protein accompany mycoplasma infection. A recent report shows that many hybridoma lines and original stocks of many myelomas lines used for cell fusion are infected with mycoplasmas (94). Various aspects of mycoplasmal effects on host cell culture have been extensively reviewed in recent years (3,4,40,50,63,76,79,118).

The incidence of infection is high and very frequently remains unnoticed (76), apparently because mycoplasmas often have no visible effect on the host cells and their growth does not produce turbidity in cell cultures even though they are present in concentrations of 10^7–10^8 colony-forming units (CFU) per milliliter of supernatant medium. The overall incidence in several recent surveys has ranged from 2.1 to 63% (4,20,76,77,79,88,94).

TABLE VI

Effects of Mycoplasmas on Cell Cultures[a]

Interference with the growth rate of cells
Inhibition of lymphocyte transformation
Stimulation of lymphocyte transformation
Induction of morphological alterations, including cytopathology
Altered DNA, RNA, and protein synthesis
Alterations of ribosomal RNA profiles
Alteration of enzyme patterns
Interference with selection of mutant mammalian cells
Induction of chromosomal aberrations
Depletion of the essential amino acid arginine from cell culture growth medium
Inhibition of virus yields
Enhancement of virus yields
Copurification of mycoplasmas with cell organelles (e.g., mitochondria)
Redistribution and modification of host cell plasma membrane antigens
Apparent reduction in tumorigenic potential of malignant cells

[a] References in text.

The reported incidence rates vary depending, in part, on the *in vitro* age of the culture, quality control practices, and the efficiency of the detection procedures (76,79). The primary sources of mycoplasmal infections are animal sera and laboratory personnel. The infected cell cultures, if not discarded or cured, later become the biggest source of infection.

Mycoplasmas can affect cell cultures in various ways, including contribution of mycoplasmal gene products, utilization of essential medium components, acidification or alkalinization of growth media, and induction of interferon production by host cells.(76)

C. Diagnosis

1. Present Techniques

a. Respiratory Mycoplasmas (*Mycoplasma pneumoniae*). A variety of methods have been described for the detection of *M. pneumoniae* and its infections (Table VII). Of these, currently three basic laboratory methods are available to routine diagnostic laboratories: isolation of the organism by culture; demonstration of the development of cold hemagglutinins; or a rise in the titer of complement-fixing (CF) antibodies in patients' sera.

Complex culture procedures, slow growth of the organism (one to several weeks), and its differentiation from other mycoplasmas are major time-consuming tasks of routine microbiology laboratories. Very few laboratories have the facilities or can afford the cost of properly implementing

TABLE VII

Methods to Detect *Mycoplasma pneumoniae* Infections

Methods	Reference
Microbiological culture	18,34,51,141
Antigen detection	
Immunofluorescence	43
Counterimmunoelectrophoresis	148
Enzyme immunoassays (EIA)	42,56
Serodiagnosis	
Mycoplasmacidal test	35
Radioimmunoprecipitation	8
Radioimmunoassay	9
Complement fixation (CF)	13,53
Metabolic inhibition	135
Tetrazolium reduction inhibition	111
Cold hemagglutination	92
Direct hemagglutination	30
Indirect hemagglutination	24,69,127

culture procedures. Alternative procedures for the direct detection of *M. pneumoniae* antigens in clinical specimens have been attempted by using conventional polyclonal antisera in immunofluorescence (43), counterimmune electrophoresis (148), and enzyme immunoassays (42,56), but with limited success. Because of the difficulties involved in the detection of *M. pneumoniae* by culture procedures or immunological techniques in clinical specimens, diagnosticians and epidemiologists have relied heavily on the serodiagnosis of diseases caused by this organism. Measurement of cold hemagglutinin levels in serum is a relatively simple, rapid, and inexpensive technique. The test, however, is positive only in ~50% of patients with *M. pneumoniae* pneumonia (29,55), and as many as 50% of the patients with cold aggluntinins in their serum do not appear to have *M. pneumoniae* infections (29). This test is best considered to be nonspecific in that the IgM antibodies that are detected are directed against 1 antigen of the human erythrocyte membrane and may occur in a few diseases not caused by *M. pneumoniae*. The test thus lacks both sensitivity and specificity.

The development of antibodies to *M. pneumoniae* is most commonly detected by CF test. Rising antibody titers become detectable during the second week after onset of illness and reach a peak during the fourth week (39). This rise in titer thus occurs too late to be of value to clinicians in planning therapy. The test is economical, simple to perform, readily adaptable in diagnostic laboratories, and is often combined with other CF tests for evidence of infection with respiratory viruses. However, it lacks specificity (61,90) and sensitivity (121). Mycoplasmacidal, radioimmunoprecipitation, radioimmunoassay, metabolic inhibition, tetrazolium reduction inhibition, and hemagglutination tests are technically demanding, and some problems in specificity or sensitivity have limited the widespread use of these procedures in routine diagnostic laboratories. An enzyme immunoassay (EIA) technique utilizing *M. pneumoniae* adhesin protein as antigen (48) for detection of IgG and IgM levels in *M. pneumoniae* patients' sera offers considerable promise; however, commercial sources for this antigen are nonexistent. The task of antigen purification and certification is beyond the capacity of diagnostic laboratories, thus limiting the evaluation of this procedure in routine serodiagnosis of *M. pneumoniae* infections.

Recent serological data have revealed that *M. pneumoniae* is capable of eliciting a complex immune response and that sera from infected rabbits or humans may contain not only *M. pneumoniae*-specific, but also cross-reacting antibodies directed against antigen(s) shared by other mycoplasmas (58,108). Predictably, false positive reactions are likely to occur in immunological procedures utilizing crude antigens or conventional polyclonal antisera. There is a need to develop reagents with high specificity and sensitivity for the rapid diagnosis of *M. pneumoniae* infections.

b. Urogenital Mycoplasmas (*Ureaplasma urealyticum, Mycoplasma hominis,* and *Mycoplasma genitalium*). Several culture and serological methods have been described for the detection of these organisms and their infections (Table VIII). Culture identification is the most widely used procedure for *U. urealyticum* and *M. hominis*. Ureaplasmas grow fast (2–5 days) and colonies are easily identified by virtue of the formation of characteristic brown accretion colonies in the presence of manganese sulfate. *Mycoplasma hominis* is a relatively slow grower and may take up to ~1 week. Speciation of *M. hominis,* however, is difficult because species-specific or serotype-specific antibodies are not available commercially. Culture media and procedures are complex, and only a very few specialized laboratories offer culture services. Therapy is generally based on the judgments of the physician. Very little is known about the culture isolation of *M. genitalium*. It is an extremely slow grower and can take as long as 50 days of incubation (143).

TABLE VIII

Methods to Detect Urogenital Mycoplasma Colonization and Infections

Organisms and methods	Reference
Ureaplasma urealyticum	
Microbiological culture	10a,51,115,127a
Serodiagnosis	
Complement fixation	85,96
Growth inhibition	116
Metabolic inhibition	65
Mycoplasmacidal test	68
Indirect hemagglutination	24
Enzyme immunoassays	7,64,149
Immunofluorescence	52,85
Mycoplasma hominis	
Microbiological culture	10a,51,127a
Serodiagnosis	
Complement fixation	96
Growth inhibition	66
Metabolic inhibition	67
Indirect hemagglutination	96
Immunofluorescence	36,96
Mycoplasmacidal test	67
Enzyme immunoassay	83
Mycoplasma genitalium	
Microbiological culture	143
Serodiagnosis	
Immunofluorescence	85

Several serological tests (Table VIII) have been described in the literature, but none is available to diagnostic laboratories in lieu of culture. Technology for detecting antibodies to genital mycoplasmas is poorly developed and serological data available thus far on these organisms are as conflicting as are their associations with human diseases. Serological cross-reactions among mycoplasmas are very common (15,58,59,71, 108,122). Poor growth of the organisms resulting in low yield of cells to be used as antigens, the requirement of complex growth media, and poor knowledge of the antigens responsible for the human immune response have hampered the development of specific and sensitive immunodiagnostic tests. Another feature that complicates *U. urealyticum* and *M. hominis* serology is the existence of multiple serotypes (65,106). Complement fixation, growth inhibition, metabolic inhibition, mycoplasmacidal, hemagglutination, and immunofluorescence tests have problems with either sensitivity or specificity. Some tests are only suitable for speciation of the organisms. Several reports for the detection of antibodies to human mycoplasmas by EIA (7,64,145,149) have recently appeared, although these studies are still preliminary and their usefulness in clinical diagnosis is questionable (45). An indirect-immunofluorescence test (85) has been recommended (51) for the serodiagnosis of *U. urealyticum* infection, but the method is complicated and requires cell culture facilities. The test has not been evaluated by diagnostic laboratories. A sensitive EIA assay using a strain that shows extensive cross-reaction with other strains has been used for the diagnosis of *M. hominis* infections; however, during the course of the disease an increase in IgG and a decrease in IgM have been found both among women with pelvic inflammatory disease and among controls (85).

Difficulties in the cultivation of urogenital mycoplasmas and a lack of standardized immunodiagnostic techniques have prompted considerable interest in the development of alternative diagnostic tools such as NA probes and MAb.

c. Cell Culture Mycoplasmas. Many different methods have been developed and proposed for the detection of cell culture mycoplasmas (40,50,63). Some microbiological, biochemical, biophysical, and microscopic detection methods are summarized in Table IX. Methodology, advantages, and disadvantages of each method have been extensively reviewed in recent years (41,76,77,79,81,109,119). Microbiological culture is definitely the method of choice; however, failure of some strains of mycoplasmas to grow on conventional culture media (20) and complex culture procedures have limited the widespread use of this procedure in the routine screening of cell cultures for mycoplasmal infection. Indirect

TABLE IX

Methods to Detect Cell Culture Mycoplasmas

Method	Reference
Microbiological culture	81
DNA fluorescent staining	20
Immunofluorescence	20
Uridine phosphorylase	62
Uridine–uracil ratio	110
Scanning electron microscopy	93
Autoradiography	123
RNA speciation	139
Mycoplasmal-mediated cytotoxicity	78

methods have been used successfully for the detection of mycoplasmas, but no single one has been shown to detect successfully all mycoplasma-infected cell cultures. Most of the indirect methods are insensitive. In order to increase sensitivity, they require indicator cell lines that are inoculated with the cell culture supernatants prior to testing. Thus cultural methods and indirect tests available thus far are complicated, time-consuming, not economical, and not attractive enough for small-scale laboratories to perform them. A rapid, economical, easy to perform, and sensitive test is urgently needed to combat the mycoplasmal cell infections that are affecting developments in basic as well as applied research. Can genetic and antibody probes replace the presently existing methods to monitor mycoplasmal cell culture infection? This is reviewed in detail in this chapter.

2. *Genetic and Antibody Probes as the Potential Future Diagnostic Tools*

Present techniques for the detection of mycoplasmas and their infections do not meet the goal of diagnostic microbiology. The goal of diagnostic microbiology is to provide methods that are rapid, sensitive, specific, economical, and easy to perform without employing sophisticated and expensive equipment.

Over the past decade, a considerable body of literature focusing on NA probes and MAb has developed, and progress in developing these tools for use in the diagnostic laboratory has generated a lot of excitement among clinical microbiologists. The excitement stems from the likelihood that these reagents can shorten the time required to identify fastidious pathogens like mycoplasmas, detect organisms directly in clinical specimens, and reduce the overall costs associated with processing specimens for the culture, isolation, and identification of these organisms.

Mycoplasmas are an extremely diverse group (54,102,124) with very low DNA sequence homology (1–4%). However, like other prokaryotes, these organisms have a significant degree of sequence homology in their rRNA genes (1,2). These studies have aided in constructing NA probes of varying specificities. Nucleic acid probes for human and cell-infecting mycoplasmas developed so far fall into three classes. The first comprises probes made of rRNA genes or fragments thereof. Because of the highly conserved nature of rRNA genes (1,2), probes of this type exhibit low specificity and are referred to as group-specific or general probes. These probes are designed to detect as many mycoplasma species as possible. The second class consists of cloned mycoplasma DNA fragments composed of sequences specific for certain species or strains. The third class of probes utilizes whole genomic DNA as probes. Because of low DNA sequence homology among mycoplasmas, the probes of this class are shown to be species-specific (44,105).

Considering the difficulties encountered in cultivating mycoplasmas from clinical specimens or cell cultures, mycoplasmologists have great hope that the recent developments in NA probes and hybridoma technology will finally permit the rapid and specific detection of these organisms. Nucleic acid probes and MAb to these organisms have recently been produced, and their use in routine diagnosis of mycoplasmal infections is just beginning. The following results address the application of these agents for the improved and rapid detection of mycoplasmas.

III. RESULTS AND DISCUSSION

A. Gene Probe Detection of *Mycoplasma pneumoniae* in Clinical Specimens

Three laboratories have reported the production of gene probes of varying sensitivity and specificity (Table X). Efficiency of MP20, MP30 (37), and pPN4 (47) probes in the detection of *M. pneumoniae* in clinical specimens has not been tested as yet. The only commercially available probe (cDNA to rRNA of *M. pneumoniae*; Gen-Probe, Inc.) has been recently tested for its efficiency in detecting *M. pneumoniae* in clinical specimens (25). A total of 116 specimens, mostly from patients with respiratory illnesses, were tested for *M. pneumoniae* by probe methods and culture. Throat swabs placed in 2 ml of a transport medium consisting of tryptic soy broth, 0.5% bovine serum albumin, and 500 IU of penicillin G per milliliter

TABLE X

Genetic Probes for the Detection of *Mycoplasma pneumoniae*

Designation and origin	Specificity	Sensitivity	Testing of infected materials	Advantages and deficiencies	References
Gen-Probe rapid diagnostic system for *Mycoplasma pneumoniae*, cDNA to rRNA (kit, Gen-Probe, San Diego, California)	Reacts with *M. pneumoniae* and *M. genitalium*	$<10^4$ *M. pneumoniae* cells	Clinical specimens	"In-solution" hybridization detects rRNA, rapid. Reacts with *M. genitalium* too	25,57,138
MP20 Synthetic oligonucleotide probe (20 bp) to rRNA	Reacts with *M. pneumoniae*	$<10^3$ Cells	Information not available	Chemically defined probe, detects rRNA	37
MP30 synthetic oligonucleotide probe (30 bp) to rRNA	Reacts with *M. pneumoniae* and *M. genitalium*	$<10^3$ Cells	Information not available	Chemically defined probe, detects rRNA	37
pPN4, DNA fragment cloned in plasmid pUC13	Specific to *M. pneumoniae*. Reactivity with bacterial species not tested	10^5 *M. pneumoniae* cells	*M. pneumoniae* clinical specimens not tested	Does not react with *M. genitalium*. Poor sensitivity	47

(51). Tracheal aspirates, auger suctions, sputa, pericardial fluids, and lung tissues were diluted (~1 : 10) in the transport medium immediately after collection or on arrival in the laboratory. The specimens, other than sputa and tissues, were vortexed briefly before being tested. Sputa were homogenized in the transport medium by being drawn and expelled several times through a 25-gauge needle. Lung tissues were minced by using sterile Pasteur pipette tips and scalpel blades. Specimens were cultured on arrival or after storage at −70°C as previously described (26,32). The probe was tested in accordance with the instructions of the manufacturer. A 1-ml volume of each specimen was centrifuged (12,000 *g* for 10 min), and the rRNA released after the lysis of cells in the sediment was allowed to hybridize with the ^{125}I-cDNA probe at 72°C for 1 hr. Hybridized cDNA–rRNA was separated from single-stranded cDNA and washed, and the radioactivity was counted in a γ counter. A sample–negative control ratio of ≥3 was taken as positive for *M. pneumoniae*. The culture and probe results for the different types of specimens tested are given in Table XI. The frequencies of detection of the organism in throat swabs, sputa, tracheal aspirates, and lung tissues seemed comparable by both tests. Of the 116 specimens studied, 44 (37.9%) were positive by culture and 47 (40.5%) were positive by probe. Of the 52 specimens positive by either method, 84.6% were positive by culture and 90.3% were positive by probe. The overall comparison of culture and probe results is presented in Table XII. The sensitivity of the probe analysis relative to culture (proportion of culture-positive specimens detected by the probe) was 39 of 44 (89%; 95% confidence interval, 76–95%). The relative specificity (proportion of

TABLE XI

Detection of *Mycoplasma pneumoniae* in Different Types of Specimens by Culture and cDNA Probe[a]

	Number (%) positive		
Specimen (*n*)	Culture	Probe	Total
Throat swab (74)	33 (45.6)	34 (45.9)	37 (50)
Sputum (17)	7 (41.1)	7 (41.1)	8 (47)
Lung biopsy or autopsy (9)	3 (33.3)	2 (22.2)	3 (33.3)
Tracheal aspirate (7)	1 (14.2)	3 (42.8)	3 (42.8)
Pleural fluid (4)	0 (0)	0 (0)	0 (0)
Pericardial fluid (3)	0 (0)	0 (0)	0 (0)
Total (116)	44 (37.9)	47 (40.5)	52 (44.8)

[a] Reproduced with permission from Dular *et al.* (25).

TABLE XII

Overall Comparison of Probe with Culture[a]

Culture result	Number of probe results	
	Positive	Negative
Positive	39	5
Negative	8	64

[a] Reproduced with permission from Dular *et al.* (25).

culture-negative specimens also negative by probe) was 64 of 72 (89%; 95% confidence interval, 89–94%). Culture and probe results agreed for 103 of 116 specimens (88.8%). Overall, 11% of the specimens showed discrepancies between the culture and probe results, which might be anticipated because the sensitivity of each is affected by different factors.

Quantitation experiments determined the lower limit of sensitivity to be $\sim 10^4$ CFU/ml (Fig. 1). The probe has also been evaluated by two other laboratories (57,138) and found to be equivalent in sensitivity and specificity to the culture of respiratory secretions. The only problem associated with this probe is its reactivity with *M. genitalium*. This organism has recently been shown to be present in throat specimens of respiratory patients (5). The role of *M. genitalium* in human respiratory disease is unknown. The probe detection method, thus, cannot differentiate between *M. pneumoniae* and *M. genitalium*. Future studies utilizing *M. pneumoniae*-specific (37,47) and *M. genitalium*-specific (47) probes or MAb (5) will provide valuable information on the prevalence of *M. genitalium* in human respiratory tract and its association with respiratory diseases, if any. The data presented here and elsewhere (57,138) show, however, that the Gen-Probe rapid detection system is sensitive and specific, and fulfills the need for the rapid detection of *M. pneumoniae* in various types of clinical specimens.

B. Gene Probe Detection of Urogenital Mycoplasmas

Whole genomic DNA and cloned DNA fragments have been prepared and shown to be specific and sensitive for the detection of urogenital mycoplasmas (Table XIII). Cloned DNA probes specific for *M. genitalium* (47,104) have not been tested on clinical specimens as yet. Their diagnostic value is therefore uncertain. To illustrate the diagnostic uses of gene

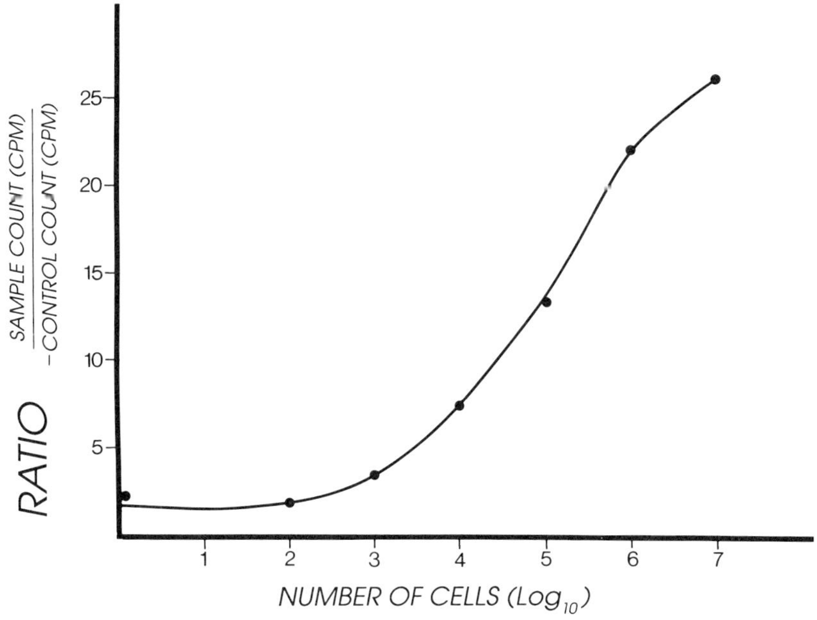

Fig. 1. Sensitivity of cDNA probe (Gen-Probe) to *Mycoplasma pneumoniae*. Mid-log phase broth culture of *M. pneumoniae* was concentrated by centrifugation (30,000 *g*, 30 min) and suspended in tryptic soy broth. Tenfold serial dilutions containing 10^7, 10^6, and 10^5, 10^4, 10^3, and 10^2 CFU/ml were tested. Tryptic soy broth served as negative control.

probes, two studies utilizing ^{32}P-labeled whole genome DNA are discussed here. Both concern the detection of *U. urealyticum, M. hominis,* and *M. genitalium* in urethral specimens of men attending a sexually transmitted disease clinic. For the detection of *U. urealyticum* and *M. hominis,* urethral specimens were collected on swabs and placed into 0.8 ml of mycoplasma transport medium (51). All specimens were tested by DNA probe for the presence of *M. hominis* and *U. urealyticum* and 144 of these specimens were tested by quantitative culture as well.

The *M. hominis* and *U. urealyticum* chromosomal DNA probes were tested against purified DNA and pure cultures of the genital organisms *M. genitalium* (G-37 and ATCC 33530), *U. urealyticum* (serovars 1–9), and *M. fermentans* (ATCC 19989). No cross-hybridization was observed between these organisms and the *M. hominis* probe, while *U. urealyticum* type 8 hybridized with all nine serovars of ureaplasma, but not with other mycoplasmas. Ten different strains of *M. hominis* were also tested and shown to hybridize with *M. hominis* probe, but not with the ureaplasma probe. A panel of bacteria often found in the urogenital tract, including

TABLE XIII

Genetic Probes for the Detection of *Ureaplasma urealyticum*, *Mycoplasma hominis*, and *Mycoplasma genitalium*

Designation and origin	Specificity	Sensitivity	Testing of infected materials	Advantages and deficiencies	References
Whole chromosomal DNA of *U. urealyticum*, type 8	Specific to *U. urealyticum*	$>10^3$ Cells	Urogenital specimens	Specific, but specimens with $<10^3$ cells will go undetected	105
Whole chromosomal DNA of *M. hominis* strain ATCC 14027	Specific to *M. hominis*	Not determined	Urogenital specimens	Specific, but insensitive	105
Whole chromosomal DNA of *M. genitalium* ATCC 33530	Specific to *M. genitalium*	Not determined	Men urethral specimen	Specific, culture data not available for comparison	44
PGN3, DNA fragment cloned in plasmid pUC13	Specific to *M. genitalium*	10^5 *M. genitalium* cells	*M. genitalium* clinical specimens not tested	Does not react with *M. pneumoniae*, poor sensitivity	47
A 256-bp DNA fragment of *M. genitalium* cloned in bacteriophage M13	Specific to *M. genitalium*	1.2×10^4 *M. genitalium* cells	*M. genitalium*, clinical specimens not tested	Does not react with bacterial species from genital tract. Reactivity with *M. pneumoniae* not tested.	104

group B streptococci, *Bacteroides, Gardnerella vaginalis, Mobiluncus,* and *Neisseria gonorrhoeae,* failed to hybridize with either probe, indicating that the whole chromosomal probes are species-specific. Data presented in Table XIV show that the DNA probe detected 9 of 16 *M. hominis* culture-positive specimens (56%) and culture detected 9 of 11 DNA probe-positive specimens (82%). The addition of probe technology to *M. hominis* culture methodology increased the total number of positive specimens by 2 (10%). For ureaplasmas, the probe detected 36 of 57 culture-positive specimens (63%), and culture detected 36 of 54 probe-positive specimens (67%). The combination of the two technologies would increase the number of positive specimens by 32%. These results suggest that the culture methodology misses a significant number of ureaplasmal isolates. The numbers for *M. hominis* are simply too small to make a judgment. The probe for *U. urealyticum* is also ineffective in specimens with an initial population of $<10^3$ CFU/ml. The only advantage of probe detection methodology for *U. urealyticum* and *M. hominis* is that the probe test will detect viable and nonviable organisms, as well as strains that are difficult to culture. However, isolation, purification, and radiolabeling of DNA are beyond the capacity of most diagnostic laboratories.

A whole genomic ^{32}P-labeled DNA probe has been recently used to demonstrate the prevalence of *M. genitalium* in men with and without urethritis (44). Urethral swabs from the patients were collected and placed immediately in 1 ml of mycoplasma transport medium (51). Specimens (100 μl samples) were applied to nitrocellulose paper and lysed. Hybridiza-

TABLE XIV

Comparison of DNA Probe and Culture for the Detection of *Mycoplasma hominis* and *Ureaplasma urealyticum* (n = 144)[a]

	M. hominis DNA probe		*U. urealyticum* DNA probe	
Culture[b]	+	−	+	−
+	9	7	36	21
−	2	126	18	69

[a] Reproduced with permission from Roberts *et al.* (105).

[b] Initial plates showing colonies were scored as positive; initial plates not showing colonies were scored as negative. χ^2 for *M. hominis* = 60.29, $p < .001$; χ^2 for *U. urealyticum* = 26.50, $p < .001$.

tion was performed at 30°C overnight, washed at room temperature, washed at 40°C, dried, and exposed to X-ray film for 5–7 days at −70°C. The probe detected as little as 50–100 pg of DNA (6×10^4–1.2×10^5 genomes) per dot blot. The probe did not react with *M. hominis, U. urealyticum, Haemophilus influenzae, Haemophilus parainfluenzae, Haemophilus ducreyi, G. vaginalis, Mobiluncus* spp., *Eikenella corrodens,* or *Bacteroides* spp.

Data presented in Table XV show that the probe detected *M. genitalium* in 30 (15%) of the men, including 11 (30%) of 37 homosexuals versus 19 (11%) of 166 heterosexuals ($p = .009$). The data suggest that *M. genitalium* is more common among homosexuals than heterosexuals. The organism was detected in urethral specimens more often among those with persistent or recurrent nongonococcal urethritis than among other clinical diagnostic groups. The prevalence of *M. genitalium* in heterosexual men resembled that of *M. hominis,* but was less than that of *U. urealyticum* in patients with no urethritis, acute gonococcal urethritis, or nongonococcal urethritis. In homosexual men the prevalence of *M. genitalium* was higher. The sensitivity of whole genomic DNA probe is slightly higher than that obtained with the ^{32}P-labeled cloned DNA probes (47,104). However, evaluation of the sensitivity of whole genomic DNA probe against patients' specimens awaits a reliable culture method.

TABLE XV

Prevalence of *Mycoplasma genitalium* by Diagnosis and Sexual Preference[a]

	Number (%) with *M. genitalium*	
Diagnosis	Heterosexual	Homosexual/bisexual
No urethritis	6/65 (9)	4/19 (21)
Acute gonococcal urethritis	2/19 (11)	1/2 (50)
Acute *Chlamydia*-positive NGU[b]	2/28 (7)	1/2 (50)
Acute *Chlamydia*-negative NGU	2/22 (9)	2/9 (22)
Persistent or recurrent NGU	7/32 (22)	3/5 (60)
Total	19/166 (11)	11/37 (30)

[a] Reproduced with permission from Hooton *et al.* (44).
[b] Nongonococcal urethritis.

The probe study described here has added a great deal to our knowledge of this newly discovered mycoplasma. Further studies to support these findings on its prevalence among diseased as well as normal persons is certainly required to understand the etiological role of this organism in urethritis.

C. Gene Probe Detection of Cell Culture Mycoplasmas

Detection of cell culture mycoplasmas has drawn maximum attention during the last two decades. In fact, the first genetic probe to be applied to the detection of mycoplasmas was used for cell culture infection (100). Since then many genetic probes have been constructed and tested for their sensitivity and specificity (Table XVI). However, very few of these probes have been tested for the detection of cell-infecting mycoplasmas. Probe pMC5 (101) has been tested against four cell-infecting mycoplasmas yielding positive results and is shown to be successful in diagnosing cell culture infections and speciation of the mycoplasmas. However, the test involves the Southern blot hybridization procedure, which may take up to a week to perform. Several other probes that have been constructed and tested for specificity and sensitivity have not been tested as yet for their application to the routine screening of cell culture mycoplasmas. The commercially available cDNA probe (Gen-Probe, Inc., San Diego, California) has been on the market for several years and has been evaluated by several laboratories for its suitability as a substitute or as an adjunct to culture isolation of mycoplasmas. The probe was first tested in 1986 (80), and positive results were obtained in 27 of 52 cell cultures tested (Table XVII). Two false negatives occurred with the use of the genetic probe. Both cell cultures were human lymphoblastoid cells propagated in serum-free media. *Mycoplasma hyorhinis* was isolated from both cell cultures. The genetic probe did not detect this infection, giving hybridization values of 8.5 and 9.3% in different assays, slightly below the 10% threshold value. The concentration of *M. hyorhinis* in these cultures was $\sim 10^6$ mycoplasmas per milliliter. This lower concentration was believed to be the cause of false negative results. The sensitivity of the assay varied with different mycoplasmas. The range of detection was from 9.3×10^3 to 2.6×10^6 CFU/ml. The currently available probe, supposed to be more sensitive, however, has failed to detect 13 of 14 mycoplasma-infected cell cultures (77). The cell cultures consisted of 3T6 cells infected separately with *M. hyorhinis, M. orale, M. arginini, M. fermentans,* and *A. laidlawii.* The probe failed to detect the presence of mycoplasmas even when the concentration was in excess of 10^7 mycoplasmas per milliliter. Most of these false negatives were in the 5–7% hybridization range.

TABLE XVI

Genetic Probes for Cell-Infecting (Cell-Contaminating) Mycoplasmas

Designation and origin	Specificity	Sensitivity	Testing of infected materials	Advantages and deficiencies	References
pMC5, a 4.8-kbp *Mycoplasma capricolum* DNA insert in pBR325 containing 23 S, 5 S, and part of 16 S rRNA genes	Nonspecific. Reacts with other prokaryotes and plant chloroplasts, no reaction with eukaryotic DNA	<1 ng mycoplasmal DNA	Cell-infecting mycoplasmas	Identifies mycoplasmas in Southern blot, detects rRNA. Lengthy procedure, up to 1 week	101
MYC-14 and MYC-25 synthetic oligonucleotide probe (14 and 25 bp, respectively) to rRNA	Group-specific, reacts to *Mycoplasma* spp. but not *Ureaplasma urealyticum*	10^4 cells	Cell-infecting mycoplasmas	Chemically defined probes. Not tested for reactivity to *Acholeplasma spp.*	37
cDNA to *Acholeplasma laidlawii* and *Mycoplasma hominis* rRNA genes (Mycoplasma Detection Kit, Gen-Probe)	Reacts with *Acholeplasma* and *Mycoplasma* spp.	9.3×10^3 to 2.6×10^6 cells. Sensitivity varies with species.	Cell-infecting mycoplasmas	Detects rRNA, rapid	49, 80

Recombinant plasmid pMCB1002 containing 5 S, 23 S, and most of 16 S rRNA gene	Reacts with several mycoplasmas	Not determined	Cell-infecting mycoplasmas	"In-solution" hybridization, lengthy procedure	40
cDNA to *M. hominis* 16 S and 23 S rRNA or pMH23 recombinant plasmid, containing 23 S rRNA gene of *M. hominis*	Reacts with mycoplasmas and other prokaryotes but not with mammalian cells	1 ng DNA by dot-blot hybridization	Infected cell cultures, mice and human arthritis specimens	Useful to detect mycoplasmas in arthritic joints	54
M13MH129, a 0.9-kbp fragment of *Mycoplasma hyorhinis* 23 S rRNA gene cloned in M13	More specific than pMC5	0.5 pg of homologous DNA or $<10^5$ mycoplasma cells	Cell-infecting mycoplasmas (not evaluated in routine screening as yet)	Specific to mycoplasmas	38
Mycoplasma hyorhinis probes, 3.6-kbp and 9.4-kbp *M. hyorhinis* DNA fragments cloned in λCh4-Mh rG1	Specific to *M. hyorhinis*	5×10^4 *M. hyorhinis* cells	Detection and identification of *M. hyorhinis* in cell cultures	Species-specific	125, 126

TABLE XVII

Results of Mycoplasma Assay of Cell Cultures with Genetic Probes[a]

Number tested	52
Number positive	27 (51.9%)
Number negative	45

[a] Reproduced with permission from McGarrity and Kotani (80).

We have recently used this probe for the detection of several cell culture lines obtained from several laboratories. Results shown in Table XVIII are disappointing. This is surprising, since other laboratories have shown the probe to be as efficient as culture (49). It should be noted, however, that the efficiency of culture isolation varies drastically from laboratory to laboratory.

IV. CONCLUSIONS

A number of gene probes have been developed and characterized in terms of their sensitivity and specificity to human and cell culture mycoplasmas. Cloned DNA fragments, whole genomic DNA, synthetic oligo-

TABLE XVIII

Comparison of cDNA Probe (Gen-Probe) for the Detection of Mycoplasmal Cell Contamination[a]

Detection methods	Number positive/number tested
Culture	15/19
Culture and DNA staining	17/19[b]
Dot–EIA	17/19[c]
Mycotect	12/19[d]
Gen-Probe	11/19

[a] R. Douma, R. Dular, B. R. Brodeur, and S. Kasatiya (unpublished data).

[b] Culture and DNA staining were carried out by standard protocols (76).

[c] Dot–EIA was performed using MYC-4 monoclonal antibody (23) and peroxidase-conjugated anti-mouse immunoglobulins.

[d] Performed as instructed by the supplier (Bethesda Research Laboratories, Gaithersburg, Maryland).

nucleotides, and conserved rRNA genes have been tested for their sensitivity, specificity, and potential application to the routine diagnosis of human pathogenic and cell-infecting mycoplasmas. The exploitation of these probes, however, has only begun. Most of the probes described here have not been tested for their diagnostic potential as yet. The commercially available cDNA probe (Gen-Probe) fulfills the long-awaited need for a rapid, sensitive, and specific diagnostic tool for the detection of *M. pneumoniae*. The probe successfully detected *M. pneumoniae* in a variety of clinical specimens from patients with respiratory infection. The major drawback of the probe is its reactivity with *M. genitalium,* which has recently been shown to be present in the human respiratory tract.

The *M. genitalium*-specific cloned probes are of interest, and their application should help mycoplasmologists study the distribution of this difficult-to-culture organism in different anatomical sites. The studies utilizing whole genomic DNA as probes for *U. urealyticum, M. hominis,* and *M. genitalium* are encouraging; however, it is beyond the capacity of many diagnostic laboratories to cultivate these organisms for the isolation of DNA. *Ureaplasma urealyticum* and *M. hominis* are fairly fast growers, and the dot-blot hybridization assay described here does not seem to offer any advantage over conventional cultural methods.

The group-specific cDNA probe (Mycoplasma T.C.11, Gen-Probe) for the detection of cell-infecting mycoplasmas is insensitive and needs improvement.

V. GENE PROBES VERSUS ANTISERA AND MONOCLONAL ANTIBODIES

Mycoplasmas have inherent antigenic similarities, and serological cross-reactions are common (15,58,59,71,108,122,150). Immunodiagnostic techniques utilizing polyclonal antisera or crude antigens suffer from a lack of sensitivity and specificity. Studies in the developments of gene probes and MAb are still in their infancy. A logical comparison between gene probes and MAb in mycoplasmology is not possible as yet. Many gene probes have been described for the detection of mycoplasmas (Section III), but very few have been tested for their diagnostic potential. Monoclonal antibodies to *M. pneumoniae, U. urealyticum, M. hominis, M. genitalium,* and several prominent cell culture-infecting mycoplasmas have been produced and characterized (Table XIX). Monoclonal antibody MP-169 shows a nonspecific reaction with media components, but has been shown to detect *M. pneumoniae* antigen in clinical specimens by Western blot immunobinding assay, a procedure that requires >6 hr and is

TABLE XIX

Monoclonal Antibodies against Human and Cell Culture Mycoplasmas

Antibody	Specificity	Reference
MP-169	*M. pneumoniae*	72,73
OC_2FS	*M. pneumoniae*	15,16
MP.1	*M. pneumoniae*	10
MYC-12	*M. pneumoniae*	23
M.43	*M. pneumoniae* (adhesin protein)	31
U.u.5B2	*U. urealyticum* (urease)	107a
PG21	*M. hominis*	14
Mg-209, Mg-124, Mg-182, Mg-229, Mg-240	*M. genitalium* (adhesin protein)	46
MH-1	*M. hyorhinis*	10
MO-1	*M. orale*	10
MA-1	*M. arginini*	10
AL-1	*A. laidlawii*	10
MS-1	*M. salivarium*	10
SFRI-Myco 1, SFRI-Myco 2	*M. fermentans*	98
MYC-4, MYC-9	*Mycoplasma, Ureaplasma,* and *Acholeplasma* spp.	23

labor-intensive and expensive (72,73). OC_2FS has been used to detect *M. pneumoniae* from enrichment culture broths by Western blot immunobinding assay (15,16). The procedure has taken from 2 to 12 days to obtain positive results. The antibody OC_2FS also reacts with *A. laidlawii* and *M. genitalium.* Antibodies MP.1 (10) and M.43 (31) have not been used for clinical diagnosis. Studies on monoclonal MYC-12 are at a very preliminary stage. Studies are currently underway and the results obtained by nitrocellulose Dot–EIA using peroxidase-conjugated anti-mouse immunoglobulins as second antibody are encouraging. Of 20 respiratory specimens tested, 16 were positive by culture, cDNA probe (Gen-Probe), and Dot–EIA (R. Dular, R. Douma, B. Brodeur, and N. C. Irvine, unpublished observation). Both gene probe and Dot–EIA take <3 hr to perform. Reagents needed to perform Dot–EIA are stable and nonisotopic.

Monoclonal antibody U.u.5B2 to *U. urealyticum* (107a) is specific to urease of this organism. The antibody has reacted to 6 serotypes tested. Other serotypes have not been tested and its application in diagnostic microbiology remains to be explored. Monoclonal antibody to *M. hominis* was tested against 14 different strains (not serotyped), but reacted with only 6 strains (14).

A panel of MAb to cell culture mycoplasmas (10,98) has been described and used successfully for the speciation of the isolates from infected cell culture. Monoclonals MYC-4 and MYC-9 react with several mycoplasma species tested, including *Ureaplasma, Mycoplasma,* and *Acholeplasma* spp. (23). Its application in the routine screening of cell cultures for mycoplasmal contamination is shown in Table XVIII. Results, though preliminary, are very encouraging.

VI. PROSPECTS FOR THE FUTURE

Because most investigators have only recently produced and characterized mycoplasma gene probes and MAb, the actual application of these reagents for the detection of these organisms has only begun. One interesting topic for investigation in the immediate future will be the testing of the efficiency of *M. pneumoniae*-specific synthetic and cloned DNA probes in the detection of this pathogen in clinical specimens. Studies utilizing *M. pneumoniae*- and *M. genitalium*-specific probes would provide supportive evidence on the presence of *M. genitalium* in oropharyngeal specimens of respiratory patients and control groups. The commercially available rapid system for the detection of *M. pneumoniae* is cross-reactive to *M. genitalium*. It was previously believed that *M. genitalium* is confined to the urogenital tract of humans, but this is not true. The organism is found in throat specimens (5), and its etiological role remains unknown. The use of *M. genitalium*- and*M. pneumoniae*-specific probes already described in the literature can be very useful in differentiating these organisms. Antigenic and genetic studies suggest that these two species have more similarities than dissimilarities (71,150). Whole genomic probes shown to be sensitive and specific for urogenital mycoplasmas have limitations. Complex culture procedures and poor yield of the cells will limit these procedures to specialized laboratories. Cloned gene sequences or genes would certainly prove to be more attractive to diagnostic laboratories. Differentiation of pathogenic and nonpathogenic serotypes of *M. hominis* and *U. urealyticum* should be given top priority in understanding the pathogenicity of these organisms. DNA hybridization tests, cleavage of genomic DNA by restriction endonucleases, and polyacrylamide gel electrophoresis of cell proteins indicate that the *U. urealyticum* serotypes fall into two genotypically distinct, but related, clusters (103). Efforts should be directed at the selection and cloning of genes specific for each of the two *U. urealyticum* clusters. Probes based on these genes will facilitate the solution of the problem of whether pathogenicity is associated with one cluster only. Similar attempts should be made for *M. hominis*.

The currently available cDNA probe (Gen-Probe) for the detection of mycoplasmal cell contamination needs improvement. Sensitive synthetic cDNA probes MYC-14 and MYC-25 (37) should be evaluated for their efficiency and made available to diagnostic laboratories for the screening of mycoplasmal infection of cell cultures.

Monoclonal antibody MYC-12 should be tested for its reactivity with *M. genitalium* and Dot–EIA procedure should be standardized. Monoclonal antibodies MYC-4 and MYC-9 are of interest because of their reactivity with a broad range of mycoplasmas. They have the potential for use as a group-specific probe for the routine screening of cell cultures for mycoplasmal contamination.

VII. SUMMARY

Mycoplasmas are pathogens of humans, animals, plants, and insects, and they frequently cause *in vitro* cell infection. They are extremely fastidious in nutritional requirement. *In vitro* cultivation of mycoplasmas requires complex and expensive media, and in many cases growth is extremely slow or fails altogether. In the past several years, major attention has been given to developing rapid, sensitive, and specific methods for detecting these organisms in clinical specimens and *in vitro* cell culture. A panel of gene probes has been developed and characterized, and some have been evaluated for their efficiency in detecting these organisms in clinical specimens and cell cultures. A cDNA probe commercially available from Gen-Probe Corporation was tested in various types of clinical specimens from respiratory patients and was found to be sensitive and specific. The DNA probe available for the detection of cell culture mycoplasmas, however, was found to be insensitive. Whole genomic DNA probes have been used to detect *U. urealyticum, M. hominis,* and *M. genitalium* in human urethral specimens. Specificity of these probes, however, is questionable. Many cloned gene sequences and synthetic oligonucleotide probes described here are of potential value as reagents for the rapid diagnosis of mycoplasmas. Monoclonal antibodies to several human and cell-infecting mycoplasmas have been described and the diagnostic application of these antibodies have been explored.

VIII. MATERIALS AND METHODS

The gene probes described here for the detection of *M. pneumoniae* and cell culture mycoplasmas are commercially available from Gen-Probe, 9020 Chesapeake Drive, San Diego, California 92123. All the reagents and reaction tubes are supplied in kit form. Detailed detection procedures are supplied with every kit. For the detection of urogenital mycoplasmas whole genomic DNA has been used. Purification and labeling of DNA are standard procedures described in most molecular biology books. The dot-blot hybridization procedure (146) is also described in several laboratory manuals.

ACKNOWLEDGMENTS

The able editorial and secretarial assistance of Gloria Tremblay is gratefully acknowledged. I thank Ann Prytula, N. C. Irvine, W. Willoughby, and S. Kasatiya for cooperation and encouragement. The opinions and assertions in this chapter are the views of the author and are not to be construed as official or as reflecting the views of the Ontario Ministry of Health.

REFERENCES

1. Amikam, D., Glaser, G., and Razin, S. (1984). Mycoplasmas (*Mollicutes*) have a low number of rRNA genes. *J. Bacteriol.* **158,** 376–378.
2. Amikam, D., Razin, S., and Glaser, G. (1982). Ribosomal RNA genes in *Mycoplasma*. *Nucleic Acids Res.* **10,** 4215–4222.
3. Barile, M. F. (1979). Mycoplasma–tissue cell interactions. *In* "The Mycoplasmas" (J. G. Tully and R. F. Whitcomb, eds.), Vol. 2, pp. 425–474. Academic Press, New York.
4. Barile, M. F., and Grabowski, M. W. (1978). Mycoplasma–cell culture–virus interactions: A brief review. *In* "Mycoplasma Infection of Cell Cultures" (G. J. McGarrity, D. Murphy, and W. W. Nichols, eds.), pp. 135–150. Plenum, New York.
5. Baseman, J. B., Dallo, S. F., Tully, J. G., and Rose, D. L. (1988). Isolation and characterization of *Mycoplasma genitalium* strains from the human respiratory tract. *J. Clin. Microbiol.* **26,** 2266–2269.
6. Biberfeld, G. (1985). Infection sequelae and autoimmune reactions in *Mycoplasma pneumoniae* infection. *In* "The Mycoplasmas" (S. Razin and M. F. Barile, eds.), Vol. 4, pp. 293–311. Academic Press, Orlando, Florida.
7. Brown, M. B., Cassel, G. H., Taylor-Robinson, D., and Shepard, M. C. (1983). Measurement of antibody to *Ureaplasma urealyticum* by an enzyme-linked immunosorbent assay and detection of antibody response in patients with nongonococcal urethrtis. *J. Clin. Microbiol.* **17,** 288–295.
8. Brunner, H., and Chanock, R. M. (1973). A radioimmunoprecipitation test for detection of *Mycoplasma pneumoniae* antibody. *Proc. Soc. Exp. Biol. Med.* **143,** 97–105.
9. Brunner, H., Schaeg, W., Bruck, W., Schummer, U., Sziegolait, D., and Schiefer, H.-G. (1978). Determination of IgG, IgM and IgA antibodies to *Mycoplasma pneumoniae* by indirect staphylococcal radioimmunoassay. *Med. Microbiol. Immunol.* **165,** 29–41.
10. Buck, D. W., Kennett, R. H., and McGarrity, G. (1982). Monoclonal antibodies specific for cell culture mycoplasmas. *In Vitro* **18,** 377–281.

10a. Caliando, J. J., and McCormack, W. M. (1983). Recovery of mycoplasmas from blood and special tissues. *In* "Methods in Mycoplasmology" (J. G. Tully and S. Razin, eds.), Vol. 2, pp. 27–35. Academic Press, New York.
11. Cassel, G. H., Clyde, W. A., Jr., and Davis, J. K. (1985). Mycoplasmal respiratory infections. *In* "The Mycoplasmas" (S. Razin and M. F. Barile, eds.), Vol. 4, pp. 65–106. Academic Press, Orlando, Florida.
12. Cassel, G. H., and Cole, B. (1981). Mycoplasmas as agents of human disease. *N. Engl. J. Med.* **304,** 80–89.
13. Chanock, R. M., James, M. D., Fox, H. H., Turner, H. C., Mufson, M. A., and Hayflick, L. (1962). Growth of Eaton PPLO in broth and preparation of complement-fixing antigen. *Proc. Soc. Exp. Biol. Med.* **110,** 884–889.
14. Christiansen, G., Andersen, H., Bikelund, S., and Freundt, E. (1987). Genomic and gene variation in *Mycoplasma hominis* strains. *Isr. J. Med. Sci.* **23,** 595–602.
15. Cimolai, N., Bryan, L. E., To, M., and Woods, D. E. (1987). Immunological cross-reactivity of *Mycoplasma pneumoniae* membrane-associated protein antigen with *Mycoplasma genitalium* and *Acholeplasma laidlawii. J. Clin. Microbiol.* **25,** 2136–2139.
16. Cimolai, N., Schryvens, A., Bryan, L. E., and Woods, D. E. (1988). Culture-amplified immunological detection of *Mycoplasma pneumoniae* in clinical specimens. *Diagn. Microbiol. Infect. Dis.* **9,** 207–212.
17. Clyde, W. A., Jr. (1979). *Mycoplasma pneumoniae* infections of man. *In* "The Mycoplasmas" (J. G. Tully and R. F. Whitcomb, eds.), Vol. 2, pp. 275–306. Academic Press, New York.
18. Clyde, W. A., Jr. (1983). Recovery of mycoplasmas from the respiratory tract. *In* "Methods in Mycoplasmology" (J. G. Tully and S. Razin, eds.), Vol. 2, pp. 9–17. Academic Press, New York.
19. Couch, R. B. (1981). *Mycoplasma pneumoniae* (primary atypical pneumonia). *In* "Principles and Practice of Infectious Diseases" (G. L. Mandoll, R. G. Douglas, and J. E. Bennett, eds.), pp. 1484–1498. Wiley, New York.
20. Del Giudice, R. A., and Hopps, H. E. (1978). Microbiological methods and fluorescent microscopy for the direct demonstration of mycoplasma infection of cell cultures. *In* "Mycoplasma Infection of Cell Cultures" (G. J. McGarrity, D. Murphy, and W. W. Nichols, eds.), pp. 57–69. Plenum, New York.
21. Del Giudice, R. A., Tully, J. G., Rose, D. L., and Cole, R. M. (1985). *Mycoplasma pirum,* sp. nov., a terminal-structured mollicute from cell cultures. *Int. J. Syst. Bacteriol.* **35,** 285–291.
22. Dienes, L., and Edsall, G. (1937). 'Observations on the L-organism of Klieneberger. *Proc. Soc. Exp. Biol. Med.* **36,** 740–744.
23. Douma, R., Dular, R., Brodeur, B. R., and Kasatiya, S. S. (1987). Monoclonal antibodies specific to *Mycoplasma pneumoniae* and cell-contaminating mycoplasmas. *Can. Assoc. Clin. Microbiol. Infect. Dis., Abstr. Annu. Meet., 55th,* p. A-27.
24. Dowdle, W. R., and Robinson, R. Q. (1964). An indirect hemagglutination test for diagnosis of *Mycoplasma pneumoniae* infection. *Proc. Soc. Exp. Biol. Med.* **116,** 947–950.
25. Dular, R., Kajioka, R., and Kasatiya, S. (1988). Comparison of Gen-Probe commercial kit and culture technique for the diagnosis of *Mycoplasma pneumoniae* infection. *J. Clin. Microbiol.* **26,** 1068–1069.
26. Dular, R., Lambert, M., Bruce, B. W., Phipps, P. H., Rossier, E., and Kasatiya, S. (1987). *Mycoplasma pneumoniae* infections in a rural setting in Canada. *Can. Med. Assoc. J.* **136,** 1271–1273.
27. Dyvig, K., and Cassell, G. H. (1987). Transposition of Gram-positive transposon Tn916 in *Acholeplasma laidlawii* and *Mycoplasma pulmonis. Science* **235,** 1392–1394.

28. Embree, J. E., Kraus, V. W., Embil, J. A., and MacDonald, S. (1980). Placental infection with *Mycoplasma hominis* and *Ureaplasma urealyticum:* Clinical correlation. *Obstet. Gynecol.* **56,** 475–481.
29. Evans, A. S., and Brobst, M. (1961). Bronchitis, pneumonitis and pneumonia in University of Wisconsin students. *N. Engl. J. Med.* **265,** 401–405.
30. Feldman, H. A., and Suhs, R. H. (1966). Serologic epidemiologic studies with *Mycoplasma pneumoniae.* 1. Demonstration of an hemagglutinin and its inhibition antibody. *Am. J. Epidemiol.* **83,** 345–356.
31. Feldner, J., Göbel, U., and Bredt, W. (1982). *Mycoplasma pneumoniae* adhesin localized to tip structure by monoclonal antibody. *Nature (London)* **298,** 765–767.
32. Fleming, C., Hodge, D., Toma, S., Dular, R., and Kasatiya, S. (1987). Isolation of *Mycoplasma pneumoniae* from respiratory tract specimens in Ontario. *Can. Med. Assoc. J.* **137,** 48–50.
33. Foy, H. M., Kenny, G. E., Conney, M. F., and Allen, I. D. (1979). Long-term epidemiology of infections with *Mycoplasma pneumoniae. J. Infect. Dis.* **139,** 681–687.
34. Freundt, E. A. (1983). Culture media for classic mycoplasmas. *In* "Methods in Mycoplasmology" (S. Razin and J. G. Tully, eds.), Vol. 1, pp. 127–135. Academic Press, New York.
35. Gale, T. L., and Kenny, G. E. (1970). Complement–dependent killing of *Mycoplasma pneumoniae* by antibody, kinetics of the reaction. *J. Immunol.* **104,** 1175–1183.
36. Gallo, D., Dupuis, K. W., Schmidt, N. J., and Kenny, G. E. (1983). Broadly reactive immunofluorescence test for measurement of immunoglobulin M and G antibodies to *Ureaplasma urealyticum* in infant and adult sera. *J. Clin. Microbiol.* **17,** 614–618.
37. Göbel, U. B., Maas, R., Haun, G., Vinga-Martins, C., and Stanbridge, E. J. (1987). Synthetic oligonucleotide probes complementary to rRNA for group and species-specific detection of mycoplasmas. *Isr. J. Med. Sci.* **23,** 742–746.
38. Göbel, U. B., and Stanbridge, E. J. (1984). Cloned mycoplasma ribosomal RNA genes for the detection of mycoplasma contamination in tissue culture. *Science* **226,** 1211–1213.
39. Grayston, J. T., Alexander, E. R., Kenny, G. E., Clark, E. R., Fremont, J. C., and MacColl, W. A. (1965). *Mycoplasma pneumoniae* infections: Clinical and epidemiological studies. *JAMA, J. Am. Med. Assoc.* **191,** 369–374.
40. Harasawa, R., Mizusawa, H., and Koshimizu, K. (1986). A reliable and sensitive method for detecting mycoplasmas in cell cultures. *Microbiol. Immunol.* **30,** 919–921.
41. Hayflick, L. (1965). Tissue cultures and mycoplasmas. *Tex. Rep. Biol. Med.* **23,** 285–303.
42. Helbig, J. H., and Witzleb, W. (1984). Enzyme-linked immunosorbent assay (ELISA) zum antigennachweis von *Mycoplasma pneumoniae. Z. Gesamte Hyg. Ihre Grenzgeb.* **30,** 106–107.
43. Hers, J. F. P. (1963). Fluorescent antibody technique in respiratory viral diseases. *Am. Rev. Respir. Dis.* **88,** 316–333.
44. Hooton, T. M., Roberts, P. L., Stamm, W. E., Roberts, M. C., Holmes, K. K., and Kenny, G. E. (1988). Prevalence of *Mycoplasma genitalium* determined by DNA probe in men with urethritis. *Lancet* **1,** 266–268.
45. Horowitz, S. A., Duffy, L., Garrett, B., Stephens, J., Davis, J. K., and Cassel, G. H. (1986). Can group- and serovar-specific proteins be detected in *ureaplasma urealyticum? Pediatr. Infect. Dis.* **5,** 325–331.
46. Hu, P. C., Schaper, U., Collier, A. M., Clyde, W. A., Jr., Horikawa, M., Huang, Y.-S., and Barile, M. F. (1987). A *Mycoplasma genitalium* protein resembling the *Mycoplasma pneumoniae* attachment protein. *Infect. Immun.* **55,** 1126–1131.
47. Hyman, H. C., Yogev, D., and Razin, S. (1987). DNA probes for detection and

identification of *Mycoplasma pneumoniae* and *Mycoplasma genitalium*. *J. Clin. Microbiol.* **25,** 726–728.

48. Jacobs, E., Fuchte, K., and Bredt, W. (1986). A 168-kilodalton protein of *Mycoplasma pneumoniae* used as antigen in a Dot-Enzyme-linked immunosorbent assay. *Eur. J. Clin. Microbiol.* **5,** 435–440.
49. Johansson, K. E., and Bölske, G. (1987). Evaluation of a DNA probe for detection of mycoplasmas in cell cultures obtained from Swedish laboratories. *Isr. J. Med. Sci.* **23,** 541 (abstr.).
50. Jurmanova, K., and Mackatkova, M. (1986). Detection of mycoplasmas in cell cultures and biologicals. *Arch. Exp. Veterinaer med.* **40,** 136–141.
51. Kenny, G. E. (1985). Mycoplasmas. *In* "Manual of Clinical Microbiology" (E. H. Lennette, A. Balows, W. J. Hausler, Jr., and H. J. Shadomy, eds.), 4th ed., pp. 407–411. Am. Soc. Microbiol., Washington, D.C.
52. Kenny, G. E. (1986). Serology of mycoplasmal infections. *In* "Manual of Clinical Laboratory Immunology" (N. R. Rose, H. Friedman, and J. L. Fahey, eds.), 3rd ed., pp. 440–445. Am. Soc. Microbiol., Washington, D.C.
53. Kenny, G. E., and Grayston, J. T. (1965). Eaton PPLO (*Mycoplasma pneumoniae*) complement-fixing antigen extraction with organic solvents. *J. Immunol.* **95,** 19–25.
54. Kingsbury, D. T. (1985). Rapid detection of mycoplasmas with DNA probes. *In* "Rapid Detection and Identification of Infectious Agents" (D. T. Kingsbury and S. Falkow, eds.), pp. 219–233. Academic Press, Orlando, Florida.
55. Kingston, J. R., Chanock, R. M., Mufson, M. A., Hellman, L. P., James, W. D., Fox, H. H., Manko, M. A., and Boyers, J. (1961). Eaton agent pneumonia. *JAMA J. Am. Med. Assoc.* **176,** 118–123.
56. Kist, M., Jacobs, E., and Bredt, W. (1982). Release of *Mycoplasma pneumoniae* substances after phagocytosis by guinea pig alveolar macrophages. *Infect. Immun.* **36,** 357–362.
57. Kontra, J., Smith, D., Karam, T., and McGregor, R. R. (1988). Preliminary evaluation of the Gen-Probe system for detection of *Mycoplasma pneumoniae*. *Am. Soc. Microbiol., Abstr. Annu. Meet., 88th,* p. 340.
58. Kotani, H., Huang, H., and McGarrity, G. J. (1987). Identification and isolation of mycoplasmas by immunobinding. *Isr. J. Med. Sci.* **23,** 752–758.
59. Kotani, H., and McGarrity, G. J. (1986). Identification of mycoplasma colonies by immunobinding. *J. Clin. Microbiol.* **23,** 783–785.
60. Krieg, N. R., and Holt, J. G., eds. (1984). "Bergey's Manual of Systematic Bacteriology," Vol. 1, pp. 740–793. Williams & Wilkins, Baltimore, Maryland.
61. Leinikki, P. O., Panzar, P., and Tykka, H. (1978). Immunoglobulin M antibody response against *Mycoplasma pneumoniae* liquid antigen in patients with acute pancreatitis. *J. Clin. Microbiol.* **8,** 113–118.
62. Levine, D. P., and Lerner, A. M. (1978). The clinical spectrum of *Mycoplasma pneumoniae* infections. *Med. Clin. North Am.* **62,** 961–978.
63. Levine, E. M., and Becker, B. G. (1978). Biochemical methods for detecting mycoplasma. *In* "Mycoplasma Infection of Cell Cultures" (G. J. McGarrity, D. G. Murphy, and W. W. Nichols, eds.), pp. 87–104. Plenum, New York.
64. Liepman, M.-F., Wattre, P., Dewilde, A., Papierok, G., and Delecour, M. (1988). Detection of antibodies to *Ureaplasma urealyticum* in pregnant women by enzyme-linked immunosorbent assay using membrane antigen and investigation of the significance of the antibodies. *J. Clin. Microbiol.* **26,** 2157–2160.
65. Lin, J.-S. (1985). Human mycoplasmal infections: Serologic observations. *Rev. Infect. Dis.* **7,** 216–231.

66. Lin, J.-S., Alpert, S., and Radnay, K. M. (1975). Combined type-specific antisera in the identification of *Mycoplasma hominis*. *J. Infect. Dis.* **131,** 727–730.
67. Lin, J.-S., and Kass, E. H. (1975a). Complement-dependent and complement-independent interactions between *Mycoplasma hominis* antibodies in vitro. *J. Med. Microbiol.* **8,** 397–404.
68. Lin, J.-S., and Kass, E. H. (1976b). Immune inactivation of T-strain mycoplasmas. *J. Infect. Dis.* **122,** 93–95.
69. Lind, K. (1968). An indirect hemagglutination test for serum antibodies against *Mycoplasma pneumoniae* using formalinised, tanned sheep erythrocytes. *Acta Pathol. Microbiol. Scand.* **73,** 459–472.
70. Lind, K., and Bentzon, M. W. (1987). Change in the epidemiological pattern of *Mycoplasma pneumoniae* infection in Denmark from 1958 to 1985. *Isr. J. Med. Sci.* **23,** 523 (abstr.).
71. Lind, K., Lindhart, B. O., Schutten, H. J., Blom, J., and Christiansen, C. (1984). Serological cross reaction between *Mycoplasma genitalium* and *Mycoplasma pneumoniae*. *J. Clin. Microbiol.* **20,** 1036–1043.
72. Madsen, R. D., Saeed, F. A., Gray, O., Fendly, B., and Coates, S. R. (1986). Species-specific monoclonal antibody to a 43,000-molecular-weight membrane protein of *Mycoplasma pneumoniae*. *J. Clin. Microbiol.* **24,** 680–683.
73. Madsen, R. D., Weinger, L. B., McMillan, J. A., Saeed, F. A., North, J. A., and Coates, S. R. (1988). Direct detection of *Mycoplasma pneumoniae* antigen in clinical specimens by a monoclonal antibody immunoblot assay. *Am. J. Clin. Pathol.* **89,** 95–99.
74. Mårdh, P.-A., and Weström, L. (1970). Antibodies against pleuropneumonia-like organisms in patients with salpingitis. *Br. J. Vener. Dis.* **46,** 390–397.
75. McCormack, W. M., and Taylor-Robinson, D. (1984). The genital mycoplasma. *In* "Sexually Transmitted Diseases" (K. K. Holmes, P. A. Mardh, P. F. Sparling, and P. J. Wiesner, eds.), pp. 408–421. McGraw-Hill, New York.
76. McGarrity, G. J. (1982). Detection of mycoplasma infection of cell cultures. *Adv. Cell Cult.* **2,** 99–131.
77. McGarrity, G. J. (1988). "International Research Program in Comparative Mycoplasmology: A Report to Working Group on Cell Culture Mycoplasmas." Coriell Institute for Medical Research, Camden, New Jersey.
78. McGarrity, G. J., and Carson, D. A. (1982). Adenosine phosphorylase-mediated nucleoside toxicity. *Exp. Cell Res.* **139,** 199–206.
79. McGarrity, G. J., and Kotani, H. (1985). Cell culture mycoplasmas. *In* "The Mycoplasmas" (S. Razin and M. F. Barile, eds.), Vol. 4, pp. 353–390. Academic Press, Orlando, Florida.
80. McGarrity, G. J., and Kotani, H. (1986). Detection of cell culture mycoplasmas by a genetic probe. *Exp. Cell Res.* **163,** 273–278.
81. McGarrity, G. J., Sarama, J., and Vanaman, V. (1979). Factors influencing microbiological assays of cell culture mycoplasmas. *In Vitro* **15,** 73–81.
82. McNaughton, R. D., Robertson, J. A., Ratzlaff, V. J., and Molberg, C. R. (1983). *Mycoplasma hominis* infection of the central nervous system in a neonate. *Can. Med. Assoc. J.* **129,** 353–354.
83. Miettinen, A., Paavonen, J., Jansson, E., and Leinikki, P. (1983). Enzyme immunoassay for serum antibody to *Mycoplasma hominis* in women with acute pelvic inflammatory disease. *Sex. Transm. Dis.* **10,** 289–293.
84. Moller, B. R. (1981). Comparison of serological tests for detection of *Mycoplasma hominis* antibodies in female grivet monkeys with experimentally induced salpingitis. *Acta Pathol. Microbiol. Scand., Sect. B* **89B,** 7–11.

85. Moller, B. R., Taylor-Robinson, D., and Furr, A. M. (1984). Serological evidence implicating *Mycoplasma genitalium* in pelvic inflammatory disease. *Lancet* **1,** 1102–1103.
86. Murray, H. W., Masur, H., Senterfit, L. B., and Roberts, R. B. (1975). The protean infections in adults. *Am. J. Med.* **58,** 229–242.
87. Naessens, A., Foulon, W., Cammu, H., Goossens, A., and Lauwers, S. (1987). Epidemiology and pathogenesis of *Ureaplasma urealyticum* in spontaneous abortion and early preterm labor. *Acta Obstet. Gynecol. Scand.* **66,** 513–516.
88. Nicklas, W., and Mauter, P. (1988). Experience with the routine monitoring of cell cultures for mycoplasma contamination. *Zentralbl. Bakteriol., Mikrobiol. Hyg., Abt. 1, Orig. A* **267,** 510–518.
89. Norcord, E., and Roux, E. (1898). Le microbe de la péri-pneumoniae. *Ann. Inst. Pasteur, Paris* **12,** 240–262.
90. Oderda, G., and Kraut, J. R. (1980). Rising antibody titer to *Mycoplasma pneumoniae* in acute pancreatitis. *Pediatrics* **66,** 305.
91. Payan, D. G., and Madoff, S. (1981). Infection of a brain abscess by *Mycoplasma hominis*. *J. Clin. Microbiol.* **14,** 571–573.
92. Peterson, O. L., Ham, T. H., and Finland, M. (1943). Cold agglutinins (autohemagglutinins) in primary atypical pneumonia. *Science* **97,** 167.
93. Phillips, D. (1978). Detection of mycoplasma contamination of cell cultures by electron microscopy. *In* "Mycoplasma Infection of Cell Cultures" (G. J. McGarrity, D. Murphy, and W. W. Nichols, eds.), pp. 105–118. Plenum, New York.
94. Polak-Vogelzang, A. A., Burgman, J., and Reijgens, R. (1987). Comparison of two methods for detection of mollicutes (Mycoplasmatales and Acholeplasmatales) in cell cultures in The Netherlands. *Antonie van Leeuwenhoek* **53,** 107–118.
95. Pönkä, A. (1979). The occurrence and clinical picture of serologically verified *Mycoplasma pneumoniae* infections with emphasis on central nervous system, cardiac and joint manifestations. *Ann. Clin. Res.* **24,** 1–60.
96. Purcell, R. H., Chanock, R. M., and Taylor-Robinson, D. (1969). Serology of the mycoplasmas of man. *In* "The Mycoplasmatales and the L-phase Bacteria" (L. Hayflick, ed.), pp. 221–264. Appleton-Century-Crofts, New York.
97. Quinn, P. A., Butany, J., Chipman, M., Taylor, J., and Hannah, W. (1985). A prospective study of microbial infection in stillbirths and early neonatal death. *Am. J. Obstet. Gynecol.* **151,** 238–249.
98. Radka, S. F., Hester, D. M., Polak-Vogelzang, A. A., and Bolhuis, R. L. H. (1984). Detection of mycoplasma contamination in lymphoblastoid cell lines by monoclonal antibodies. *Hum. Immunol.* **9,** 111–116.
99. Razin, S. (1985). Molecular biology and genetics of mycoplasmas (mollecutes). *Microbiol. Rev.* **49,** 419–455.
100. Razin, S., Gross, M., Wormser, M., Pollack, Y., and Glaser, G. (1984). Detection of mycoplasmas infecting cell cultures by DNA hybridization. *In Vitro* **20,** 404–408.
101. Razin, S., Harasawa, R., and Barile, M. F. (1983). Cleavage patterns of the mycoplasma chromosome, obtained by using restriction endonucleases, as indicator of genetic relatedness among strains. *Int. J. Syst. Bacteriol.* **33,** 201–206.
102. Razin, S., Tully, J. G., Rose, D. L., and Barile, M. F. (1983). DNA cleavage patterns as indicators of genotypic heterogeneity among strains of *Acholeplasma* and *Mycoplasma* species. *J. Gen Microbiol.* **129,** 1935–1944.
103. Razin, S., and Yogev, D. (1986). Genetic relatedness among *Ureaplasma urealyticum* serotypes (serovars). *Pediatr. Infect. Dis.* **5,** 300–304.
104. Risi, G. F., Jr., Martin, D. H., Silberman, J. A., and Cohen, J. C. (1988). A DNA probe

for detecting *Mycoplasma genitalium* in clinical specimens. *Mol. Cell. Probes* **2,** 327–335.
105. Roberts, M. C., Hooton, M., Stamm, W., Holmes, K. K., and Kenny, G. E. (1987). DNA probes for the detection of mycoplasmas in genital specimens. *Isr. J. Med. Sci.* **23,** 618–620.
106. Robertson, J. A., and Stemke, G. W. (1982). Expanded serotyping scheme for *Ureaplasma urealyticum* strains isolated from humans. *J. Clin. Microbiol.* **15,** 873–878.
107. Robinson, L. B., Wichelhausen, R. K., and Roizman, B. (1956). Contamination of human cell cultures by pleuropneumonialike organisms. *Science* **124,** 1147–1148.
107a. Saada, A.-B., Deutsch, V., and Kahane, I. (1988). Interaction of monoclonal antibody with the urease of *Ureaplasma urealyticum. FEMS Microbiol.* **55,** 187–190.
108. Sasaki, T., Bonissol, C., Stoiljkovic, B., and Ito, K. (1987). Demonstration of cross-reactive antibodies to mycoplasmas in human sera by ELISA and immunoblotting. *Micobiol. Immunol.* **31,** 639–648.
109. Schneider, E. L., and Stanbridge, E. J. (1976). Comparison of methods for the detection of mycoplasmal contamination of cell cultures: A review. *In Vitro* **11,** 20–34.
110. Schneider, E. L., Stanbridge, E. J., and Epstein, C. J. (1974). Incorporation of ^{3}H-uridine and ^{3}H-uracil into RNA: A simple technique for the detection of mycoplasma contamination of cultured cells. *Exp. Cell Res.* **84,** 311–318.
111. Senterfit, L. B., and Jensen, K. E. (1966). Antimetabolic antibodies to *Mycoplasma pneumoniae* measured by tetrazolium reduction inhibition. *Proc. Soc. Exp. Biol. Med.* **122,** 786–790.
112. Sethi, K. K. (1972). On the incidence of mycoplasma contamination in cell cultures. *Zentralbl. Bakteriol., Mikrobiol. Hyg., Abt. 1, Orig. A* **219,** 550–554.
113. Shaw, D. R., and Lim, I. (1988). Extragenital *Mycoplasma hominis* infection: A report of two cases. *Med. J. Aust.* **148,** 144–145.
114. Shepard, M. C. (1954). The recovery of pleuropneumonialike organisms from Negro men with and without nongonococcal urethritis. *Am. J. Syph., Gonorrhea, Vener. Dis.* **38,** 113–124.
115. Shepard, M. C. (1983). Culture media for ureaplasmas. *In* "Methods in Mycoplasmology" (S. Razin and J. G. Tully, eds.), Vol. 1, pp. 137–146. Academic Press, New York.
116. Shepard, M. C., Lunceford, C. D., Ford, D. K., Purcell, R. H., Taylor-Robinson, D., Razin, S., and Black, F. T. (1974). *Ureaplasma urealyticum* gen. nov., sp. nov.: Proposed nomenclature for the human T (T-strain) mycoplasmas. *Int. J. Syst. Bacteriol.* **24,** 160–171.
117. Simberkoff, M. S., and Toharsky, B. (1976). Mycoplasmemia in adult patients. *JAMA J. Am. Med. Assoc.* **236,** 2522–2524.
118. Stanbridge, E. J., and Doersen, C.-J. (1978). Some effects that mycoplasmas have upon their infected host. *In* "Mycoplasma Infection of Cell Cultures" (G. J. McGarrity, D. G. Murphy, and W. W. Nichols, eds.), pp. 119–134. Plenum, New York.
119. Stanbridge, E. J., and Katayama, C. (1978). Principles of morphological and biochemical methods for the detection of mycoplasma contamination of cell cultures. *In* "Mycoplasma Infection of Cell Cultures" (G. J. McGarrity, D. G. Murphy, and W. W. Nichols, eds.), pp. 57–69. Plenum, New York.
120. Stanbridge, E. J., and Reff, M. E. (1979). The molecular biology of mycoplasmas. *In* "The Mycoplasmas" (M. F. Barile and S. Razin, eds.), Vol. 1, pp. 157–185. Academic Press, New York.
121. Steinberg, P., White, R. J., Flud, S. L., Gutekunst, R. R., Chanock, R. M., and

Senterfit, L. B. (1969). Ecology of *Mycoplasma pneumoniae* infections in marine recruits at Paris Island, South Carolina. *Am. J. Epidemiol.* **89,** 62–73.

122. Stemke, G. W., and Robertson, J. A. (1981). Modified colony indirect epifluorescence test for serotyping *ureaplasma urealyticum* and an adaptation to detect common antigenic specificity. *J. Clin. Microbiol.* **14,** 582–584.
123. Studzinski, G. P., Gierthy, J. R., and Cholon, J. J. (1973). An autoradiographic screening test for mycoplasmal contamination of mammalian cell cultures. *In Vitro* **8,** 466–472.
124. Sugino, W. M., Wek, R. C., and Kingsburry, D. T. (1980). Partial nucleotide sequence similarity within species of *Mycoplasma* and *Acholeplasma. J. Gen Microbiol.* **121,** 333–338.
125. Taylor, M. A., Wise, K. S., and McIntosh, M. A. (1984). Species-specific detection of *Mycoplasma hyorhinis* using DNA probes. *Isr. J. Med. Sci.* **20,** 778–780.
126. Taylor, M. A., Wise, K. S., and McIntosh, M. A. (1985). Selective detection of *Mycoplasma hyorhinis* using cloned genomic DNA fragments. *Infect. Immun.* **47,** 827–830.
127. Taylor, P. (1979). Evaluation of an indirect haemagglutination kit for the rapid serological diagnosis of *Mycoplasma pneumoniae* infections. *J. Clin. Pathol.* **32,** 280–283.

127a. Taylor-Robinson, D. (1983). Recovery of mycoplasmas from the genitourinary tract. *In* "Methods in Mycoplasmology" (J. G. Tully and S. Razin, eds.), Vol. 2, pp. 19–26. Academic Press, New York.

128. Taylor-Robinson, D. (1984). Mycoplasma infection of the human urogenital tract with particular reference to nongonococcal urethritis. *Ann. Microbiol.* (Paris) **135A,** 129–134.
129. Taylor-Robinson, D. (1985). Mycoplasmal and mixed infections of the human male urogenital tract and their possible complications. *In* "The Mycoplasmas" (S. Razin and M. F. Barile, eds.), Vol. 4, pp. 27–63. Academic Press, Orlando, Florida.
130. Taylor-Robinson, D., Csonka, G. W., and Prentice, M. J. (1977). Human intraurethral inoculation of ureaplasmas. *Q. J. Med.* [N. S.] **46,** 309–326.
131. Taylor-Robinson, D., Furr, P. M., and Webster, A. D. B. (1986). *Ureaplasma urealyticum* in the immunocompromised host. *Pediatr. Infect. Dis.* **5,** 236–238.
132. Taylor-Robinson, D., and McCormack, W. M. (1979). Mycoplasmas in human genitourinary infections. *In* "The Mycoplasmas" (J. G. Tully and R. F. Whitcomb, eds.), Vol. 2, pp. 308–366. Academic Press, New York.
133. Taylor-Robinson, D., and McCormack, W. M. (1980). The genital mycoplasmas. *N. Engl. J. Med.* **302,** 1003–1010, 1063–1067.
134. Taylor-Robinson, D., and Munday, P. E. (1988). Mycoplasmal infection of the female genital tract and its complications. *In* "Genital Tract Infection in Women" (M. J. Hare, ed.), pp. 228–247. Churchill-Livingstone, Edinburgh and London.
135. Taylor-Robinson, D., Purcell, R. H., Wong, D. C., and Chanock, R. M. (1966). A color test for the measurement of antibody to certain mycoplasma species based upon the inhibition of acid production. *J. Hyg.* **64,** 91–104.
136. Taylor-Robinson, D., Tully, J. G., Furr, P. M., Cole, R. M., Rose, D. L., and Hanna, N. F. (1981). Urogenital mycoplasma infection of man: A review with observations on a recently discovered mycoplasma. *Isr. J. Med. Sci.* **17,** 524–530.
137. Taylor-Robinson, D., Tully, J. G., and Barile, M. F. (1985). Urethral infection in male chimpanzees produced experimentally by *Mycoplasma genitalium. Br. J. Exp. Pathol.* **66,** 95–101.
138. Tilton, R. C., Dias, F., Kidd, H., and Ryan, R. W. (1988). DNA probe versus culture for detection of *Mycoplasma pneumoniae* in clinical specimens. *Diagn. Microbiol. Infect. Dis.* **10,** 109–112.

139. Todaro, G. J., Aaronson, S. A., and Rands, E. (1970). Rapid detection of mycoplasma-infected cell cultures. *Exp. Cell Res.* **65,** 256–257.
140. Tuazon, C. U., and Murray, H. W. (1983). Atypical pneumonias. *In* "Respiratory Infections: Diagnosis and Management" (J. E. Pennington, ed.), pp. 251–267. Raven Press, New York.
141. Tully, J. G. (1981). Laboratory diagnosis of *Mycoplasma pneumoniae* infections. *Isr. J. Med. Sci.* **17,** 644–647.
142. Tully, J. G., and Taylor-Robinson, D. (1986). Taxonomy and host distribution of ureaplasmas. *Pediatr. Infect. Dis.* **5,** S292–S295.
143. Tully, J. G., Taylor-Robinson, D., Cole, R. M., and Rose, D. L. (1981). A newly discovered mycoplasma in the human urogenital tract. *Lancet* **1,** 1288–1291.
144. Tully, J. G., Taylor-Robinson, D., Rose, D. L., Furr, P. M., Graham, C. E., and Barile, M. F. (1986). Urogenital challenge of primate species with *Mycoplasma genitalium* and characteristics of the infection induced in chimpanzees. *J. Infect. Dis.* **153,** 1046–1054.
145. Turunen, H., Leinikki, P., and Jansson, E. (1982). Serological characterization of *Ureaplasma urealyticum* by enzyme-linked immunosorbent assay (ELISA). *J. Clin. Pathol.* **35,** 439–443.
146. Wahl, G. M., Stern, M., and Stork, G. M. (1979). Efficient transfer of large DNA fragments from agarose gel to diazobenzyloxymethyl paper and rapid hybridization by using dextran sulfate. *Proc. Natl. Acad. Sci. U.S.A.* **76,** 3683–3687.
147. Waites, K. B., Crouse, D. T., Nelson, K. G., Rudd, P. T., Canupp, K. C., Ramsey, C., and Cassell, G. H. (1988). Chronic *Ureaplasma urealyticum* and *Mycoplasma hominis* infections of central nervous system in preterm infants. *Lancet* **1,** 17–21.
148. Wiernik, A., Jarstrand, C., and Tunevall, G. (1978). The value of immunoelectroosmophoresis (IEOP) for etiological diagnosis of acute respiratory tract infections due to pneumococci and *Mycoplasma pneumoniae*. *Scand. J. Infect. Dis.* **10,** 173–176.
149. Wiley, C. A., and Quin, P. A. (1984). Enzyme-linked immunosorbent assay for detection of specific antibodies to *Ureaplasma urealyticum* serotypes. *J. Clin. Microbiol.* **19,** 421–426.
150. Yogev, D., and Razin, S. (1986). Common deoxyribonucleic acid sequences in *Mycoplasma genitalium* and *Mycoplasma pneumoniae* genomes. *Int. J. Syst. Bacteriol.* **36,** 426–436.

16

Detection of TEM β-Lactamase Genes Using DNA Probes

KEVIN J. TOWNER
Department of Microbiology and PHLS Laboratory
University Hospital
Nottingham, England

I. INTRODUCTION

Numerous reports describing the use of DNA probes for the clinical diagnosis of infectious diseases have now appeared in the scientific literature and these have been comprehensively reviewed in this volume and elsewhere (54). Such probe-based tests permit the rapid screening of large numbers of samples and have the major potential advantage that they can be used to detect specific microorganisms without primary isolation and growth. Similar tests can also be used to characterize antibiotic-resistance determinants present in bacteria and have become increasingly widely used for epidemiological studies (53) where, under certain circumstances, they may offer considerable advantages compared to the more conventional biochemical and/or physical tests used to characterize resistance gene products. The fact that probe-based tests can be used to detect specific genes without primary isolation and growth means that this technology has the potential for enabling the antimicrobial susceptibility of an infecting organism to be rapidly determined directly from crude patient specimens, with obvious concomitant clinical advantages.

On a world-wide basis the most important compounds used to treat infections caused by gram-negative bacteria are the β-lactam antibiotics, comprising the numerous different penicillin and cephalosporin derivatives. Clinically significant resistance to these compounds in gram-negative bacteria largely results from the production of β-lactamase enzymes (often coded for by genes on plasmids or transposons), which cleave the β-lactam ring of penicillins and cephalosporins to produce compounds lacking antibacterial activity (51). Of the many β-lactamases classified to date, the TEM class enzymes have consistently proved to be the most prevalent types coded for by transmissible plasmids (31,34,41,45,46). This chapter describes the different probes that have been developed to test for the presence of TEM β-lactamase genes and the way in which they can be used directly to detect resistant bacteria in clinical specimens.

An important consideration in the design of a hybridization test is the choice of a suitable label for the DNA probe. Most previous epidemiological applications have used radioactive labels directly incorporated into the probe, normally by the process of nick translation (40). Many routine microbiology laboratories lack the facilities and expertise to handle highly radioactive probe labels, and attempts have consequently been made to develop nonradioactive methods of labeling probes. The production of a colored end-product by some nonradioactive detection systems potentially allows the results from large numbers of clinical samples to be automatically assessed in a quantitative manner. This chapter therefore also includes details of a possible approach to an automated assessment of hybridization results using a computer-controlled image analysis system.

II. BACKGROUND

A. Clinical Importance of Plasmid-Encoded β-Lactamases

Since the late 1950s the development of numerous semisynthetic penicillin and cephalosporin derivatives has resulted in the introduction of a large number of clinically effective β-lactam antibiotics that inhibit enzymes involved in bacterial cell wall biosynthesis and cell division. High-level clinically significant resistance to β-lactam antibiotics in gram-negative bacteria, particularly bacterial species involved in hospital-acquired infection, has historically been due predominantly to production by bacteria of β-lactamase enzymes that inactivate these antibiotics (31). In gram-negative bacteria these enzymes are normally encoded by plasmid-carried genes (31). More than 30 such plasmid-encoded enzymes and variants have now been described (24,31,32,48).

In an attempt to circumvent this resistance problem, a number of broad-spectrum "β-lactamase-stable" cephalosporins were introduced during the 1980s. These antibiotics were considered to be of particular use for treating life-threatening infections caused by gram-negative bacteria such as the Enterobacteriaceae and *Pseudomonas aeruginosa*. Initially it seemed that significant resistance to these valuable new antibiotics was only encountered in bacterial genera that possessed inducible chromosomal β-lactamases (25); however, a number of outbreaks of infection due to the emergence of conjugative plasmids coding for β-lactamases active against these new antibiotics were soon also reported (4,20,38,47).

The rapid spread of β-lactam resistance between different species of gram-negative bacteria has already been shown to be largely due to the interchange of conjugative plasmids (39). Further dissemination can then be helped by the process of transposition (19), which is known to be responsible for the spread of certain genes coding for β-lactamase production between different groups of plasmids (23,39,42). The emergence of transferable resistance to the newer cephalosporin antibiotics therefore exemplifies the major risk of spread of such resistance to otherwise sensitive strains, with a consequent reduction in the effectiveness of these valuable antibiotics.

Hospital-acquired infection is a leading cause of mortality in hospitalized patients. Attempts to reduce the incidence of such infection are crucially dependent on detailed epidemiological investigation of clinical isolates, including studies of the epidemiology of their antibiotic resistance genes. For this reason it is important to have a reliable and rapid method of identifying β-lactamases that is capable of dealing economically with the large number of samples generated by detailed epidemiological investigations.

B. Conventional Methods for Classifying β-Lactamases

β-Lactamases are widespread in the microbial world and, with few exceptions, β-lactamase activity has been detected in most strains of bacteria tested (5). The various classification schemes and terminology of β-lactamases have been reviewed previously (50), but it is likely that classification schemes will continue to be revised as the molecular characteristics of these enzymes are gradually elucidated.

As stated in the previous section, most clinically relevant β-lactamases encountered in gram-negative bacteria are plasmid-encoded. The variety of methods used to identify these enzymes include substrate profiles, inhibition studies, isoelectric points, molecular mass determinations, production of antisera, and relative substrate affinity values (17,28). The most commonly used method involves the analytical isoelectric focusing of cell extracts in parallel with protein standards and β-lactamase reference enzymes (29–31,46). This technique can be performed either on polyacrylamide gels (30) or cellulose acetate membranes (11). Unfortunately the technique, while it has been extremely useful and reliable over the years for the identification of plasmid-encoded β-lactamases, does not permit the rapid analysis of the large number of clinical isolates required for epidemiological studies, primarily because a protein extract must be prepared separately for each β-lactamase-producing strain studied. In addition, there are now an increasing number of β-lactamases that have very similar isoelectric points (31) and that consequently require a combination of the other biochemical tests mentioned earlier for full identification. This is even more inappropriate for routine clinical use, and it increasingly seems that DNA probes may offer the most promise for studying the epidemiology and distribution of known β-lactamase genes, in addition to monitoring the emergence of "new" β-lactamase types.

C. Distribution and Types of Plasmid-Encoded β-Lactamases

The fact that genes coding for β-lactamase production occur on plasmids and transposons means that different β-lactamase types originally found in only one genus of bacteria will eventually appear in other groups. This process is accelerated by the widespread use of β-lactam antibiotics, which causes the incidence of resistance to rise locally and then spread worldwide. Table I illustrates the types and prevalence of different β-lactamases, as identified by isoelectric focusing, found in clinical isolates

TABLE I

Major Groups of Plasmid-Encoded β-Lactamases Identified in Clinical Isolates of Ampicillin-Resistant *Escherichia coli* from Different Parts of the World

		Percentage of strains with β-lactamase type[a]					
Region	Reference	TEM-1 only	TEM-2 only	TEM-1/TEM-2 + others	OXA Group only	PSE Group only	Others
Brazil	34	72	NT[b]	6	10	2	9
England	46	76	4	2	5	0	13
France	34	76	NT	0	10	0	0
Indonesia	31	82	NT	0	9	9	0
South Africa	31	49	NT	6	11	0	0
Spain	41	91	0	3	2	0	3
Thailand	31	93	NT	7	0	0	0
United States	34	79	NT	0	3	0	0

[a]Percentages of strains encoding a chromosomal enzyme only are not shown.
[b]NT, Not tested.

of ampicillin-resistant *Escherichia coli* obtained from different parts of the world. In each of these surveys of resistant *E. coli* strains, the TEM-1 β-lactamase was by far the commonest, with its close relative TEM-2 also sometimes encountered. In other species this may not always be the case; indeed other surveys have been carried out that indicate that the various plasmid-determined β-lactamases have different distributions according to bacterial genus (16,28,33,34,45). It should also be expected that the prevalence of different β-lactamases will vary in time and place when particular strains became epidemic (31). This may often be in response to changes in the use of antibiotics.

As mentioned earlier, the introduction of "β-lactamase-stable" "third generation" cephalosporin antibiotics resulted in the recognition of new plasmid-encoded β-lactamases, now termed TEM-3 to TEM-7 (48), which are active against these new antibiotics. These novel enzymes seem to be structural variants of the TEM-1 β-lactamase and it has been shown that spontaneous mutants of TEM enzymes can be obtained *in vitro* that have increased activity against third-generation cephalosporins (48). It therefore seems that TEM-1 and the other closely related members of the TEM family will together continue to compose the most clinically important and prevalent group of plasmid-encoded β-lactamases found in gram-negative bacteria. The remainder of this chapter is therefore devoted to describing the DNA probes that have been constructed for distinguishing the TEM group from other β-lactamases, the usefulness of these probes for distinguishing β-lactamase variants within the TEM family, and the possible ways in which these probes can be used with nonradioactive probe-labeling and detection systems to examine directly crude clinical samples obtained from patients. DNA probes have also been described that recognize β-lactamases belonging to groups outside of the TEM family (2,3,12,24,35,37), but these will not be considered further in this chapter.

III. RESULTS AND DISCUSSION

A. DNA Probes for TEM β-Lactamase Genes and Their Specificity

1. Probes Constructed from Plasmids

Attempts to construct probes for TEM β-lactamase genes from preexisting plasmids have largely utilized the well-characterized 4.36-kb chimeric plasmid pBR322, which carries the structural gene for TEM-1 β-lactamase. The complete nucleotide sequence of this plasmid has been determined (49), and the structure of the TEM-1-encoding region indicating relevant restriction sites is shown in Fig. 1. Purified pBR322 DNA is readily available from a large number of commercial suppliers and it therefore serves as a convenient starting point for probe construction. The fragments of pBR322 that have been successfully utilized as probes for TEM β-lactamase genes are described in the following subsections.

a. One-Kilobase *Eco*RI–*Hin*fI Fragment. The first successful attempt to develop a gene probe for TEM β-lactamase genes (9) involved sequential digestion of pBR322 with *Eco*RI and *Hin*fI. This yielded 11 fragments, the largest of which was 1 kb in size and formed a potential probe (probe A in Fig. 1) that encompassed most of the β-lactamase structural gene; however, the probe also contained ~200 bp extraneous to the structural gene. The probe was initially tested (9) against seven other β-lactamase classes (Table II) and was found to hybridize only with strains producing the TEM-1, TEM-2, or OXA-2 β-lactamases. The cross-hybridization between the genes encoding TEM-1 and TEM-2 was to be expected as these genes differ in sequence by only a single base pair (1,49), corresponding to a single amino acid difference at position 14 in the mature proteins. The cross-hybridization observed with OXA-2 was, however, surprising as the nucleotide sequence of the OXA-2 gene (10) indicates that there is no homology between TEM-1 and OXA-2.

The range of β-lactamase classes tested against this probe has subsequently been extended (Table II) in two further studies (6,18). These

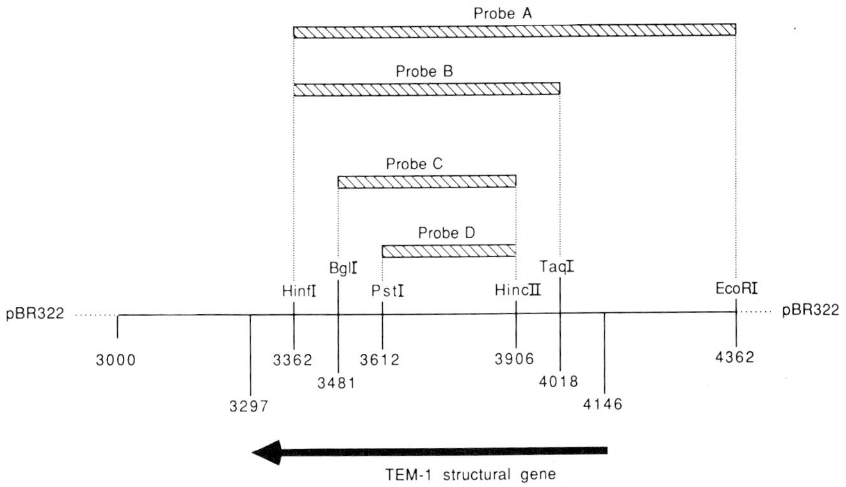

Fig. 1. Structure of the TEM-1-encoding region of pBR322 showing relevant restriction sites and extent of DNA fragments forming probes.

TABLE II

Hybridization Results Obtained in Different Studies Using TEM Probes Derived from pBR322[a]

β-Lactamase type[b]	Probe A[c] (9)	(18)	(6)	Probe B[d] (36)	Probe C[e] (24)	Probe D[f] (48)
TEM-1	+	+	+	+	+	+
TEM-2	+	+	+	+	+	
TEM-3						+
TEM-4						+
TEM-5						+
TEM-6						+
TLE-1			−		+	
OXA-1	−		−	−	−	
OXA-2	+	−	−	−	−	
OXA-3	−	−	−	−	−	
OXA-4		−	−		−	
OXA-5			−		−	
OXA-6			−		−	
OXA-7			−		−	
PSE-1		−	−	−	−	
PSE-2		−	−	−	−	
PSE-3	−	−	−	−	−	
PSE-4	−	−	−	−	−	
HMS-1	−	−	−	−	−	
SHV-1	−	−	−	−	−	
SAR-1			−			
BRO-1			−			
ROB-1			−		−	
LCR-1			−		−	
AER-1			−		−	
CARB-3					−	
CARB-4			−		−	
NPS-1			−			
N-29			−			
N-3			−			
CEP-1					−	
CEP-2			−			

[a]Numbers in parentheses are references. Blank combinations were not tested in the studies listed.

[b]Additional β-lactamases not tested in any of the listed studies include TEM-7, TLE-2, LXA-1, OHIO-1, and SHV-2.

[c]One kilobase *Eco*RI–*Hin*fI fragment.

[d]656-Base pair *Hin*fI–*Taq*I fragment.

[e]424-Base pair *Hin*cII–*Bgl*I fragment.

[f]298-Base pair *Hin*cII-*Pst*I fragment.

additional studies confirmed the cross-hybridization observed between the TEM-1 and TEM-2 genes, but, using different OXA-2 encoding plasmids from that originally tested (9), were unable to demonstrate cross-hybridization with OXA-2. The original reaction may therefore have been due to either a silent copy of the TEM-1 gene on the plasmid tested or a nonspecific hybridization reaction involving the probe region external to the structural gene.

Two of these studies (9,18) used probe DNA labeled with ^{32}P by the process of nick translation, followed by conventional autoradiography as a subsequent means of detecting hybridized sequences. In contrast, the third study (6) demonstrated that probe DNA labeled with biotin-11-dUTP could be successfully used in spot hybridization tests, provided that the filters used as a support were incubated at 37°C for 1 hr in a solution of proteinase K (0.5 mg/ml) prior to hybridization. Subsequent detection was by means of a commercially available streptavidin–polyalkaline phosphatase-based detection system that generated a colored end-product. The advantages of this approach include probe stability, safety of use, lack of disposal problems, and the rapidity of colorimetric detection compared with autoradiography.

b. 656-Base Pair *Hin*fI–*Taq*I Fragment. This fragment is the largest of the 16 fragments obtained by a *Hin*fI–*Taq*I double digest of pBR322 DNA. It is completely internal to the TEM-1 structural gene and contains none of the secretion signal sequence found in the 1-kb *Eco*RI–*Hin*fI probe described earlier. The 656-bp *Hin*fI–*Taq*I fragment (probe B in Fig. 1) was used (36) to probe the β-lactamase genes indicated in Table II. Hybridization was observed only with TEM-1 and TEM-2; in addition, this probe is internal to probe A and can therefore be expected to give a negative reaction with all the additional β-lactamase genes shown in Table II to give a negative reaction with probe A.

c. 424-Base Pair *Hin*cII–*Bgl*I Fragment. This fragment, internal to the TEM-1 structural gene (probe C in Fig. 1), has also been used as a probe (24). Cross-hybridization was observed only with TEM-2 and, curiously, TLE-1 (Table II). In contrast, no hybridization between TEM-1 and TLE-1 was reported using probe A (6). TLE-1 resembles TEM-1 in substrate profile and reactions with inhibitors, but differs in its isoelectric point and enzyme banding pattern on flat-bed electrofocusing (32). The original strain of *E. coli* producing TLE-1 was shown (32) to carry two plasmids: pMG204a coding for TEM-1, and pMG204b coding for TLE-1. One possible explanation for the anomalous result is that the TEM-1 gene may have transposed to pMG204b, with the concomitant generation of an apparently positive hybridization result. This probe can also be expected to give a negative reaction with all of the additional β-lactamase genes shown in Table II to give a negative reaction with probe A.

d. 298-Base Pair *Hinc*II–*Pst*I Fragment. This intragenic fragment of the TEM-1 β-lactamase gene (probe D in Fig. 1) has been used to examine the relationship between TEM-1 and the new broad-spectrum enzymes TEM-3 to TEM-6 (48). In each case hybridization was obtained under high-stringency conditions between the probe and purified DNA from all strains producing the new enzymes.

2. *Synthetic Probes*

The probes described in the previous section are efficient at distinguishing the TEM family from other β-lactamase groups, but are not effective in distinguishing different β-lactamases within the TEM family. Using stringent hybridization conditions and synthetic oligonucleotides it is now considered possible to detect single-base pair point mutations in a given sequence (14). A 15-mer oligonucleotide probe (5′-ATGATGAGCACTTTT-3′) has been synthesized (3) that corresponds to an internal sequence of the structural gene. This probe distinguishes TEM-1 and TEM-2 from other groups of β-lactamases, but does not distinguish between TEM-1 and TEM-2. However, two oligonucleotides have also been synthesized (36) in which the central bases correspond to the one amino acid difference between TEM-1 and TEM-2 (1,49). Using high-stringency conditions, the TEM-1 oligonucleotide (5′-CCCAACTGATCTTCA-3′) reacts only with the TEM-1 gene, while the TEM-2 oligonucleotide (5′-CCCAACTTATCTTCA-3′) reacts only with the TEM-2 gene (36). In both cases these oligonucleotides correspond to the noncoding strand.

The novel broad-spectrum β-lactamases (TEM-3 to TEM-7) which have emerged in response to the use of third-generation cephalosporins also seem to be close structural variants of either TEM-1 or TEM-2. In the case of TEM-3 it has been determined that there are two substitutions between TEM-3 and TEM-2, and three substitutions between TEM-3 and TEM-1 (48). It therefore seems probable that it will be possible to synthesize oligonucleotides that can distinguish between all of the members of the TEM family when sufficient DNA sequence data becomes available. The synthesis of oligomers with a mismatch toward the center of the sequence increases the extent of discrimination because of an increased destabilizing effect. It should also be noted that the temperature of hybridization and washing conditions may need to be determined empirically for each oligomer, taking into account the G + C and A + T content (36).

3. *Use of Probes for Epidemiological Studies and Comparison with Isoelectric-Focusing Results*

If DNA probes are to be used on a routine clinical basis, it is first necessary to evaluate their sensitivity and specificity in direct comparison with the conventional methods currently in use. In the case of TEM β-lactamase probes this means comparing their performance with the "gold standard" of analytical isoelectric focusing.

An initial study (18) used isoelectric focusing and molecular hybridization with the 1-kb TEM probe derived from pBR322 (probe A in Fig. 1) to screen for TEM β-lactamase production in 328 ampicillin- or ticarcillin-resistant bacterial isolates representing 11 gram-negative genera. Bacteria were transferred to the surface of nitrocellulose filters placed on MacConkey agar plates and, after an incubation period of 18–24 hr at 37°C, the cells were lysed *in situ* and their DNA denatured. Following hybridization with the radioactively labeled probe and subsequent autoradiography, the TEM gene was detected by hybridization in 174 isolates (53%). This compared with 166 (51%) that produced a β-lactamase identified as TEM by isoelectric focusing. Apparent false-negative hybridization results were seen with 8 strains, while 16 isolates of various types were probe-positive but did not produce a TEM β-lactamase detectable by isoelectric focusing. The results obtained by both methods were concordant in 92.7% of the entire sample.

A similar, but slightly smaller study (12), using a radioactively labeled 424-bp TEM probe (probe C in Fig. 1), compared DNA hybridization with isoelectric focusing in 122 gram-negative isolates resistant to ampicillin, ticarcillin, or both. A total of 64 strains produced either TEM-1 or TEM-2 β-lactamase as determined by isoelectric focusing; each of these was also positive using the TEM probe. In contrast to the previous results (18), no false positives were found with the 424-bp TEM probe, perhaps because the probe used was entirely intragenic or the sample tested was somewhat smaller.

A third study (35) used the TEM-1 oligonucleotide described in Section III,A,2, to examine a total of 114 β-lactamase-producing gram-negative bacteria belonging to at least 16 species. This synthetic probe was again labeled with radioactivity and then used in colony hybridization experiments to probe the DNA liberated from bacteria grown on a nylon membrane. A total of 44 strains were determined by isoelectric focusing to produce the TEM-1 enzyme, either alone or in combination with other β-lactamases, and each of these strains was also detected using the TEM-1 oligonucleotide probe. The probe did not hybridize with any of the bacteria shown by isoelectric focusing to produce the TEM-2 enzyme alone, thereby further demonstrating the specificity of the oligonucleotide probe (36). The probe did, however, hybridize with 5 strains that could not be shown to produce TEM-1 by isoelectric focusing. The results were thus 96% concordant using this synthetic probe.

In all three of the studies just described, the correlation between hybridization and isoelectric focusing was extremely good, if not always complete. Isoelectric focusing suffers from the major drawbacks of the limited number of isolates that can be tested per day and the increasing number of enzymes that have extremely similar isoelectric points. DNA probe methodology can cope with considerably larger sample numbers and, using appropriate probes, it is possible to distinguish between different β-lactamases having the same or similar isoelectric points. The various probes used do seem to generate occasional false positives, but it is not yet clear whether this is due to the (possibly significant) presence of "silent" or nonexpressed TEM genes, or whether DNA sequences are encountered in clinical bacteria that are structurally related to TEM genes, but that are unrelated to β-lactamase production.

A major drawback with DNA probe technology in the clinical laboratory is the use of radioactivity, but it has now been shown that biotin-labeled probes can also be used to detect TEM β-lactamase genes (6,27). A colony hybridization method has been used (13) to examine biotin and ^{32}P-labeled TEM probes for sensitivity and specificity in identifying the type of β-lactamase made by 103 clinical isolates. The results obtained were compared with those generated by isoelectric focusing. The ^{32}P-labeled probe (probe C in Fig. 1) correctly identified all TEM-1-encoding strains, and no false positive or false negative findings were obtained. The same probe labeled with biotin was less specific and generated ~20% false positive and false negative results using a colony hybridization method, but was specific when purified plasmid DNA was used as the target. In contrast, a separate study (6) obtained excellent specificity using a biotin-labeled TEM probe (probe A in Fig. 1) to examine dot blots of broth cultures known to produce a variety of different β-lactamases. It therefore seems that the type of sample and the way in which it is prepared is of prime importance if biotin-labeled probes are to be used successfully. It is also possible to label oligonucleotide probes with biotin (8), but this has not yet been reported in conjunction with TEM β-lactamase probes. Further developments in nonradioactive methods of labeling probes should do much to encourage the spread of DNA probe technology into clinical laboratories.

B. Use of Probes for Direct Examination of Patient Specimens

1. Application of a Nonradioactive Detection System

Most studies using nonradioactively labeled DNA probes have involved either (1) isolation and growth in pure culture of infecting organisms or (2) purification of plasmid or whole-cell DNA prior to hybridization with the probe. There have been very few studies in which probes carrying a nonradioactive label have been used in conjunction with crude clinical specimens, particularly when looking for antibiotic resistance.

A rare example of such an application demonstrated (43) that alkaline phosphatase-conjugated 26-bp oligonucleotide probes could be used to detect heat-labile (LT) and heat-stable (ST) enterotoxigenic *E. coli* in direct stool blots, but that they were less sensitive than the same probes labeled with radioactivity. This method involves covalently linking alkaline phosphatase directly to the C-5 position of the thymidine bases contained in the oligonucleotide by means of a 12-atom spacer arm (15) and is a particularly effective nonradioactive labeling method for short synthetic probes.

Of the various alternatives available for labeling larger DNA fragments, the use of biotin as a modifying group seems, to date, to have gained widest acceptance. Subsequent detection uses a streptavidin–alkaline phosphatase conjugate to generate an insoluble colored precipitate in the presence of a suitable dye (21,22). It was initially demonstrated (6) that the 1-kb *Eco*RI–*Hin*fI pBR322 fragment could be labeled with biotin and used to detect TEM β-lactamase genes in purified cultures of bacteria. In an extension of this work (7), fresh urine samples were obtained from randomly selected patients with proven urinary tract infection and filtered directly onto hydrated nitrocellulose membranes using a microfiltration apparatus to produce a 12×8 dot-blot array. Thus 96 urine samples could be screened simultaneously. Following hybridization procedures, detection of a positive result was by means of a commercially available kit incorporating a streptavidin–alkaline phosphatase conjugate. Full details of the methods used are given in Section VIII. A study using this protocol (7) has shown that bacteria coding for TEM β-lactamase production could be readily recognized in crude urine samples at cell densities $\geq 10^5$ cells per spot (equivalent to 10^6 cells/ml infected urine). The hybridization of cells contained in either urine or saline did not significantly affect the results.

Figure 2 shows part of a typical dot blot of infected urines after hybridization and application of the detection system. Of the 40 infected urines on the filter, 23 were probe-positive and 17 were probe-negative. It

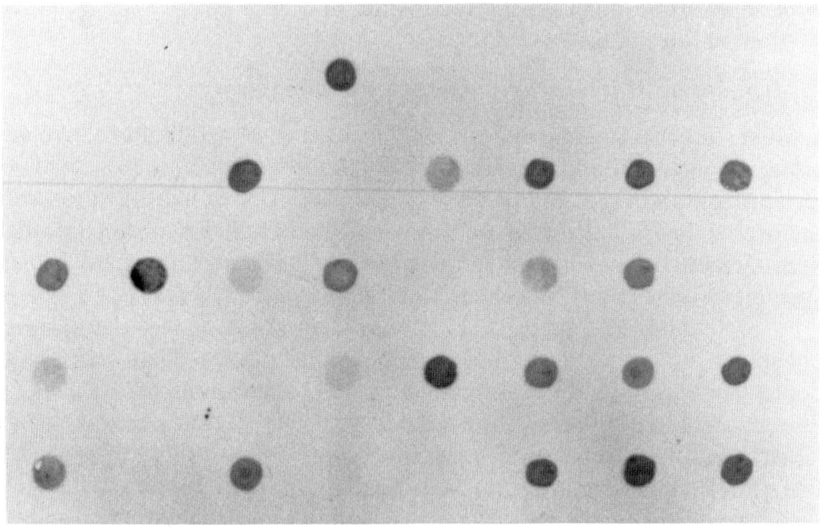

Fig. 2. Dot blot of infected urines following hybridization with the TEM probe. Of the 40 infected urines on the filter, 23 were probe-positive and 17 were probe-negative.

should be noted that there were differences in the intensity of coloration of the positive dots (see Sections III,B,2 and IV).

Treatment of filters with proteinase K prior to hybridization is an essential step in this protocol. In the absence of this treatment all samples gave an apparently positive hybridization signal (7), possibly due to a nonspecific reaction of the streptavidin component of the detection system with protein or other components contained in the infected urines.

2. Automated Assessment of Hybridization Results

a. A Computer-Controlled Image Analysis System. The generation of hybridization results in the fixed 12 × 8 array produced by the microfiltration apparatus suggested that it might be possible to design a machine capable of reading the intensity of hybridization at each dot-blot position. As an initial step we decided to utilize an existing image analysis system (Mastascan 2; Mast Laboratories) designed to evaluate multipoint-inoculated antibiotic susceptibility tests. The hybridization membranes were held in position by a specially constructed baseplate, and new software was written so that the Mastascan computer could assess the 96 spots in the 12 × 8 format produced on each membrane. The system generated a digital value corresponding to the intensity of the hybridization signal at each position. This value was computed by taking nine "internal" intensity readings in a 3 × 3 matrix *within* the position of each spot. An additional four "external" readings were taken around each position. Any variation in the illumination or the staining intensity of the membrane was then compensated for by computing the average of the nine "internal" readings minus the average of the four "external" readings.

b. Sensitivity and Reproducibility of Results Generated Using the System. The sensitivity of the Mastascan system and its ability to distinguish between "positive" and "negative" bacterial cells was tested by using a series of cell dilutions of either *E. coli* J53.2 or *E. coli* J53.2 (pBR322) to contaminate noninfected urine. Figure 3 shows a typical

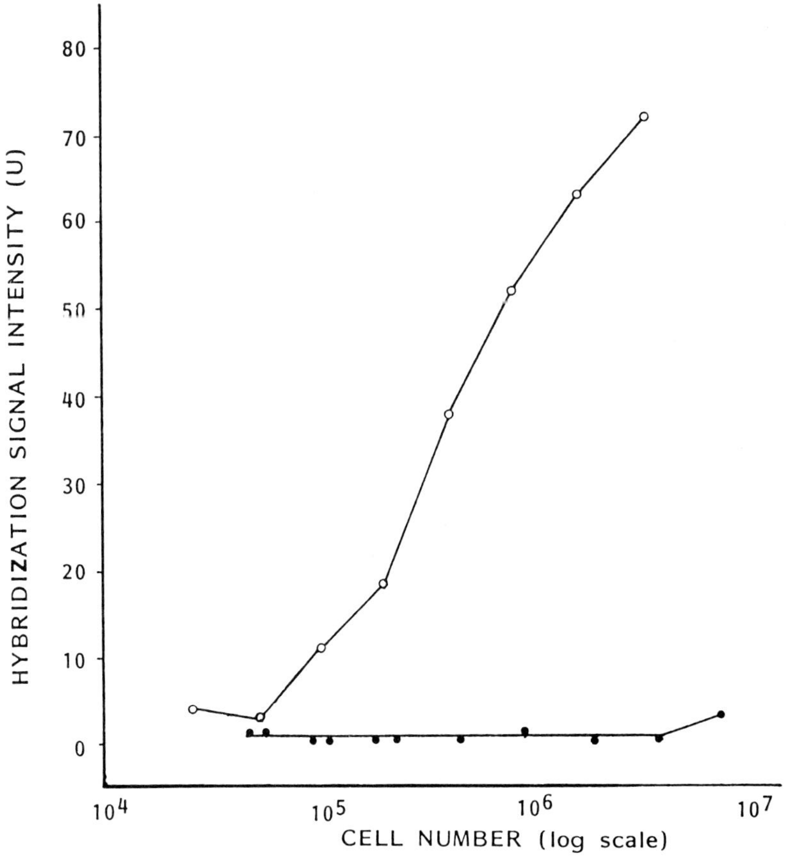

Fig. 3. Relationship between the number of cells filtered per spot (logarithmic scale) and the signal intensity obtained with the Mastascan system following hybridization with the TEM probe. ○——○, J53.2 carrying pBR322; ●——●, J53.2 alone.

example of the digital hybridization values generated by the Mastascan system plotted against cell numbers. The results indicated that the negative control (J53.2 alone) generated a slightly elevated hybridization "intensity" at high cell concentrations, but that a positive hybridization result could be readily distinguished by the system (as with the naked eye) at cell concentrations exceeding 10^5 cells per spot.

The reproducibility of the system was tested using a standard inoculum of 3×10^6 cells per 100-μl spot (7). When duplicate blots of the positive and negative control strains were repeatedly screened using the system, the positive controls generated a mean reading of 34.2 units ($n = 32$; SD = 5.6), while the negative controls gave a reading of 3.7 units ($n = 32$; SD = 1.8).

c. Comparison of Hybridization Results Obtained Using the Mastascan System with Isoelectric Focusing. The results obtained in the previous subsection suggested that the Mastascan system might offer an approach to assessing rapidly and objectively the large number of hybridization results generated in clinical and epidemiological studies. In a further validation of the system, the results generated by the Mastascan system were compared (7) with those obtained after isolation of the infecting organism, followed by conventional sensitivity tests and isoelectric focusing procedures. A total of 81 infected urine samples were included in the study. The results obtained by conventional and DNA hybridization methodology are compared in Table III. Only four anomalous results were obtained. Three of these were apparently probe false positives. One of

TABLE III

Comparison of Results Obtained from 81 Infected Urine Samples (a) Using the TEM Probe Directly, Followed by Assessment with the Mastascan System; and (b) Following Isolation of the Infecting Organism, Conventional Antimicrobial Susceptibility Testing, and Isoelectric Focusing[a]

	Conventional methodology			
	Ampicillin^R			
	TEM$^+$	TEM$^-$	Ampicillin^S	Mixed
TEM Probe +	14	3[b]	0	3
TEM Probe −	1[c]	9	44	7

[a]Results shown are the number of strains in each category.
[b]Probe "false positive."
[c]Probe "false negative" (see text).

these seemed to be due to excessive urine pigmentation, but the remaining two may have been due to TEM β-lactamase genes that were not being expressed. Similar results were described in Section III,A,3. One probe false negative was also obtained, and it seemed that this result was due to the presence of insufficient cell numbers in the original infected urine. Thus in this rather limited sample there was an overall good correlation between the results obtained by conventional methods and those assessed by the Mastascan computer.

IV. CONCLUSIONS

Several groups of investigators have now described DNA probes that can be used successfully to detect TEM β-lactamase-encoding genes in clinical isolates of bacteria. These probes can be either specific fragments obtained by restriction endonuclease digestion of known plasmid molecules or short synthetic oligonucleotides. In all cases these probes seem to be efficient at distinguishing β-lactamase-encoding genes belonging to the TEM family from β-lactamase genes belonging to unrelated groups. When compared with conventional analytical isoelectric focusing there is a high concordance between the results obtained with the two methodologies.

It is more difficult to distinguish between genes *within* the TEM family. Although synthetic oligonucleotides have been described that distinguish between TEM-1 and TEM-2, it seems that the synthesis of further probes for discrimination between the other enzymes within the TEM family must await the elucidation of detailed DNA sequence data for the more novel enzymes. Short synthetic oligonucleotides can hybridize much faster than probes derived from plasmid fragments, because the rate of hybridization is in part determined by the sequence complexity and length. They are also considerably easier to produce and purify. However, oligonucleotide probes are harder to label, perhaps carrying only one reporter group per molecule, and might turn out to be less sensitive when used to probe clinical specimens directly (see below).

It has been demonstrated that either radioactive or nonradioactive labels can be used in conjunction with probes for TEM β-lactamase genes. To date it seems that radioactive probes are slightly more sensitive, but the use of a nonisotopic detection system has many advantages over techniques incorporating labels such as ^{32}P. These advantages include probe stability, safety of use, lack of disposal problems, and the rapidity of colorimetric detection compared with autoradiography. The introduction of methods allowing the nonisotopic detection of hybridization, such as those based on the high-affinity binding of biotin and streptavidin, therefore offers the possibility of introducing DNA hybridization tests for antibiotic-resistance genes (in general) into routine clinical microbiology laboratories lacking the facilities or expertise to handle highly radioactive probe labels.

At present, DNA probe methodology has been used primarily for epidemiological purposes, where it has provided useful information regarding the evolutionary relationships between different antibiotic-resistance genes and a means of following their dissemination. However, perhaps the main potential advantage of introducing these methods into routine diagnostic laboratories devolves from the fact that the technique can be used to detect specific genes without necessitating primary isolation and growth of an organism. DNA hybridization technology therefore provides the ability to detect directly certain types of antibiotic resistance expressed by pathogenic microorganisms contained in crude clinical samples. This ability would be particularly valuable for organisms that are hard to isolate, slow growing, or otherwise difficult to cultivate.

If hybridization tests for specific antibiotic-resistance genes are to be introduced into routine laboratories, it will eventually be necessary for labeled DNA probes to become available in kit form. The nonradioactive hybridization procedures described in this chapter are simple to perform and consist essentially of a series of incubation and wash steps. It may be possible to design a machine to automate these procedures and hence minimize the technician time required. As described earlier, it has already proved possible to modify an existing automated system, which can now be used to quantify rapidly the intensity of the hybridization signal obtained following direct probing of large numbers of clinical samples.

Apart from diagnostic clinical uses, the availability of an image analysis system to quantify hybridization signals has a number of potentially interesting research applications. The intensity of the signal obtained is directly related to the number of "positive" cells, presumably reflecting the number of copies of the target gene sequence present. The technique may therefore be extremely useful for applications such as studies of plasmid metabolism, particularly those in which changes in plasmid copy numbers or curing resulting from exposure to varying concentrations of different antibiotics need to be determined.

V. GENE PROBES VERSUS ANTISERA AND MONOCLONAL ANTIBODIES

While it is certainly feasible to produce antisera and monoclonal antibodies (MAb) capable of detecting the products of specific antibiotic-resistance genes (including TEM β-lactamase genes), this is not an area that has received much attention. Antisera and MAb recognize gene products rather than the genes themselves. This may be a disadvantage in screening for antibiotic resistance because pathogenic bacteria in clinical lesions are often growing extremely slowly and may switch on their resistance mechanisms only in response to antibiotic treatment. In this respect there would be few advantages over conventional antibiotic susceptibility tests. The major advantage of DNA probes is that they can be used to screen clinical samples directly for *potential* antibiotic resistance, even if the resistance gene is not being expressed at the time of testing.

A more useful application of MAb in this field might be in *combination* with DNA hybridization. Using a biotin-labeled probe for hybridization, a detection system could be designed based on a high-affinity antibiotin MAb conjugated to a variety of detectable molecules, with secondary reagents designed to maximize the potential limits of detection. Such a system could potentially overcome the nonspecific binding problems encountered between clinical specimens and the biotin–streptavidin system. These areas are currently the subject of detailed investigations.

VI. PROSPECTS FOR THE FUTURE

Most methods described in the literature for routine DNA probe-based assays are merely extensions of methods used in research laboratories. The hybridization protocols described in this chapter are no exception to this general rule. The requirements for a routine assay are somewhat different from those of a research technique, and there are therefore a number of areas that should be investigated to see if the hybridization approach can be made more suitable for routine clinical screening of bacteria for antibiotic resistance.

First, it is important that work continues in an attempt to increase the sensitivity of nonradioactive detection systems. Such systems are still at an early stage of development and it is likely that optimal detection systems and conditions have yet to be formulated. It can be anticipated that it will be possible to develop tests that are several orders of magnitude more sensitive than that described in this chapter. Improvements might include the introduction of sandwich hybridizations (55) combined with a move to a hybridization format using a liquid rather than a solid support (52), the combination of a DNA hybridization format with a MAb-based detection system (see Section V), or the design of detection systems incorporating "enzyme cascades" in which the reaction product of one enzyme is amplified by a second (or third) enzyme.

Second, the use of DNA probes as a direct means of detecting antibiotic resistance in clinical specimens will depend on the generation of probes or probe sets that cover all of the mechanisms of resistance to a particular antibiotic. While the TEM family of β-lactamases may account for much of the clinically important β-lactam resistance found in gram-negative bacteria, there are many other types of plasmid-mediated β-lactamases (see Section II,C). It is clear from the results discussed in this chapter that it is possible to develop probes that are specific for a particular resistance gene. In this way it should be possible to produce gradually a set of probes covering all of the known plasmid-mediated β-lactamases, perhaps each with a different type of nonradioactive probe label for easy identification. An alternative might involve determining a DNA sequence common to all β-lactamases and using this as a "universal probe" for β-lactam resistance (3).

Finally, an example of an approach to automated assessment of hybridization results has been briefly described in this chapter. Other automated formats, both for assessing hybridization results and for performing the actual hybridization reaction and washing sequences, are likely to be developed in the near future. It seems certain that these developments, coupled with improvements in probes, nonradioactive detection systems, and practical utility in a nonresearch situation, will lead to an increasing application of this technology by routine diagnostic laboratories for determining the antimicrobial susceptibility of pathogenic bacteria in clinical specimens.

VII. SUMMARY

Plasmid-encoded β-lactamases belonging to the TEM family are responsible for the majority of resistance to β-lactam antibiotics encountered in clinical isolates of gram-negative bacteria. DNA probes are available that distinguish the TEM family from other β-lactamase groups. Results obtained are comparable with those obtained by conventional analytical isoelectric focusing. Progress has been made in producing synthetic oligonucleotides that distinguish between β-lactamases within the TEM family, but DNA sequence data for some of the newer enzymes are not yet available. Probes can be radioactively labeled, but the use of nonradioactive labels such as biotin offers the possibility of introducing these tests into routine clinical microbiology laboratories lacking the facilities or expertise to handle highly radioactive probe labels. Using a biotin-labeled probe it is possible to detect bacteria carrying TEM β-lactamase genes in infected urine samples. Large numbers of hybridization tests can be rapidly and objectively assessed using a computer-controlled image analysis system. Improvements to these methods should lead to increasing use of hybridization technology for antimicrobial susceptibility testing of bacteria by routine diagnostic laboratories.

VIII. MATERIALS AND METHODS

This section describes a simple protocol (using commercially available materials) in which the 1-kb probe can be used to detect resistant bacteria directly in crude urine samples.

A. Preparation and Labeling of Probe DNA

Purified plasmid pBR322 DNA (Bethesda Research Laboratories, BRL) was double-digested with restriction endonucleases *Eco*RI and *Hin*fI, used according to the manufacturer's recommendations (Boehringer-Mannheim). The 1-kb fragment was separated on a 0.7% agarose gel, collected by electrophoretic elution and purified using an Elutip-d column (Schleicher and Schuell). Nick translation was used to label the purified fragment with biotin-11-dUTP (BRL), again according to the manufacturer's instructions. The labeled fragment was separated from unincorporated label by ethanol precipitation and resuspended in Tris–EDTA buffer (10 m*M* Tris–1 m*M* EDTA, pH 8.0).

B. Sample Treatment

Portions of fresh urine samples (0.1 ml) were filtered directly onto hydrated nitrocellulose membranes (Schleicher and Schuell), using a Bio-Dot microfiltration apparatus (Bio-Rad Laboratories) to produce a 12 × 8 dot-blot array. Filters were air-dried, after which DNA was liberated from cells contained in the urine, denatured, and bound to the membrane using standard methods for colony hybridizations (26). Prior to hybridization procedures, filters were hydrated in a solution comprising 2× SSC (1× SSC is 0.15 *M* sodium chloride–0.015 *M* sodium citrate, pH 7.0) and 0.1% sodium dodecyl sulfate (SDS). The filters were then soaked for 1 hr at 37°C in a solution of proteinase K (Sigma) dissolved in 2× SSC–0.1% SDS at a concentration of 0.5 mg/ml. Finally the filters were washed three times (each for 5 min) in 2× SSC.

C. Hybridization Procedures

Prehybridization and hybridization conditions were essentially similar to those recommended for radioactively labeled probes (44), except that the concentration of formamide was reduced to 45%, and 0.1% SDS was added to the prehybridization and hybridization buffers. A probe concentration of ~0.1 μg/ml of hybridization solution was used (0.1 ml solution per square centimeter of nitrocellulose membrane) and hybridization was routinely carried out at 42°C overnight. Detection of a positive hybridization result was by means of a BluGENE detection system kit (BRL), used as recommended by the manufacturer. This kit is similar to that used in the original study of broth cultures (6), but employs a streptavidin–alkaline phosphatase conjugate and simpler washing procedures, resulting in a reduced detection time (3 hr).

ACKNOWLEDGMENTS

The work described in this chapter was partially funded by a research grant from the Trent Regional Health Authority. I would additionally like to thank my colleagues in Nottingham and elsewhere for many stimulating discussions on the use of DNA probes for antibiotic susceptibility testing. I am particularly indebted to my research assistant, Ian Carter, for performing many of the nonradioactive hybridization experiments, Neil Pearson for adapting the Mastascan image analysis system, and Linda Bowering for preparing the manuscript so efficiently.

REFERENCES

1. Ambler, R. P., and Scott, G. K. (1978). Partial amino acid sequence of penicillinase encoded by *Escherichia coli* plasmid R6K. *Proc. Natl. Acad. Sci. U.S.A.* **75,** 3732–3736.
2. Bisessar, U., and James, R. (1988). Molecular cloning of the *Shv-1* β-lactamase gene and construction of an *Shv-1* hybridization probe. *J. Gen. Microbiol.* **134,** 835–840.
3. Boissinot, M., Mercier, J., and Lévesque, R. C. (1987). Development of natural and synthetic DNA probes for OXA-2 and TEM-1 β-lactamases. *Antimicrob. Agents Chemother.* **31,** 728–734.
4. Brun-Buisson, C., Legrand, P., Philippon, A., Montravers, F., Ansquer, M., and Duval, J. (1987). Transferable enzymatic resistance to third-generation cephalosporins during nosocomial outbreaks of multiresistant *Klebsiella pneumoniae*. *Lancet* **2,** 302–306.
5. Bush, K., and Sykes, R. B. (1984). Interaction of β-lactam antibiotics with β-lactamases as a cause for resistance. *In* "Antimicrobial Drug Resistance" (L. E. Bryan, ed.), pp. 1–31. Academic Press, Orlando, Florida.
6. Carter, G. I., Towner, K. J., and Slack, R. C. B. (1987). Detection of TEM β-lactamase genes by non-isotopic spot hybridization. *Eur. J. Clin. Microbiol.* **6,** 406–409.
7. Carter, G. I., Towner, K. J., Pearson, N. J., and Slack, R. C. B. (1989). Use of a non-radioactive hybridization assay for direct detection of gram-negative bacteria carrying TEM β-lactamase genes in infected urine. *J. Med. Microbiol.* **28,** 113–117.
8. Chu, B. C. F., and Orgel, L. E. (1985). Detection of specific DNA sequences with short biotin-labeled probes. *DNA* **4,** 327–331.
9. Cooksey, R. C., Clark, N. C., and Thornsberry, C. (1985). A gene probe for TEM β-lactamases. *Antimicrob. Agents Chemother.* **28,** 154–156.

10. Dale, J. W., Godwin, D., Mossakowska, D., Stephenson, P., and Wall, S. (1985). Sequence of the OXA-2 β-lactamase: Comparison with other penicillin-reactive enzymes. *FEBS Lett.* **191,** 39–44.
11. Eley, A., Ambler, J., and Greenwood, D. (1983). Isoelectric focusing of β-lactamases on cellulose acetate membranes. *J. Antimicrob. Chemother.* **12,** 193–196.
12. Huovinen, S., Huovinen, P., and Jacoby, G. A. (1988). Detection of plasmid-mediated β-lactamases with DNA probes. *Antimicrob. Agents Chemother.* **32,** 175–179.
13. Huovinen, S., Huovinen, P., and Jacoby, G. A. (1988). Reliability of biotinylated DNA probes in colony hybridization: Evaluation of an improved colony lysis method for detection of TEM-1 β-lactamase. *Mol. Cell. Probes* **2,** 83–85.
14. Itakura, K., Rossi, J. J., and Wallace, R. B. (1984). Synthesis and use of synthetic oligonucleotides. *Annu. Rev. Biochem.* **53,** 323–356.
15. Jablonski, E., Moomaw, E. W., Tullis, R. H., and Ruth, J. L. (1986). Preparation of oligodeoxyribonucleotide–alkaline phosphatase conjugates and their use as hybridization probes. *Nucleic Acids Res.* **14,** 6115–6128.
16. Jacoby, G. A., Sutton, L., and Medeiros, A. A. (1980). Plasmid-determined β-lactamase of *Pseudomonas aeruginosa*. *In* "Current Chemotherapy and Infectious Disease" (J. D. Nelson and C. Grassi, eds.), pp. 769–771. Am. Soc. Microbiol., Washington, D.C.
17. James, R. (1983). Relative substrate affinity index values: A method for identification of beta-lactamase enzymes and prediction of successful beta-lactam therapy. *J. Clin. Microbiol.* **17,** 791–798.
18. Jouvenot, M., Deschaseaux, M. L., Royez, M., Mougin, C., Cooksey, R. C., Michel-Briand, Y., and Adessi, G. L. (1987). Molecular hybridization versus isoelectric focusing to determine TEM-type β-lactamases in gram-negative bacteria. *Antimicrob. Agents Chemother.* **31,** 300–305.
19. Kleckner, N. (1981). Transposable elements in prokaryotes. *Annu. Rev. Genet.* **15,** 341 404.
20. Knothe, H., Shah, P., Krcmery, V., Antal, M., and Mitsuhashi, S. (1983). Transferable resistance to cefotaxime, cefoxitin, cefamandole and cefuroxime in clinical isolates of *Klebsiella pneumoniae* and *Serratia marcescens*. *Infection* **6,** 315–317.
21. Langer, P. R., Waldrop, A. A., and Ward, D. C. (1981). Enzymatic synthesis of biotin-labeled polynucleotides: Novel nucleic acid affinity probes. *Proc. Natl. Acad. Sci. U.S.A.* **78,** 6633–6637.
22. Leary, J. J., Brigati, D. J., and Ward, D. C. (1983). Rapid and sensitive colorimetric method for visualising biotin-labeled DNA probes hybridized to DNA or RNA immobilized on nitrocellulose: Bio-blots. *Proc. Natl. Acad. Sci. U.S.A.* **80,** 4045–4049.
23. Levesque, R. C., and Jacoby, G. A. (1988). Molecular structure and interrelationships of multiresistance β-lactamase transposons. *Plasmid* **19,** 21–29.
24. Levesque, R. C., Medeiros, A. A., and Jacoby, G. A. (1987). Molecular cloning and DNA homology of plasmid-mediated β-lactamase genes. *Mol. Gen. Genet.* **206,** 252–258.
25. Lindberg, F., and Normark, S. (1986). Contribution of chromosomal β-lactamases to β-lactam resistance in Enterobacteria. *Rev. Infect. Dis.* **8,** Suppl. 3, S292–S304.
26. Maniatis, T., Fritsch, E. F., and Sambrook, J. (1982). "Molecular Cloning: A Laboratory Manual," pp. 312–315. Cold Spring Harbor Lab., Cold Spring Harbor, New York
27. Martel, A. Y., Gosselin, P., Ouellette, M., Roy, P. H., and Bergeron, M. G. (1987). Isolation and molecular characterization of β-lactamase-producing *Haemophilus parainfluenzae* from the genital tract. *Antimicrob. Agents Chemother.* **31,** 966–968.
28. Matthew, M. (1979). Plasmid-mediated β-lactamases of gram-negative bacteria. Properties and distribution. *J. Antimicrob. Chemother.* **5,** 349–358.

29. Matthew, M., and Harris, A. M. (1976). Identification of β-lactamases by analytical isoelectric focusing: Correlation with bacterial taxonomy. *J. Gen. Microbiol.* **96,** 55–67.
30. Matthew, M., Harris, A. M., Marshall, M. J., and Ross, G. W. (1975). The use of analytical isoelectric focusing for detection and identification of β-lactamases. *J. Gen. Microbiol.* **88,** 169–178.
31. Medeiros, A. A. (1984). β-lactamases. *Br. Med. Bull.* **40,** 18–27.
32. Medeiros, A. A., Cohenford, M., and Jacoby, G. A. (1985). Five novel plasmid-determined β-lactamases. *Antimicrob. Agents Chemother.* **27,** 715–719.
33. Medeiros, A. A., Gilleece, E. S., and O'Brien, T. F. (1981). Distribution of plasmid-type β-lactamases in ampicillin-resistant salmonellae from humans and animals in the United States. *In* "The Molecular Biology, Pathogenicity and Ecology of Bacterial Plasmids" (S. B. Levy, R. C. Clowes, and E. L. Koenig, eds.), p. 634. Plenum, New York.
34. Medeiros, A. A., Ximenez, J., Blickstein-Goldworm, K., O'Brien, T. F., and Acar, J. (1980). β-lactamases of ampicillin-resistant *Escherichia coli* from Brazil, France and the United States. *In* "Current Chemotherapy and Infectious Diseases" (J. D. Nelson and C. Grassi, eds.), pp. 761–762. Am. Soc. Microbiol., Washington, D.C.
35. Ouellette, M., Paul, G. C., Philippon, A. M., and Roy, P. H. (1988). Oligonucleotide probes (TEM-1, OXA-1) versus isoelectric focusing in β-lactamase characterization of 114 resistant strains. *Antimicrob. Agents Chemother.* **32,** 397–399.
36. Ouellette, M., Rossi, J. J., Bazin, R., and Roy, P. H. (1987). Oligonucleotide probes for the detection of TEM-1 and TEM-2 β-lactamase genes and their transposons. *Can. J. Microbiol.* **33,** 205–211.
37. Ouellette, M., and Roy, P. H. (1986). Analysis by using DNA probes of the OXA-1 β-lactamase gene and its transposon. *Antimicrob. Agents Chemother.* **30,** 46–51.
38. Petit, A., Sirot, D. L., Chanal, C. M., Sirot, J. L., Labia, R., Gerbaud, G., and Cluzel, R. A. (1988). Novel plasmid-mediated β-lactamase in clinical isolates of *Klebsiella pneumoniae* more resistant to ceftazidime than to other broad-spectrum cephalosporins. *Antimicrob. Agents Chemother.* **32,** 626–630.
39. Richmond, M. H., Bennett, P. M., Choi, C-L., Brown, N., Brunton, J., Grinsted, J., and Wallace, L. (1980). The genetic basis of the spread of β-lactamase synthesis among plasmid-carrying bacteria. *Philos. Trans. R. Soc. London, Ser. B* **289,** 349–359.
40. Rigby, P. J. W., Dieckmann, M. A., Rhodes, C., and Berg, P. (1977). Labelling deoxyribonucleic acid to high specific activity *in vitro* by nick-translation with DNA polymerase. *Int. J. Mol. Biol.* **113,** 237–251.
41. Roy, C., Foz, A., Segura, C., Tirado, M., Fuster, C., and Reig, R. (1983). Plasmid-determined β-lactamases identified in a group of 204 ampicillin-resistant Enterobacteriaceae. *J. Antimicrob. Chemother.* **12,** 507–510.
42. Saunders, J. R. (1984). Genetics and evolution of antibiotic resistance. *Br. Med. Bull.* **40,** 54–60.
43. Seriwatana, J., Echeverria, P., Taylor, D. N., Sakuldaipeara, T., Changchawalit, S., and Chivoratanond, O. (1987). Identification of enterotoxigenic *Escherichia coli* with synthetic alkaline phosphatase-conjugated oligonucleotide DNA probes. *J. Clin. Microbiol.* **25,** 1438–1441.
44. Silhavy, T. J., Berman, M. L., and Enquist, L. W. (1984). "Experiments with Gene Fusions," pp. 191–195. Cold Spring Harbor Lab., Cold Spring Harbor, New York.
45. Simpson, I. N., Harper, P. B., and O'Callaghan, C. H. (1980). Principal β-lactamases responsible for resistance to β-lactam antibiotics in urinary tract infections. *Antimicrob. Agents Chemother.* **17,** 929–936.
46. Simpson, I. N., Knothe, H., Plested, S. J., and Harper, P. B. (1986). Qualitative and

quantitative aspects of β-lactamase production as mechanisms of β-lactam resistance in a survey of clinical isolates from faecal samples. *J. Antimicrob. Chemother.* **17,** 725–737.
47. Sirot, D., Sirot, J., Labia, R., Morand, A., Courvalin, P., Darfeuille-Michaud, A., Perroux, R., and Cluzel, R. (1987). Transferable resistance to third-generation cephalosporins in clinical isolates of *Klebsiella pneumoniae:* Identification of CTX-1, a novel β-lactamase. *J. Antimicrob. Chemother.* **20,** 323–334.
48. Sougakoff, W., Goussard, S., Gerbaud, G., and Courvalin, P. (1988). Plasmid-mediated resistance to third-generation cephalosporins caused by point-mutations in TEM-type penicillinase genes. *Rev. Infect. Dis.* **10,** 879–884.
49. Sutcliffe, J. G. (1978). Nucleotide sequence of the ampicillin resistance gene of *Escherichia coli* plasmid pBR322. *Proc. Natl. Acad. Sci. U.S.A.* **75,** 3737–3741.
50. Sykes, R. B. (1982). The classification and terminology of enzymes that hydrolyze β-lactam antibiotics. *J. Infect. Dis.* **145,** 762–765.
51. Sykes, R. B., and Matthew, M. (1976). The β-lactamases of gram-negative bacteria. *J. Antimicrob. Chemother.* **2,** 115–157.
52. Syvänen, A.-C., and Korpela, K. (1986). Detection of ampicillin and tetracycline resistance genes in uropathogenic *Escherichia coli* strains by affinity-based nucleic acid hybrid collection. *FEMS Microbiol. Lett.* **36,** 225–229.
53. Tenover, F. C. (1986). Studies of antimicrobial resistance genes using DNA probes. *Antimicrob. Agents Chemother.* **29,** 721–725.
54. Tenover, F. C. (1988). Diagnostic deoxyribonucleic acid probes for infectious diseases. *Clin. Microbiol. Rev.* **1,** 82–101.
55. Wolf, H., Leser, U., Haus, M., Gu, S. Y., and Pathmanathan, R. (1986). Sandwich nucleic acid hybridization: A method with a universally labeled probe for various specific tests. *J. Virol. Methods* **13,** 1–8.

17

SIA Technology for Probing Microbial Genes

ROBERT J. JOVELL, EVERLY CONWAY de MACARIO, AND ALBERTO J. L. MACARIO

Wadsworth Center for Laboratories and Research
New York State Department of Health, and
School of Public Health
State University of New York
Albany, New York

I. INTRODUCTION

Despite advances in the techniques for nucleic acid (NA) hybridization, quantification is still largely unresolved. Improvements have been obtained using new membranes with enhanced binding capacity via ionic or covalent attachment, for example. We have also witnessed improvements in more specific, less expensive, and easier to use nonradiolabeled NA probes that offer diagnostic advantages over some immunological and culturing techniques. A variety of hardware for the manipulation of these refined membranes and probes is also available. Yet, conspicuously absent is a practical system for accurately measuring and/or quantifying the hybrids formed, other than the human eye.

We have recently developed a system for measuring hybridization of enzyme-labeled gene probes to microliter amounts of target DNA immobilized onto transparent circles placed on a glass slide. This system is based on a modification of the slide immunoenzymatic assay (SIA) originally developed for quantification of antigens and antibodies (9,10) (U.S. Patent no. 4,682,891). It will be referred to as the SIA–DNA system.

The new system described here was designed to quantify hybridization of an enzyme-labeled oligonucleotide probe to specific base sequences present in a DNA sample in pure form, in a cell lysate or *in situ* (whole cells). The model used to calibrate the system comprises: (1) an enterotoxigenic *Escherichia coli* (ETEC) strain with a plasmid that contains an inserted DNA fragment coding for a heat-labile toxin (LT), which serves as the target DNA sequence; and (2) a commercially available alkaline phosphatase-labeled oligonucleotide DNA probe with complementary sequences to the target sequence.

II. BACKGROUND

Fundamental elements of currently available mixed-phase DNA-based assay systems include (1) a labeled, gene-specific probe that can form stable hybrids with target sequences and yield a detectable signal; (2) a solid support upon which nucleic acids can be immobilized without compromising their capacity for reacting with incoming probe; and (3) a method for measuring specific hybrid formation.

DNA probes have been traditionally constructed by radiolabeling double-stranded DNA fragments ~500–1500 nucleotides in length by nick translation (28,34). DNA is nicked with pancreatic DNase I, thereby generating a 3′-hydroxyl terminus to which deoxynucleotide 5′-[^{32}P] triphosphate residues may be joined via *E. coli* DNA polymerase I.

Such probes, as a rule, are expensive, short-lived (14-day half-life for ^{32}P), hazardous to manipulate, and difficult to dispose of safely. In addition, nick-translated probes in solution have a natural tendency to rehybridize with their own complementary strands prior to reaching the sample target NA.

A. Development of Nonradioactive Nucleic Acid Probes

Progress in the development of nonradioactive NA probes has been fueled by a growing desire to reduce dependence on radiolabeled materials without compromising the quality of detection systems. Sensitivity of enzyme-linked gene probes is now thought to approximate that of radioactive probes, since many enzyme molecules can theoretically occupy the same position held by a single radionuclide (3–6,16,23,24,26).

Manuelidis *et al.* (26) used DNA probes labeled with biotin–nucleotides via nick translation to hybridize *in situ* with polynucleotide sequences within genes on *Drosophila* polytene chromosomes. Horseradish peroxidase (HRP)-conjugated antibiotin antibodies were used to bind to the hybrids. Visualization under light and electron microscopy (LM, EM) was achieved by adding substrate, hydrogen peroxide with chromogen 3,3′-diaminobenzidine (DAB). Resolution by this method was reported to be greater than that achieved by conventional LM or EM.

Goltz *et al.* (17) used a probe labeled with biotin for *in situ* hybridization to specific sequences in human papilloma virus (HPV) DNA in a specimen that had been immobilized onto a microscope slide. A streptavidin–HRP complex was added to bind to the biotin label linked to the hybridized DNA probe. A colored precipitate formed after addition of substrate, which was visualized under LM. Results were obtained in ~1 hr. Other tests were conducted with biotin-labeled probes specific for sequences found in herpes simplex virus (HSV) types I and II DNA, and with probes for *Chlamydia trachomatis* (16,22).

Buffone *et al.* (3) used biotin-labeled DNA probes for detection of specific cytomegalovirus DNA sequences immobilized onto nitrocellulose (NC). Hybrids were treated with complexes of avidin–biotinylated alkaline phosphatase, which bound strongly to the biotin label linked to the probe–sample sequence complex. Visualization of the hybrid complex was achieved by addition of substrates 5-bromo-4-chloro-3-indolyl phosphate (BCIP) and nitro blue tetrazolium (NBT), which were converted to an insoluble product by alkaline phosphatase and precipitated onto the NC membrane.

Cardullo *et al.* (5) described the use of fluorescently labeled oligodeoxynucleotides (ODNT) in a process called nonradiative fluorescence resonance energy transfer (FRET) to study NA hybridization. Hybridization with ODNT requires donor and acceptor fluorophores, with overlapping excitation and emission spectra, to be in sufficiently close proximity to allow the excited-state energy of the donor molecule (situated on either the probe or the target DNA sequence) to be transferred by a resonance dipole-induced dipole interaction to the neighboring acceptor fluorophore. The FRET process is dependent on molecular distances, and has been proposed as a useful means for detection of NA hybridization in solution, rather than in a mixed-phase assay.

Burns *et al.* (4) developed a methodology for antenatal sex determination on small numbers of uncultured cells through *in situ* hybridization using biotinylated DNA probes specific for the Y chromosome (pHY2.1). Hybridization was visualized by addition of either a streptavidin–HRP complex that bound to the biotin label linked to the hybrid complex, or a goat antibiotin antibody followed by a HRP-conjugated rabbit anti-goat IgG. In both methods, reaction visualization was achieved by addition of substrate, hydrogen peroxide and the chromogen DAB. The signal generated was amplified by silver precipitation.

Langenberg *et al.* (22) performed *in situ* DNA hybridization using a biotinylated probe to detect HSV DNA in lesion specimens from 118 episodes of recurrent genital herpes. Detection of successful hybridization was developed by adding an avidin–biotinylated HRP complex that bound to the biotin label linked to the hybrid complex. A chromogen–substrate solution of 3-amino-9-ethylcarbazole and hydrogen peroxide was added and later examined under LM for cells displaying characteristic red nuclear staining. Results were compared to conventional cell culture methods and, though not as sensitive as cell culturing, the DNA probe attained an overall sensitivity of 92% in comparison to cell culturing. However, *in situ* hybridization tests were simple to perform and could be completed in 2–2.5 hr, whereas viral isolation from cell culture requires tissue culture as well as cytopathology skills and can take as long as 5 days.

Chevrier *et al.* (6) used acetylaminofluorene (AAF)-labeled genomic DNA probes to identify and classify *Campylobacter* strains by immobilizing DNA extracted from the cells onto NC and compared results to those obtained using ^{32}P-labeled probes. Anti-AAF monoclonal antibody (MAb) was allowed to bind to the AAF linked to the hybrid complex, followed by addition of alkaline phosphatase-labeled sheep anti-mouse IgG. The entire complex was visualized by formation of dark-blue precipitates after addition of substrates NBT and BCIP.

Colony hybridization (18,35) using a DNA probe labeled directly with peroxidase has been applied to detect sequences coding for the entire heat-labile enterotoxin (LT^+) A subunit. This activates adenylate cyclase activity and a portion of the B binding subunit. Hybridizations were visualized on the filters by addition of chromogenic enzyme substrate solution containing hydrogen peroxide.

B. Membranes for Immobilization of Nucleic Acids

An important advance in modern NA detection technology has been the introduction of mixed-phase hybridization. In the past, hybridization experiments were performed with both probe and target DNA in solution. Problems arose as some of the single-stranded probe and target DNA reannealed with themselves rather than hybridizing with each other, resulting in decreased sensitivity. This problem was resolved, in part, by introducing a solid support upon which denatured target DNA or RNA is immobilized in readiness for hybridization with incoming probe.

Immobilization is routinely achieved through the noncovalent binding of sample DNA or RNA to a NC membrane. The mechanism involved in the attachment of nucleic acids to this material is yet unclear, but probably depends on a combination of hydrophobic interactions, hydrogen bonding, and salt bridges (2,15,24,37–40).

In 1961, Hall and Spiegelman (19) reported the annealing of RNA and DNA in solution. Quantification of the RNA–DNA hybrids in solution was, however, confounded by competition with the re-formation of DNA–DNA complexes in the same solution. A logical alternative was to immobilize the denatured DNA irreversibly onto a solid support with subsequent hybridization on the same support. This would effectively prevent reannealing of the single DNA strands.

In 1963, Nygaard and Hall (30,31) observed that after a mixture of free RNA and RNA–DNA complexes flowed through a NC membrane filter, the free RNA passed through without retention while the RNA–DNA complexes bound to the membrane. The NC membrane filter promised to be the solid support upon which hybridization of labeled RNA to unlabeled, immobilized DNA could proceed.

Gillespie and Spiegelman in 1965 (15) used ^{3}H-DNA to examine how well DNA remained bound to NC by monitoring the ^{3}H counts per minute. They went on to detect DNA–RNA hybridization by immobilizing denatured DNA onto NC membranes and hybridizing complementary ^{32}P-RNA to the membrane-fixed DNA. The NC membrane retained both the denatured DNA and hybridized RNA. Unpaired RNA could be rinsed away.

In 1975, Southern (38) introduced a method for transferring fragments of DNA from agarose gels to NC membranes (''Southern'' transfer). Once immobilized on the membrane, the DNA fragments were allowed to hybridize with radioactive RNA. Those DNA fragments containing transcribed sequences, complementary to the RNA, appeared as sharp bands on autoradiographs. Meinkoth and Wahl (28) reported optimal transfer of single-stranded DNA fragments of lengths of ~1000 bp. Smaller fragments are commonly resolved on polyacrylamide gels and electroeluted onto NC (12,25,40).

In 1979, Renart *et al.* (33) transferred bands consisting of protein antigens that had been separated in polyacrylamide–agarose composite gels to diazobenzyloxymethyl paper, where they became coupled covalently ("Western" transfer). The transferred bands were revealed by autoradiograph after sequential addition of unlabeled specific antibody and ^{125}I-labeled protein A. Towbin and Gordon (40) reported the transfer of proteins in an electric field with noncovalent binding to NC.

In 1983, Bresser *et al.* (2) reported the selective immobilization of mRNA onto NC ("Northern" transfer). mRNA bound to NC while DNA and rRNA did not, in 12.2 molal NaI at 25°C or below.

Unfractionated NA sequence detection and quantification ("spot" hybridization) on NC does not require electrophoresis or transfer (12,25,39). Greater sensitivity than blotting is attained as a result of the presence of both degraded and intact sequences, which, when made cumulatively available to the probe, yield a strengthened hybridization signal. Further sensitivity has been achieved through deproteinization of the nucleic acid, since protein actively competes for available binding sites on the NC (39,40).

Today, a variety of specialized transfer membranes are available as solid supports for nucleic acids and/or protein immobilization. Positively charged nylon membranes enhance immobilization of proteins and low molecular weight nucleic acids; diazotized membranes effect covalent binding of RNA, DNA, and proteins; ion exchange membranes such as diethylaminoethyl (DEAE) membranes provide anionic immobilization, and cellulose membranes effect cationic immobilization.

C. Slide Immunoenzymatic Assay and SIA–DNA System

The slide immunoenzymatic assay (SIA) was originally designed as an antibody- and/or antigen-capture system (7,9,10). The central component is a glass slide with small circles, or reaction areas, delimited by a hydrophobic material. These circles serve as transparent solid supports for immobilization of proteins and other analytes. Sequential addition of appropriate reagents to these immobilized analytes allow for *in situ* reactions and for direct, spectrophotometric quantification of the analytes. A variety of chromogenic substrates were shown to be suitable for measuring the presence of enzyme-conjugated primary, or secondary antibodies. The intensity of color is measured directly on the slide using a vertical-beam spectrophotometer.

The SIA–DNA system incorporates the primary concept of SIA, that is, immobilization of an analyte onto a light-transparent, small circular solid surface for the purpose of analysis. In the case of SIA–DNA, however, the purpose is the detection and measurement of hybridization of NA probes to target base sequences rather than antigen–antibody binding.

Enzyme-labeled oligonucleotides are allowed to hybridize to complementary base sequences on the sample's single-stranded DNA, which has been immobilized onto the small circle on the SIA–DNA slide. Last-step addition of a drop of chromogenic substrate solution to the circle enables visualization of a successful hybridization by generating a color of measurable intensity. Absorbance values are obtained by intersecting each circle with the light beam of a vertical-beam spectrophotometer.

To ensure that the analytes remain on the reaction circle until completion of the assay procedure, the surface of the SIA slide circles had to be modified. This was achieved by coating the surface of the circles with a positively charged material, capable of withstanding exposure to pH extremes.

III. RESULTS AND DISCUSSION

A. Modus Operandi, Samples, Negative Controls, Instruments

Microliter volumes of test and control samples containing nucleic acids, that is, whole bacteria or their lysates, or purified DNA were applied to individual circles of an SIA–DNA slide (Fig. 1a). This slide was designed to have a "3×8" array of circles of 3 mm in diameter that would accept and keep as drops 5–20 μl of sample. The array conforms to that of a standard microtitration plate. A combination of four SIA–DNA slides inserted into a newly designed SIA adapter align to form a pattern equal to an entire 96-well microtitration plate. A thin layer of hybrophobic material around the circles serves to contain test samples and reactants throughout the hybridization process. The glass surface of the reaction circles was treated with a lysine preparation that remains permanently bound (see Section VII). The circle, thus positively charged, became a suitable stage for adhesion of sample nucleic acids, whether in pure form, in cell lysates, or *in situ* (whole cells). From start to finish, all steps of the procedure, including the final-step application of a hybrid-detecting color substrate, were performed on each sample anchored on its circle.

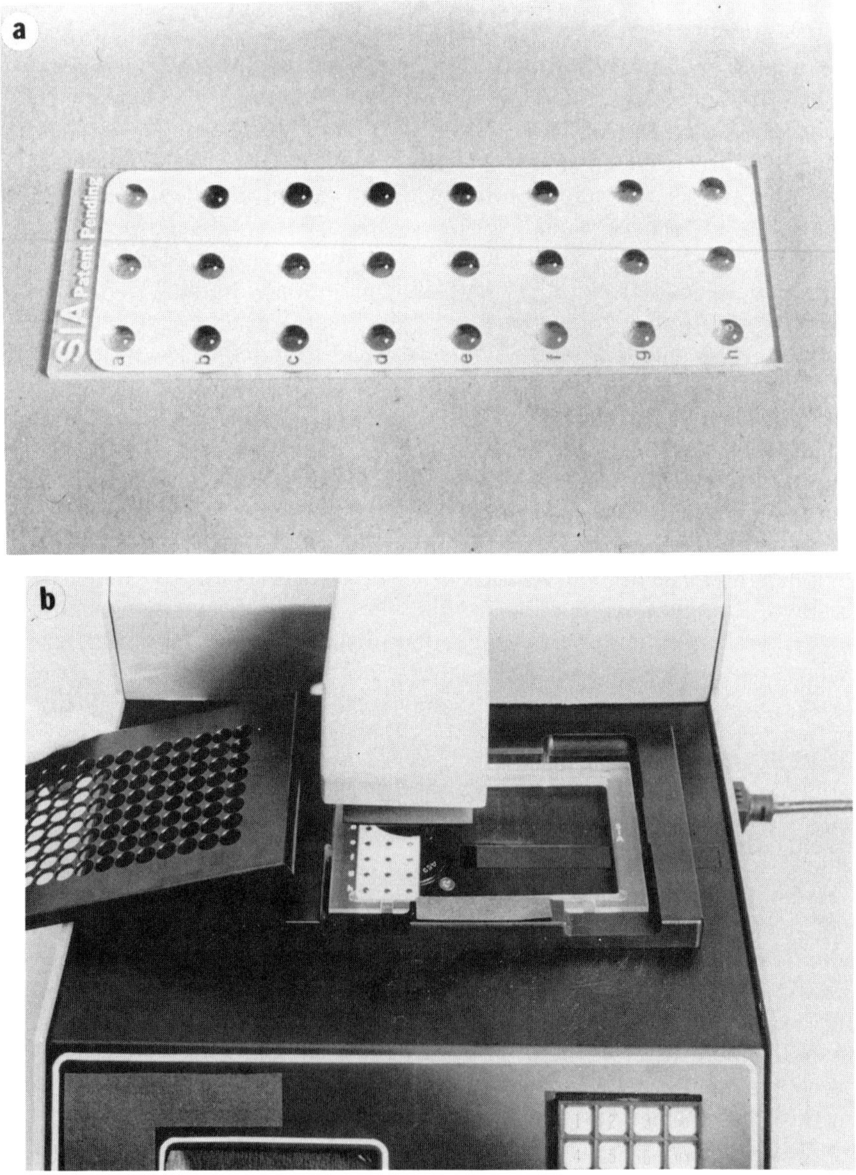

Fig. 1. (a) A standard glass slide used in the SIA–DNA system is shown having a "3 × 8" array of reaction circles containing drops of sample. A thin layer of hydrophobic material around the reaction circles serves to contain test samples and reactants throughout the procedure. (b) An SIA–DNA slide is shown in its SIA adapter, which is inserted into the microtitration plate holder of a standard vertical-beam spectrophotometer in preparation for absorbance readings. Four SIA–DNA slides inserted into the SIA adapter align to form a pattern of circles equal to a 96-well microtitration plate. Each sample drop is intersected with the vertical light beam and color intensity is recorded. The slide can be returned to a humid chamber to continue incubation and perform subsequent readings.

After suitable incubation to allow color development of the substrate, the SIA–DNA slide in the SIA adapter was inserted into a standard unmodified vertical-beam spectrophotometer (Fig. 1b). Each sample drop was allowed to intersect the instrument's vertical light beam for individual quantification of color intensity. Sample and negative control absorbance values were recorded. Negative control readings, which were subtracted from the sample readings in the data shown in this report, were similar for experiments involving purified DNA, bacterial cell lysates or whole cells (AM = 0.03; SD = 0.02; n = 59).

Several readings of the same slide can be done at intervals, returning the slide to a humid chamber between readings. The reaction proceeds undisturbed by the manipulations. In the case of most chromogenic substrates, presence of color and, to a degree, relative intensity can also be determined visually without the use of instrumentation.

B. Identification and Quantification of LT^+ Gene Sequences in Purified Plasmid DNA and in Cell Lysates

1. Reproducibility

Tests were performed on 3 different days using a commercially available alkaline phosphatase-conjugated oligonucleotide probe to detect complementary sequences in purified LT^+ EWD299 plasmid DNA samples. Plasmid DNA samples were shown on agarose gel electrophoresis to comprise a mixture of supercoiled, nicked, and linear DNA, and concentrations were adjusted for testing 1 mg DNA/ml (i.e., 10 μg DNA per SIA–DNA reaction circle). Test conditions were as described in Section VII, with a last-step addition of chromogenic substrate p-nitrophenyl phosphate (PNPP). A single SIA–DNA slide housed all samples run on a single day. Absorbance values (A_{410}) for all samples were recorded after incubation with substrate. Figure 2 shows the results (AM = 1.52; SD = 0.105; n = 10).

Tests were similarly performed to probe for LT^+ gene sequences in ETEC cell lysates with the same oligonucleotide probe utilized for the purified plasmid DNA samples just described. Ten-microliter volumes of an ETEC cell suspension containing 6.65×10^6 cells/ml (i.e., 6.65×10^4 cells per circle) were added to the circles, lysed by addition of NaOH and neutralized with ammonium acetate before addition of probe, as described in Section VII. Figure 2 shows A_{410} readings obtained from positive lysate sample hybridizations (AM = 0.99; SD = 0.082; n = 10).

Result reproducibility was satisfactory. CV% mean was 2.18 and 2.62, and percentage sample deviation was 5.50 and 6.46 for assays with purified DNA and cell lysates, respectively.

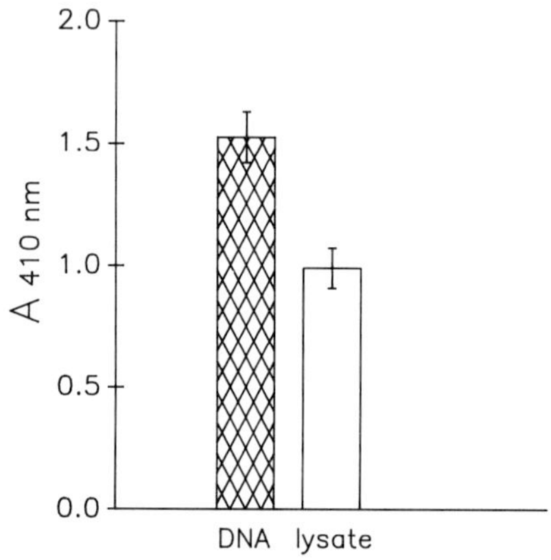

Fig. 2. Reproducibility results using quantitative SIA–DNA system. Mean absorbance values (±SD) are represented by the crosshatched and open bars for purified DNA and bacterial cell lysate, respectively.

2. *Sensitivity*

Tests using the same oligonucleotide probe as already described were performed to assess sensitivity. For this, a series of decreasing amounts of purified LT^+ plasmid DNA was used. Titrations were performed on 3 different days under the same conditions stated before. DNA concentrations ranged from 1 mg to 10 ng/ml (i.e., 10 μg to 1 pg DNA per circle). Test conditions were as described in Section VII. Figure 3 shows mean A_{410} values (n = 3) for each concentration of plasmid DNA measured at 2 hr and at 16 hr after addition of substrate. When readings were done 2 hr after substrate addition 100 pg DNA per circle could be readily measured, while after 16 hr, levels as low as 1 pg DNA per circle could be detected.

Sensitivity was also assayed using a series of lysates of decreasing numbers of LT^+ ETEC cells. Titrations were performed on 3 different days. Lysates were prepared from cell suspensions ranging in concentration from 6.65×10^6 to 6.65×10^1 cells/ml (i.e., from 6.65×10^4 cells per circle down to dilutions in which 1 or no cells per circle were expected). Negative controls were done using the same concentrations of LT^- *E. coli* cells. Test conditions were as described in Section VII. Figure 4 shows mean A_{410} values (n = 3) measured 16 hr after substrate addition. A positive hybridization signal could be detected in circles with 60 cells.

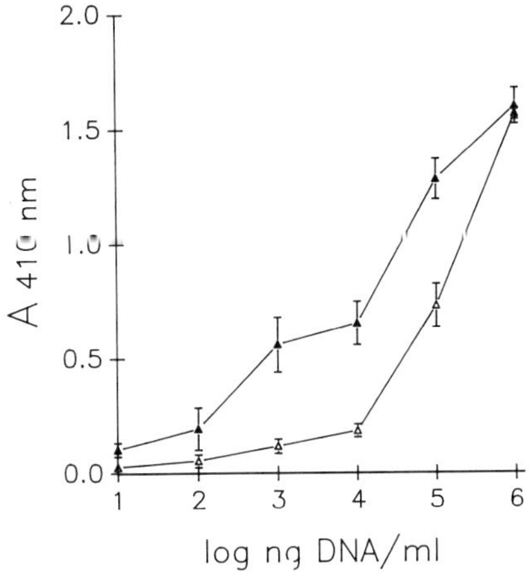

Fig. 3. Sensitivity of the SIA–DNA system for measuring hybridization of probe over a range of purified target DNA concentrations. Mean absorbance values (±SD) measured at 2 (△) and 16 hr (▲) after addition of substrate are shown.

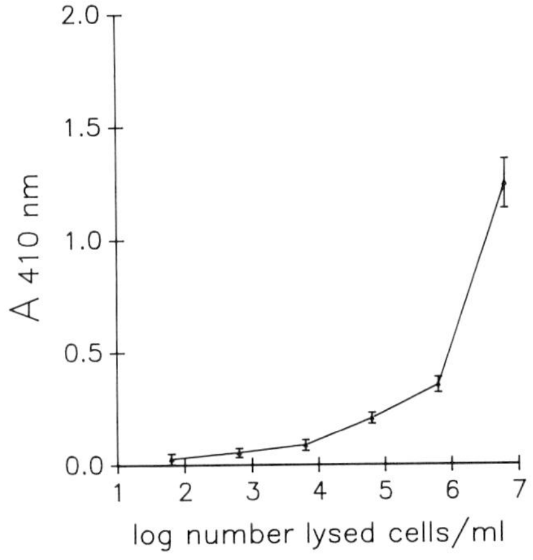

Fig. 4. Example of results obtained with the SIA–DNA system for measuring hybridization of probe with target DNA in bacterial cell lysates over a range of bacterial cell concentrations. Mean absorbance values (±SD) were measured 16 hr after addition of substrate.

C. Quantitative Measurement of LT^+ Gene Sequences in Whole Cells

Experiments were performed to assess the sensitivity of the SIA–DNA system applied to identification of gene sequences using whole cells anchored on the slide circles. Appropriate LT^+ ETEC cells were used. A range of cell concentrations was tested from 6.65×10^5 to 6.65×10^0 cells/ml (i.e., from 6.65×10^3 cells per circle down to dilutions in which 1 or no cells per circle were expected). Test conditions were similar to those used for measurement of hybridization in cell lysates with one modification: the cell-lysing reagent was diluted 4-fold. This effectively prevented cell lysis yet was sufficient to permeabilize the bacterial cell walls and simultaneously denature resident plasmid DNA. Probe was added, incubated at 50°C for 15 min, and rinsed as before. Substrate PNPP was added and drops were measured for color intensity 16 hr later. Values (data not shown) were slightly lower compared to those obtained with the cell lysates, but nonetheless indicated that target DNA had been denatured and made available to incoming labeled probe.

Of the three methods (i.e., detection of gene sequences in purified DNA, in DNA of cell lysate, or in whole-cell DNA *in situ*), the cell lysate method has potential advantages for clinical applications. Quantification of specific, diagnostic hybrids without the need for DNA extraction and purification would be desirable. *In situ* hybrid quantification is less sensitive than the cell lysate method yet requires approximately equal effort.

For research purposes, quantification of specific sequences in sample nucleic acids may be more appropriately performed with extracted and purified DNA samples. Detection sensitivity is greater in purified NA samples. Enzymatically digested NA fragments would be well suited for hybridization analysis and quantification using the SIA–DNA system.

For sample specimens having a limited number of cells available for probing, the enhanced sensitivity of the SIA–DNA system may be helpful. Detection of single-copy genes in as few as 5×10^4 cells using a ^{32}P-labeled probe has been reported (14). The SIA–DNA system detected specific NA sequences in lysates of 6×10^1 cells.

IV. CONCLUSIONS

The introduction of the solid-support system described here, allowing direct, automated measurement of probe–target DNA hybrid formation, is an improvement over existing technology. The SIA–DNA system uses a solid phase other than nitrocellulose or similar membrane supports for immobilization of nucleic acids. The support consists of a transparent glass slide that can firmly bind sample nucleic acids and other charged analytes in small, positively charged circular surfaces, and retain these analytes throughout the duration of the assay. The system uses microliter volumes of sample and reagents. The procedure can be completed in ~2 hr for purified DNA samples and in somewhat longer times for assays with whole cells or lysate thereof. The system is nonradioisotopic, needs little experience to perform, and is suited to direct sample measurement in a standard, unmodified vertical-beam spectrophotometer.

The SIA–DNA system is potentially applicable to the diagnosis of viral infections (3,21,22) and of other diseases caused by bacteria such as those discussed in other chapters in this book, and by other agents such as parasites. The system is also potentially useful for hybridizations involving RNA (2,12,15,19,25,28,30,31), for parentage testing (36), and for identification and quantification of gene sequences in laboratory research.

V. PROSPECTS FOR THE FUTURE

Enhancement of the SIA–DNA system's sensitivity will likely be achieved by incorporating antibodies, labeled and unlabeled, into the signal amplification process. Many of the signal amplification compounds used in today's immunological assays can be applied to the SIA–DNA system with few modifications. For example, hybrids formed using biotinylated probes could be reacted with antibiotin antibodies, which, in turn, would be allowed to react with enzyme-labeled second antibodies. A variety of enzymes, which can be quantified using chromogenic substrates, are available for use as signal-emitting labels. The antibody detection label could be a fluorochrome such as fluorescein, instead of an enzyme, for *in situ* demonstration of hybridization. Fluorogenic substrates, such as 4-methylumbelliferone, can also be utilized and quantified by a fluorometer (13).

Still other antibody-based amplification techniques could be used to increase the signal generated by the SIA–DNA system. The peroxidase–antiperoxidase (PAP) reagent used in the SIA–PAP for amplifying antibody–antigen reactions (8) would be applicable.

Other improvements are likely to come with the use of new signal amplification compounds that will enable the detection of minimal amounts of DNA. Biotinylated DNA probes hybridized to complementary DNA sequences of bacterial or viral pathogens immobilized on the SIA–DNA slide can potentially be detected with addition of a variety of enzyme-labeled streptavidin complexes. Since the streptavidin molecule has multiple biotin-binding sites, further signal amplification could be achieved by addition of biotinylated enzymes. Chromogenic substrates can then be added and color intensities recorded instrumentally.

VI. SUMMARY

The slide immunoenzymatic assay (SIA), originally designed to detect antigens and antibodies in microliter-volume samples, was modified to measure DNA hybridization. This SIA–DNA system uses as solid support a glass slide with small circular reaction areas instead of the commonly known nitrocellulose membranes. It was standardized using target DNA (from enterotoxigenic LT^+ *E. coli*) in purified form, in a bacterial cell lysate, or in whole cells. The target DNA was anchored onto the small circles on the glass slide. The surface of the circle was precoated with a positively charged material. The immobilized DNA was denatured to yield single strands directly on the circles and then it was probed with an alkaline phosphatase-labeled probe. A chromogenic substrate was used to measure reaction intensity with an SIA slide adapter and a standard vertical-beam spectrophotometer. LT^- cells or lysates were used as negative controls. Absorbance values were recorded showing reproducible results with satisfactory sensitivity.

VII. MATERIALS AND METHODS

A. Bacteria

Enterotoxigenic *E. coli* (ETEC) and nonenterotoxigenic *E. coli* were obtained from the American Type Culture Collection (ATCC, Rockville, Maryland). The ETEC ATCC strain 37218 has a 6.8-kb ampicillin-resistance (Ap^r) plasmid (EWD299) constructed to contain the gene expressing a heat-labile enterotoxin (LT^+) (11,29). A nonenterotoxigenic *E. coli,* ATCC strain 33849, was the same strain used as DNA vector for distribution of the LT^+ plasmid just mentioned. It is void of the plasmid and was used as the negative control strain (LT^-).

LT^+ strain 37218 was grown at 37°C overnight on ampicillin-containing Luria-Bertani (LB) agar plates (final concentration 50 μg ampicillin/ml). Cells were harvested from the plate and suspended in water to give an $OD_{660} = 2.0$, yielding $\sim 6 \times 10^6$ cells per milliliter. LT^- strain 33849 was grown on LB agar plates without ampicillin and harvested similarly.

B. Plasmid DNA Preparation

To obtain a supply of purified target DNA, large-scale amplification of EWD299 LT^+ plasmid DNA in ATCC strain 37218 was performed according to standard methods (1,12,25). Isolation and purification of the amplified plasmids was performed as described in Promega, Biological Research Products 1988/89 Catalogue and Applications Guide (Transcription Systems, p. 14). Presence of DNA in the reaction circle was controlled using ethidium bromide (20,32,41). A stock preparation with a concentration of 1.0 mg LT^+ DNA/ml was used in this study.

C. DNA Probe

An alkaline phosphatase-labeled oligonucleotide probe specific for the ETEC LT^+ gene sequence (NEN Research Products, DuPont, Boston, Massachusetts) was used. The probe, 15–35 bases in length, with alkaline phosphatase covalently linked to the C-5 position of a thymidine base through a spacer arm, was diluted 1 : 50 in hybridization buffer and a 10-μl volume applied to each circle. The slide was incubated 15 min at 50°C and then rinsed first in 0.15 *M* sodium chloride–0.015 *M* sodium citrate buffer (1× SSC) with 1% sodium dodecyl sulfate (SDS) for 5 min at 40°C, then in 1× SSC for 2 min, and finally in water for 2 min at room temperature (23°C).

D. SIA–DNA Slides

Glass SIA slides (Cel-Line Inc., Newfield, New Jersey) were designed by the authors to configure a pattern of 24 individual reaction circles, each 3 mm in diameter, surrounded by a thin layer of hydrophobic material, which serves to confine reactants within the circles (7,9,10). Circle surfaces were coated with a clear, permanent application of a lysine–acetate solution containing L-lysine (27), methanol, butyl acetate, ethyl acetate, and nitrocellulose toluene solubilized in acetone.

E. SIA–DNA System Procedures

1. Bacterial Cell Lysates

a. On-Slide Lysate Preparation. A 10-μl volume of bacterial cell suspension (LT^+ or LT^- *E. coli*) was added to a lysine-coated circle on the SIA–DNA slide. To this drop was added a 5-μl volume of freshly prepared lysing–denaturing solution (0.25 *M* NaOH–0.1 *M* NaCl), after which the slide was incubated in a humid chamber at room temperature (23°C) for 30 min. A 10-μl volume of neutralizing solution (0.3 *M* ammonium acetate) was added to this lysate and the mixture allowed to dry. The slide was heat-fixed, water-rinsed, and air-dried. To minimize nonspecific binding of labeled DNA probe to the circle's charged surface, a 10-μl volume of 1 mg/ml of nonhomologous DNA from methanogenic bacterium (*Methanosarcina mazei*) was added, air-dried, and rinsed. A 10-μl volume 1 : 50 dilution of labeled oligonucleotide probe was added to each circle and the SIA–DNA slide was incubated in a humid chamber at 50°C for 15 min. The slide was rinsed 1 min in 1× SSC buffer at 40°C, then 1 min in 1× SSC, and finally in distilled water 1 min, all at room temperature. On drying, each reaction circle was covered with a 15 μl drop of substrate PNPP (1 mg/ml) and the slide incubated in a humid chamber at room temperature. Color intensity in each drop was measured using a standard vertical-beam spectrophotometer equipped with a 410-nm filter (MiniReader II, Dynatech, Inc., Chantilly, Virginia). LT^- *E. coli* strain ATCC 33849 was used as the negative control.

2. *Purified Plasmid DNA*

Ten-microliter volumes of LT^+ or LT^- purified plasmid DNA (1 mg/ml) were added to lysine-coated reaction circles on SIA–DNA slides. Five-microliter volumes of freshly prepared lysing–denaturing solution (0.25 *M* NaOH–0.1 *M* NaCl) were added to each liquid sample drop and the slide was incubated in a humid chamber at room temperature (23°C) for 15 min. Ten-microliter volumes of 0.3 *M* ammonium acetate neutralizing reagent were added and the mixtures were allowed to air-dry onto the reaction circles. Samples were hybridized with probe, rinsed, and covered with substrate as described before. Titrations were done using 10 μl of DNA samples ranging from 10 ng to 1 mg DNA/ml.

3. *Whole Cells*

Ten-microliter volumes of an ETEC LT^+ cell suspension were added to reaction circles of an SIA–DNA slide. Permeabilization of the bacterial walls and denaturation of plasmid DNA inside the cells was achieved by addition of 5 μl volumes of 0.1 *M* NaOH–0.02 *M* NaCl to the drops of sample on the circles. Fifteen minutes later, 5 μl volumes of 1 *M* ammonium acetate were added to each drop and the drops were allowed to dry. Probe was added, incubated, rinsed, and substrate PNPP added as before.

ACKNOWLEDGMENTS

We thank James S. Swab for his help in preparing the manuscript. The portion of the work involving preparation and use of DNA from *Methanosarcina mazei* was partially supported by grant No. 706IERBEA85 from GRI-NYSERDA-NY Gas.

REFERENCES

1. Birnboim, H. C., and Doly, J. (1979). A rapid alkaline extraction procedure for screening recombinant plasmid DNA. *Nucleic Acids Res.* **7,** 1513–1523.
2. Bresser, J., Hubbell, H. R., and Gillespie, D. (1983). Biological activity of mRNA immobilized on nitrocellulose in NaI. *Proc. Natl. Acad. Sci. U.S.A.* **80,** 6523–6527.
3. Buffone, G. J., Schimbor, C. M., Demmler, G. J., Wilson, D. R., and Darlington, G. J. (1986). Detection of cytomegalovirus in urine by nonisotopic DNA hybridization. *J. Infect. Dis.* **154,** 163–166.
4. Burns, J., Chan, V. T. W., Jonasson, J. A., Fleming, K. A., Taylor, S., and McGee, J. O'. D. (1985). Sensitive system for visualizing biotinylated DNA probes hybridized *in situ:* Rapid sex determination of intact cells. *J. Clin. Pathol.* **38,** 1085–1092.
5. Cardullo, R. A., Agrawal, S., Flores, C., Zamecnik, P. C., and Wolf, D. E. (1988). Detection of nucleic acid hybridization by nonradiative fluorescence resonance energy transfer. *Proc. Natl. Acad. Sci. U.S.A.* **85,** 8790–8794.
6. Chevrier, D., Larzul, D., Megraud, F., and Guesdon, J.-L. (1989). Identification and classification of *Campylobacter* strains by using nonradioactive DNA probes. *J. Clin. Microbiol.* **27,** 321–326.
7. Conway de Macario, E., Jovell, R. J., and Macario, A. J. L. (1987). Multiple solid-phase system for storage of dry ready-for-use reagents and efficient performance of immunoenzymatic and other assays. *J. Immunol. Methods* **99,** 107–112.
8. Conway de Macario, E., Jovell, R. J., and Macario, A. J. L. (1987). Slide Immunoenzymatic Assay (SIA): Improving sensitivity to measure antibodies when samples are very small and dilute, and antigen is scarce. J. Immunoassay **8,** 283–295.
9. Conway de Macario, E., Macario, A. J. L., and Jovell, R. J. (1983). Quantitative Slide Micro-Immunoenzymatic Assay (Micro-SIA) for antibodies to particulate and non-particulate antigens. *J. Immunol. Methods* **59,** 39–47.
10. Conway de Macario, E., Macario, A. J. L., and Jovell, R. J. (1986). Slide immunoenzymatic assay (SIA) in hybridoma technology. *In* "Methods in Enzymology" (J. J. Langone and H. van Vunakis, eds.), Vol. 121, Part I, pp. 509–525. Academic Press, Orlando, Florida.
11. Dallas, W. S., Gill, D. M., and Falkow, S. (1979). Cistrons encoding *Escherichia coli* heat-labile toxin. *J. Bacteriol.* **139,** 850–858.
12. Davis, L. G., Dibner, M. D., and Battey, J. F. (1986). "Basic Methods in Molecular Biology." Am. Elsevier, New York.
13. Downs, T. R., and Wilfinger, W. W. (1983). Fluorometric quantification of DNA in cells and tissue. *Anal. Biochem.* **131,** 538–547.
14. Feddersen, R. M., and Van Ness, B. G. (1989). Single-copy gene detection requiring minimal cell numbers. *BioTechniques* **7,** 44–49.
15. Gillespie, D., and Spiegelman, S. (1965). A quantitative assay for DNA–RNA hybrids with DNA immobilized on a membrane. *J. Mol. Biol.* **12,** 829–842.
16. Goltz, S. P., Donegan, J. J., Yang, H.-L., Pollice, M., Todd, J. A., Molina, M. M., Victor, J., and Kelker, N. (1989). The use of nonradioactive DNA probes for rapid diagnosis of sexually transmitted bacterial infections. This volume, Chapter 1.
17. Goltz, S. P., Todd, J. A., and Yang, H.-L. (1987). A rapid DNA probe test for detecting HPV in biopsy specimens. *Am. Clin. Prod. Rev.* **6,** 16–19.
18. Guilfoyle, D. E. (1988). The use of DNA colony hybridization for the identification of pathogenic microorganisms. *Am. Biotechnol. Lab.* **6,** 6–11.
19. Hall, B. D., and Spiegelman, S. (1961). Sequence complementarity of T2-DNA and T2-specific RNA. *Proc. Natl. Acad. Sci. U.S.A.* **47,** 137–146.

20. Karsten, U., and Wollenberger, A. (1977). Improvements in the ethidium bromide method for direct fluorometric estimation of DNA and RNA in cell and tissue homogenates. *Anal. Biochem.* **77,** 464–470.
21. Keller, G. H., and Khan, N. C. (1988). Identification of HIV sequences using nucleic acid probes. *Am. Clin. Lab.* **7,** 10–15.
22. Langenberg, A., Smith, D., Brakel, C. L., Pollice, M., Remington, M., Winter, C., Dunne, A., and Corey, L. (1988). Detection of herpes simplex virus DNA from genital lesions by *in situ* hybridization. *J. Clin. Microbiol.* **26,** 933–937.
23. Langer, P. R., Waldrop, A. A., and Ward, E. C. (1981). Enzymatic synthesis of biotin-labeled polynucleotides: Novel nucleic acid affinity probes. *Proc. Natl. Acad. Sci. U.S.A.* **78,** 6633–6637.
24. Leary, J. J., Brigati, D. J., and Ward, D. C. (1983). Rapid and sensitive colorimetric method for visualizing biotin-labeled DNA probes hybridized to DNA or RNA immobilized on nitrocellulose: Bio-blots. *Proc. Natl. Acad. Sci. U.S.A.* **80,** 4045–4049.
25. Maniatis, T., Fritsch, E. F., and Sambrook, J. (1982). "Molecular Cloning: A Laboratory Manual." Cold Spring Harbor Lab., Cold Spring Harbor, New York.
26. Manuelidis, L., Langer-Safer, P. R., and Ward, D. C. (1982). High resolution mapping of satellite DNA using biotin-labeled DNA probes. *J. Cell Biol.* **95,** 619–625.
27. Mazia, D., Schatten, G., and Sale, W. (1975). Adhesion of cells to surfaces coated with polylysine. *J. Cell Biol.* **66,** 198–200.
28. Meinkoth, J., and Wahl, G. (1984). Hybridization of nucleic acids immobilized on solid supports. *Anal. Biochem.* **138,** 267–284.
29. Moseley, S. L., and Falkow, S. (1980). Nucleotide sequence homology between the heat-labile enterotoxin gene of *Escherichia coli* and *Vibrio cholerae* deoxyribonucleic acid. *J. Bacteriol.* **144,** 444–446.
30. Nygaard, A. P., and Hall, B. D. (1963). A method for the detection of RNA–DNA complexes. *Biochem. Biophys. Res. Commun.* **12,** 98–104.
31. Nygaard, A. P., and Hall, B. D. (1964). Formation and properties of RNA–DNA complexes. *J. Mol. Biol.* **9,** 125–142.
32. Pearse, A. G. E. (1985). "Histochemistry: Theoretical and Applied. Analytical Technology," Vol. 2, p. 639. Churchill-Livingstone, New York.
33. Renart, J., Reiser, J., and Stark, G. R. (1979). Transfer of proteins from gels to diazobenzyloxymethyl-paper and detection with antisera: A method for studying antibody specificity and antigen structure. *Proc. Natl. Acad. Sci. U.S.A.* **76,** 3116–3120.
34. Rigby, P. W. J., Dieckmann, M., Rhodes, C., and Berg, P. (1977). Labeling deoxyribonucleic acid to high specific activity *in vitro* by nick translation with DNA polymerase I. *J. Mol. Biol.* **113,** 237–251.
35. Romick, T. L., Lindsay, J. A., and Busta, F. F. (1987). A visual DNA probe for detection of enterotoxigenic *Escherichia coli* by colony hybridization. *Lett. Appl. Microbiol.* **5,** 87–90.
36. Rose, S. D., and Keith, T. P. (1988). Application of DNA probes in parentage testing. *Am. Clin. Lab.* **7,** 16–20.
37. Seed, B. (1982). Attachment of nucleic acids to nitrocellulose and diazonium-substituted supports. *In* "Genetic Engineering: Principles and Methods" (J. K. Setlow and A. Hollaender, eds.), Vol. 4, pp. 91–102. Plenum, New York.
38. Southern, E. M. (1975). Detection of specific sequences among DNA fragments separated by gel electrophoresis. *J. Mol. Biol.* **98,** 503–517.
39. Towbin, H., and Gordon, J. (1984). Immunoblotting and dot immunobinding current status and outlook. *J. Immunol. Methods* **72,** 313–340.

40. Towbin, H., Staehelin, T., and Gordon, J. (1979). Electrophoretic transfer of proteins from polyacrylamide gels to nitrocellulose sheets: Procedure and some applications. *Proc. Natl. Acad. Sci. U.S.A.* **76**(9), 4350–4354.
41. Waring, M. J. (1961). Structural requirements for the binding of ethidium to nucleic acids. *Biochim. Biophys. Acta* **114**, 234–244.

Index

A

B

D

E

H

I

T

U

V

W